Ecological Studies, Vol. 157

Analysis and Synthesis

Edited by

I.T. Baldwin, Jena, Germany
M.M. Caldwell, Logan, USA
G. Heldmaier, Marburg, Germany
O.L. Lange, Würzburg, Germany
H.A. Mooney, Stanford, USA
E.-D. Schulze, Jena, Germany
U. Sommer, Kiel, Germany

Ecological Studies

Volumes published since 1997 are listed at the end of this book.

Springer
Berlin
Heidelberg
New York
Hong Kong
London
Milan
Paris
Tokyo

M.G.A. van der Heijden   I.R. Sanders (Eds.)

Mycorrhizal Ecology

1st Edition 2002, 2nd Printing 2003

With 59 Figures, 3 in Color, and 16 Tables

Dr. Marcel G.A. van der Heijden
Free University of Amsterdam
Department of Systems Ecology
Faculty of Earth and Life Sciences
De Boelelaan 1087
1081 HV Amsterdam
The Netherlands

Dr. Ian R. Sanders
University of Lausanne
Institute of Ecology
Biology Building
1015 Lausanne
Switzerland

Cover illustration: Marcel G.A. van der Heijden

ISSN 0070-8356
ISBN 3-540-00204-9 Springer-Verlag Berlin Heidelberg New York

Originally published as hardcover edition in 2002 (ISBN 3-540-42407-5)
Library of Congress Cataloging-in-Publication Data applied for

A catalog record for this book is available from the Library of Congress
Bibliographic information published by Die Deutsche Bibliothek
Die Deutsche Bibliothek lists this publication in the Deutsche Nationalbibliographie;
detailed bibliographic data is available in the Internet at http://dnb.ddb.de

This work is subject to copyright. All rights are reserved, whether the whole or part of the material is concerned, specifically the rights of translation, reprinting, reuse of illustrations, recitation, broadcasting, reproduction on microfilm or in any other way, and storage in data banks. Duplication of this publication or parts thereof is permitted only under the provisions of the German Copyright Law of September 9, 1965, in its current version, and permissions for use must always be obtained from Springer-Verlag. Violations are liable for prosecution under the German Copyright Law.
Springer-Verlag Berlin Heidelberg New York
a company of BertelsmannSpringer Science+Business Media GmbH

http://www.springer.de

© Springer-Verlag Berlin Heidelberg 2002, 2003
Printed in Germany

The use of general descriptive names, registered names, trademarks, etc. in this publication does not imply, even in the absence of a specific statement, that such names are exempt from the relevant protective laws and regulations and therefore free for general use.

Production: Friedmut Kröner, Heidelberg
Cover design: *design & production* GmbH, Heidelberg
Typesetting: Kröner, Heidelberg

31/3150 YK – 5 4 3 2 1 0 – Printed on acid free paper

Preface

Plants collaborate with many micro-organisms in the rhizosphere to form mutualistic associations. One of the best examples is the mycorrhizal symbiosis between plants and fungi. Here, fungi support plants with mineral nutrients and other services and the fungi, in turn, receive photosynthates from the autotrophic plants. Mycorrhizal associations are common in almost all ecosystems and 80 % of all land plants associate with these mutualistic soil fungi. There is an increasing awareness among biologists, ecologists and mycologists that mycorrhizal associations need to be considered in order to understand the ecology and evolution of plants, plant communities and ecosystems. In the last decade, many advances and breakthroughs have been made in mycorrhizal ecology. We aim to summarise these advances in this Volume, with special emphasis given to the ecological function of the mycorrhizal symbiosis.

This Volume is divided into six sections. The first section gives an introduction to the mycorrhizal symbiosis and discusses the progress that has been made in understanding the ecological function of this association. The second section deals with the eco-physiology of mycorrhizal plants. It also covers the influence of global changes on the symbiosis. The third section discusses the influences of mycorrhizal fungi on biodiversity and ecosystem functioning. It also discusses factors that influence the diversity and structure of mycorrhizal fungal communities. The fourth section shows the multitrophic nature of the mycorrhizal symbiosis. It shows that mycorrhizal fungi interact with individuals and communities of different trophic levels including plants, herbivores, insects and other soil fungi. The fifth section takes a more evolutionary approach to looking at host specificity between plants and mycorrhizal fungi. Finally, the last section gives a summary of this Volume.

We aim to bring together these major topics in mycorrhizal ecology and this also reflects our personal interest in integrating the study of the mycorrhizal symbiosis within the much wider field of plant ecology. Moreover, the field of mycorrhizal ecology has expanded greatly in recent years. It was, therefore, not possible to include all disciplines and choices had to be made. In addition, during the preparation of this Volume, several exciting findings in our field were published. Several authors have included these findings in their contributions but of course not all could be covered. We aimed to cover these last-minute breakthroughs in Chapte 17, "Mycorrhizal Ecology: Synthesis and Perspectives".

The editors are thankful to many persons who contributed to this Volume. First, we would like to thank Dieter Czechlik and Andrea Schlitzberger from Springer Verlag for their invitation to start thisVolume. We would also like to thank Prof. O.L. Lange, the Series Editor of Ecological Studies. I am especially thankful to Rien Aerts, Head of the Department of Systems Ecology at the Vrije Universiteit Amsterdam, for his continuous support and for giving me the opportunity to accomplish this exciting project (MvdH). Many thanks also to David Read for his help, suggestions and contribution to thisVolume. Each of the chapters was peer-reviewed by two independent referees. We are, therefore, grateful to all the people who reviewed the chapters. Last, but not least, MvdH would like to thank his closest friends and relatives for their encouragement, support and interest in his work. IS thanks the Swiss National Science Foundation for financial support of a professorial fellowship during the period in which this Volume was edited.

December 2001

Marcel G.A. van der Heijden,
Amsterdam, The Netherlands

Ian R. Sanders,
Lausanne, Switzerland

Foreword: Mycorrhizae and "Joined-Up" Ecology

In the second half of the 20th century, rapid progress was achieved in many fields of biological research by the development of specialist approaches. Ecology was no exception to this trend and much has been achieved by the concentrated attack that is possible when groups of scientists co-operate to address well-defined objectives.

As the end of the last century approached, it became apparent that to serve the best interests of ecological research, this reductionist agenda would require more sophisticated development than was necessary in neighbouring biological sciences such as genetics, molecular biology and biomedical science. In particular, it became necessary to remind ourselves that the ultimate objectives of ecology are to understand the assembly and dynamic properties of communities, ecosystems, landscapes and biomass. To achieve this "scaling-up", I propose that the most significant challenge for ecologists at the present time is to continue to exploit the many current opportunities for penetration into detailed mechanisms but simultaneously to actively promote integration of sub-disciplines within ecological research. Mycorrhizal ecology is uniquely positioned and qualified to provide a lead in this difficult task. Quite clearly, as the expert authors of the various chapters in this Volume explain, their subject has been an interdisciplinary pursuit throughout its history with important and continuing involvement of plant physiologists, mycologists, soil scientists, animal ecologists and, more recently, molecular biologists and geneticists. Equally important, since its inception, mycorrhizal ecology has recognised the diversity of phenomena represented by the various associations between mycorrhizal fungi and vascular plants and the very different consequences of this diversity for the survival and functioning of plant communities and ecosystems in contrasted environments.

Ever since my first encounter with Jack Harley, I have been aware that a motivating force for mycorrhizal ecologists and their subterranean allies has been the corporate and deeply held conviction that research below ground has been a neglected and undervalued part of ecology. The exciting achievements and expanding perspectives reviewed in this Volume are a vindication of the long campaign to direct sharp intelligence and adequate funding to the many unsolved problems that lie below the soil surface.

With the case for the importance of mycorrhizal ecology now firmly established in terrestrial studies, what will be the future role of research on this topic? This Volume provides several pointers to particular developments in both fundamental and applied ecology. In addition to these, however, I would not be surprised to see a more general effect on the conduct of ecological research and the development of ecological theory. For the purposes of management and conservation, it has become essential to develop an array of dynamic models that are driven by the biology of component organisms and capture the functioning of ecosystems of widely contrasting types. As ecologists of many persuasions converge upon the ecosystem as a subject for modelling and as a vehicle for advice to policy makers, the pivotal position of dominant plant species and their mycorrhizal associates, as mediators of ecosystem productivity and biogeochemistry, is likely to be widely recognised. I am confident that in such circumstances the continuation of a research tradition embracing in situ observation, experimental manipulation, quantitative biochemistry and mechanistic interpretation will assure an influential role for mycorrhizal experts of the future.

J.P. Grime

Contents

Section A: Introduction

1   Towards Ecological Relevance –
Progress and Pitfalls in the Path Towards
an Understanding of Mycorrhizal Functions in Nature . . .   3
D.J. READ

1.1	Summary	3
1.2	Introduction	4
1.3	The Platform Provided by Reductionist Approaches	4
1.4	Approaches to Examination of Mycorrhizal Function at the Community Level	8
1.4.1	The Microcosm Approach	10
1.4.2	The Field Approach	21
1.5	Conclusion	23
References		24

Section B: Ecophysiology, Ecosystems Effects and Global Change

2   Carbon and Nutrient Fluxes Within
and Between Mycorrhizal Plants   33
S.W. SIMARD, D. DURALL, M. JONES

2.1	Summary	34
2.2	Introduction	34
2.3	Ecto-mycorrhiza and Nutrient Uptake	35
2.3.1	Quantitative Effects of Ecto-mycorrhiza on Nutrient Uptake by Plants	35
2.3.2	Mechanisms of Increased Nutrient Uptake	39
2.3.3	Transport of Nutrients Within the Ecto-mycorrhizal Fungal Mycelium	43
2.4	Carbon Flux in Ecto-mycorrhizal Plants	44

2.4.1	Mechanism of Carbon Exchange Between Symbionts	44
2.4.2	Quantification of Belowground C Partitioning	45
2.4.3	Biological and Physical Effects on Belowground Carbon Partitioning	46
2.5	Carbon and Nutrient Transfers Between Plants	47
2.5.1	Hyphal Links Between Plants	47
2.5.2	Carbon and Nutrient Transfer Between Plants	50
2.5.3	Factors Regulating Carbon and Nutrient Transfer	55
2.6	Conclusions on the Ecological Significance of Interplant Transfer	59
References		61

3 Function and Diversity of Arbuscular Mycorrhizae in Carbon and Mineral Nutrition ... 75
I. Jakobsen, S.E. Smith, F.A. Smith

3.1	Summary	75
3.1	Introduction	76
3.2	Carbon Nutrition and Growth of the Fungus	76
3.3	Mineral Nutrition of the Fungus	79
3.4	Nutrition of the Plant: The Autotrophic Symbiont	82
3.5	Mycorrhizal Responsiveness	84
3.6	From Individual Symbioses to Plant Communities	87
References		88

4 Foraging and Resource Allocation Strategies of Mycorrhizal Fungi in a Patchy Environment ... 93
P.A. Olsson, I. Jakobsen, H. Wallander

4.1	Summary	93
4.2	Introduction	94
4.3	The Mycorrhizal Mycelium	98
4.3.1	Structures and Growth Rates	98
4.3.2	Density of the Mycelium Front	99
4.3.3	Energy Storage	100
4.4	Foraging of the Ecto-mycorrhizal Mycelium	100
4.4.1	Response to Inorganic Nutrients	100
4.4.2	Response to Organic Nutrient Sources	101
4.4.3	Response to Plant Roots	102
4.5	Foraging of the Arbuscular Mycorrhizal Mycelium	102
4.5.1	Response to Inorganic Nutrients	102

4.5.2	Response to Organic Material and Compounds	103
4.5.3	Response to Plant Roots	104
4.6	Differences in Foraging Strategies Between Arbuscular Mycorhizal and Ecto-mycorrhizal Fungi	104
4.7	Models and Theories on the Foraging of Mycorrhizal Fungi	105
4.8	Conclusions and Future Perspectives	109
References		110

5	**The Role of Various Types of Mycorrhizal Fungi in Nutrient Cycling and Plant Competition** R. Aerts	117
5.1	Summary	117
5.2	Introduction	118
5.3	Soil Nutrient Sources	120
5.4	Different Mycorrhizal Types Can Tap Different Soil Nutrient Sources	122
5.5	Vascular Plant ^{15}N Natural Abundance as an Indicator for the Type of Mycorrhiza-Mediated Plant Nutrition?	124
5.6	A Conceptual Model for Mycorrhizal Impacts on Plant Competition and Coexistence	125
5.7	Effects of Increased Nitrogen Deposition	127
5.8	Plant-Soil Feedbacks: Mineralization or 'Organicisation'?	129
5.9	Future Challenges	130
References		131

6	**Global Change and Mycorrhizal Fungi** M. Rillig, K.K. Treseder, M.F. Allen	135
6.1	Summary	135
6.2	Introduction	136
6.2.1	What Is Global Change?	136
6.2.2	Rationale for Considering Mycorrhizae in Global Change Biology	138
6.3	Global Change: Multiple Factors	139
6.3.1	Elevated Atmospheric CO_2	139
6.3.2	Nitrogen Deposition	143
6.3.3	Altered Precipitation and Temperature	144
6.3.4	Ozone	147
6.3.5	UV Radiation	148
6.4	Global Change Factor Interactions	149

6.5	Long-Term Versus Short-Term Responses	150
6.6	Global Change and Effects on Symbionts	151
6.7	Conclusions	152
References		153

Section C: Biodiversity, Plant and Fungal Communities

7   Diversity of Ecto-mycorrhizal Fungal Communities in Relation to the Abiotic Environment   163
S. Erland, A.F.S. Taylor

7.1	Summary	163
7.2	Introduction	164
7.3	Natural Abiotic Factors Influencing Ecto-mycorrhizal Community Structure	172
7.3.1	Edaphic Factors	172
7.3.2	Soil Organic Matter and Spatial Heterogeneity	173
7.3.3	Moisture	175
7.3.4	pH	176
7.3.5	Temperature	177
7.3.6	Wildfire	177
7.4	Effects of Pollution and Forest Management Practices on Ecto-mycorrhizal Fungal Communities	179
7.4.1	Elevated CO_2	179
7.4.2	Ozone	180
7.4.3	Heavy Metals	180
7.4.4	N-Deposition and Fertilisation	181
7.4.5	Acidification	184
7.4.6	Liming	185
7.4.7	Wood Ash Application and Vitality Fertilisation	186
7.5	Conclusions	187
7.6	Future Progress	192
References		193

8	Genetic Studies of the Structure and Diversity of Arbuscular Mycorrhizal Fungal Communities	201
	J.P. Clapp, T. Helgason, T.J. Daniell, J.P.W. Young	
8.1	Summary	202
8.2	Introduction	202
8.2.1	The Importance of Community Structure in Arbuscular Mycorrhizal Fungi	202
8.2.2	Molecular Identification of Arbuscular Mycorrhizal Fungi .	203
8.3	Strategies for Identifying Arbuscular Mycorrhizal Fungi Using Molecular Markers	205
8.3.1	Methods Available	205
8.3.2	Primers Developed for Analysis of Arbuscular Mycorrhizal Fungal Ribosomal Genes: A Short History	206
8.4	Investigations of Arbuscular Mycorrhizal Fungal Communities	208
8.4.1	Primers that Target Single Families or Species	208
8.4.2	Primers Intended to Target all Glomalean Fungi	209
8.4.3	Alternative Approaches to Molecular Characterisation of Arbuscular Mycorrhizal Fungi	212
8.5	Comparison of Molecular and Morphological Data	212
8.5.1	Phylogeny and Taxonomy	213
8.5.2	Field Data	213
8.6	The Genetic Organisation of Arbuscular Mycorrhizal Fungi	215
8.6.1	Heterogeneous Sequences in Spores and Cultures	215
8.6.2	"Contaminant" Sequences in Spores and Cultures	217
8.6.3	How Is the Sequence Heterogeneity Organised?	218
8.6.4	How Is Sequence Heterogeneity Maintained?	219
8.6.5	How Does This Affect Analysis of Field Sequences?	220
8.7	Future Prospects	221
References		221

9	Diversity of Arbuscular Mycorrhizal Fungi and Ecosystem Functioning	225
	M. Hart, J.N. Klironomos	
9.1	Summary	225
9.2	Introduction	226
9.3	Linking Biodiversity and Ecosystem Function	227
9.4	Arbuscular Mycorrhizal Fungal Diversity and Ecosystem Functioning	229
9.4.1	Arbuscular Mycorrhizal Fungal Networks	229

9.4.2	Functional Specificity	230
9.4.3	Differential Effects	231
9.4.4	Community Effects	231
9.5	Succession of Arbuscular Mycorrhizal Fungi	237
9.6	Conclusions	239
References		239

10	Arbuscular Mycorrhizal Fungi as a Determinant of Plant Diversity: In Search for Underlying Mechanisms and General Principles	243
	M.G.A. van der Heijden	
10.1	Summary	243
10.2	Introduction	244
10.3	Arbuscular Mycorrhizal Fungi as a Determinant of Plant Diversity	245
10.3.1	Importance of Plant Species Composition and Nutrient Availability	245
10.3.2	Mycorrhizal Dependency	249
10.3.3	Underlying Mechanisms and Explaining Models	252
10.4	Ecological Significance of Arbuscular Mycorrhizal Fungal Diversity	255
10.4.1	Influence on Plants and Plant Communities	255
10.4.2	Thoughts on Underlying Mechanisms	256
10.4.3	Mycorrhizal Species Sensitivity	258
10.5	Conclusions	260
References		261

11	Dynamics Within the Plant – Arbuscular Mycorrhizal Fungal Mutualism: Testing the Nature of Community Feedback	267
	J.D. Bever, A. Pringle, P.A. Schultz	
11.1	Summary	268
11.2	Introduction	268
11.3	Mutual Interdependence of Plant and Arbuscular Mycorrhizal Fungal Growth Rates Cause Feedback Dynamics	270
11.3.1	Positive Feedback	271
11.3.2	Negative Feedback	272
11.4	Identifying Positive Versus Negative Feedback: Complimentary Approaches	274

11.4.1	Interdependence of Plant and Fungal Population Growth Rates: A Mechanistic Approach	274
11.4.2	Plant Response to Manipulated Arbuscular Mycorrhizal Fungal Communities: A Phenomenological Approach	275
11.5	Evidence for Interdependence of Plant and Fungal Population Growth Rates	276
11.5.1	Specificity of Plant Response to Arbuscular Mycorrhizal Fungal Species	277
11.5.2	Specificity of Arbuscular Mycorrhizal Fungal Response to Plant Species	278
11.6	Testing Feedback Between Plant and Arbuscular Mycorrhizal Fungal Communities	279
11.6.1	Evaluation of Feedback Using the Mechanistic Approach	279
11.6.2	Evaluation of Feedback Using the Phenomenological Approach	283
11.6.3	A Comparison of Mechanistic and Phenomenological Approaches to Testing Feedback	285
11.7	Implications and an Extension of the Feedback Framework	286
11.7.1	Negative Feedbacks and the Evolutionary Maintenance of Mutualism	287
11.7.2	Spatial Structure and the Dynamics of Feedback	288
11.8	Conclusion	289
References		290

Section D: Multitrophic Interactions

| 12 | Mycorrhizae-Herbivore Interactions: Population and Community Consequences | 295 |

G.A. GEHRING, T.G. WHITHAM

12.1	Summary	295
12.2	Introduction	296
12.3	Effects of Aboveground Herbivory on Mycorrhizal Fungi	297
12.3.1	Population Level Responses	297
12.3.2	Community Level Responses	299
12.3.3	Carbon Limitation as a Mechanism of Herbivore Impacts on Mycorrhizae	301
12.3.4	Conditionality in Mycorrhizal Responses to Herbivory	302
12.4	Effects of Mycorrhizal Fungi on the Performance of Herbivores	305

12.4.1	Patterns of Interaction	305
12.4.2	Mechanisms of Mycorrhizal Impacts on Herbivores	307
12.4.3	Not All Mycorrhizae Are Equal	308
12.5	Ecological and Evolutionary Implications	311
12.5.1	Herbivore Effects on Mycorrhizae	311
12.5.2	Mycorrhizae Effects on Herbivores	312
12.5.3	Combined Effects	314
12.6	Suggestions for Future Research	315
References		316

13 Actions and Interactions of Soil Invertebrates and Arbuscular Mycorrhizal Fungi in Affecting the Structure of Plant Communities ... 321
A.C. GANGE, V.K. BROWN

13.1	Summary	321
13.2	Introduction	322
13.3	Soil Invertebrate Groups	323
13.3.1	Earthworms	324
13.3.2	Nematodes	325
13.3.3	Mites	327
13.3.4	Insects	328
13.3.5	Other Invertebrates	333
13.4	Field Studies	333
13.5	Conclusions	340
References		341

14 Interactions Between Ecto-mycorrhizal Fungi and Saprotrophic Fungi ... 345
J.R. LEAKE, D.P. DONNELLY, L. BODDY

14.1	Summary	346
14.2	Introduction	346
14.2.1	The Importance of Ecto-mycorrhizal and Saprotrophic Fungi in Forest Soils	347
14.2.2	Structural and Functional Similarities Between Mycelia of Saprotrophic and Ecto-mycorrhizal Fungi	348
14.3	Competition for Nutrients Between Ecto-mycorrhizal and Saprotrophic Fungi	353

14.3.1	Evidence of Saprotrophic Nutrient Mobilising Activities of Ecto-mycorrhizal Fungi	354
14.3.2	The Effect of Short-Circuiting of the N and P Cycles by Ecto-mycorrhizal Fungi on the Activities of Saprotrophic Fungi	356
14.4	Interactions Between Ecto-mycorrhizal and Saprotrophic Fungi Observed in Microcosms Containing Natural Soil	359
14.4.1	Transfers of P Between Interacting Mycelia of Ecto-mycorrhizal and Saprotrophic Fungi	359
14.4.2	The Effect of Interaction with Saprotrophic Fungi on Growth and Carbon Allocation in Ecto-mycorrhizal Mycelium	360
14.4.3	The Effect of Interaction with Ecto-mycorrhizal Mycelium on the Growth and Morphology of Mycelial Cords of the Saprotrophic Fungus *Phanerochaete velutina*	364
14.5	Conclusions	366
References		367

Section E: Host Specificity and Co-evolution

15	Mycorrhizal Specificity and Function in Myco-heterotrophic Plants	375
	D.L. TAYLOR, T.D. BRUNS, J.R. LEAKE, D.J. READ	
15.1	Summary	375
15.2	Introduction	376
15.3	Evidence for Specificity in Myco-heterotrophs	377
15.3.1	Overview of Specificity in the Orchidaceae	378
15.3.2	Overview of Specificity in the Monotropoideae	393
15.4	Influences on Specificity	395
15.4.1	Local Distribution of Fungi	395
15.4.2	Habitat and Genetic Influences on Specificity	396
15.4.3	Ontogenetic Influences on Specificity	397
15.5	Evolution of Specificity	399
15.6	Fungal Trophic Niches and Mycorrhizal Carbon Dynamics	399
15.7	Conclusions and Future Goals	405
References		407

16	Specificity in the Arbuscular Mycorrhizal Symbiosis	415
	I.R. SANDERS	

16.1	Summary	416
16.2	Introduction	416
16.3	Definitions of Specificity	417
16.4	Why Is Specificity in the Mycorrhizal Symbiosis Ecologically Interesting?	417
16.5	Theoretical Considerations on the Evolution of Specificity in Mutualistic Symbioses	419
16.6	Why Do We Not Already Know Whether Specificity Exists?	421
16.7	Evidence Supporting a Lack of Specificity	423
16.7.1	Arbuscular Mycorrhizal Fungi Have a Broad Host Range	423
16.7.2	A Systematic Perspective	424
16.7.3	Repeatable Patterns of Arbuscular Mycorrhizal Fungal Community Structure	425
16.7.4	Physiological Evidence	425
16.8	Evidence Supporting Specificity	425
16.8.1	Arbuscular Mycorrhizal Fungal Effects on Plant Fitness	425
16.8.2	Plant Species Effects on Fungal Fitness	426
16.9	What Information Is Missing and Which Experiments Are Needed?	428
16.9.1	Arbuscular Mycorrhizal Fungal Benefit from Specific Hosts	428
16.9.2	Reciprocal Benefit Between Plant Species and Fungal Species	428
16.9.3	A Genetic Basis for Specificity in Plants	429
16.9.4	A Genetic Basis for Specificity in Arbuscular Mycorrhizal Fungi	430
16.9.5	The Importance of the Hyphal Network for Understanding Specificity	433
16.10	Conclusions	434
References		434

Section F: Conclusions

17	Mycorrhizal Ecology: Synthesis and Perspectives	441
	M.G.A. VAN DER HEIJDEN, I.R. SANDERS	

17.1	Introduction	441
17.2	Ecophysiology, Ecosystem Effects and Global Change	442

17.3	Biodiversity, Plant and Fungal Communities	446
17.4	Multitrophic Interactions	449
17.5	Host Specificity and Co-evolution	450
17.6	Conclusions	452
References		454

Subject Index 457

Taxonomic Index 465

Contributors

AERTS, RIEN
   Institute of Ecological Science, Department of Systems Ecology, Vrije Universiteit, De Boelelaan 1087, NL-1081 HV Amsterdam, The Netherlands, e-mail: aerts@bio.vu.nl

ALLEN, MICHAEL F.
   Center for Conservation Biology, University of California, Riverside, California 92521, USA, e-mail: mallen@ucrac1.ucr.edu

BODDY, L.
   Cardiff School of Biosciences, University of Wales, Cardiff CF1 3TL, UK, e-mail: BoddyL@cardiff.ac.uk

BEVER, JAMES D.
   Department of Biology, Indiana University, Bloomington, Indiana 47405, USA, e-mail: jbever@indiana.edu

BROWN, VALERIE K.
   Centre for Agri-Environmental Research (CAER)
   Department of Agriculture, University of Reading, Earley Gate, P.O. Box 237, Reading RG6 6AR, UK,
   e-mail: v.k.brown@reading.ac.uk

BRUNS, T.
   University of California, Berkeley, Department of Plant and Microbial Biology, 111 Koshland Hall, Berkeley, California 94720, USA, e-mail: boletus@garnet.berkeley.edu

CLAPP, JUSTIN P.
   Department of Biosciences, University of Kent, P.O. Box 228, Canterbury, Kent, CT2 7YW, UK, e-mail: j.p.clapp@ukc.ac.uk

DANIELL, TIM J.

Department of Biology, University of York, P.O. Box 373, York, YO10 5YW, UK, e-mail: t.daniell@scri.sari.ac.uk. *Present address*: Plant Soils & Environment Division, Scottish Crop Research Institute, Invergowrie, Dundee, DD2 5DA, UK

DONNELLY, D.P.

Department of Animal & Plant Sciences, University of Sheffield, Sheffield S10 2TN, UK, e-mail: d.donnelly@shef.ac.uk

DURALL, DANIEL D.

Biology Department, Okanagan University College, Kelowna, British Columbia, V1 V 1V7, Canada, e-mail: DDurall@notes.okanagan.bc.ca

ERLAND, SUSANNE

Department of Microbial Ecology, Lund University, Ecology Building, 223 62 Lund, Sweden, e-mail: susanne.erland@mbioekol.lu.se

GANGE, ALAN C.

School of Biological Sciences, Royal Holloway, University of London, Egham, Surrey TW20 0EX, UK, e-mail: a.gange@rhul.ac.uk

GEHRING, CATHERINE A.

Department of Biological Sciences, Northern Arizona University, Flagstaff, Arizona 86011, USA, e-mail: Catherine.Gehring@NAU.EDU

GRIME, J.P.

Unit of Comparative Plant Ecology, Department of Animal & Plant Sciences, University of Sheffield, Sheffield S10 2TN, UK, e-mail: j.p.grime@sheffield.ac.uk

HART, MIRANDA M.

Department of Botany, University of Guelph, Guelph, ON, N1G 2W1, Canada, e-mail: mhart@uoguelph.ca

HELGASSON, THORUNN

Department of Biology, University of York, P.O. Box 373, York, YO10 5YW, UK, e-mail: th7@york.ac.uk

JAKOBSEN, IVER

Plant-Microbe Symbioses, Plant Biology and Biogeochemistry Department, Risø National Laboratory, 4000 Roskilde, Denmark, e-mail: iver.jakobsen@risoe.dk

List of Contributors

JONES, MELANIE

Biology Department, Okanagan University College, Kelowna, British Columbia, V1 V 1V7, Canada, e-mail: mjones@okuc02.okanagan.bc.ca

KLIRONOMOS, J.K.

Department of Botany, University of Guelph, Guelph, ON, Canada, N1G 2W1, e-mail: jklirono@uoguelph.ca

LEAKE, J.R.

Department of Animal & Plant Sciences, University of Sheffield, Sheffield S10 2TN, UK, e-mail: j.r.leake@sheffield.ac.uk

OLSSON, PÅL AXEL

Department of Microbial Ecology, Ecology Building, Lund University, 223 62 Lund, Sweden, e-mail: Pal_Axel.Olsson@mbioekol.lu.se

PRINGLE, ANNE

Department of Biology, Duke University, Durham, North Carolina 27708, USA, e-mail: anne.pringle@duke.edu

READ, D.J.

Department of Animal & Plant Sciences, University of Sheffield, Sheffield S10 2TN, UK, e-mail: d.j.read@sheffield.ac.uk

RILLIG, MATTHIAS C.

Division of Biological Sciences, The University of Montana, Missoula, Montana 59812, USA, e-mail: matthias@selway.umt.edu

SANDERS, IAN R.

University of Lausanne, Institute of Ecology, Bâtiment de Biologie, 1015 Lausanne, Switzerland, e-mail: Ian.Sanders@ie-bsg.unil.ch

SCHULTZ, PEGGY ANN

Department of Biology, Indiana University, Bloomington, Indiana 47405, USA, e-mail: pschultz@bio.indiana.edu

SIMMARD, SUZANNE W.

Kamloops Forest Region, British Columbia Ministry of Forests, Kamloops, British Columbia V2 C 2T7, Canada,
e-mail: Suzanne.Simard@gems9.gov.bc.ca

SMITH, S.E.

Department of Soil and Water and Centre for Plant Root Symbioses, The University of Adelaide, South Australia, 5005, Australia, e-mail: ssmith@waite.adelaide.edu.au

SMITH, F.A.

Department of Environmental Biology and Centre for Plant Root Symbioses, The University of Adelaide, South Australia, 5005, Australia, e-mail: andrew.smith@adelaide.edu.au

TAYLOR, ANDY F.S.

Department of Forest Mycology and Pathology, Swedish Agricultural University, Box 7026, 750 07, Uppsala, Sweden, e-mail: andy.taylor@mykopat.slu.se

TAYLOR, D.L.

University of California, Berkeley, Department of Integrative Biology, 3060 Valley Life Sciences Building, Berkeley, California 94720, USA, e-mail: dltaylor@socrates.Berkeley.edu

TRESEDER, KATHLEEN K.

Department of Biology, University of Pennsylvania, Philadelphia, Pennsylvania 19104, USA, e-mail: treseder@mail.ucr.edu

VAN DER HEIJDEN, MARCEL G.A.

Department of Systems Ecology, Vrije Universiteit, De Boelelaan 1087, 1081 HV Amsterdam, The Netherlands, e-mail: Heijden@bio.vu.nl

WALLANDER, HÅKAN

Department of Microbial Ecology, Ecology Building, Lund University, 223 62 Lund, Sweden, e-mail: hakan.wallander@mbioekol.lu.se

WHITHAM, THOMAS G.

Department of Biological Sciences and Merriam-Powell Center for Environmental Research, Northern Arizona University, Flagstaff, Arizona 86011, USA, e-mail: Thomas.Whitham@nau.edu

YOUNG, J. PETER W.

Department of Biology, University of York, P.O. Box 373, York, YO10 5YW, UK, e-mail: jpy1@york.ac.uk

Section A:

Introduction

1 Towards Ecological Relevance – Progress and Pitfalls in the Path Towards an Understanding of Mycorrhizal Functions in Nature

D.J. READ

Contents

1.1	Summary	3
1.2	Introduction	4
1.3	The Platform Provided by Reductionist Approaches	4
1.4	Approaches to Examination of Mycorrhizal Function at the Community Level	8
1.4.1	The Microcosm Approach	10
1.4.2	The Field Approach	21
1.5	Conclusion	23
References		24

1.1 Summary

The major achievement of the first hundred years of research on the mycorrhizal symbiosis is the observation that the symbiosis is almost universally present in natural communities of terrestrial plants. However, studies of the functional characteristics of the mycorrhizal associations have used, for the most part, reductionist approaches, and the role of the symbiosis in the dynamics of terrestrial plant communities has been largely overlooked. This chapter, along with others contained in the book, describes attempts made so far to place the mycorrhizal function in the broader context. The strengths and weaknesses of reductionist approaches to investigation of mycorrhizal function are assessed and the overriding need to recognise and tackle the inherent complexity of plant and microbial communities is seen as a fundamental prerequisite for progress towards ecological relevance. Two distinct pathways are seen to have the potential to facilitate this progress, one involving 'microcosm', the other 'field' approaches. The relative advantages and disadvantages of each approach is examined and the features of experimental design which will enhance the potential to obtain ecologically meaningful outputs are considered in detail. It is concluded that, in addition to the requirement for greater sophistication in our experimental approaches, there is a need for more effective collaboration with specialists in related disciplines, most notably soil chemists, bacteriologists, micro-faunists and those dealing with other fungal groups, if we are to gain an appreciation of the status of the mycorrhizal symbiosis in the larger context of ecosystem function.

1.2 Introduction

It is heartening as we enter the second century of research on mycorrhiza that attention is increasingly focused on the functions of the symbiosis in the natural and semi-natural systems in which it evolved. This volume, while encapsulating the spirit of the new adventure, also provides two further opportunities. It enables us to assess the strength of the platform from which we launch into this challenging area and to identify which experimental approaches might provide the most realistic evaluation of the roles played by mycorrhizae in natural communities.

1.3 The Platform Provided by Reductionist Approaches

It can justifiably be claimed that a major achievement of the first hundred years of research on the mycorrhizal symbiosis was to force recognition, amongst an initially largely sceptical scientific community, that the symbiosis was almost universally present in natural communities of terrestrial plants. Over this period, much emphasis has been placed on activities of the cataloguing kind. Six major types of mycorrhiza have been recognised (Harley 1959; Harley and Smith 1983; Smith and Read 1997; Read 1998), so that the basic structural and functional attributes of each can be summarised (Fig. 1.1). The taxonomic position of the fungi forming these types is understood, at least in broad terms, and the pattern of their distribution amongst plant families (Trappe 1987) and species (Harley and Harley 1987) is known.

Recently, the application of molecular methods to analysis of the taxonomic and phylogenetic relationships within the major groups of mycorrhizal fungi (e.g. White et al. 1990; Simon et al. 1992) has yielded valuable new insights. For instance, exciting discoveries have been made to describe host specificity and coevolution between plants and mycorrhizal fungi based on molecular methods (Taylor et al., Chap. 15, this Vol.). These approaches also promise to reveal much about mycorrhizal community structure in the field (Clapp et al., Chap. 8, this Vol.). Unfortunately, progress towards understanding how mycorrhizae influence the functions of plant and microbial commu-

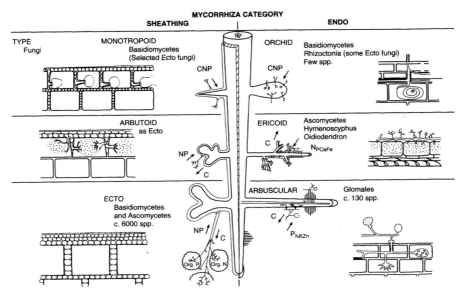

Fig. 1.1. The diagnostic structural features of the six recognised types of mycorrhiza. Two basic categories are designated, one in which the root surface is sheathed in a fungal mantle (*SHEATHING*), and one lacking a mantle but in which hyphae proliferate internally (*ENDO*). The defining structures of each type are fungal pegs (*MONOTROPOID*), Hartig net and intracellular penetration (*ARBUTOID* – also seen in the subtype 'ectendo'), Hartig net, mantle, external mycelial network (*ECTO*), peloton (*ORCHID*), hyphal complexes in hair roots (*ERICOID*) and arbuscules or hyphal coils (*ARBUSCULAR*). The most important nutrient acquisition pathways are shown. *C* Carbon, *N* nitrogen, *P* phosphorus, *Org* organic, *K* potassium, *Zn* zinc, *Ca* calcium, *Fe* iron. Predominant directions of flux of each element are indicated by *arrows*. (Modified from Read 1998)

nities in nature has not matched that achieved in the descriptive areas. There have been serious attempts, largely based on extrapolation of results obtained in laboratory studies, to suggest what these functions might be in ecosystems but they consist, in essence, of hypothetical scenarios that remain to be tested (see e.g. Allen 1991; Read 1991). We find ourselves in this unsatisfactory situation largely because, faced with the complexity of natural systems, the response of many experimentalists has been to adopt reductionist approaches. Experimental designs were simplified to the extent that their ecological relevance was compromised. We now require an evaluation of the extent to which such approaches have provided understanding of mycorrhizal function in the real world

Some useful insights have undoubtedly been gained. Thus, the sheer number of experiments carried out under 'controlled' conditions enable us to make firm predictions about the processes whereby nutrient supplies to individual roots, as well as to single plants, usually grown for short periods of time in small containers, are augmented. Details of nutrient uptake and supply by ecto-mycorrhizal fungi (EMF) and arbuscular mycorrhizal fungi (AMF) are given by Simard et al. (Chap. 2, this Vol.) and Jakobsen et al. (Chap. 3, this Vol.) respectively. However, the results produced by most laboratory and fild experiments describe only the potential of the symbiosis. We remain largely ignorant of the extent to which this potential is expressed in nature. Insofar as excision of the root removes the major vegetative components of both the plant and fungal partners, while containerisation of individual plants or species removes interactions between them, neither approach can provide a platform from which to establish a meaningful assault upon questions of function at the ecosystem level.

Reductionism has afflicted mycorrhizal research in other ways. Not only has there been an overemphasis upon studies of the nutritional role of the symbiosis but also a preoccupation with its involvement in phosphorus (P) supply to the plant. At its height, this enabled review articles describing nutrient supply and demand to mycorrhizal plants to deal exclusively with the phosphate ion (see, for example, Koide 1991). One consequence of these preoccupations is that mycorrhizae have become defined as 'symbioses in which an external mycelium of a fungus supplies soil-derived nutrients to a plant' (Smith and Read 1997). While not being necessarily inaccurate, such definitions carry with them the implication that absorption is the key, if not the only, function of the symbiosis. A broader view emphasising evolutionary perspectives, recognises that gene transfer is the key driver in natural selection and considers the symbiosis in terms of its contribution to the reproductive success or 'fitness' of the partners. Viewed in this way, a mycorrhiza is more appropriately defined as 'a structure in which a symbiotic union between a fungus and the absorbing organs of a plant confers increases of fitness on one or both partners' (Read 1999). This definition is compatible with the emerging recognition of the multifunctional nature of the symbiosis (Newsham et al. 1995a).

At the community level we are aware that mycorrhizae may express a number of potential attributes both of nutritional and broader kinds. Amongst those perceived to be of likely importance to the plant partner are the abilities to mobilise a range of nutrients from complex substrates (Aerts., Chap. 5 and Leake et al., Chap. 14, this Vol.), the provision of resistance to disease (Carey et al. 1992), to the effects of naturally occurring climatic stresses like drought (Allen and Allen 1986) and to the impacts of pollutants (Meharg and Cairney 2000). At the same time AMF and most EMF appear to be obligately dependent upon their autotrophic associates for the resources necessary to produce reproductive propagules. The result of this broader perception of the nature of the mycorrhizal condition is, that its importance for both plant and fungus can indeed go beyond that of absorption to make wider contributions to survival, fecundity and hence 'fitness' of the partners (Read 1997, 1999).

Greater awareness of these more fundamental biological attributes of the symbiosis is timely because it should enable them to be considered when ecosystem scale experiments are being designed and analysed. It leads to an emphasis on the need to examine responsiveness of the mycorrhizal partners over their full life cycles or at least over critical parts thereof, in the presence of the environmental variables likely to be imposed upon the systems in nature. Since these variables themselves differ at global (e.g. latitudinal) and local (e.g. edaphic) scales, it can be predicted that selection will have favoured distinctive attributes in the plant and fungal partners under each environmental circumstance. These attributes have been readily recognised by plant ecologists and used to define biomes (Odum 1971) or communities (Ellenberg 1988) based upon suites of largely phenological characteristics seen above ground.

That similar selective forces have moulded mycorrhizal fungal communities is indicated by the changes observed at the global scale in the extent of occurrence of arbuscular mycorrhizal, ecto-mycorrhizal and ericoid mycorrhizal categories of the symbiosis in different biomes (Read 1991). While these shifts are relatively easy to detect, there is every likelihood that effects at the more local scales will influence selection of genotypes within each category of the symbiosis but, to date, we have been less aware of this possibility. Small scale changes of community structure, for example of the kind enabling identification of particular phyto-sociological units within broader groupings, sensu Braun-Blanquet (1928) and Rodwell (1991), may well be linked to, even driven by, distinctive genotypes of mycorrhizal fungi. The recent demonstration (van der Heijden et al. 1998a,b) that genotypic diversity within AM fungal communities can have major impacts upon the performance and compo-

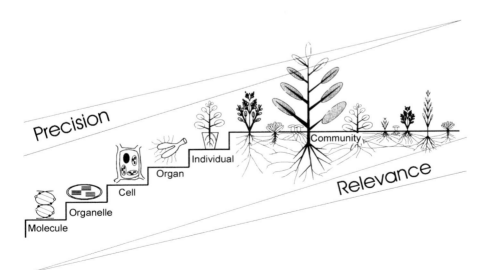

Fig. 1.2. Studies of the mycorrhizal symbiosis can be carried out at a number of levels. As we move up the hierarchy of complexity there is a likelihood that our observations have greater relevance but at the expense of precision. (Modified from Körner 1993)

sition of plant assemblages is strongly suggestive of such effects (for further information see Hart and Klironomos, Chap. 9 and van der Heijden, Chap. 10, this Vol.). Clearly, such observations are of direct relevance to issues of conservation and biodiversity below and above ground which are in urgent need of further investigation.

The 'retreat' from the challenge posed by complexity is not a unique feature of mycorrhizal research. The problems posed by the need to scale up from readily tractable simple questions to the multifactorial issues that determine ecosystem function have been acknowledged by ecologists (e.g. O'Neill et al. 1986). Körner (1993) recognised a hierarchy of complexities along which, as studies advance from the molecular, cellular or organ levels to those at the community scale, precision was inevitably lost, the great compensation being that gains in ecological relevance were obtained (Fig. 1.2). In the context of mycorrhizal research, the 'wood-wide web' of symbiotic mycelium which forms a network connecting individual species, genera, even families of plants in natural communities (Simard et al., Chap. 2, this Vol.; Read 1997) provides a good example of the complexity which prevails in the upper echelons of the hierarchy, but which is inevitably overlooked by reductionist approaches. The need to recognise such complexity and to incorporate it into experimental design will be emphasised through the remainder of the present chapter and throughout this book.

1.4 Approaches to Examination of Mycorrhizal Function at the Community Level

It is evident from Fig. 1.2 that as we advance to ask questions about the role of mycorrhiza in ecosystem processes, the complexity required of our experimental designs will, of necessity, become progressively greater. Some, the reductionists, would say that they become so great as to render the questions unanswerable but such counsels of despair may derive as much from fear of the logistics of ecosystem scale experiments as from problems inherent in the science itself. Those who latterly have ventured into the new and challenging domain have done so by two different routes which can be designated 'microcosmic' (Sect. 1.4.1 sensu Grime et al. 1987; van der Heijden et al. 1998a,b) and 'field' (Sect. 1.4.2 sensu Fitter 1986). These mirror approaches, also only recently developed, taken by ecologists to investigate the 'rules of association' in plant communities assimilating the information gained from both approaches to produce broader pictures. What is evident is that if we are to investigate complex ecosystem processes by either route, we should do so in such a way as to maximise relevance and to minimise the number of confounding factors. With this requirement in mind some of the more and less desirable features of 'microcosm' and 'field' approaches to the question of mycorrhizal function in ecosystems are considered below, and summarised in Table 1.1.

Table 1.1. Some of the major advantages and disadvantages of experimentation in microcosms and in the field, with special attention to issues of concern to those investigating the mycorrhizal symbiosis

Microcosms Advantages	Disadvantages	Field Advantages	Disadvantages
Enable control or manipulation of physical variables e.g. above ground: irradiance (quality and quantity), temperature, watering regime; below ground: soil temperature, water potential.	Absence of natural stochasticity – no realistic simulation of extreme events or fluctuations e.g. of water and nutrient supply.	The only truly representative circumstance. The 'yard stick' against which possible 'relevance' of any alternative must be judged.	Difficulty of investigating below-ground phenomena without disturbing the essential advantages of balance.
Control or manipulation of physicochemical aspects of substrates e.g. patchiness, inputs and throughputs of nutrients (budgets), rooting volume and depth.	Absence (generally) of seasonality.	Relevant climatic conditions.	Susceptibility of experimental sites to natural disasters e.g. fire, epidemic herbivore or pathogen attack.
Control of plant species or genotype composition and density.	Generally short term i.e. tend to examine what are, in essence, events representative of early rather than late successional or equilibrium environment.	Realistic soil structure. Physical and chemical heterogeneity of soil environment is maintained.	Problems of identifying single factor effects under circumstances where experimental manipulations will influence physical and biological components
Ability to recover entire plant root systems.	Emphasis upon early (developmental) rather than late (reproductive) phases of plant life cycle.	Qualitatively and quantitatively meaningful populations of all categories of organisms.	Difficulty in harvesting whole root systems or determining what proportion has been recovered.
	Entire life cycles rarely examined.		
Some ability to manipulate microbial and mesofaunal composition through pre-sterilisation, inoculation, containment and separation e.g. by use of selective mesh sizes.	Restricted to small (usually herbaceous) plants, unsuitable for large longer living species.	Presence of naturally selected communities of locally relevant mycorrhizal symbionts – often including mixtures of functional types (e.g. AM with orchid, ecto with ericoid etc).	Since we lack ecologically sensitive methods enabling us to selectively remove mycorrhizal (or other microbial) species or groups difficulties in identifying their influence arise.
	Use (generally) spores or fragmented hyphae rather than the mycelial network as units of inoculum i.e. simulate disturbed rather than stable environments.		
Ability to carry out comparative analyses of mycorrhizal and non-mycorrhizal situations under near identical (control) conditions.	Uncertainty about associated microbial population particularly effects of containment on relative sizes of populations. Uniformity of substrates (e.g. removal of woody debris) can change representation of key functional groups.	Potentially amenable to prosecution of long-term experiments or observations, in which entire life cycles can be followed.	
Ability to investigate attributes of known species and genotypes of mycorrhizal fungi in axenic, monoxenic and mixed associations with selected combinations of 'host' plants.	Experiments involving 'addition' of biological components normally require pre-sterilisation of soil. Effects of this are large and difficult to measure.		
	Larger fauna e.g. earthworms, insect larvae usually absent; – effects of this upon physical (e.g. comminution) and chemical (e.g. digestive) processing of substrates not simulated.		

1.4.1 The Microcosm Approach

1.4.1.1 Desirable Features of Microcosm Experiments

In a biological context a microcosm can be defined as a self-contained compartment which enables communities of organisms to be re-assembled, manipulated or observed under prescribed conditions. The alternative terms 'mesocosm' and 'macrocosm' are often used in attempts to indicate relative sizes, but since all such compartments are small when compared with real ecosystems, the generic use of the word 'microcosm' seems appropriate.

Microcosms have been extensively used for the experimental study of ecological systems (Beyer and Odum 1993), but construction of the assemblage to be investigated requires considerable care. Having selected a target ecosystem, a number of considerations each enhancing the potential for 'relevance' (Fig. 1.1) arise. These can be categorised (Table 1.2) into below-ground and above-ground compartments.

1.4.1.2 The Below-Ground Compartment

1.4.1.2.1 Selection of Soil

The soil should be representative of the type supporting the 'target' community in the field. This apparently 'common sense' requirement has in fact been widely overlooked by many of the pioneers of ecosystem reconstruction. Several, e.g. Lawton (1995), Fraser and Grime (1998, 1999), Buckland and Grime (2000), have resorted to an extreme reduction by employing simply 'builders-sand', this being subsequently referred to as 'soil'. No doubt they had their own reasons for making such a drastic simplification and these are beyond the scope of this treatment. The vital point is that this approach should not become accepted as a model for use in mycorrhizal studies. One of the most important features to emerge from studies at the individual plant level is that soil characteristics, particularly the inseparable aspects of quality of source and quantity of supply of nutrients, broadly determine which category of mycorrhiza dominates in the ecosystem (Read 1991). At the finer scale it influences the responsiveness of the individual plant or species (Frank 1894; Murdoch et al. 1967) to mycorrhizal colonisation. In any event, since most ecologists would accept that species composition of natural plant communities is determined by parent material-soil-climate interactions and microbiologists would acknowledge that these factors exercise selective effects upon the whole microbial community, it seems advisable to be sensitive from the outset about choice of growth medium.

Table 1.2. Attributes of microcosm design which will enhance relevance in studies of mycorrhizal involvement in community and ecosystem processes

The below-ground compartment – soils, microbes and microfauna	The above-ground compartment – plants and herbivores
a) Use of soil in which physico-chemical characteristics, particularly pH and nutrient supplies are qualitatively and quantitatively representative of those prevailing in the community in question.	f) Nomination of important, ecologically well characterised and widely distributed plant community type.
b) Selection of soil type which can be subjected to sterilisation (to facilitate comparisons of non-mycorrhizal (NM) 'control' with mycorrhizal (M) communities) without major changes in these physico-chemical attributes.	g) Inclusion of key individual species and functional groups e.g. in grassland: grasses (AM) (C_3 or C_4 where relevant to system), sedges (NM), herbs (AM responsive and unresponsive), legumes (rhizobial and AM), orchids (OM), small shrubs (AM, some EM or ericoid mycorrhizal), bryophytes. In tropical forest: trees (AM or EM), legumes, understories (AM and EM). In boreal forest: trees (EM), understories (EM, AM or ericoid mycorrhizal). Community to be initially reconstructed in a phenotypically sensitive manner e.g. tillers replanted and seeds introduced in regeneration niches.
c) Selection of genotypes of the mycorrhizal symbionts known to be present in the community in question. Consideration of interactions between categories of symbionts e.g. arbuscular mycorrhizal (AM) vs. orchid mycorrhizal (OM) or ecto-mycorrhizal (EM) vs. ericoid mycorrhizal. Inclusion of 'other' root symbionts.	h) Members of each group to be employed in the form of genotypes selected from the community in question.
d) Provision of inoculum in a manner representative of that prevailing in nature i.e. a pre-established mycelial network in grassland or forest communities, spores or root fragments for disturbed early successional environments or agricultural monocultures.	i) Addition of herbivores for grazing regimes – ecologically realistic ranges of species and meaningful numbers of individuals.
e) Reconstruction of ecologically relevant microbial communities in M and NM microcosms by addition of all major functional groups i.e. microflora and microfauna, including mycovores and herbivores with population structures and sizes appropriate for plant-soil system under investigation.	j) Supplementary imposition upon communities reconstructed in this way of environmental perturbations e.g. seasonality, drought, nutrient enrichment

1.4.1.2.2 Awareness of Effects of Sterilisation upon Soils

Since, in most experiments, the aim will be to compare community dynamics in the presence and absence of mycorrhizal inoculum there will be a need for sterilisation and this brings particular problems. Both autoclaving and irradiation change the physico-chemical nature of soil, effects upon organic constituents being particularly marked (Bowen and Cawse 1964; Stribley et al. 1975). One response to this problem has been to use naturally occurring soil of very low organic matter content, which can be sterilised with minimal impact upon the nutrient dynamics. Fortunately, many natural ecosystems in sandy soils support highly mycotrophic systems and their soils can be used as 'models' to gain some insights into mycorrhizal effects. Grime et al. (1987) used calcareous dune sand as a substrate upon which to reconstruct a community representative of that occurring widely on sandy substrates. The pH, base-status and nutrient regime (e.g. P largely in the form of insoluble shelly material, N as NO_3) after autoclaving are such that extrapolation to the natural habitat is possible. An alternative approach (van der Heijden et al. 1998b, Exp. 1) is to use an inert quartz sand as the bulk substrate but to enhance the 'relevance' by adding a proportion of sterilised natural soil from the field site in question. Preliminary manipulation of ratios of inert to natural substrates may enable the side-effects of sterilisation to be minimised while the physico-chemical relevance gained by natural substrate addition is maximised.

While these treatments can yield ecologically meaningful substrates for studies of AM systems, the difficulties facing those wishing to experiment with the highly organic substrates in which ericoid and ecto-mycorrhizal fungi proliferate are more daunting. Since it is the quality of the nutrient resource which on the one hand is most sensitive to the sterilisation techniques, and on the other the factor determining susceptibility to microbial exploitation, there is a need to resolve this issue. In the absence of a fungicide capable of selectively removing ecto- or ericoid mycorrhizal fungi, or a non-destructive heat or irradiation treatment, studies to date have examined the abilities of these fungi to exploit unsterilised natural substrates like litter (Bending and Read 1995; Perez-Moreno and Read 2000). While selective exploitation of the resources contained in these materials by mycorrhizal fungi can be demonstrated, the contribution of other microbial fractions to the mobilisation process is difficult to evaluate. An alternative approach to this question is to grow representative tissues aseptically and then to supply these as necromass to axenically produced mycorrhizal seedlings (Kerley and Read 1988). The deficiency here is that the organic residues are not physico-chemically identical to those which have been subjected to natural decomposition processes in soil litter horizons.

1.4.1.2.3 Selection of Symbiont Species and Genotypes

There is increasing evidence that the soil, as well as the plant, can exert selective effects upon fungal genotypes in the main types of mycorrhiza. Amongst the soil factors which can cause shifts in AMF population structure at the species level are water content (Anderson et al. 1984), pH (Porter et al. 1987), temperature (Koske 1987) and organic content (Johnson et al. 1991). In combination, such effects are likely to produce populations that are peculiar to a given soil type. In a study of the selective effects imposed by soil upon AMF community structure, Johnson et al. (1992) showed that of 12 abundant species, the occurrence of 6 was strongly influenced by soil type. There is thus much to suggest that the population of microbial symbionts will be moulded by local environmental conditions in the same way as is the plant population seen above ground and that, particularly in stable natural and semi-natural environments, particular genotypic traits will have been selected at the intra-specific level over many generations. The fact that disturbance appears to lead, on the one hand, to severe losses of genetic diversity and, on the other, to selection of a distinctive, perhaps ruderal, genotypes of mycorrhizal fungi (Helgason et al. 1998), indirectly lends support to the view that AM populations are specially adapted to particular local conditions.

Recognition of the likelihood that AMF populations may be edaphically selected is important. It runs counter to a prevailing view, perhaps largely based upon the low levels of speciation seen in the Glomales, that genetic diversity is low. The demonstration that there is extensive functional specialisation in the plant-fungus partnership (Klironomos 2000) should lead us to be sensitive about selections of fungal genotypes for plant community experiments (see c in Table 1.2). While it may be appropriate, indeed ideal, for studies of molecular or physiological processes to use 'model' organisms, the trend in ecosystem studies should be away from use of ecologically irrelevant genotypes obtained from culture collections, towards selection of those isolated from the ecosystem being investigated. Since some of these may be poorly represented in the local spore population the use, as initial inoculum, of root segments freshly cut from the plant species in question has much to recommend it. Parallel taxonomic characterisation of these AM communities, using conventional or molecular methods as appropriate, is clearly desirable. Genetic characterisation will in turn provide the potential for determination of the extent of intra-specific specialisation within AM communities. Access to these populations will enable experimental investigation of the relative importance of soil type on the one hand, and plant genotype on the other, in selecting for genetic differentiation in populations of the fungal symbionts.

Ecto-mycorrhizal fungal communities develop most intensively in the organic residues of their autotrophic partners (Read 1991). While there is evidence that on a given soil, interspecific differences in the quality of litter produced by individual plant species can have profound influences upon the structure of the mycorrhizal community (Conn and Dighton 2000), we know little about the selective effects of sub-surface soil conditions. At the large scale there are, of course, differences between the species composition of ecto-mycorrhizal fungal communities of acidic and basic soils. Though ecto-mycorrhizal forests are characteristically associated with acidic soils, plant communities dominated by ecto-mycorrhizal plants e.g. *Pinus* or *Fagus* spp. occur widely on calcareous substrates. Distinctive naturally selected species or genotypes of ecto-mycorrhiza fungi appear to play a key role in enabling their host plants to tolerate the particular nutritional constraints of such soils (Lapeyrie 1990).

The enormous variability of physiological properties seen in the dicaryotised progeny of individual ecto-mycorrhizal species, e.g. *Hebeloma cylindrosporum* (Debaud et al. 1995) demonstrates a profligate potential at the intra-specific level for production of genotypes capable of dealing with distinctive soil conditions. Indeed it may be their inherently greater rate of genetic turnover that enables ecto-mycorrhizal symbionts to play such an important role in providing tolerance in their host plants to adverse soil properties (Meharg and Cairney 2000). Here, as in the case of AM fungi, it is evident that EMF species and genotypes to be used in studies of nutrition, or of tolerance to environmental extremes, must be selected with due regard for the possibility of specificity to habitat types as well as to host.

There are yet other scales of resolution, not so far considered in experimental studies, involving interactions between the fungi forming different categories of mycorrhiza. Within AM dominated calcareous grassland, for example, individual members of the Orchidaceae which are hosts to basidiomycetous (usually *Rhizoctonia*) mycorrhizae (OM), or clones of shrubs such as *Helianthemum* that form ecto-mycorrhiza (EM), often co-occur. The dynamic interactions between these very distantly related fungal groups may play a key role in determining the structure of the plant community seen above ground (see also g in Table 1.2).

At a coarser level still, communities in which the dominants have one type of mycorrhiza, for example the EM plant species of boreal forest, may have an understorey of plants with another, in this case often ericoid (ERM) type. The presence of one type of mycorrhiza may be inhibitory to that of another. Field observations (Kovacic et al. 1984) have suggested that under ecto-mycorrhizal *Pinus* spp the development of tree or herbaceous seedlings of species which are normally colonised by AM fungi is inhibited. Conversely, diversity of herbaceous species that are normally hosts to AM fungi in the Great Smoky Mountains, showed a strong positive relationship to the abundance of trees capable of forming AM (Newman and Reddell 1988). Observations of these kind provide circumstantial evidence that the mycorrhizal mycelial network may play an influential role, negative or positive in its effects, upon recruitment into the community. There is some evidence for spatial segregation of the different types of mycorrhizal root with respect to soil horizons (see Read

1991), but the basis of such effects and their involvement in determining the observed community structure remains to be investigated. The nature of these interactions in small plants could be examined using microcosms but this is one of those areas of research where, for larger plants at least, field studies, perhaps aided by use of rhizoscopes, rhizotrons or root windows, may be the only realistic way forward.

One further consideration should be made by those who recognise the need to reconstruct a balanced population of root symbionts. Any researcher who spends time carefully observing field grown roots will soon become aware of the presence of fungi which are not readily classified into one of the known functional groups of mycorrhizal organisms. Despite the widespread occurrence of what Peyronel (1924) referred to as 'dual infection' we have, probably for reasons of convenience, largely ignored these fungi (Read 1992). Though some of them, most notably the so called dark-septate (DS) forms have received some attention recently (Jumponnen and Trappe 1998), many are less readily visible even in 'stained' roots and their status remains unclear. Experiments in which plant responses to 'conventional' mycorrhizal fungi are re-examined in the presence and absence of other widely occurring symbionts would take us a further step towards relevance.

1.4.1.2.4 Types of Inoculum

It has become the custom, perhaps again largely for reasons of convenience, to use spores or soil as sources of inoculum for test plants being grown in pot experiments. Using this approach, rates of colonisation, which can be critical determinants of plant response, are inherently variable, depending as they must in part upon the concentration of inoculum used. It is an essential prerequisite of an experiment designed to enquire about the involvement of AM fungi in determining the 'rules of assembly' of plant communities, that the nature of the infection process be taken into account. In cultivated agricultural land and in post-disturbance early successional environments, spores and other fractionated point sources of inoculum may indeed constitute the only major source of inoculum, so it becomes entirely appropriate to use inoculum of these kinds in reconstructing such environments. However, one of the most ecologically relevant advances in the understanding of the mycorrhizal condition over recent years has been recognition of the importance of mycelial networks both as sources of inoculum and as mechanisms for nutrient capture and distribution (Read et al. 1985). These are seen to facilitate early colonisation of compatible seedlings (Birch 1986), their recruitment into pre-established plant communities (Grime et al. 1987) as well as the exclusion of incompatible species (Francis and Read 1994, 1995).

Clearly, if the regeneration niche, which is so important a determinant of incorporation into established communities (Grubb 1977) is to be meaningfully recreated in experiments, it should not only contain a mix of relevant fungal genotypes but these should be present in pre-established mycelial networks. Such networks or 'webs' (Read 1998) can be constructed by pre-planting of a sub-set of inoculated individuals representative of the dominant component in the ecosystems concerned. Examples of such approaches as applied to study of AM networks are seen in Grime et al. (1987) and to ecto-mycorrhizal systems in Finlay and Read (1986a,b).

1.4.1.2.5 Reconstruction of 'Free-Living' Microbial Communities

It has been widely conceded that while soil sterilisation is necessary to facilitate the production of mycorrhiza-free 'control' plants, the procedure brings with it the need to re-establish a background flora of non-symbionts in the soil (see Koide and Li 1989). Unfortunately it has become the convention in research in mycorrhiza to deal with this requirement simply by adding a soil solution that has been passed through a filter paper or sieve, the pore sizes of which are assumed to enable transmission of bacteria while excluding all but the smallest fungal or animal propagules. Even the most basic knowledge of the structure of the soil microbial community indicates that this concession represents little more than a token gesture towards reality. There is now good evidence both that the component which is reintroduced to sterilised soil, and that which is excluded by filtration can exert significant impacts upon any experiments designed to investigate responses to mycorrhizal colonisation. The addition of a sievate of micro-organisms to pasteurised soil supporting mycorrhizal or non-mycorrhizal species has been shown to exert important but distinctive effects upon the two categories of plant (Hetrick et al. 1988; Kitt et al. 1988). The mycorrhizal response of C3 and C4 has been shown to be affected by the presence of specific soil micro-organisms (Hetrick et al. 1989, 1990). Results of these kinds alert us to the importance of the bacterial component of the microbial community. What of the rest?!

While studies of relationships between mycorrhizal mycelial networks and the bacterial population of soils are in their infancy, it is already becoming clear in the case of ecto-mycorrhizal systems that the mycorrhizosphere and the bulk soil can each have a bacterial flora which is distinctive in terms both of species composition and function (Timonen et al. 1998; Heinonsalo et al. 2000). One of the key questions to confront new generations of researchers will concern the extent and nature of the involvement of bacteria in the processes of root colonisation and nutrient mobilisation in natural ecosystems.

Our increasing awareness of the potential of mycorrhizal fungi to act as 'bio-control' agents, in the field (Newsham et al. 1995b) as well as in pots, brings into focus the likelihood that there will be interactions between the different functional groups of fungi in soil (Leake et al., Chap. 14, this Vol.). This emphasises the need to include specific fungal components in inocula to be applied to soil microcosms. The notion that sterilised soil can be left to become spontaneously re-infected from the air spora, dominated as it is by a select group of ruderal saprophytes, has the virtue of convenience but not of ecological sense. Any microbial inoculum should include a fungal component which is both qualitatively (species composition) and quantitatively (inoculum potential) representative of that known to occur in the soil in question. The identification of appropriate functional groups (potential pathogens, wood and litter decomposers, saccharophiles) would be a critical part of this selection process. This follows also from the observation that mycorrhizal fungi interact with saprotrophic fungi (Fig. 14.3; Chap. 14). To deal with the issue of population size it may be necessary to pre-inoculate soils with the required genotypes and, incubate them for a period sufficient to enable equilibria to develop prior to addition of mycorrhizal and non-mycorrhizal plants.

Of course, the same considerations apply to the soil fauna, which can have a big impact on mycorrhizal functioning (see Gange and Brown, Chap. 13, this Vol.). Much has been made of the possibility that fungivores, especially collembolans, might reduce the effectiveness of mycorrhizal fungi in the field by causing damage to their mycelial networks (Warnock et al. 1982; Fitter and Sanders 1992). While the experiments upon which these conclusions were based can themselves be criticised for being unrealistic, for example in relation to their failure to provide dietary choice, they nonetheless recognised the possible importance of a previously ignored biotic component. Recent studies providing a choice of food sources have suggested not only that collembolans preferentially feed on non-mycorrhizal fungi in the rhizosphere (see Gange 2000), but also that the presence of these animals can have positive effects upon plant growth (Harris and Boerner 1990; Lussenhop 1996). A possible mechanism for such effects is suggested by Gange and Bower (1996) who have shown that at moderate densities collembolans can enhance mycorrhizal colonisation. The complexity of trophic interactions is further emphasised by their observation that the growth stimulation arises not as a result of enhanced nutrient status but because foliar tissues of AM-infected plants are more resistant to grazing by chewing insects. Herbivorous insects can also exert direct influences on mycorrhizal functioning and, through these effects, change the structure of the mycorrhizal community as a whole (Gehring and Whitham, Chap. 12, this Vol.).

The issue of population size is highlighted by a microcosm study of EMF responses to collembolan grazing (Ek et al. 1994). While at high densities *Onychiurus armatus* impeded hyphal growth, at low densities it stimulated growth of the fungus *Paxillus involutus* and increased mycorrhizal colonisation.

Setälä (1995), Setälä et al. (1999) and Laakso and Setälä (1999) have pioneered analysis of the consequences of reintroduction of microfaunal groups for responses of ecto-mycorrhizal plants grown in microcosms. The conceptual advance embodied in their experimental designs lies in the recognition of food-chain hierarchies which, in the final analysis, will determine the nature and extent of mycophagy in soils. Inclusion of qualitatively and quantitatively realistic populations of invertebrates in mycorrhizal and non-mycorrhizal microcosms might be best achieved using the pre-incubation approach recommended to enable the development of fungal populations.

1.4.1.3 Above Ground Attributes

1.4.1.3.1 Nomination of Important and Widely Recognised Plant Community Types

Understanding of the contribution of mycorrhiza to ecosystem function is likely to be advanced most effectively by concentrating initially upon plant communities of well defined structure and wide geographic distribution. Amongst these, it will be most rewarding to investigate assemblages in which, for nutritional or other reasons, maximal impacts of mycorrhizal colonisation can be hypothesised. Pioneers of microcosm study have largely adopted this approach, their efforts having been devoted to representative communities of calcareous grassland (Grime et al. 1987; van der Heijden et al. 1998a,b), old field succession (van de Heijden et al. 1998b; Klironomos 2000) and tall grass prairies (Hetrick et al. 1988, 1989). These largely herbaceous communities offer the most tractable experimental materials but other types of ecosystem, which are likely to be equally responsive to the mycorrhizal symbiosis, are in urgent need of examination.

Amongst these, species-rich tropical rain forests and contrasting species-poor boreal forest systems, which constitute the main sinks of carbon in the terrestrial biosphere, are prime candidates for study. An initial reaction to the notion of experimental reconstruction of such systems may be one of trepidation but tropical rain forest assemblages have been reconstructed in macrocosms (Körner and Arnone 1992). This study demonstrated the importance of the below-ground compartment of tropical forests as the ultimate destination of much of the carbon fixed by the autotrophs. However, while providing a tantalising glimpse of 'reality', it was ultimately frustrating in that it failed to include fungal symbionts or indeed, 'real' soil. It nonetheless provides a model which, while being incomplete, demonstrates the potential for more realistic approaches. A smaller scale microcosm experiment by Perry et al. (1989), using pine species representative of boreal forests, examined the influence of different combinations of ecto-mycorrhizal fungi upon the outcome of competitive interactions between young trees. Strong impacts of fungal symbiont species were demonstrated

1.4.1.3.2 Key Individuals, Functional Groups and Phenotypes of Plants

Currently, debate concerning the contribution of plant genomic diversity to ecosystem properties is polarised around the question of whether it is the number of species or merely the presence of key functional types which determines ecosystem function (Huston 1997). Experiments on the role of biodiversity in ecosystem function have been subject to particular criticism on the basis that as numbers of species are progressively raised to give greater diversity in microcosms or plots so, inevitably, the chances of including a species with critical functional attributes (e.g. the ability to fix nitrogen) will increase. This feature, referred to as a 'sampling effect' will lead to changes in properties of the system which are attributable to that key species, not to diversity per se. Pioneering studies of the effects of fungal symbiont diversity on above ground attributes of plant communities (van der Heijden et al. 1998b) have been criticised on these grounds (Wardle 1999). While responses to such criticism have been robust (e.g. van der Heijden et al. 1999), the issue is one which should be at the forefront of thinking when experiments on diversity and ecosystem function are being designed.

It is clearly essential to include not only species representative of the dominant and subordinate phenotypes of the community in question, but also to consider the components in terms of their functional characteristics. Thus in grassland, while the dominant and subordinate grasses are essential components so also, in many systems, will be non-mycorrhizal 'graminoids' of genera such as *Carex, Juncus* and *Luzula*. Appropriate mixtures of C_3 and C_4 species may be required. This is especially important in tall grass prairies or savannahs where such species are abundant. Of the forbs, legumes are sufficiently distinctive in functional terms to require special consideration. Their contribution to ecosystem function has been well characterised in agricultural systems where emphasis has been placed upon the N fixing rhizobial symbioses. However, since effectiveness of N fixation is determined by P availability, the rhizobial and AM systems are likely to function in an interactive way and one of the main contributions of AM to ecosystem functions in grassland may be through its indirect impacts upon the ability of legumes to fix N. Accurate representation of the nature and extent of the leguminous component of any natural vegetation system is thus a prerequisite of any ecosystem reconstruction.

Preliminary analysis of the assemblages to be investigated can provide evidence of the relative contribution, for example, of species which are recruited to the community by resprouting or seeding, those which are clonal, and those which are perennial, biennial and annual. In this context, it is of interest that the growth form of clonal plants can be changed by mycorrhizal colonisation (Streitwolf-Engel et al. 1997), a feature which will influence availability of space for other species. Having classified the assemblage in this way, common sense dictates that the plants be reassembled as growth forms which represent as closely as possible those seen in nature. Unless the aim is to simulate primary succession where seed or very small vegetative propagules may be appropriate, the most ecologically relevant approach to reconstruction is to introduce species which reproduce vegetatively using representative units, e.g. tillers, runners, bulbs, and to use seed only where it is ecologically meaningful to do so. To those communities such as grasslands in which a component of the population is recruited annually from seed, the natural event can be simulated by pre-establishment of clonal dominants growing in the mycorrhizal or non-mycorrhizal condition and sowing of seed into the active turf (Grime et al. 1987). Clearly, phenotypic characteristics vary enormously within and between communities and no prescription for reconstruction can deal adequately with such differences. Suffice it to say that relevance can be enhanced by giving due consideration to these attributes prior to attempting any reconstruction, and that recourse to sowing of seed mixtures, except in the case of annual communities, will not provide ecological understanding.

As indicated earlier (see c in Table 1.2) some grassland and forest communities contain more than one type of mycorrhizal symbiosis. While reductionist approaches have enabled us to characterise ecto-mycorrhizal, ericoid mycorrhizal, and orchid mycorrhizal symbioses as being functionally distinct from those of arbuscular mycorrhizal plants (see Fig. 1.1), we have yet to embark upon investigation of the extent to which interactions between the different mycorrhizal symbionts determine the recruitment and survival of their respective host species in communities. As consciousness of the conservational and amenity values of diversity increases, so pressure upon us to investigate these interactions will grow.

1.4.1.3.3 Considerations Relating to Genotypic and Phenotypic Characteristics of the Plants in the Community

In any natural or semi-natural community the genotypic characteristics of the plants as well as of the fungi (see c in Table 1.2) will be the product of selective forces, imposed by the local environment, over many generations. Relevance can be enhanced by recognising this 'historical' perspective from the outset. Plants grown from propagules collected at the site in question are more likely to provide an ecologically realistic response to introduction of mycorrhizal inoculum than are those grown from material produced under different circumstances. Of particular concern is the trend, based upon 'convenience' for ecologist to use seed purchased from so called 'wild flower' nurseries, where plants are grown in monocultures, often over many generations, and under relatively fertile conditions. Such material, while it may be of value for amenity and some conservation purposes, is likely to be sufficiently distinct in genotypic terms not to produce representative responses in communities reconstructed to reflect particular local conditions.

1.4.1.3.4 Effects of Other Functional Groups upon Responses of Communities to Mycorrhizal Colonisation

Since above-ground herbivory is well known to exercise selective effects upon plant community development (Fretwell 1977, 1987), it is important to include carefully balanced populations of appropriate shoot feeders in any analysis of mycorrhizal impacts. This need is further emphasised by recognition that a primary determinant of palatability is tissue nutrient composition which is known frequently to be enhanced by mycorrhizal colonisation. There is also evidence that if herbivory is sufficiently intensive to reduce assimilatory surface area, mycorrhizal colonisation can itself be strongly inhibited (Gehring and Whitham, Chap. 12, this Vol.). Simulations of the large scale effects of grazing upon mycorrhizal and non-mycorrhizal communities can be achieved by clipping (Grime et al. 1987). A beneficial side-effect of this treatment is that it yields material that can be analysed for assessment of the impacts of mycorrhiza upon biomass production and nutrient quality in the community. A recent study (Fraser and Grime 1999) of the impacts of herbivory and fertility on a synthesised plant community might serve as a model for reconstruction of above-ground herbivore communities. Unfortunately, as in so many cases where a particular component of the ecosystem is targeted, overall 'relevance' in this study is reduced by the lack of attention to the below-ground compartment where again 'builders sand' (but with mycorrhizal inoculum!) was employed as a growth medium.

1.4.2 The Field Approach

1.4.2.1 Studies of Mycorrhizal Function in the Field

Because mycorrhizal fungi are of almost ubiquitous occurrence in terrestrial ecosystems any analysis of their function in the field requires the application of treatments enabling them to be removed so that non-mycorrhizal 'control' plots are available for comparison. The performance of the community can then be assessed with and without their presence.

A number of field scale methods for near complete sterilisation of soil are available for those wishing to examine successional events starting from bare soil (Stroetman et al. 1994). Amongst these, fumigation with methyl-bromide (MB) or MB-chloropicrin mixtures (Marx et al. 1991) has been most widely employed. However, all such methods are drastic and impose serious side effects particularly upon organic soils. A more sensitive procedure involving exposure of soil to direct sunlight, sometimes referred to as 'solarisation' has been widely used in tropical systems (Birch 1958) and can be employed to prepare soils for mycorrhizal inoculation (Marx et al. 1991).

The impacts of mycorrhiza upon successional events are perhaps best examined by careful analysis of the processes of recolonisation following natural 'sterilisation' events. Using this approach, Allen (1987, 1988) evaluated the factors enabling the introduction of AM propagules to 'sterile' volcanic ash and carried out assessment of the roles of fungal symbionts in facilitating plant community development after the Mount St Helens eruption. Studies of the same type have enabled some appreciation of the role of ecto-mycorrhizal fungi in recolonisation of soils after glacial retreat (Trappe 1988) and fire (Visser 1995).

For those wishing to investigate the function of AM fungi 'in situ' within established late successional or climax plant communities the choice of 'sterilisation' treatments is very restricted. In the ideal world a fungicide which precisely targets the mycorrhizal fungus or fungi in question without having 'side-effects' upon the plants, would be used. Since no such compound exists, pioneers of field investigation have resorted to the use of fungicides which are partially selective. Of these, the most popular has been benomyl (l-butyl-carbomyl)-2-(benzimadazole) (carbonic acid), a compound which until recently was widely available for use in control of fungal pathogens of soil under the name 'Benlate'. This fungicide has been shown in field studies to reduce the amounts of AM colonisation in alpine (Fitter 1986), temperate (McGonigle and Fitter 1988) and prairie (Hartnett and Wilson 1999) grasslands, as well as in plants of a woodland ecosystem (Merryweather and Fitter 1996). The consequences of reduced colonisation have been small in the case of the cool grasslands but large in the prairie and woodland studies. Factors which may determine the extent of effect of the fungicide are duration of exposure and monitoring, inherent responsiveness of the species making up the community, and any side effects which the chemical may have.

The advantages of long-term assessment are as evident in this as in any other experimental analysis of ecosystem response to an environmental variable. Further, the slower the rate of turnover in the system, the longer will be the period required for any changes to be detected. This may help to explain the apparent lack of response to benomyl recorded by Fitter (1986) in the alpine grassland. In this context, experiments of the kind carried out by Hartnett and Wilson (1999) in the tall-grass Konza prairie are likely to be the most instructive. They followed the response of the community to a twice yearly application of benomyl over 5 years. The treatment was shown to reduce AM colonisation to less than 25% of that observed in untreated plots. As a result of exposure to benomyl, the C_4 grasses which naturally dominate this system were suppressed at the expense of subordinate C_3 grasses and forbs. The overall effect of this suppression was an increase in both species richness and diversity in plots with reduced mycorrhizal colonisation. These effects became more pronounced with time. This experiment also highlights the impact of the inherent responsiveness of the constituent species in the community (for an extensive discussion on mycorrhizal responsiveness, see Chap. 10, this Vol.). Earlier pot experiments (Hetrick et al. 1986) had shown that the C_4 species were more responsive to colonisation by AM fungi than their C_3 or forb associates. While the ecological significance of these effects observed under reductionist conditions were a matter only for speculation, the field experiment allows us to infer that the C_4 grasses are dependent for their natural dominance upon mycorrhizal colonisation.

While such inferences can be made, the issues concerning the extent of specificity and of side effects of fungicide application cannot be ignored, the more so because if they are present their effects would also presumably increase over time. There are some reassuring observations, unfortunately mostly based upon pot studies. Fitter and Nichols (1988) showed that reduced colonisation in benomyl-treated plants was achieved without any detectable effect of the treatment on availability of water-soluble phosphorus (P). Likewise, Hartnett and Wilson (1999) cite as evidence that the primary impact of benomyl lay in suppression of mycorrhizal colonisation, the observation that its effects could be reversed by increasing the availability of P, and that decreases in growth of obligately mycorrhizal prairie plants occurred rather than the increases, which would be predicted if the greatest effects of the fungicides were upon pathogenic or other antagonistic fungal species.

Despite such claims, careful evaluation of the side effects of fungicides applied to natural vegetation (Paul et al. 1989) has emphasised their lack of specificity. In this era of increasing consciousness of the multi-functionality of mycorrhizal fungi, and especially of awareness of their role in protecting plants against pathogenic fungi (Newsham et al. 1994), it is clearly inappropriate to ignore the possibility of impacts upon these other functional groups.

Other fungicides have been shown to reduce mycorrhizal activity. Thus thiobendazole has been used to reduce *Rhizoctonia*-type colonisation in orchids (Alexander and Hadley 1984) while rovral (Rhone-Poulenc) in which the active ingredient is iprodione has been applied to give control of AM fungi (Gange and Brown 1992). Again, the extent of specificity of these agents is unknown.

1.5 Conclusion

The long and difficult climb towards understanding of the impacts of mycorrhiza upon the species composition and dynamics, above and below ground, of plant communities is just beginning. This volume demonstrates both the strengths and the weaknesses of the position from which we launch into the future. A strength may be that we have much precise information about mycorrhizal function under simplified conditions. The weakness, on the other hand, is that we have, as yet, little reliable information about the extent to which these functions are expressed under relevant, essentially multi-factorial circumstances of the kind that prevail in nature.

Two factors in addition to reductionism, have contributed to retard progress to date. The first is simply that the essentially short-term structure of science funding, be it for graduate or post doctoral students, even thematic programmes, is not compatible with the need to investigate processes at the ecosystem level many of which, by their very nature, are essential long term phenomena. The second, which may be more amenable to resolution, is that progress towards understanding of ecosystem level processes requires interdisciplinary collaboration. Reductionism has done more to produce isolation of mycorrhizal researchers then it has to foster links with other disciplines. The failure of many of us (for which, admittedly, we are not solely to blame!) to collaborate effectively with soil scientists, bacteriologists, micro-faunists, and even with mycologists whose interests lie in saprophytic or pathogenic domains has left us in a position where we know little of the extent to which mycorrhizae contribute to linkages or hierarchies in food chains. Of the wood-wide web we have, belatedly, become aware but even this, so far, has been perceived as a closed circuit involving only one functional group. Our failure to communicate leaves us with a lamentable lack of knowledge of the physico-chemical nature of the substrates through which the mycorrhizal web extends and of the importance of our chosen group of microbes relative to others, like bacteria, in processes such as nutrient mobilisation and transport. This book addresses big questions. They are so big that they will require effective symbioses between ourselves and other specialist groups if they are to be answered adequately.

References

Alexander C, Hadley G (1984) The effect of mycorrhizal infection of *Goodyera repens* and its control by fungicide. New Phytol 97:391–400
Allen EB, Allen MF (1986) Water relations of xeric grasses in the field: interactions of mycorrhizae and competition. New Phytol 104:559–571
Allen MF (1987) Re-establishment of mycorrhizas on Mount St Helens: migration vectors. Trans Br Mycol Soc 88:413–417
Allen MF (1988) Re-establishment VA of mycorrhizae following severe disturbance: comparative patch dynamics of a shrub desert and a subalpine volcano. Proc R Soc Edinb 94B:63–71
Allen MF (1991) The ecology of mycorrhizae. Cambridge Univ Press, Cambridge
Anderson RC, Liberta AE, Dickman LA (1984) Interaction of vascular plants and vesicular-arbuscular mycorrhizal fungi across a soil moisture-nutrient gradient. Oecologia (Berl) 64:111–117
Bending GD, Read DJ (1995) The structure and function of the vegetative mycelium of ecto-mycorrhizal plants. V. The foraging behaviour of ecto-mycorrhizal mycelium and the translocation of nutrients from exploited organic matter. New Phytol 130:401–409
Beyer RJ, Odum HT (1993) Ecological microcosms. Springer, Berlin Heidelberg New York
Birch HF (1958) The effect of soil drying on humus decomposition and nitrogen availability. Plant Soil 10:9–31
Birch CPD (1986) Development of VA mycorrhizal infection in seedlings in semi-natural grassland turf. In: Gianinazzi-Pearson V, Gianinazzi S (eds) Physiological and genetical aspects of Mycorrhizae. INRA, Paris, France, pp 233–237

Bowen HJM, Cawse PA (1964) Effects of ionizing radiation on soils and subsequent crop growth. Soil Sci 97:252–259

Braun-Blanquet J (1928) Pflanzensoziologie. Grundzüge der Vegetations Kunde. Springer, Berlin Heidelberg New York

Buckland SM, Grime JP (2000) The effects of trophic structure and soil fertility on the assembly of plant communities: a microcosm experiment. Oikos 91:336–352

Carey PD, Fitter AH, Watkinson AR (1992) A field study using the fungicide benomyl to investigate the effect of mycorrhizal fungi on plant fitness. Oecologia 90:550–555

Conn C, Dighton J (2000) Litter quality influences on decomposition, ecto-mycorrhizal community structure and mycorrhizal root surface acid phosphatase activity. Soil Biol Biochem 32:489–496

Debaud JC, Marmeisse R, Gay F (1995) Intra specific genetic variation in ecto-mycorrhizal fungi. In: Varma A, Hock B (eds) Mycorrhiza: structure, function, molecular biology and biotechnology. Springer, Berlin Heidelberg New York, pp 79–113

Ellenberg H (1988) Vegetation ecology of central Europe. Cambridge Univ Press, Cambridge

Ek H, Sjögren M, Arnebrant K, Söderström B (1994) Extramatrical mycelial growth, biomass allocation and nitrogen uptake in ecto-mycorrhizal systems in response to collembolan grazing. Appl Soil Ecol 1:155–169

Finlay RD, Read DJ (1986a) The structure and function of the vegetative mycelium of ecto-mycorrhizal plants. I. Translocation of ^{14}C-labelled carbon between plants interconnected by a common mycelium. New Phytol 103:143–156

Finlay RD, Read DJ (1986b) The structure and function of the vegetative mycelium of ecto-mycorrhizal plants. II. The uptake and distribution of phosphorus by mycelium interconnecting host plants. New Phytol 103:157–165

Fitter AH (1986) Effect of benomyl on leaf phosphorus concentration in alpine grassland: a test of mycorrhizal benefit. New Phytol 103:767–776

Fitter AH, Nichols R (1988) The use of benomyl to control fungal infection by vesicular-arbuscular mycorrhizal fungi. New Phytol 110:201–206

Fitter AH, Sanders IR (1992) Interactions with the soil fauna. In: Allen MF (ed) Mycorrhizal functioning. Chapman and Hall, London, pp 333–354

Francis R, Read DJ (1994) The contribution of mycorrhizal fungi to the determination of plant community structure. Plant Soil 159:11–25

Francis R. Read DJ (1995) Mutualism and antagonism in the mycorrhizal symbiosis, with special reference to impacts on plant community structure. Can J Bot 73 [Suppl]: 1301–1309

Frank A (1894) Die Bedeutung der Mykorrhiza-pilze für die gemeine Kiefer. Forst Centralbl 16:185–190

Fraser LH, Grime JP (1998) Top-down control and its effects upon the biomass and composition of three grasses at high and low soil fertility in outdoor microcosms. Oecologia 113:239–246

Fraser LH, Grime JP (1999) Interacting effects of herbivory and fertility on a synthesized plant community. J Ecol 87:514–525

Fretwell SD (1977) The regulation of plant communities by the food chains exploiting them. Perspect Biol Med 20:169–185

Fretwell SD (1987) Food chain dynamics: the central theory of ecology? Oikos 50:291–301

Gange AC (2000) Disruption of arbuscular mycorrhizal mutualism by Collembola – fact or fantasy? Trends Ecol Evol 15:369–372

Gange AC, Brown VK (1992) Interactions between soil-dwelling insects and mycorrhizas during early plant succession. In: Read DJ, Lewis DH, Fitter AH, Alexander IJ (eds) Mycorrhizas in ecosystems. CAB International, Wallingford, pp 177–182

Gange AC, Bower E (1996) Interactions between insects and mycorrhizal fungi. In: Gange AC, Brown VK (eds) Multitrophic interactions in terrestrial systems. Blackwell Science, London, pp 115–132

Grime JP, Mackey JML, Hillier SH, Read DJ (1987) Floristic diversity in a model system using experimental microcosms. Nature 328:420–422

Grubb PJ (1977) The maintenance of species-richness in plant communities: the importance of the regeneration niche. Bio Rev 52:107–145

Harley JL (1959) The Biology of Mycorrhiza. Leonard Hill, London

Harley JL, Harley EL (1987) A check-list of mycorrhiza in the British Flora. New Phytol 105:1–102

Harley JL, Smith SE (1983) Mycorrhizal Symbiosis. Academic Press, London

Harris KK, Boerner REJ (1990) Effects of belowground grazing by collembola on growth, mycorrhizal infection, and P uptake of *Geranium robertianum*. Plant Soil 129:203–210

Hartnett C, Wilson GWT (1999) Mycorrhizae influence plant community structure and diversity in tall grass prairie. Ecology 80:1187–1195

Heinonsalo J, Jorgensen KS, Haahtela K, Sen R (2000) Effects of *Pinus sylvestris* root growth and mycorrhizosphere development on bacterial carbon source utilization and hydrocarbon oxidation in forest and petroleum-contaminated soils. Can J Microbiol 46:451–464

Helgason T, Daniell TJ, Husband R, Fitter AH, Young JPY (1998) Ploughing up the wood-wide web? Nature 392:431

Hetrick BAD, Kitt DG, Wilson GWT (1986) The influence of phosphorus fertilisation, drought, fungus species and non-sterile soil on mycorrhizal growth responses in tall grass prairie plants. Can J Bot 64:1199–1203

Hetrick BAD, Kitt DG, Wilson GWT (1988) Mycorrhizal dependence and growth habit of warm-season and cool-season tall-grass prairie species. Can J Bot 66:1376–1380

Hetrick BAD, Wilson GWT, Hartnett DC (1989) Relationship between mycorrhizal dependence and competitive ability of two tall grass prairie grasses. Can Jo Bot 67:2608–2615

Hetrick BAD, Wilson GWT, Todd TC (1990) Differential responses of C3 and C4 grasses to mycorrhizal symbiosis, phosphorus fertilization, and soil microorganisms. Can J Bot 68:461–467

Huston MA (1997) Hidden treatments in ecological experiments: re-evaluating the ecosystem function of biodiversity. Oecologia 110:449–460

Johnson NC, Pfleger FL, Crookston RK, Simmons SR, Copeland PJ (1991) Vesicular-arbuscular mycorrhizae respond to corn and soybean cropping history. New Phytol 117:657–663

Johnson NC, Tilman D, Wedin D (1992) Plant and soil controls on mycorrhizal fungal communities. Ecology 73:2034–2042

Jumpponen A, Trappe JM (1998) Dark septate endophytes: a review of facultative biotrophic root-colonizing fungi. New Phytol 140:295–310

Greenberger J, Alan H, Grinstein A (1976) Solar heating by polyethylene mulching for the control of diseases caused by soil-borne pathogens. Phytopathology 66:683–688

Kerley SJ, Read DJ (1988) The biology of mycorrhiza in the Ericaceae XX. Plant and mycorrhizal necromass as nitrogenous substrates for the ericoid mycorrhizal fungus *Hymenscyphus ericae* and its host. New Phytol 109:473–481

Kitt DG, Hetrick BAD, Wilson GWT (1988) Relationship of soil fertility to suppression of the growth response of mycorrhizal big bluestem in nonsterile soil. New Phytol 109:473–481

Klironomos J (2000) Host-specificity and functional diversity among arbuscular mycorrhizal fungi. In: Bell CR, Brylinsky M, Johnson-Green P (eds) Microbial biosystems: new frontiers. Proceedings of the 8th international symposium on microbial ecology. Society for Microbial Ecology, Halifax, Canada, pp 845–851

Koide R (1991) Nutrient supply, nutrient demand and plant response to mycorrhizal infection. New Phytol 117:365–386

Koide RT, Li M (1989) Appropriate controls for vesicular-arbuscular mycorrhizal research. New Phytol 111:35–44

Körner C (1993) Scaling from species to vegetation: the usefulness of functional groups. In: Schultze ED, Moony HA (eds) Biodiversity and ecosystem function. Springer, Berlin Heidelberg New York, pp 117–140

Körner C, Arnone J III (1992) Responses to elevated carbon dioxide in artificial tropical ecosystems. Science 257:1672–1675

Koske RE (1987) Distribution of VA mycorrhizal fungi along a latitudinal temperature gradient. Mycologia 79:55–68

Kovacic DA, St John TV, Dyer MI (1984) Lack of vesicular-arbuscular mycorrhizal inoculum in a ponderosa pine forest. Ecology 65:1755–1759

Laakso J Setälä H (1999) Sensitivity of primary production to changes in the architecture of belowground food webs. Oikos 87:57–64

Lapeyrie F (1990) The role of ecto-mycorrhizal fungi in calcareous soil tolerance by symbiocalcicole woody-plants. Ann Sci For 47:579–589

Lawton J (1995) Ecological experiments with model systems. Science 269:328–331

Lussenhop J (1996) Collembola as mediators of microbial symbiont effects upon soybean. Soil Biol Biochem 28:363–369

Marx DH, Ruehle JL, Cordell CE (1991) Methods for studying nursery and field response of trees to specific ecto-mycorrhiza. Methods Microbiol 23:383–411

McGonigle TP, Fitter AH (1988) Ecological consequences of arthropod grazing on VA mycorrhizal fungi. Proc R Soc Edinb B94:25–32

Merryweather J, Fitter A (1996) Phosphorus nutrition of an obligately mycorrhizal plant treated with the fungicide benomyl in the field. New Phytol 132:307–311

Meharg AA, Cairney JWG (2000) Co-evolution of mycorrhizal symbionts and their hosts to metal contaminated environments. Adv Ecol Res 30:69–112

Murdoch CL, Jackobs JA, Gerdemann JW (1967) Utilisation of phosphorus sources of different availability by mycorrhizal and non-mycorrhizal maize. Plant Soil 27:329–334

Newman EI, Reddell P (1988) Relationship between mycorrhizal infection and diversity in vegetation: evidence from the Great Smoky Mountains. Funct Ecol 2:259–262

Newsham KK, Fitter AH, Watkinson AR (1994) Root pathogenic and arbuscular mycorrhizal fungi determine fecundity of asymptomatic plants in the field. J Ecol 82:805–814

Newsham KK, Fitter AH, Merryweather JW (1995a) Multifunctionality and biodiversity in arbuscular mycorrhizas. Tree 10:407–411

Newsham KK, Fitter AH, Watkinson AR (1995b) Arbuscular mycorrhiza protect an annual grass from root pathogenic fungi in the field. J Ecol 83:991–1000

Odum EP (1971) Fundamentals of ecology. Saunders, Philadelphia, USA

O'Neill RV, DeAngelis DL, Waide JB, Allen TFH (1986) A hierarchical concept of ecosystems. Monogr Popul Biol 23. Princeton Univ Press, Princeton

Paul ND, Aryes PG, Wyness LE (1989) On the use of fungicides for experimentation in natural vegetation. Funct Ecol 3:759–769

Perez-Moreno J, Read DJ (2000) Mobilization and transfer of nutrients from litter to tree seedlings via the vegetative mycelium of ecto-mycorrhizal plants. New Phytol 145:301–309

Perry DA, Margolis H, Choquette C, Molina R, Trappe JM (1989) Ecto-mycorrhizal mediation of competition between coniferous tree species. New Phytol 112:501–511

Peyronel B (1924) Prime recherche sulle micorize endotrofiche e sulla microflora radicola normale delle fanergame. Riv Biol 5:463–485

Porter WM, Robson AD, Abbott LK (1987) Field survey of the distribution of vesicular-arbuscular mycorrhizal fungi in relation to soil pH. J Appl Ecol 24:659–662

Read DJ (1991) Mycorrhizas in ecosystems. Experientia 47:376–391

Read DJ (1992) Experimental simplicity versus natural complexity in mycorrhizal systems. In: Fontana A (eds) Fungi, plants and soil. National Council for Research, Turin, Italy, pp 75–104

Read DJ (1997) The ties that bind. Nature 388:517–518

Read DJ (1998) Mycorrhiza – the state of the art. In: Varma A, Hock B (eds) Mycorrhiza: structure, function, molecular biology and biotechnology, 2nd edn. Springer, Berlin Heidelberg New York

Read DJ (1999) The ecophysiology of mycorrhizal symbioses with special reference to impacts upon plant fitness. In: Press MC, Scholes JD, Barker MG (eds) Physiological plant ecology. Blackwell Science, Oxford

Read DJ, Francis R, Finlay RD (1985) Mycorrhizal mycelia and nutrient cycling in plant communities. In: Fitter AH, Atkinson D, Read DJ, Usher MB (eds) Ecological interactions in soil. Blackwell Scientific, Oxford, pp 193–217

Rodwell JS (1991) British plant communities, vol 1–5. Cambridge Univ Press, Cambridge

Setälä H (1995) Growth of birch and pine seedlings in relation to grazing by soil fauna on ecto-mycorrhizal fungi. Ecology 76:1844–1851

Setälä H, Kulmala P, Mikola J, Markkola AM (1999) Influence of ecto-mycorrhiza on the structure of detrital food webs in pine rhizosphere. Oikos 87:113–122

Simon L, Lalonde M, Bruns TD (1992) Specific amplification of 18S fungal ribosomal genes from vesicular-arbuscular endomycorrhizal fungi colonizing roots. Appl Environ Microbiol 58:291–295

Smith SE, Read DJ (1997) Mycorrhizal symbiosis. Academic Press, San Diego

Streitwolf-Engel R, Boller T, Wiemken A, Sanders IR (1997) Clonal growth traits of two Prunella species are determined by co-occurring arbuscular mycorrhizal fungi from a calcareous grassland. J Ecol 85:181–191

Stribley DP, Read DJ, Hunt R (1975) The biology of mycorrhiza in the Ericaceae. V. The effect of mycorrhizal infection, soil type and partial soil sterilisation (by γ irradiation) on growth of Cranberry (*Vaccinium macrocarpon* Ait). New Phytol 75:119–130

Stroetmann I, Kampfer P, Dott W (1994) Efficiency of different methods for sterilization of different soil types. Zentralbl Hyg Umweltmed 195:111–120

Timonen S, Jorgensen KS, Haahtela K, Sen R (1998) Bacterial community structure at defined locations of *Pinus sylvestris Suillus bovinus* and *Pinus sylvestris Paxillus involutus* mycorrhizospheres in dry pine forest humus and nursery peat. Can J Microbiol 44:499–513

Trappe JM (1987) Phylogenetic and ecologic aspects of mycotrophy in the angiosperms from an evolutionary standpoint. In: Safir GR (ed) Ecophysiology of VA Mycorrhizal plants. CRC Press, Boca Raton, pp 5–25

Trappe JM (1988) Lessons from alpine fungi. Mycologia 80:1–10

van der Heijden MGA, Boller T, Wiemken A, Sanders IR (1988a) Different arbuscular mycorrhizal fungal species are potential determinants of plant community structure. Ecology 79:2082–2091

van der Heijden MGA, Klironomos JN, Ursic M, Moutoglis P, Streitwolf-Engel R, Boller T, Wiemken A, Sanders IR (1998b) Mycorrhizal fungal diversity determines plant biodiversity, ecosystem variability and productivity. Nature 396:69–72

van der Heijden MGA, Klironomos JN, Ursic M, Moutoglis P, Streitwolf-Engel R, Boller T, Wiemken A, Sanders IR (1999) "Sampling effect", a problem in biodiversity manipulation? A reply to David A Wardle. Oikos 87:408–410

Visser S (1995) Ecto-mycorrhizal fungal succession in jack pine stands following wildfire. New Phytol 129:389–401

Wardle DA (1999) Is "sampling effect" a problem for experiments investigating biodiversity-ecosystem function relations? Oikos 87:403–407

Warnock AJ, Fitter AH, Asher MB (1982) The influence of a springtail, *Folsomia candida* (Insecta, Collembola) on the mycorrhizal association of leek, *Allium porrum* and the vesicular-arbuscular endophyte, *Glomus fasciculatus*. New Phytol 90:283–292

White TJ, Bruns T, Lee S, Taylor J (1990) Amplification and direct sequencing of fungal ribosomal RNA genes for phytogenetics. In: Innis MA, Gelfand DM, Sninsky JJ, White TJ (eds) PCR protocols: a guide to methods and applications. Academic Press, London, pp 315–233

Section B:

Ecophysiology, Ecosystem Effects and Global Change

2 Carbon and Nutrient Fluxes Within and Between Mycorrhizal Plants

SUZANNE W. SIMARD, MELANIE D. JONES, DANIEL M. DURALL

Contents

2.1	Summary	34
2.2	Introduction	34
2.3	Ecto-mycorrhiza and Nutrient Uptake	35
2.3.1	Quantitative Effects of Ecto-mycorrhiza on Nutrient Uptake by Plants	35
2.3.2	Mechanisms of Increased Nutrient Uptake	39
2.3.3	Transport of Nutrients Within the Ecto-mycorrhizal Fungal Mycelium	43
2.4	Carbon Flux in Ecto-mycorrhizal Plants	44
2.4.1	Mechanism of Carbon Exchange Between Symbionts	44
2.4.2	Quantification of Belowground C Partitioning	45
2.4.3	Biological and Physical Effects on Belowground Carbon Partitioning	46
2.5	Carbon and Nutrient Transfers Between Plants	47
2.5.1	Hyphal Links Between Plants	47
2.5.2	Carbon and Nutrient Transfer Between Plants	50
2.5.3	Factors Regulating Carbon and Nutrient Transfer	55
2.6	Discussion of the Ecological Significance of Interplant Transfer	59
References		61

2.1 Summary

Mycorrhizal fungi are involved in the uptake of nutrients in exchange for C from host plants, and possibly in the transfer of C and nutrients between plants. Ecto-mycorrhizal fungi (EMF) increase uptake rates of nutrients by a variety of mechanisms, including increased physical access to soil, changes to mycorrhizosphere or hyphosphere chemistry, and alteration of the bacterial community in the mycorrhizosphere. They influence mycorrhizosphere chemistry through release of organic acids and production of enzymes. Movement of nutrients within an ecto-mycorrhizal (EM) mycelial network, as well as exchange of C and nutrients between symbionts, appear to be regulated by source-sink relationships. Estimates of the quantity of plant C partitioned belowground (to roots and EMF) varies widely (40–73%) depending on the methodology used and ecosystem studied, and is affected by several factors such as the identity of plant and fungal species, plant nutrient content, and EM age.

There is considerable evidence for EM and arbuscular mycorrhizal (AM) hyphal links between plants of the same or different species, from direct observation, specificity studies and molecular DNA analysis. Many studies provide supporting evidence for C, N and P transfer through hyphal links, and others indicate that transfer also occurs through soil pathways. Evidence exists for the regulation of interplant transfer by plant source-sink relationships for assimilate, fixed N or P. However, mycorrhizal fungi also appear to influence transfer through mycelial source-sink relationships, fungal species differences, or degree of colonisation. Recent studies indicate that patterns of interplant C transfer differ between EM and AM systems. Controversy remains over the importance of interplant carbon and nutrient transfer to plant community and ecosystem dynamics. Reasons for this include inadequate demonstration of net transfer as well as the low number of field studies.

2.2 Introduction

Two major plant processes are affected by the relationship between individual plants and their ecto-mycorrhizal fungal (EMF) associates: (1) nutrient uptake, which is often increased due to nutrient supply by the fungi; and (2) partitioning of C within the plant, with some ultimately being transferred to the EMF. The exchange of C and nutrients between symbionts may not be simultaneous (Jones et al. 1991) and the relative demands of each symbiont on the other will depend on the environment in which the association is occurring (Zhou and Sharik 1997). We begin this chapter with a discussion of C and nutrient exchange between an individual plant and its EMF, but it is important to remember that a single fungal mycelium can be associated with several plants (Molina et al. 1992). This increases the complexity of the potential C and nutrient fluxes in plant communities. In the second part of this chapter we review the evidence for the association of individual mycelia with several plants and how mycorrhizal fungi may facilitate C and nutrient fluxes amongst them. In this latter section, we include references to arbuscular mycorrhiza (AM), because much important work on interplant C and nutrient transfer has been carried out on these associations. Our review ends with a discussion of possible ecological implications of interplant transfer, including potential consequences for interplant competition, fungal ecology, and ecosystem productivity. This review builds on those of Newman (1988), Miller and Allen (1992) and Smith and Read (1997), and thus, concentrates on studies conducted during the past 5–10 years.

2.3 Ecto-mycorrhiza and Nutrient Uptake

The fine lateral roots of many woody plants are heavily colonised by EMF. Because fine roots are the major sites of nutrient uptake, ecto-mycorrhiza may be viewed as the main nutrient-absorbing organs in these plant species. Although some woody species absorb considerable amounts of nutrients through the apices of rapidly growing, non-mycorrhizal coarse roots (Marschner 1996 and references therein), given the overall surface area of fine versus coarse roots in soil (Vogt et al. 1982), ecto-mycorrhiza are undoubtedly important in the overall nutrition of woody plants. Ecto-mycorrhizal fungi influence uptake into roots because, in order to reach root tissues, nutrients must either travel within fungal hyphae into the Hartig net (the structure that forms the interface between the plant and the fungus), or travel through the apoplast of the fungal mantle (Harley and Smith 1983).

2.3.1 Quantitative Effects of Ecto-mycorrhiza on Nutrient Uptake by Plants

Ecto-mycorrhizal plants often have higher nutrient contents (mol nutrient plant^{-1}) than non-mycorrhizal plants (Table 2.1), especially when grown in soils with low nutrient availability (Jones et al. 1990; Bâ et al. 1999). In soils where one or more nutrients is at a high level, ecto-mycorrhiza formation is frequently suppressed (Jones et al. 1990; Dupponois and Bâ 1999). It is interesting that literature on growth responses to EMF focuses on increased nutrient uptake, whereas that on plant growth promoting rhizobacteria lists production of growth-stimulating compounds such as phytohormones, and

Table 2.1. Ratios of nutrient contents, concentrations and specific uptake rates between ecto-mycorrhizal and non-mycorrhizal plants of the same species in recent 'pot' experiments. Nutrients were supplied in solution or in insoluble organic or inorganic forms

Element	Form supplied	Substrate	Container	Plant species	Fungal symbiont	Content per plant EM/NM ratio	Concentration g^{-1} tissue EM/NM ratio	Specific uptake rate[a] EM/NM ratio	Reference
N	NH_4^+ in P-limiting nutrient solution	Perlite	70-ml syringe	*Pinus sylvestris*	*Paxillus involutus* *Suillus luteus* *Thelephora terrestris*	–	NSD	2 2 5	Van Tichelen et al. (1999)
N	Commercial complete nursery fertiliser solution	Peat/vermiculite	110-ml nursery container	*Picea mariana*	*Hebeloma crustiniforme* *Laccaria bicolor*	No difference in plant biomass	1.1–1.2	–	Quoreshi and Timmer (1998)
N	20–25 mg l^{-1} NH_4^+-N in complete nutrient solution	Sand	1.5-l pot	*Pinus sylvestris*	*Pisolithus arhizus*	1.7	NSD	3 µg N tip^{-1} day^{-1}	Högberg (1989)
N	NO_3^- (10 mol m^{-3})	Solution	Plastic growth pouches	*Pinus pinaster*	*H. cylindrosporum*	NSD	NSD	NSD	Scheromm et al. (1990)
N	Protein (bovine serum albumin)	Perlite	250-ml flask	*Betula pendula* *Picea mariana* *Picea sitchensis* *Pinus contorta*	*H. crustuliniforme*	22.3 9.8 3.2 5.1	24.5 4.0 2.4 2.2	–	Abuzinadah and Read (1986)
N	Protein (bovine serum albumin)	Perlite	250-ml flask	*B. pendula*	*H. crustuliniforme* *Amanita muscaria* *P. involutus*	49.2 27.9 17.6	3.0 2.9 5.2	b	Abuzinadah and Read (1989)
N	65 µM Pi	Perlite	70-ml syringe	*Pinus sylvestris*	*P. involutus* *S. luteus* *T. terrestris*	–	NSD	10 4 3	Van Tichelen et al. (1999)
P	50 µM Pi	Sand	163-ml container	*Pinus rigida*	*Pisolithus tinctorius*	~2.3	NSD	–	Cumming (1993)
P	$AlPO_4$ in P-free nutrient solution	Sand	163-ml container	*Pinus rigida*	*Pisolithus tinctorius*	~5.5	1.65	–	Cumming (1993)

Table 2.1. (Continued)

Element	Form supplied	Substrate	Container	Plant species	Fungal symbiont	Content per plant EM/NM ratio	Concentration g^{-1} tissue EM/NM ratio	Specific uptake rate[a] EM/NM ratio	Reference
P	Inositol hexaphosphate	Sand	163-ml container[a]	Pinus rigida	Pisolithus tinctorius	~2.2	NSD	–	Cumming (1993)
P	4 mg kg^{-1} extractable Pi	Soil	6.5-cm pot	Salix viminalis	L. proxima T. terrestris	2.9 2.1	NSD	[b]	Jones et al. (1990)
P	Apatite	Sand/peat	9×9×9 cm pot	P. sylvestris	S. variegatus Unknown fungus	1.5–2.7	NSD?	–	Wallander (2000)
K	Commercial complete nursery fertiliser solution	Peat/vermiculite	110-ml nursery container	Picea mariana	Hebeloma crustiniforme Laccaria bicolor	no difference in plant biomass	1.1–1.2	–	Quoreshi and Timmer (1998)
K	Biotite	Sand/peat	5×5×5-cm pots	P. sylvestris	S. variegatus P. involutus	NSD	NSD	–	Wallander and Wickman (1999)
K	Not determined	Non-sterilised but EM-free savannah soil	Bags containing 2 kg of soil	Afzelia africana	Scleroderma spp. Unknown isolate	NSD in plant biomass	1.4–2.1	–	Bâ et al. (1999)
S	SO_4^{2-}	Sterilised soil	[a]	Quercus robur	L. laccata	8.3, based on difference in xylem loading	–	NSD	Seegmüller et al. (1996)
S	SO_4^{2-}	Perlite/peat	5-l pots	Fagus sylvatica	P. involutus L. laccata	–	–	0.8 0.6	Kreuzwieser (1997) as cited in Rennenberg (1999)
Zn	0.3 μM Zn^{2+}	Perlite	Petri plates	P. sylvestris	S. bovinus Unknown isolate	1.2 1.1	~1.4 ~1.3	–	Bücking and Heyser (1994)

Ratios are presented only when differences between EM and NM plants were statistically significant; NSD, no significant difference between EM and NM plants for this variable; –, means that this could not be determined from the manuscript.
[a] μmol g^{-1} root unless otherwise specified.
[b] There was no net uptake of the element by NM plants.

interactions with other soil microbes as being equally important (Kloepper 1993). Ecto-mycorrhizal fungi can also stimulate plant growth through non-nutritional effects. For example, EMF can stimulate plant growth solely through the production of auxins (Karabaghli-Degron et al. 1998) or by stimulating expression of auxin-regulated plant genes (Martin et al. 1997). In addition, EMF influence the community of rhizosphere microflora (see below). Damage from root pathogens may be reduced in EM plants (Duchesne et al. 1987). Therefore, it is important to consider whether EMF increase plant growth under low nutrient conditions due to increased nutrient uptake, or whether increased nutrient accumulation in EM plants occurs due to growth stimulation by non-nutritional effects of the fungi.

Experiments that maintain similar growth rates between EM and nonmycorrhizal (NM) plants provide results that are relevant to this question. Quoreshi and Timmer (1998) exposed *Picea mariana* seedlings to exponential increases in N, P and K fertiliser such that ecto-mycorrhizae had no effect on seedling growth rate. Nevertheless, accumulation of N, P and K was 5–18% higher and resulted in higher concentrations in EM plants. Van Tichelen et al. (1999) found that specific uptake rates (uptake per g of root per unit time) of P were up to ten-fold higher and those of ammonium were up to five-fold higher in EM than NM *Pinus sylvestris* plants, even though relative growth rates were the same. These two studies confirm that nutrient uptake can be higher in EM plants independent of growth stimulation or alterations in root/shoot ratio.

In many soils, only a small proportion of the total soil nutrient content is readily available to plants. For example, in most soils P, Cu and Zn move very slowly towards roots because they have low solubilities or adsorb strongly to soil particles (Marschner 1995). Other nutrients, such as more complex forms of organic N (e.g. proteins), cannot be used as nutrient sources by NM roots of some plants (Abuzinadah et al. 1986; Turnbull et al. 1996). Compared with NM plants of the same species, EM plants take up organic nutrients and ammonium-N more effectively (Table 2.1). This is especially true for N in complex organic forms (Abuzinadah and Read 1986; Turnbull et al. 1996; see also Sect. 2.3.2.2). Rather than affecting specific uptake rates of S, participation in the ecto-mycorrhizal symbiosis enhances xylem loading (Rennenberg 1999), possibly increasing the amount of S returned to roots for protein synthesis. Little work has been done comparing rates of Ca and Mg uptake in EM and NM plants, although EMF hyphae can transport these nutrients to plants (Melin and Nilsson 1955; Jentschke et al. 2000).

Most studies of nutrient uptake by EM plants, including some of those cited in this chapter, have been carried out in pots containing sterile media. Results from these studies must be interpreted cautiously because the root density may be unrealistically low or high, the microbial community will differ from that under natural conditions, the plants are typically not competing with other plants, and because tree seedlings must be used for such experiments.

Furthermore, only a small proportion of EMF species have been used in these studies (Table 2.1) because most grow very slowly or are difficult to culture. Nevertheless, even within the small number of investigated fungi, it is obvious that substantial physiological differences exist. Therefore, different EMF should be expected to influence tree nutrition in different ways, and any comments in this chapter on the function of ecto-mycorrhizae should be viewed as generalisations only. For all these reasons, it is essential that more investigations of ecto-mycorrhizal nutrition such as those of Turnbull et al. (1996), Wallenda and Read (1999), Arocena and Glowa (2000), and Conn and Dighton (2000) will be conducted in situ in the field. These types of experiments are necessary to fully understand the functioning of ecto-mycorrhiza under natural conditions.

2.3.2 Mechanisms of Increased Nutrient Uptake

2.3.2.1 Increased Physical Access Via Extramatrical Hyphae

Ecto-mycorrhizae produce prodigious amounts of hyphae, which extend from the surface of the mantle into the soil, and are responsible for much of the nutrient uptake in EM plants (Brandes et al. 1998). Hyphae may do this by providing a larger total surface area for absorption, by providing absorptive surfaces beyond nutrient depletion zones, by expressing transporters with lower K_m values, or by accessing smaller soil pores than can be penetrated by roots. Estimates of external hyphae range from 300 m per metre root of mycorrhiza in sterilised sandy soil (Jones et al. 1990) to 1000–8000 m per metre root in non-autoclaved organic soil (Read and Boyd 1986). Based on these estimates and assuming an average hyphal diameter of 3 µm and an average mycorrhiza (root + mantle) diameter of 400 µm, hyphae produce up to a 60-fold increase in surface area. This may explain increased inflow rates (uptake per unit root length) or specific uptake rates observed in the hydroponic or semi-hydroponic studies described above, but the usefulness of this increased surface area for nutrient uptake from soil depends on its distribution. For molecules with low mobility relative to the demand by the plant, a zone of depletion forms around the root or mycorrhiza and proliferation of the absorptive area within this zone provides no benefit. Nutrient uptake by hyphae from beyond the depletion zone is likely to be significant for P and ammonium, but is unlikely to be important for Ca or Mg (Marschner 1996).

In order for the extension of mycorrhizal hyphae beyond the depletion zone to be important to plant nutrient uptake, two conditions must be met. First, these hyphae must not only translocate nutrients back to the root, but must do so at rates faster than the nutrient would otherwise move through the soil. Estimates of C and nutrient translocation through EMF hyphae range from 7.5 mm h^{-1} to 20 cm h^{-1} (e.g. Finlay and Read 1986a; Timonen et al. 1996). These rates exceed expected diffusion rates within the hyphae, which in turn, are much greater than movement through the soil for nutrients of low soil mobility (Finlay and Read 1986b; Timonen et al. 1996). Second, root density must be low enough that depletion zones from adjacent roots do not overlap. Most comparisons of nutrient uptake by NM and EM plants have been done in pots with single plants. Growth differences tend to be highest once colonisation is well established. Later in the experiment, when root densities increase, differences in plant growth and nutrient uptake have been seen to diminish or the effect may even become reversed (Jones et al. 1991). The effectiveness of the external hyphae for the uptake of poorly soluble nutrients will also be reduced if depletion zones around hyphae overlap, as detected by Pedersen et al. (1999). It is important to understand how common this is under natural conditions. The use of stable or radioisotope tracers to study uptake by EM plants in situ in the field (e.g. Näsholm et al. 1998) would help to answer this question and such experiments are certainly required.

Until recently, difficulty in producing ecto-mycorrhiza in hydroponics has limited study of the kinetics of uptake systems of ecto-mycorrhiza to excised roots from the field. Such studies have indicated that kinetics differ substantially amongst ecto-mycorrhiza formed by different fungi (Wallenda and Read 1999). Using a semi-hydroponic system with a relative addition rate of 3 % day^{-1}, Van Tichelen and Colpaert (2000) were able to compare P uptake kinetics in EM and NM *Pinus sylvestris*. They found that the K_m values for the high affinity P_i uptake system were lower and the V_{max} was up to seven-fold higher in root systems colonised by *Paxillus involutus*, compared with NM roots. This indicates that phosphorus could be taken up more efficiently at lower concentrations (a lower K_m value) and that there was a higher maximum uptake capacity for phosphorus (a higher V_{max}) when EMF were present. Enhanced uptake kinetics for methylamine, an ammonium analogue, has also been demonstrated in ecto-mycorrhiza (Javelle et al. 1999). These characteristics would be expected to influence total influx rates into plant roots when EMF hyphae extend beyond depletion zones or into soil microsites with high nutrient concentrations in the soil solution.

Because roots are primarily restricted to soil macropores (Hatano et al. 1988), EMF hyphae could provide access to additional nutrients by growing into micropores and penetrating soil aggregates. Access by hyphae to the interior of soil aggregates would be significant because the concentration of water-soluble nutrients is higher in their interior than in surrounding water films (Hildebrand 1994). Wilpert et al. (1996) and Schack-Kirchner et al. (2000) found only 8–30 % of the hyphal length within soil aggregates in two forest soils. Of these hyphae, 70 % were restricted to the outer 50 µm of the aggregates, possibly due to low rates of oxygen diffusion. These results suggest that hyphae and roots have similar distributions in soil, but an alternate interpretation is possible. Schack-Kirchner et al. (2000) define macropores as those with diameters greater than 10 μm, but we would not expect pores as small as 10 μm to be occupied by roots. Thus, entry of hyphae into these smallest macropores would indeed provide access to nutrients otherwise not available to roots. Moreover, Jongmans et al. (1997) and Van Breemen et al. (2000) provide visual evidence that EMF might gain unique access to P, K, Ca and Mg by penetrating feldspar and hornblende particles in the E horizon of podzols.

2.3.2.2 Changes to the Chemistry of the Mycorrhizosphere or Hyphosphere

There is good evidence that EMF can alter the chemistry of the mycorrhizosphere by increasing the rate of weathering of soil minerals and supply the released nutrients to the plant. Ecto-mycorrhiza-enhanced release of Fe from biotite (Watteau and Berthelin 1994), P from apatite (Wallander et al. 1997a), and Ca from calcium phosphates (Lapeyrie et al. 1990) have been observed. The evidence for the significant enhancement of K uptake from biotite and microcline is weak, although increased fungal biomass seemed to increase weathering of these minerals (Wallander and Wickman 1999; Wallander 2000). Minerals such as mica and chlorite appear to be transformed at higher rates in the ecto-mycorrhizosphere of *Abies lasiocarpa* compared with the rhizosphere of NM roots (Arocena et al. 1999), resulting in higher cation concentrations (Arocena and Glowa 2000). This is important because it suggests that EMF not only increase uptake from the soil solution, but can also increase the capacity of the soil to supply nutrients.

The increased weathering may be due to the release of organic acids by EMF (Van Breemen et al. 2000). For example, the production of Ca oxalate crystals by EMF has been associated with the breakdown of Ca phosphate granules (Cromack et al. 1979; Griffiths et al. 1994; Chang and Li 1998). Arocena et al. (1999) attributed lowered pH in the ecto-mycorrhizosphere to the release of organic acids by the fungi. Although organic acids such as oxalic, formic, lactic, citric and malic acids are common components of the exudates of both NM and EM roots (Jones 1998), concentrations are generally correlated with levels of fungal biomass in ecto-mycorrhiza (Wallander and Wickman 1999; Wallander 2000). A recent model predicts that organic acid secretion can be important in P solubilisation and uptake (Kirk 1999). This means that ecto-mycorrhiza and their external hyphae have the potential to weather minerals both within and beyond the rhizosphere.

Ecto-mycorrhizal fungi also access soil nutrients through the production of enzymes that degrade complex organic compounds (Bending and Read 1995a; Conn and Dighton 2000). This mechanism is especially important for organic N and P, but has the potential to be important for organic S as well. In forest soils, where most ecto-mycorrhiza are found, a high percentage of N occurs as complex organic N (Qualls et al. 1991; Michelsen et al. 1998; see

Aerts, Chap. 5, this Vol.). Although there is increasing evidence that roots of many plants, including NM plants, can absorb simple organic N such as amino acids (Kielland 1994; Näsholm et al. 1998), EM plants can absorb these at faster rates than NM plants of the same species (Chalot and Brun 1998). Moreover, EMF can absorb N from more complex forms of organic N, such as proteins (Abuzinadah and Read 1989; Chalot and Brun 1998). Natural abundance of ^{15}N tends to be higher in EM than in arbuscular or NM plants; this could be due to isotopic fractionation during transfer from fungus to plant or it could be further evidence that EM plants accumulate N from organic N sources (Table 5.1 in Chap. 5; Michelsen et al. 1998). Both EM and NM roots secrete phosphatases (Dighton 1983), and enzyme activities per gram of root or mycorrhiza appear to be inversely related to the P status of the plant or fungus (Cumming 1993). Higher enzyme activities in the vicinity of mycorrhizal roots may be because of the increased surface area for secretion provided by the hyphae (Marschner 1996).

Ecto-mycorrhizal fungi are often considered less effective at decomposing organic matter than saprotrophic fungi because they have limited abilities to degrade recalcitrant organic compounds such as lignin and phenolics (Durall et al. 1994a; Bending and Read 1997). Nevertheless, specific EMF can produce phosphomonoesterase activities comparable to, or higher than, those of decomposer fungi (Dighton 1991 and references therein; Colpaert and Van Laere 1996). Moreover, Leake et al. (Chap. 14, this Vol.) show that competitive interactions for nutrients occur between ecto-mycorrhizal fungi and saprotrophic fungi. Gramss et al. (1999) found that some EMF produced enzyme levels comparable to wood decomposers or terricolous fungi. Observing enzyme production in the laboratory does not establish that these enzymes would function under the low temperatures and pHs of soils where many EM plants grow. Nevertheless, evidence is accumulating that extracellular hydrolytic enzymes produced by EMF retain their activities at low temperatures (Tibbett et al. 1998) and pHs (Botton and Chalot 1995), and when bound to clay particles (Leprince and Quiqampoix 1996). Moreover, Bending and Read (1995b) detected elevated enzyme activities in organic matter from the forest floor fermentation horizon following colonisation by *Paxillus involutus*. In parallel experiments, *Suillus bovinus* and *Thelephora terrestris* translocated N and P from the same substrate, suggesting that EMF hyphae do indeed access organic forms of N and P using hydrolytic enzymes (Bending and Read 1995a).

2.3.2.3 Influence on Rhizosphere Bacterial Populations

EMF may also influence nutrient uptake indirectly through their effects on bacterial populations in the rhizosphere. Bacteria have been observed on, and within, the mantles of ecto-mycorrhiza and are especially numerous on the surfaces of fine hyphae at the mycelial front (Nurmiaho-Lassila et al. 1997).

The numbers and diversity of the bacterial community depend on the location within the mycorrhizal system, on the fungal species (Schelkle et al. 1996; Nurmiaho-Lassila et al. 1997) and on the chemistry of the soil (Olsson et al. 1996; Olsson and Wallander 1998). Given that bacteria are so commonly associated with ecto-mycorrhiza, it is surprising that some studies indicate that EMF hyphae may have little or no effect on the size of bacterial populations in soils, or even cause decreases of 20–50 % (Olsson et al. 1996). In the study by Nurmiaho-Lassila et al. (1997), bacterial colonisation was heaviest on non-mycorrhizal long roots.

Any change in bacterial populations in the mycorrhizosphere or hyphosphere would be expected to influence nutrient availability and uptake by the mycorrhiza. Bacteria undoubtedly compete with roots for organic nutrients (Kaye and Hart 1997), but may also chelate nutrients for uptake by roots or mycorrhizal fungi (Paul and Clark 1989). Some plant growth-promoting rhizobacteria stimulate tree growth only in the presence of mycorrhiza (Shishido et al. 1996a,b). In addition, N-fixing bacteria appear to be associated with specific types of ecto-mycorrhiza (Li et al. 1992).

2.3.3 Transport of Nutrients Within the Ecto-mycorhizal Fungal Mycelium

Nutrient movement within a mycelial network can be complex because hyphae in different parts of the system have different physiologies (Cairney and Burke 1996). Translocation patterns also differ amongst fungi (Olsson and Gray 1998). Understanding the mechanisms that control nutrient movement is essential for predicting the effects that plants will have on transport in a mycelial network connecting several plants. Within a mycelium, C tends to be directed toward active, nutrient-acquiring regions (Fig. 14.3, Chap. 14; Bending and Read 1995b; Cairney and Burke 1996; Leake et al. 1999). When a fungus colonises a new root, this region may initially act as a sink for C and nutrients (Finlay and Read 1986b). The developing ecto-mycorrhiza will rapidly become a source of C for the fungus, rather than a sink, although the C supplied by the plant decreases as the mycorrhiza ages (Durall et al. 1994b). It is easy to envision that a section of a mycelium associated with one plant could act as a source of C, which is then translocated to a section that is a C sink during active colonisation of a second plant.

The source and sink concept fits the model proposed by Cairney (1992) on how nutrients may be translocated through rhizomorphs (multi-hyphal linear aggregates emanating from a root); however, the mechanisms of nutrient translocation within fungal hyphae are still being debated (Cairney 1992; Jennings 1995; Olsson and Gray 1998). Jennings (1995) lists three possible mechanisms for translocation within hyphae: diffusion, use of a contractile system, and pressure-driven bulk flow. The pressure gradient could be generated by differences in osmolarity at the two ends of the hyphae or by evaporation of water. Cairney (1992) discusses the special situation of translocation within rhizomorphs. He proposes that bulk flow is the major mechanism of transport away from the mycorrhiza within the rhizomorph vessel hyphae. Nearer the mycelial front, C would be loaded into the cytoplasm of smaller diameter hyphae and carried further by cytoplasmic streaming. Basipetal transport of mineral nutrients toward the root via vessel hyphae of rhizomorphs would occur by mass flow generated by a water potential gradient (Cairney 1992). According to this model, differences in transpiration rate between trees connected to the same EM mycelium would influence the direction and rate of mineral nutrient transport toward their roots. To test the mass flow hypothesis, Timonen et al. (1996) manipulated transpiration rates of EM plants and found no impact on translocation rates within hyphae. Jennings (1995) emphasises that this type of evidence is not sufficient to eliminate the possibility of bulk flow in the transpiration stream. In addition, it is important to remember that mass flow can also be caused by differences in pressure between different parts of the mycelium caused by loading or unloading of solutes. In addition to mass flow, several mechanisms of active translocation have been proposed, including cytoplasmic streaming (Finlay and Read 1986b) and peristalsis in tubules that connect vacuoles within and between cells (Ashford et al. 1994). Additional information on hyphal foraging and resource allocation within ecto-mycorrhizal fungi mycelium is given by Olsson et al. (Chap. 4, this Vol.).

2.4 Carbon Flux in Ecto-mycorrhizal Plants

Ecto-mycorrhizal fungi rely on their plant symbionts for most of their fixed C (Smith and Read 1997). Although net C flux within the symbiosis is from plant to fungus, there is increasing evidence that C also moves from fungus to plant (Cullings et al. 1996; Ek 1997; Simard et al. 1997a). The focus of this section is on the mechanism and quantity of C flux from plant to EMF, and on key factors affecting it.

2.4.1 Mechanism of C Exchange Between Symbionts

Several mechanisms have been proposed to explain the movement of C from plant symbiont to EMF (Smith and Read 1997). In one of these, Salzer and Hager (1991, 1993), Schaeffer et al. (1995), and Hampp and Schaeffer (1995) propose that C is transported to the root apoplast as sucrose and is subsequently hydrolysed to fructose and glucose by plant invertase. The fungus preferentially absorbs glucose because high levels of glucose inhibit the uptake of fructose by the fungus. In turn, the relatively high level of fructose inhibits invertase activity preventing further hydrolysis of sucrose. As the level of glucose falls, the fungus begins to absorb fructose, which removes the inhibition of invertase activity. This mechanism appears to rely on the fungus converting glucose and fructose to fungal metabolites, such as glycogen, trehalose and mannitol, thereby maintaining a concentration gradient from plant to EMF, and allowing for continued uptake of glucose and fructose by the fungus (Smith et al. 1969; Jordy et al. 1998). However, there are still gaps in our understanding of this mechanism (Smith and Read 1997). Recent findings show that increased monosaccharide concentrations at the fungus/plant interface upregulate monosaccharide protein transporters in the fungus (Rieger et al. 1992; Nehls et al. 1998). This may result in enhanced photoassimilate sequestration by the fungus that will trigger additional photoassimilate supply by the host to the fungus and will ultimately increase rates of photosynthesis (Hampp et al. 1999). This provides a mechanism by which mycorrhization can cause increased photosynthetic rates (Jones and Hutchinson 1988; Rygiewicz and Andersen 1994; Colpaert et al. 1996).

It is proposed that most movement of C from root to fungus is by way of sugar transport (Smith and Smith 1990; Smith and Read 1997); however, there is evidence of bi-directional movement of amino acids between symbionts (Lewis 1976). Carbon in amino acids is thought to usually move from fungus to root by way of the glutamine/glutamate shuttle across the symbiont interface (Smith and Smith 1988), but it also may move as glutamine from root to fungus (Lewis 1976).

2.4.2 Quantification of Belowground C Partitioning

Estimates of the quantity of C partitioned to belowground components and from plant symbiont to EMF have been quite variable, which is not surprising given the wide variability in methodology and growing conditions among studies. Field estimates of total belowground partitioning to both roots and ecto-mycorrhizal fungi of trees were 40–73% of assimilated C (Edwards and Harris 1977; Persson 1978). In laboratory studies, EM plants partitioned 3–36% more fixed C to belowground components than did non-mycorrhizal plants (Reid et al. 1983; Rygiewicz and Andersen 1994; Durall et al. 1994c). In field studies, C demand values by the EMF have ranged from 10–50% of net primary production (Fogel and Hunt 1979; Vogt et al. 1982). Field studies are likely to underestimate C used by EMF because it is difficult to accurately measure the C demand of fruiting bodies and external hyphae supported by a tree or a community of trees (Rygiewicz and Andersen 1994; Tinker et al. 1994; Smith and Read 1997). By contrast, laboratory studies can overestimate C flux to the EMF by not taking into account differences in size or rates of root turnover between mycorrhizal and non-mycorrhizal plants (Durall et al. 1994c; Tinker et al. 1994). Although estimates of C flux from EM plants in the field are wide-ranging, there are two estimates from different parts of the world that appear to be in agreement. Finlay and Söderström (1989) accounted for turnover of EM roots and the production of external mycelium, fungal mantle and sporocarps, and calculated 15% of C flux to EMF tissue in a Swedish pine forest. Vogt et al. (1982) estimated a similar C flux in an *Abies amabilis* forest, but they did not take into account the production and turnover of external mycelium or EM roots.

The partitioning of C to different belowground pools, including plant tissue, fungal tissue, exudates and respiration has been estimated in laboratory studies (Reid et al. 1983; Durall et al. 1994c; Rygiewicz and Anderson 1994). Durall et al. (1994c) found no difference in the portion of belowground C in respiration, belowground biomass or soil in EM versus NM willow. Others found the differences in belowground C partitioning between EM and non-mycorrhizal plants largely due to fungal respiration (Rygiewicz and Anderson 1994; Colpaert 1996; Ek 1997). Ek (1997), studying ecto-mycorrhizal birch, found EMF external hyphae received 20–29% of shoot net assimilation, and 43–64% of the C allocated to the mycelium was respired. In terms of exudates, Leyval and Berthelin (1993) found that mycorrhizal *Fagus sylvatica* exuded more sugars and amino acids than non-mycorrhizal plants. A recent report suggests that ecto-mycorrhiza with well developed mantles release fewer soluble exudates than those with underdeveloped mantles (Priha et al. 1999). This C efflux to the rhizosphere has been shown to affect C and N cycling as well as the numbers and biomass of microbes in the rhizosphere (Bradley and Fyles 1995).

2.4.3 Biological and Physical Effects on Belowground C Partitioning

Plant factors that may affect belowground C partitioning include identity of the plant and fungal species, nutrient content of the EM plant (Wallenda et al. 1996; Ek 1997), photosynthetic rate (Colpaert et al. 1996), plant age and root age (Cairney and Alexander 1992). Factors external to the plant include: rhizoplane community structure (Timonen et al. 1998); soil animal grazing of EMF (Setälä 1995); defoliation/herbivory (Kytöviita and Sarjala 1997; Rossow et al. 1997); and elevated CO_2 (Norby et al. 1987; Hodge 1996; Rouhier and Read 1998; Rillig et al., Chap. 6, this Vol.). In this section we focus on two of these factors: nutrient content of the plant and root age.

Carbon flux as affected by plant nutrient status can be analysed using cost:benefit analysis (Koide and Elliott 1989). Jones et al. (1991, 1998) viewed incremental C allocated belowground as the cost of ecto-mycorrhiza to the plant and incremental P uptake as the benefit. Jones et al. (1991) found that this cost:benefit ratio increased with time since establishment of the association. This would be expected to be even more extreme in a field situation where C demand will increase with time due to sporocarp production. In a comparison between AM and EM plants, Jones et al. (1998) calculated the P uptake efficiency (P uptake/C partitioned belowground) and found both EM and AM plants were more efficient than NM plants at acquiring P per unit of C allocated belowground. Other approaches to cost:benefit analysis include: incorporating nutrients other than P (Ek 1997); non-nutritional benefits (Colpaert et al. 1996); expressing both cost and benefit in C units (Tinker et al. 1994); or viewing cost:benefit with respect to the mycorrhizal fungus instead of the plant.

Although laboratory studies using ^{14}C tracers have typically found greater accumulation of ^{14}C in younger EM roots than older ones, factors other than chronological age influence the amount of ^{14}C partitioned to ecto-mycorrhiza (Cairney et al. 1989; Cairney and Alexander 1992; Durall et al. 1994b). For example, Durall et al. (1994b) found more ^{14}C photosynthate allocated to 90-day-old *Hebeloma crustuliniforme* mycorrhiza than to *Laccaria laccata* roots of the same age. In addition, light and turgid roots (i.e. morphologically young, sensu Downes et al. 1992; Cairney and Alexander 1992) were allocated more ^{14}C photosynthate than dark, shrivelled (morphologically old) ecto-mycorrhiza, irrespective of their chronological age. A related factor often overlooked is the rejuvenation of morphologically "old" ecto-mycorrhiza by colonisation of the same or different EMF (Durall et al. 1994b), potentially creating a more powerful sink for C. Ecto-mycorrhizal rejuvenation is frequently observed on field grown roots, and its importance to root turnover needs further investigation.

2.5 Carbon and Nutrient Transfers Between Plants

2.5.1 Hyphal Links Between Plants

2.5.1.1 Direct Evidence for Links Between Chlorophyllous Plants

Both AMF and, to a lesser degree, EMF species can associate with multiple hosts (Molina et al. 1992), and mycorrhizal hyphae emanating from the roots of established plants frequently colonise new roots of adjacent plants (Brownlee et al. 1983; Read et al. 1985). For these reasons, it is commonly believed that most mycorrhizal plants are interconnected by a common mycorrhizal network (CMN) at some point in their life history (Newman 1988). However, few plant pairs or communities have been examined to find out if links commonly exist in nature, despite their potential importance for plant community and ecosystem dynamics.

Conclusive evidence for hyphal links has been reported for a small number of AM (Hirrel and Gerdemann 1979; Heap and Newman 1980; Warner and Mosse 1983; Read et al. 1985) and EM plants (Read et al. 1985; Arnebrant et al. 1993) using direct observation and radio-isotope tracings in transparent laboratory containers. Field observations have been rare due to the delicate, microscopic and cryptic nature of hyphal links (Newman 1988), with the exception of two studies of EM linking coniferous trees (Trappe and Fogel 1977; Miller et al. 1989). More recently, Newman et al. (1994) identified links microscopically among different field soil-inoculated grass, legume and woody AM plant species by following hyphal connections and hyphal penetration of plant roots. In spite of these direct observations, hyphal links have never been enumerated within a system. Degree of colonisation and number of links between plants probably affects variability in C and nutrient transfer (Miller and Allen 1992), but these factors have not been correlated.

2.5.1.2 Specificity Evidence for Links Between Chlorophyllous Plants

The potential for interspecific hyphal links has been inferred in several studies examining host plant specificity using morphological and molecular methods to identify mycorrhiza. Some studies indicate high potential for a CMN to link different plant species in mixed communities, and low potential in others. Field and greenhouse studies have shown that many EMF associated with *Pseudotsuga menziesii* or *Pinus contorta* also associate with several species of coniferous trees, deciduous trees and deciduous shrubs (Read et al. 1985; Smith et al. 1995; Jones et al. 1997; Simard et al. 1997b; Horton and Bruns 1998; Varga 1998; Durall et al. 1999; Horton et al. 1999; Jumpponen 1999; Kranabetter et al. 1999; Massicotte et al. 1999a; Hagerman et al. 2001). More specifically, field and dual culture bioassays have shown that *Rhizopogon* species, as well as other EMF, associated with *P. menziesii* (Horton et al. 1999; Hagerman et al. 2000) or *P. ponderosa* (Molina et al. 1997) also associated with secondary *Tsuga heterophylla* and *P. contorta* hosts (Molina and Trappe 1994; Massicotte et al. 1994; Molina et al. 1999), as well as *Arctostaphylos* or *Arbutus* plants, forming EM anatomy with EM hosts and arbutoid anatomy with ericaceous hosts (Molina et al. 1997; Hagerman et al. 2000). Several laboratory and field studies have shown that the number of shared EM has been greater when plants were grown in a mixture rather than alone (Massicotte et al. 1994; Molina et al. 1997; Simard et al. 1997b,c), possibly because of C supplements via hyphal links from well-inoculated to less compatible or C-poor hosts. High potential for hyphal links was inferred in most of these studies, with the exceptions of *Alnus rubra* or *A. viridis* associated with *P. mensiezii* or *P. contorta*, respectively, where few EMF morphotypes were shared between hosts (Miller et al. 1991; Varga 1998). *Paxillus involutus* has been shown to link and facilitate N transfer between *A. glutinosa* and *P. contorta* (Arnebrant et al.

1993), as well as between *A. incana* and *P. sylvestris* (Ekblad and Huss-Danell 1995), but the functional status of *Paxillus–Alnus* associations has been highly variable (Pritsch et al. 1997; Massicotte et al. 1999b).

Even where morphological studies indicate high linkage potential, it has been recognised that genotypic variation within a fungal species commonly associated with different plant species could preclude development of hyphal links. Some studies have applied polymerase chain reaction–restriction fragment length polymorphism (PCR-RFLP) analysis of the internal transcribed spacer (ITS) region of rDNA to examine genetic variation within mycorrhizal morphotypes (see Erland and Taylor, Chap. 7, this Vol., for use of these methods for studying EMF communities). Several of these studies found that separation of mycorrhiza on the basis of morphology matched reasonably well with that of DNA for the most commonly shared morphotypes (Bonello et al. 1998; Varga 1998; Horton et al. 1999; Jumpponen 1999), whereas some others showed morphotyping to underestimate genetic diversity in EMF and EMF communities (Perotto et al. 1996; Kårén and Nylund 1997; Wallander et al. 1997b; Liu et al. 1998). The possibility of high genotypic variability within a population demonstrates the need to develop species-specific fungal primers (Egger 1995) and to conduct somatic incompatibility tests to confirm potential for functional hyphal links.

Arbuscular mycorrhizal fungi are generally thought to have little, or no, host specificity (Molina et al. 1992; see also Sanders, Chap. 16, this Vol.) and, therefore, to have high potential to form a CMN. However, some studies show that hosts exposed to a mix of AM fungi are preferentially colonised by certain strains, suggesting host-endophyte or ecological specificity (McGonigle and Fitter 1990; Dhillion 1992; Zhu et al. 2000). Temporal differences in colonisation have also been reported in grass and clover systems, possibly due to changing environmental conditions or changing levels of competition over time (Sanders and Fitter 1992; Zhu et al. 2000). Colonisation by AM fungi has also been influenced by the presence of other plant species grown in a mixture (Zhu et al. 2000). These factors may help explain variation in linkage and nutrient transfer among AM plants.

2.5.1.3 Achlorophyllous Plants

The earliest evidence for hyphal linkages and C transfer involved conifers and mycoheterotrophic plants in the subfamily Montropoideae (Bjorkman 1960). By the use of PCR-RFLP techniques, several mycoheterotrophic plants have been shown to form endomycorrhizal associations with fungi that simultaneously form EM with surrounding trees (Fig. 15.3 in Taylor et al., Chap. 15, this Vol.; Cullings et al. 1996; Taylor and Bruns 1997, 1999; Bidartondo et al. 2000; McKendrick et al. 2000a). In a recent study, McKendrick et al. (2000b) has provided the first experimental confirmation that growth of the myco-heterotrophic orchid, *Corallorhiza trifida,* was sustained by C directly received from a neighbouring autotrophic tree (*Betula* or *Salix*) through linked fungal mycelia of a shared symbiont (see Taylor et al., Chap. 15, this Vol.).

The relationship between mycoheterotrophic species and their narrow range of closely related EMF hosts is an extreme specialisation (Zelmer and Currah 1995; Cullings et al. 1996; Kretzer et al. 2000), and contrasts with photosynthetic plants, which form mycorrhizae with diverse EMF (Molina et al. 1992). The relationship has been termed epiparasitic because the achlorophyllous plants receive all of their fixed C from green plants through their common EMF associates, thus cheating the EM mutualism, rather than through a direct parasitic connection (Taylor and Bruns 1997). Nevertheless, a net cost to the green plant or fungal associate has not been shown (Leake 1994). In a recent study, Bidartondo et al. (2000) provide evidence for a mutualistic aspect to the relationship between *Sarcodes sanguinea* and the EM association of *Rhizopogon ellenae* and *Abies magnifica.* They found that the distribution of EMF was highly clumped in rootballs of *Sarcodes,* and suggested that *Sarcodes* plants either established within pre-existing EMF clumps or, more likely, they stimulated clump formation. The authors suggested that *R. ellenae* benefits from the stimulation by colonising a larger proportion of *Abies* roots relative to its EMF competitors, and this in turn benefits the specialised *Sarcodes* by providing a large stable C source (Bidartondo et al. 2000). *Abies* may benefit from its association with *Sarcodes* because the small plants act as inexpensive intermediaries that extend the distance of hyphal exploration and nutrient uptake (Miller and Allen 1992). These studies on mycoheterotrophic plants provide conclusive evidence that interplant C transport occurs through hyphal networks.

2.5.2 Carbon and Nutrient Transfer Between Plants

Many isotope labelling studies have shown that mycorrhizal colonisation can facilitate transfer of C, N and P between host plants. Although it has been established that materials can move between mycorrhizal plants, several issues remain under lively debate, including whether (1) transfer is bi-directional, (2) there is net transfer in one direction, (3) net transfer is large enough to affect plant physiology and ecology, (4) transferred material moves from fungal into receiver plant tissues, and (5) transfer occurs through hyphal links and/or soil pathways that involves leakage and uptake by roots or hyphae of adjacent plants. These issues have important implications for the ecological significance of interplant C and nutrient transfer.

2.5.2.1 Carbon Transfer

Simard et al. (1997a,d) provided evidence for bi-directional C transfer between EM plants in laboratory and field experiments involving dual (^{13}C-^{14}C) pulse-labelling of *P. menziesii* and *Betula papyrifera* seedlings. Carbon transfer to *P. menziesii* was balanced with that to *B. papyrifera* in the laboratory and first year field experiment (i.e. zero net transfer), but during the second field year *P. menziesii* experienced a net C gain. Net gain by *P. menziesii* averaged 6% of C isotope uptake through photosynthesis, with more in deep shade (10%) than in full or partial sun (3–4%). Label was detected in receiver plant shoots in all experiments, indicating movement of transferred C out of fungi and into plant tissues. In the field, equidistant AM *Thuja plicata* seedlings absorbed <1% and 18% of transferred isotope in the first and second years, respectively, suggesting that most C was transferred through the EM hyphal pathway, but that some was also transferred through soil pathways (e.g. in fungal and root exudates, root-respired CO_2, or sloughed fungal and root cells). In the laboratory experiment, transfer via hyphal versus soil pathways could not be distinguished, possibly because the variability in transfer exceeded the power of our design, and possibly because interconnecting hyphae anastomosed in severed root boxes. Based on the field and laboratory results, Simard et al. (1997a,d) concluded that interplant C transfer was a complex and variable process that involved hyphal as well as soil pathways. One of the shortcomings of these experiments was that the presence and functional status of hyphal connections were not unequivocally demonstrated using radiography, ultrastructure and/or molecular tests. Instead, their presence was inferred from EM shared compatibility studies (Jones et al. 1997; Simard et al. 1997b; Sakakabara et al., unpubl. data), direct observation of hyphal links (laboratory only), and EM versus AM isotope uptake (field only).

Recent criticisms of Simard et al. (1997a,d) by Robinson and Fitter (1999) and Fitter et al. (1999) have helped highlight some other important limitations of these studies, but some criticisms may be unfounded. Firstly, Robinson and Fitter (1999) pointed out that a large amount of C (18%) was absorbed by AM *T. plicata*, suggesting the occurrence of a soil transfer pathway through release and capture of exudates, and possibly negating the primacy of a hyphal transfer pathway. We have never disputed that root exudation is a significant transfer pathway. However, the occurrence of a hyphal transfer pathway is suggested in these experiments (Simard et al. 1997a), as well as several others (e.g. Francis and Read 1984), because much greater transfer occurs when mycorrhizal links are present or possible. Future studies investigating C and nutrient transfer should include quantification of hyphal and root lengths of the test plants to help ascertain the relative importance of uptake from the soil. Secondly, Robinson and Fitter (1999) argue that the studies show one-way rather than bi-directional transfer, based on the nearly ten-fold larger isotope transfer from *B. papyrifera* to *P. menziesii* than vice versa in the second year field experiment, and the possibility that received isotope lies within the error calculation limits. However, Simard (1995) showed that isotope received by both tree species was well above detection and calculation error limits, providing unequivocal evidence for two-way transfer. Furthermore, large amounts of isotope were transferred in both directions in both the first year field and laboratory studies (Simard et al. 1997a,d), exceeding one-way transfer considered significant in some other systems (e.g. Read et al. 1985; Finlay and Read 1986a). We disagree that our bi-directional C transfer results invoke a parasitic relationship nor have we inferred that the plant-fungus-plant system was mutualistic; rather, we state that the net competitive effect of one species on another should not be predicted without an understanding of interplant transfer, including the magnitude and direction, through hyphal and soil pathways. Thirdly, Robinson and Fitter (1999) suggested that [$^{13}CO_2$] far exceeded [$^{14}CO_2$] in our field labelling chambers, possibly altering C assimilation, allocation and interplant flux patterns. Their suggestion was based on the higher detection thresholds for ^{13}C as well as the isotope concentration differences that occurred in the labelling chambers (1.4% $^{12+13}CO_2$ versus 0.03% $^{12+14}CO_2$) in the laboratory study of Simard et al. (1997d). In contrast to the laboratory study, however, out planted *B. papyrifera* and *P. menziesii* in the field studies were pulsed with similar $^{12+13}CO_2$ (0.05%) and $^{12+14}CO_2$ (0.03%) concentrations (Simard et al. 1997e). The difference in isotope concentrations in the field labelling chambers resulted in the same ^{13}C and ^{14}C tissue allocation patterns (Simard 1995). This finding was supported by another study by Simard et al. (1997e), who found no effect of increasing $^{12+13}CO_2$ concentrations from 0.04 to 0.05% on ^{13}C allocation patterns. Interestingly, Fitter et al. (1998) also found that elevated CO_2 concentrations (0.07 versus 0.03%) had no effect on the occurrence or extent of C transfer between AM plants. We do agree with Robinson and Fitter (1999) and Fitter et al. (1999) that future field experiments on interconnection and transfer should include: (1) physical barriers to prevent hyphal contact between EM plants, (2) full reciprocity of all labelling treatments, (3) quantification of donor root specific activities, (4) longer-term labelling and chase periods, (5) identification of transferred C compounds, and (6) examination of both plant and fungal regulatory factors.

Recent studies are providing increasing evidence that the pattern of interplant C transfer in AM systems differs from that in EM systems. Although several studies have demonstrated C movement from one AM plant to another (Francis and Read 1984; Martins 1993; Watkins et al. 1996), none have demonstrated net transfer (Perry 1999), and most have found little (Francis and Read 1984; Read et al. 1985) or no movement of transferred C from roots to shoots (Waters and Borowicz 1994; Watkins et al. 1996; Graves et al. 1997; Fitter et al. 1998), possibly because it remained associated with fungal tissue (Fitter et al. 1999). In a well designed laboratory experiment that used changes in natural abundance of ^{13}C to quantify C transfer between C_3 (*P. lanceolata*) and C_4 (*C. dactylon*) plants linked in a CMN, as well as different sized mesh barriers to distinguish AM from soil transfer pathways, Watkins et al. (1996) found that one-way transfer to *C. dactylon* varied from 0 to 41%, with most <10% (transfer in the opposite direction could not be calculated). They also found that transfer between potentially linked plants far exceeded that of unlinked plants, most transferred C remained in the root/fungal tissues of *C. dactylon*, but most transferred C moved into both roots and shoots of *P. lanceolata*. Fitter et al. (1998) further tested the fate of transferred C by examining the regrowth of clipped receiver plants. Their results confirmed that all transferred C remained in root/fungal structures, and suggested that it moved into fungal storage vesicles instead of hyphae. Graves et al. (1997) similarly used changes in ^{13}C to quantify C transfer, but instead of using C_3 and C_4 plants, they fumigated sections of mycorrhizal and non-mycorrhiza C_4 *F. ovina* turf with ^{13}C-depleted air over a 1-week period. Results from this laboratory experiment corroborated their other studies, in that a substantial amount of C (41%) was transferred from donor roots to linked neighbours, none was transferred to unlinked neighbours, and most transferred C remained in root/fungal structures. In a separate study, Martins (1993) demonstrated that almost 50% of C received by AM plants was through hyphal links with AM neighbours, whereas the rest was due to hyphal or root release and uptake from the soil, indicating a complex transfer pathway that involves both hyphal and soil pathways. These experiments have greatly improved our understanding of transferred C in AM systems. However, further work is still required to (1) accurately quantify the amounts of C transferred, (2) confirm whether two-way or net transfer occurs in AM systems, (3) determine whether AM links are functional in transfer, (4) distinguish between C transferred into fungal versus root tissue, and (5) ascertain the extent and variability of transfer in AM systems by analysing a greater variety of plant species that differ in their C physiology. The fact that transfer among a large number of plant species needs to be investigated follows also from Fig. 10.3. This figure shows that the amount of interplant carbon transfer increases when the mycorrhizal dependency of a particular plant species increases (van der Heijden, Chap. 10, this Vol.). To our knowledge, similar research is lacking on C or nutrient transfer among plants interconnected by ericoid mycorrhiza.

2.5.2.2 Nutrient Transfer

Transfer of mineral nutrients (P and N) between intact, live plants has been repeatedly demonstrated in the field (Chiariello et al. 1982; Walter et al. 1996) and laboratory (Ritz and Newman 1986; Haystead et al. 1988; Frey and Schuepp 1992), and has been enhanced by the presence of inter- or intra-specific mycorrhizal links (e.g. Hamel and Smith 1991). Bi-directional transfer of

^{32}P and of ^{15}N has been shown to occur, particularly when donor and receiver plants were colonised by AMF; however, transfer has been nearly equivalent in both directions (Johansen and Jensen 1996). In some studies, one-way N transfer from N_2-fixing to non-N_2-fixing AM plants has been insignificant or too small to affect plant nutrient status (e.g. Giller et al. 1991; Frey and Schuepp 1992; Ikram et al. 1994; Ekblad and Huss-Danell 1995), whereas in others it has been substantial (e.g. Bethlenfalvay et al. 1991; Martin et al. 1995; Haystead et al. 1988; Mårtensson et al. 1998). Substantial N flux has also been observed between fertilised and unfertilised plants (Eissenstat 1990). Bethlenfalvay et al. (1991) found that N transfer was accompanied by significant transfer of other nutrients. Of two studies that examined N-transfer between EM plants (N_2-fixing alder and non-N_2-fixing pine), one found no significant transfer (Ekblad and Huss-Danell 1995), whereas Arnebrant et al. (1993) found up to 20 % of N in pine was derived from alder fixation and subsequent transfer.

The magnitude of N transfer appears to depend on mycorrhizal colonisation, soil nutrient status, and physiological status of the plants (Ikram et al. 1994; Martin et al. 1995), but these factors have been studied little. For example, N transfer in AM systems has been increased by AM colonisation (Haystead et al. 1988; Hamel and Smith 1991; Hamel et al. 1991a,b; Ledgard 1991), as well as by specific N limitations (Ekblad and Huss-Danell 1995), P limitations (Martin et al. 1995), or fertilisation (Bethlenfalvay et al. 1991). Hyphal links have been shown to be directly involved in interplant N transfer (e.g. Bethlenfalvay et al. 1991; Hamel and Smith 1991; Arnebrant et al. 1993); however, some studies could not distinguish between hyphal and soil pathways (Eason et al. 1991; Ikram et al. 1994; Ekblad and Huss-Danell 1995). In these latter studies, the role of mycorrhizae may have been to improve uptake efficiencies and to reduce losses and immobilisation.

Phosphorus has been shown to transfer between plants much less freely than N, possibly because of lower plant demand, less leakage out of roots or, where transfer is via the soil solution, because of lower soil mobility (Eissenstat 1990; Newman et al. 1992; Johansen and Jensen 1996). One-way P transfer has commonly been very small (Ritz and Newman 1986), or either too slow or too small to significantly affect the receiver plant's nutrient status (Eason et al. 1991; Newman and Eason 1993; Ikram et al. 1994), except possibly under very deficient P conditions (Ritz and Newman 1984). For example, Newman and Eason (1993) found that P transfer between mycorrhizal plants was slower than the life-span of leaves or tillers, while transfer between tillers within a plant was faster than their life-span. Several of these studies found movement of P into receiver shoots (e.g. Ikram et al. 1994; Johansen and Jensen 1996; Walter et al. 1996), and one found that it increased with time according to phenologies of receiver species (Walter et al. 1996). Arbuscular mycorrhizal links were often important in facilitating P transfer (Newman 1988; Newman and Eason 1993), but soil was sometimes also an important pathway (Newman and Ritz 1986; Ikram et al. 1994; Walter et al. 1996). Results of these studies suggest that the magnitude and rate of P transfer between living plants were too small to be of major ecological importance.

In contrast to intact living plant pairs, there has been rapid, enhanced nutrient transfer from the dying, mineralising roots of stressed (manual-, insect- or herbicide-defoliated) donor plants to neighbouring non-stressed plants, either through AM or ericoid mycorrhizal links (Eason and Newman 1990; Eason et al. 1991; Johansen and Jensen 1996), or through soil pathways (Ta and Faris 1987; Hamel et al. 1991a,b; Ikram et al. 1994) or possibly both (Ritz and Newman 1985; Bethlenfalvay et al. 1996). In most studies, N transfer has been much greater than P transfer, and may be large enough to affect plant nutrition (Eissenstat 1990). For example, Giller et al. (1991) found up to 35% of maize N was transferred from bean when both plants had been severely defoliated by insects. Nutrients captured from dying roots have been shown to increase several-fold if the dying and living roots shared the same mycorrhizal fungi, but the involvement of direct hyphal links has sometimes been inconclusive (Eason et al. 1991; Ikram et al. 1994; Johansen and Jensen 1996). Evidence in favour of soil pathways comes from Eason (1988), who found that stimulation of soil microbial activity reduced P transfer, implying that most P in dying roots first entered the soil solution before its transfer to receiver plants (Eason et al. 1991). Rapid shifts in competition due to reduced root growth of clipped plants may also account for the shifts in plant nutrient status (Eissenstat 1990). The benefits of N transfer, and particularly P transfer, to plant performance has been variable in these studies, ranging from enhancement to decline in plant nutrition or yield.

2.5.3 Factors Regulating Carbon and Nutrient Transfer

Most work on factors that regulate transfer has focused on plant control through source-sink relationships for photosynthate, N or P, but recently, attention has also been paid to possible fungal controlling factors, such as degree and direction of mycorrhizal colonisation.

2.5.3.1 Source-Sink Regulation of C Transfer

The role of source-sink relationships in regulating C transfer has been studied by experimentally altering sink strength or, less commonly, source strength of connected plants. Shading of receiver plants has increased transfer in AM (Francis and Read 1984; Read et al. 1985) and EM systems (Finlay and Read 1986a; Simard et al. 1997a). Shading has been suggested to increase sink strength of receiver plants by reducing net photosynthesis and assimilate supply to receiver roots, thereby increasing the assimilate concentration gradient between plants. In another study, Waters and Borowicz (1994) altered sink strength of AM-linked plants by clipping one of the partners, and found that C flowed away from clipped (stressed or putatively sink) plants toward unclipped plants; the opposite of that expected. They reasoned, however that clipping plants increased the labile C concentration in the roots (i.e. turned them into source plants), thereby increasing the diffusion gradient for C out of roots into mycorrhizal fungi, and into connected, neighbouring unclipped plants. In contrast to sink strength manipulations, Fitter et al. (1998) found that increasing source strength by growing plants in enriched CO_2 environments had no effect on the occurrence or extent of C transfer. Other experiments examined transfer between plants that naturally differed in C physiology. For example, Simard et al. (1997a) found net transfer when both *B. papyrifera* and *P. menziesii* species were fully illuminated. This corresponded to greater net photosynthetic rate and greater allocation of assimilated C to roots of *B. papyrifera* than *P. menziesii* (Simard et al. 1997e). Grime et al. (1987) observed apparent one-way transfer of ^{14}C from canopy-dominant *Festuca ovina* to understorey herbs and grasses, which they suggested occurred through a CMN along a natural source-sink gradient for assimilate, and may have accounted for increased diversity and more even biomass distribution among plants (see Hart and Klironomous, Chap. 9; van der Heijden, Chap. 10, this Vol.). This work was later criticised because net transfer and the transfer pathway were not examined (Bergelson and Crawley 1988; Fitter et al. 1999), and interplant variation in C physiology and AM dependency were not adequately demonstrated. Indeed, only one of these studies demonstrated net transfer, leaving in question the regulatory role of source-sink relationships in interplant C transfer in most mycorrhizal systems.

The form of C transferred between interconnected plants is important for understanding the mechanism of interplant transfer, and how C compounds are used by receiver plants. Fungal-specific compounds converted from plant sugars appear to feed an endogenous fungal pool of glucose, which provides C skeletons for assimilation of ammonium into amino acids (Lewis and Harley 1965; Martin et al. 1987; Söderström et al. 1988). Amino acids have been shown to pass from mycorrhizal fungi into host plant tissues (Finlay et al. 1989; Arnebrant et al. 1993; Cliquet and Stewart 1993), whereas sugars have not (Smith and Smith 1990). A mechanism for interplant C transfer has been proposed wherein sugar or organic N is transported from the donor plant to the fungus, and organic N moves from the fungus to the receiver plant (Smith and Smith 1990; Smith and Read 1997). Net flow of organic N could occur along a hydrostatic pressure gradient from high to low assimilate and N concentrations, where ammonium is also assimilated from the soil pool (Martin et al. 1987). In Simard et al. (1997a), a greater contribution of assimilate to the CMN and higher foliar N concentrations of *B. papyrifera* could then result in net C transfer in amino acids to *P. menziesii* (Smith and Read 1997). Although there exists a theoretical mechanism for mycorrhizal-facilitated interplant transfer along C-N gradients, little work has been done on the organic compounds involved, or on alternative transfer mechanisms.

2.5.3.2 Source-Sink Regulation of Nutrient Transfer

There is considerable evidence in support of source-sink regulation of interplant nutrient (N, P) flux. However, some evidence is equivocal or contrary. Evidence for source-sink regulation exists where the direction, magnitude or rate of transfer were dependent on physiological (Frey and Schüepp 1992; Walter et al. 1996) and/or nutrient differences between linked plants (Bethlenfalvay et al. 1991; Newman et al. 1992; Frey and Schüepp 1993). Tissue nutrient concentrations differ between plants, for example that are senescing versus living, fertilised versus not fertilised, N_2-fixing versus non-N_2-fixing, mycorrhizal versus non-mycorrhizal, annual versus perennial, physiologically young versus old, C_3 versus C_4, or deciduous versus coniferous (Newman et al. 1992; Walter et al. 1996; Simard et al. 1997a; Mårtensson et al. 1998). The strongest evidence for source-sink regulation of N and P transfer comes from studies where donor plants were stressed by shoot removal or defoliation, leading to senescence of donor roots, increased labile nutrient concentrations and impaired sink strength, followed by rapid nutrient transfer to nutrient-poorer receiver roots linked by a CMN (see Sect. 2.5.2.2.).

By contrast, there remains greater uncertainty regarding source-sink regulation of nutrient transfer between living plants. For example, the N or P status of nutrient-poor receiver plants usually improved when grown in association with nutrient-rich plants, either because of fertilisation, nodulation or interspecific variation (e.g. Brown et al. 1992; Arnebrant et al. 1993; Walter et al. 1996; Mårtensson et al. 1998). In some cases, however, receiver nutrient status was unaffected (Ikram et al. 1994; Ekblad and Huss-Danell 1995; Johansen and Jensen 1996). For example, very little N transfer occurred where donors were N-limited (Ritz and Newman 1986), receivers were N-sufficient (Frey and Schüepp 1992), or other nutrients were limiting (Ekblad and Huss-Danell 1995). Ekblad and Huss-Danell (1995) also found that N transfer from alder to pine did not differ significantly from zero, even though transfer was greatest where pine was N-starved and alder was fixing N_2 maximally. Brown et al. (1992) found transfer in the opposite direction to that expected; where the AM symbiosis increased C flux from maize to soybean, reduced cob biomass, and stimulated nodule activity of soybean. This may have been due to a diversion of C away from corn to soybean to meet the soybeans' high energy requirements for N_2 fixation. Walter et al. (1996) found large differences in received P among C_3 and C_4 plants, which they attributed to differences in each species' abilities to access neighbouring rhizosphere P rather than the magnitude of tissue nutrient differences. Other plant factors, such as shading of receiver plants (Ikram et al. 1994) or relative mass of donors and receivers

(Chiariello et al. 1982; Walter et al. 1996), appear to play little role in regulating P or N transfer.

Several factors interact with source-sink relations to affect the magnitude of nutrient transfer. Firstly, the mode of donor N enrichment (e.g. fertiliser versus nodules) appears to be important, possibly because of variable enrichment of different N compounds, plant tissues or plant cells. For example, greater N transfer to maize from N-fertilised (non-nodulated) than nodulated soybean in Bethlenfalvay et al. (1991) may be explained by higher root/shoot N concentration in non-nodulated than nodulated soybean (after Martin et al. 1991). Secondly, soil nutrient status appears to affect the magnitude of N transfer, which may be indirectly related to mycorrhizal infection or N_2 fixation. For example, N transfer from soybean to maize was generally greater in soils limited by N or P, where mycorrhizal infection and N_2 fixation tended to be greatest (Martin et al. 1995). Ekblad and Huss-Danell (1995) also found greatest N_2 fixation in alder and greatest N transfer to pine in low N conditions, but it was further enhanced by P fertilisation, presumably because high nodule demands for P were better met. In contrast to these studies, Giller et al. (1991) found that N transfer from beans under severe N limitations did not result in improved growth or N yield of maize. Thirdly, participation in the mycorrhizal symbiosis affects the magnitude of N transfer from N_2-fixers to non-fixers in variable ways. In several studies it enhanced transfer (see Sect. 2.5.2.2.), but in others it reduced transfer, possibly due to reduced root excretion or enhanced root re-absorption of N by the N_2 fixer (Hamel et al. 1992). Fourthly, microbial activity appears to reduce N transfer from soybean to maize, probably due to rapid utilisation of nitrogenous root exudates (Hamel et al. 1991a). Earthworms have also been shown to reduce N transfer from decomposing or living clover roots to accompanying wheat, possibly by disrupting root contact and putative hyphal links through burrowing and feeding (Schmidt and Curry 1999).

2.5.3.3 Fungal Regulation of Transfer

There is increasing evidence that mycorrhizal fungi play a role in controlling the direction, magnitude and rate of C or nutrient transfer between linked plants, particularly in AM systems. Although the ability of the fungi to sequester and provide nutrients to linked plants at different rates and amounts has largely been unexplored (Miller and Allen 1992), different characteristics of the CMN have recently been shown to influence interplant transfer. For example, Mårtensson et al. (1998) found great variation in N transfer from bean to chicory among different linking isolates of *Glomus* spp., which they suggested reflected differences in the capacity of isolates to initiate a plant trigger for N transfer. In other studies, hyphal links have been shown to enhance herbicide injury to weeds, possibly because AM fungi adjusted their own source-sink relations in favour of sink-driven fluxes (Bethlenfalvay et al. 1996; Rejon et al. 1997). Several studies provide evidence that the magnitude of transfer is affected by the degree of physical root/rhizosphere overlap or frequency of mycorrhizal links (Johansen and Jensen 1996; Walter et al. 1996; Simard et al. 1997a). By contrast, Watkins et al. (1996) and Fitter et al. (1998) found little relationship between C transfer and the degree or direction of AM colonisation, and little evidence that colonisation was related to number or functioning of links. Additional fungal factors that appear to play a role in transfer include identity of the fungal species, seasonal or phenological variation in mycorrhizal activity (Frey and Schüepp 1992; Hamel et al. 1992) as well as differential mycotrophy of different plants in mixtures (Hamel and Smith 1991; Zhu et al. 2000).

Robinson and Fitter (1999) and Fitter et al. (1998) argue strongly for fungal regulation of transfer based on their laboratory results that received C remained in plant roots or, more probably, AM fungal tissues. They suggest that C transfer is dictated by fungal C demands and growth dynamics, and that plants are simply habitat patches for the fungi. As a result, C transfer would be only weakly related to the demands of the plants, and strongly related to assimilation and storage of C by the fungus, resulting in a large variation in C transfer among plants. Wilkinson (1999) argues, however that a single explanation (i.e. mycocentric versus phytocentric) is inappropriate, and that transfer in "mycorrhizal networks is best explained by a plurality of mechanisms". Our view is that both mycorrhizal fungi and plants influence transfer, and that the relative degree of control exercised by each symbiotic partner varies among systems. In this chapter, we reviewed evidence for fungal control of nutrient translocation through source-sink relationships and fungal physiological triggers. We also reviewed evidence that plants influence the direction and magnitude of interplant transfer through differences in physiology and demands for C and nutrients. Transfers also appear to be influenced by numerous environmental and biotic factors. Our review of the literature underscores the complexity of factors that control C and nutrient transfer within plant communities, and therefore the challenge in predicting the consequences of transfer on plant community and ecosystem dynamics in nature.

2.6 Discussion of the Ecological Significance of Interplant Transfer

The potential implications of mycorrhizal links and interplant C or nutrient transfer for ecosystem structure and dynamics are numerous, and have been enumerated in various review papers (Newton 1988; Miller and Allen 1992; Perry et al. 1992; Amaranthus and Perry 1994; Zobel et al. 1997), as well as publications of individual experiments. Recent studies provide supporting evidence for some of these ecosystem implications. However, most evidence still suffers from the inability to conclusively demonstrate bi-directional or net transfer, the involvement of hyphal links, movement of transferred C or nutrients into plant receiver tissues, or that the magnitude, timing and/or duration of transfer is sufficient to affect measures of plant fitness (Newman 1988; Fitter et al. 1999). Some argue that the pathway of transfer (hyphal versus soil) is irrelevant from a practical or economic viewpoint (Bethlenfalvay et al. 1991; Brown et al. 1992; Mårtensson et al. 1998), but also acknowledge that an understanding of the mechanisms is fundamental to management of plants and mycorrhizal fungi for specific objectives. It is also arguable whether movement into plant shoot tissues is of key importance to functional significance of transfer. Perry (1999) points out that transferred C remaining in fungal tissues within the root is still a "subsidy to the nutrient gathering system of the receiver plant".

Some of the potential implications of mycorrhizal links and interplant transfer for ecosystems are listed below, with reference to supporting or contradictory studies:

1. Assists seedling establishment near mature plants by allowing seedlings to become colonised more rapidly or to tap into an established CMN supported by other plants. This has been demonstrated in EM (e.g. Borchers and Perry 1990; Simard et al. 1997c; Horton et al. 1999; Bidartondo et al., 2000), AM (e.g. Eissenstat and Newman 1990) and mixed (e.g. Jumpponen 1999) plant communities. It raises interesting questions about guild-formation among different plant and fungal species (Perry et al. 1992). In a recent study in a tropical Camaroon rain forest, EM inoculation by conspecific adults had variable effects on seedling establishment (Newberry et al. 2000), generating many questions regarding the role played by EM in promoting species aggregation or monodominance in tropical forests (Torti and Coley 1999).
2. Assists recovery of species following disturbance, which is particularly important where disturbance behaviour is unpredictable. One example is western North American forests, which frequently regenerate to mixed species stands following periodic and variable-intensity fires. Some studies provide evidence for and/or discuss fungal linkages through time (Perry et al. 1989a, 1992; Horton and Bruns 1998; Horton et al. 1999).
3. Reduces competitive dominance and promotes species diversity by allowing C or nutrients to directly flow from sufficient to deficient plants, resulting in a more even distribution of C or nutrients. Several studies provide supporting evidence for reduced competition and/or increased diversity where plants are inoculated with the same mycorrhizal fungi (Grime et al. 1987; Perry et al. 1989b; Gange et al. 1993; Moora and Zobel 1996). By contrast, other studies provide evidence that suggests competition may increase (Eissenstat and Newman 1990; Eason et al. 1991; Hartnett et al.

1993; Waters and Borowicz 1994; Bethlenfalvay et al. 1996; Rejon et al. 1997) or becomes unbalanced due to differences in mycotrophy (Hartnett et al. 1993; Zobel and Moora 1995; Moora and Zobel 1996). In none of these studies, however, was the presence of functional hyphal links or bi-directional transfer demonstrated. More detailed competition studies that adequately address interplant transfer could test the common assumption that plants compete for resources as physiologically independent organisms.

4. Affects fungal nutrient demands, source-sink relationships, foraging and uptake, and therefore affects fungal ecology. The influence of fungal physiologies on nutrient transfer to plants has been studied to some degree (e.g. Hamel et al. 1992; Bethlenfalvay et al. 1996; Rejon et al. 1997; Mårtensson et al. 1998; Fitter et al. 1998), but the implications of nutrient transfers on fungal ecology have yet to be explored.
5. Reduces nutrient losses from ecosystems by keeping more nutrients in biomass, therefore increasing productivity. This is particularly evident with preferential cycling of nutrients from dying to living plants. Reduced nutrient loss from experimental systems has been demonstrated by Hamel et al. (1991a) and others.
6. Increases the productivity, stability and sustainability of ecosystems. This has been inferred in several papers (e.g. Grime et al. 1987; Perry et al. 1989b; Read 1994; Simard et al. 1997a), but never experimentally tested.

The potential implications of severing hyphal links following management activities, such as ploughing, fertilisation, herbicide application, fungicide application or conversion to non-native plant species, are numerous and poorly understood (Perry et al. 1989b; Amaranthus and Perry 1994; Helgason et al. 1998; Simard 1999). Research on C and nutrient transfers reflects a growing interest in their implications for ecological function as well as how such co-operative interactions evolve. Two publications, one by Wilkinson (1998) and the other by Perry (1998), present thoughtful discussions on the evolutionary ecology of mycorrhizal networks.

References

Abuzinadah RA, Read DJ (1986) The role of proteins in the nitrogen nutrition of ecto-mycorrhizal plants. III. Protein utilization by *Betula*, *Picea* and *Pinus* in mycorrhizal association with *Hebeloma crustuliniforme*. New Phytol 103:507–514

Abuzinadah RA, Read DJ (1989) The role of proteins in the nitrogen nutrition of ecto-mycorrhizal plants. V. Nitrogen transfer in birch (*Betula pendula*) grown in association with mycorrhizal and non-mycorrhizal fungi. New Phytol 112:61–68

Abuzinadah RA, Finlay RD, Read DJ (1986) The role of proteins in the nitrogen nutrition of ecto-mycorrhizal plants. II. Utilization of protein by mycorrhizal plants of *Pinus contorta*. New Phytol 103:495–506

Amaranthus MP, Perry DA (1994) The functioning of mycorrhizal fungi in the field: linkages in space and time. Plant Soil 159:133-140
Arnebrant K, Ek H, Finlay RD, Söderström B (1993) Nitrogen translocation between *Alnus glutinosa* (L.) Gaertn. seedlings inoculated with Frankia sp. and *Pinus contorta* Doug. ex Loud seedlings connected by a common ecto-mycorrhizal mycelium. New Phytol 130:231-242
Arocena JM, Glowa KR (2000) Mineral weathering in ectomycorrhizosphere of subalpine fir (*Abies lasiocarpa* (Hook.) Nutt.) as revealed by soil solution composition. For Ecol Manage 133:61-70
Arocena JM, Glowa KR, Massicotte HB (1999) Chemical and mineral composition of ectomycorrhizosphere soils of subalpine fir (*Abies lasiocarpa* (Hook) Nutt. in the Ae horizon of a luvisol. Can J Soil Sci 79:25-35
Ashford AE, Ryde S, Barrow KD (1994) Demonstration of a short chain polyphosphate in *Pisolithus tinctorius* and the implications for phosphorus transport. New Phytol 126:239-247
Bâ AM, Sanon KB, Duponnois R, Dexheimer J (1999) Growth response of *Afzelia africana* Sm. seedlings to ecto-mycorrhizal inoculation in a nutrient-deficient soil. Mycorrhiza 9:91-95
Bending GD, Read DJ (1995a) The structure and function of the vegetative mycelium of ecto-mycorrhizal plants V. The foraging behaviour of ecto-mycorrhizal mycelium and the translocation of nutrients from exploited organic matter. New Phytol 130:401-409
Bending GD, Read DJ (1995b) The structure and function of the vegetative mycelium of ecto-mycorrhizal plants VI. Activities of nutrient mobilizing enzymes in birch litter colonized by *Paxillus involutus* (Fr.) Fr. New Phytol 130:411-417
Bending GD, Read DJ (1997) Lignin and soluble phenolic degradation by ecto-mycorrhizal and ericoid mycorrhizal fungi. Myco Res 101:1348-1354
Bergelson JM, Crawley MJ (1988) Mycorrhizal infection and plant species diversity. Nature 334:282
Bethlenfalvay GJ, Reyes-Solis MG, Camel SB, Ferrera-Cerrato R (1991) Nutrient transfer between the root zones of soybean and maize plants connected by a common mycorrhizal mycelium. Physiol Plant 82:423-432
Bethlenfalvay GJ, Schreiner RP, Mihara KL, McDaniel H (1996) Mycorrhiza, biocides, and biocontrol. 2. Mycorrhizal fungi enhance weed control and crop growth in a soybean-cocklebur association treated with the herbicide bentazon. Appl Soil Ecol 3:205-214
Bidartondo MI, Kretzer AM, Pine EM, Bruns TD (2000) High root concentration and uneven ecto-mycorrhizal diversity near *Sarcodes sanguinea* (Ericaceae): a cheater that stimulates its victims? Amer J Bot 87:1783-1788
Bjorkman E (1960) *Monotropa hypopitys* L. - an epiparasite on tree roots. Phys Plant 13:308-327
Bonello P, Bruns TD, Gardes M (1998) Genetic structure of a natural population of the EMF *Suillus pungens*. New Phytol 138:533-542
Borchers SL, Perry DA (1990) Growth and ecto-mycorrhiza formation of Douglas-fir seedlings grown in soils collected at different distances from pioneering hardwoods in southwest Oregon clear-cuts. Can J For Res 20:712-721
Botton B, Chalot M (1995) Nitrogen assimilation: enzymology in ecto-mycorrhiza. In: Hock B, Varma A (eds) Mycorrhiza: structure, function, molecular biology and biotechnology. Springer, Berlin Heidelberg New York, pp 325-363
Bradley RL, Fyles JW (1995) Growth of paper birch (Betula papyrifera) seedlings increases soil available C and microbial acquisition of soil-nutrients. Soil Biol Biochem 27:1565-1571
Brandes B, Godbold DL, Kuhn AJ, Jentschke G (1998) Nitrogen and phosphorus acquisition by the mycelium of the ecto-mycorrhizal fungus *Paxillus involutus* and its effect on host nutrition. New Phytol 140:735-743

Brown MS, Ferrera-Cerrato, Bethlenfalvay GJ (1992) Mycorrhiza-mediated nutrient distribution between associated soybean and corn plants evaluated by the Diagnosis and Recommendation Integrated System (DRIS). Symbiosis 12:83–94

Brownlee C, Duddridge JA, Malibari A, Read DJ (1983) The structure and function of ecto-mycorrhizal roots with special reference to their role in forming inter-plant connections and providing pathways for assimilate and water transport. Plant Soil 71:433–443

Bücking H, Heyser W (1994) The effect of ecto-mycorrhizal fungi on Zn uptake and distribution in seedlings of *Pinus sylvestris* L. Plant Soil 167:203–212.

Cairney JWG (1992) Translocation of solutes in ecto-mycorrhizal and saprotrophic rhizomorphs. Myco Res 96:135–141

Cairney JWG, Alexander IJ (1992) A study of spruce (*Picea sitchensis* (Bong.) Carr.) ecto-mycorrhizas. II Carbohydrate allocation in ageing *Picea sitchensis/Tylospora fibrillosa* (Burt.) Donk ecto-mycorrhizas. New Phytol 122:153–158

Cairney JWG, Burke RM (1996) Physiological heterogeneity within fungal mycelia: an important concept for a functional understanding of the ecto-mycorrhizal symbiosis. New Phytol 134:685–695

Cairney JWG, Ashford AE, Allaway WG (1989) Distribution of photosynthetically fixed carbon within root systems of *Eucalyptus pilularis* plants ecto-mycorrhizal with *Pisolithus tinctorius*. New Phytol 112:495–500

Chalot M, Brun A (1998) Physiology of organic nitrogen acquisition by ecto-mycorrhizal fungi and ecto-mycorrhizas. FEMS Microbiol Rev 22:21–44

Chang TT, Li CY (1998) Weathering of limestone, marble, and calcium phosphate by ecto-mycorrhizal fungi and associated microorganisms. Taiwan J For Sci 13:85–90

Chiariello N, Hickman JC, Mooney HA (1982) Endo-mycorrhizal role for interspecific transfer of phosphorus in a community of annual plants. Science 217:941–943

Cliquet JB, Stewart GR (1993) Ammonia assimilation in *Zea mays* L. infected with a vesicular-arbuscular fungus *Glomus fasciculatum*. Plant Physiol 101:685–671

Colpaert JV, Van Laere A (1996) A comparison of the extracellular enzyme activities of two ecto-mycorrhizal and leaf-saprotrophic basidiomycete colonizing beech leaf litter. New Phytol 134:133–141

Colpaert JV, Van Laere A, Van Assche JA (1996) Carbon and nitrogen allocation in ecto-mycorrhizal and non-mycorrhizal *Pinus sylvestris* L. seedlings. Tree Phys 16:787–793

Conn C, Dighton J (2000) Litter quality influences on decomposition, ecto-mycorrhizal community structure and mycorrhizal root surface acid phosphatase activity. Soil Biol Biochem 32:489–496

Cromack K Jr, Sollins P, Grausten WC, Speidel K, Todd AW, Spycher G, Li CY, Todd RL (1979) Calcium oxalate accumulation and soil weathering in mats of the hypogeous fungus *Hysterangium crassum*. Soil Biol Biochem 11:463–468

Cullings KW, Szaro TM, Bruns TD (1996) Evolution of extreme specialization within a lineage of ecto-mycorrhizal epiparasites. Nature 379:63–66

Cumming JR (1993) Growth and nutrition of nonmycorrhizal and mycorrhizal pitch pine (*Pinus rigida*) seedlings under phosphorus limitation. Tree Physiol 13:173–187

Dighton J (1983) Phosphatase production by mycorrhizal fungi. Plant Soil 71:455–462

Dighton J (1991) Acquisition of nutrients from organic resources by mycorrhizal autotrophic plants. Experentia 47:362–369

Dhillion SS (1992) Evidence for host mycorrhizal preference in native grassland species. Mycol Res 96:359–362

Downes GM, Alexander IJ, Cairney JWG (1992) A study of spruce (*Picea sitchensis* (Bong.) Carr.) ecto-mycorrhizas. I. Morphological and cellular changes in mycorrhizas formed by *Tylospora fibrillosa* (Burt.) Donk and *Paxillus involutus* (Batsch ex Fr.) Fr. New Phytol 122:141–152

Duchesne LC, Peterson RL, Ellis BE (1987) Interactions between the ecto-mycorrhizal fungus *Paxillus involutus* and *Pinus resinosa* induces resistance to *Fusarium oxysporum*. Can J Bot 66:558-562

Dupponois R, Bâ AM (1999) Growth stimulation of *Acacia mangium* Willd. by *Pisolithus* sp. in some Senegalese soils. For Ecol Manag 119:209-215

Durall DM, Todd AW, Trappe JM (1994a) Decomposition of 14C-labelled substrates by ecto-mycorrhizal fungi in association with Douglas-fir. New Phytol 127:725-729

Durall DM, Marshall JD, Jones MD, Crawford R, Trappe JM (1994b) Morphological changes and photosynthate allocation in ageing *Hebeloma crustuliniforme* (Bull.) Quel. and *Laccaria bicolor* (Maire) Orton mycorrhizas of *Pinus ponderosa* Dougl. ex. Laws. New Phytol 127:719-724

Durall DM, Jones MD, Tinker PB (1994c) Allocation of 14C-carbon in ecto-mycorrhizal willow. New Phytol 128:109-114

Durall DM, Jones MD, Wright EF, Kroeger P, Coates KD (1999) Species richness of ecto-mycorrhizal fungi in cutblocks of different sizes in the Interior Cedar-Hemlock forests of northwestern British Columbia: sporocarps and ecto-mycorrhiza. Can J For Res 29:1322-1332

Eason WR (1988) The cycling of phosphorus from dying roots including the role of mycorrhizas. PhD Thesis, University of Bristol

Eason WR, Newman EI (1990) Rapid cycling of nitrogen and phosphorus from dying roots of *Lolium perenne*. Oecologia 82:432-436

Eason WR, Newman EI, Chuba PN (1991) Specificity of interplant cycling of phosphorus: the role of mycorrhizas. Plant Soil 137:267-274

Edwards NT, Harris WF (1977) Carbon cycling in mixed deciduous forest floor. Ecology 58:431-437

Egger KN (1995) Molecular analysis of ecto-mycorrhizal fungal communities. Can J Bot 73 [Suppl]:S1415-S1422

Eissenstat DM (1990) A comparison of phosphorus and nitrogen transfer between plants of different phosphorus status. Oecologia 82:342-347

Eissenstat DM, Newman EI (1990) Seedling establishment near large plants: effects of vesicular-arbuscular mycorrhizas on the intensity of plant competition. Funct Ecol 4:95-99

Ek H (1997) The influence of nitrogen fertilization on the carbon economy of *Paxillus involutus* in ecto-mycorrhizal association with *Betula pendula*. New Phytol 135:133-142

Ekblad A, Huss-Danell K (1995) Nitrogen fixation by *Alnus incana* and nitrogen transfer from *A. incana* to *Pinus sylvestris* influenced by macronutrients and ecto-mycorrhiza. New Phytol 131:453-459

Finlay RD, Read DJ (1986a) The structure and function of the vegetative mycelium of ecto-mycorrhizal plants. I. Translocation of 14C-labelled carbon between plants interconnected by a common mycelium. New Phytol 103:143-156

Finlay RD, Read DJ (1986b) The structure and function of the vegetative mycelium of ecto-mycorrhizal plants. II. The uptake and distribution of phosphorus by mycelial strands interconnecting host plants. New Phytol 103:157-165

Finlay RD, Söderström B (1989) Mycorrhizal mycelia and their role in soil and plant communities. In: Clarholm M, Bergström (eds) Ecology of arable land, perspectives and challenges. Development in plant and soil sciences, vol 39. Kluwer Academic Publishers, Dordrecht, pp 139-148

Finlay RD, Ek H, Odham G, Söderström B (1989) Uptake, translocation and assimilation of nitrogen from ^{15}N-labelled ammonium and nitrate sources by intact ecto-mycorrhizal systems of *Fagus sylvatica* infected with *Paxillus involutus*. New Phytol 113:47-55

Finlay RD, Frostegärd A, Sonnerfeldt AM (1992) Utilisation of organic and inorganic nitrogen sources by ecto-mycorrhizal fungi in pure culture and in symbiosis with *Pinus contorta* Dougl. ex Loud. New Phytol 131:443–451

Fitter AH, Graves JD, Watkins NK, Robinson D, Scrimgeour C (1998) Carbon transfer between plants and its control in networks of arbuscular mycorrhizas. Funct Ecol 12:406–412

Fitter AH, Hodge A, Daniell TJ, Robinson D (1999) Resource sharing in plant-fungus communities: did the carbon move for you? Tree 14:70–71

Fogel R, Hunt G (1979) Fungal and arboreal biomass in a western Oregon Douglas-fir ecosystem : distribution patterns and turnover. Can J For Res 9:245–256

Francis R, Read DJ (1984) Direct transfer of carbon between plants connected by vesicular-arbuscular mycorrhizal mycelium. Nature 307:53–56

Frey B, Schüepp H (1992) Transfer of symbiotically fixed nitrogen from berseem (*Trifolium alexandrinum* L.) to maize via vesicular-arbuscular mycorrhizal hyphae. New Phytol 122:447–454

Frey B, Schüepp H (1993) The role of vesicular-arbuscular (VA) mycorrhizal fungi in facilitating inter-plant nitrogen transfer. Soil Biol Biochem 25:651–658

Gange AC, Brown VK, Sinclair LM (1993) Vesicular-arbuscular mycorrhizal fungi: a determinant of plant community structure in early succession. Funct Ecol 7:616–622

Giller KE, Ormesher J, Awah FM (1991) Nitrogen transfer from *Phaesolus* bean to intercropped maize measured using ^{15}N-enrichment and ^{15}N-isotope dilution methods. Soil Biol Biochem 4:339–346

Gramss G, Ziegenhagen D, Sorge S (1999) Degradation of soil humic extract by wood- and soil- associated fungi, bacteria, and commercial enzymes. Micro Ecol 37:140–151

Graves JD, Watkins NK, Fitter AH, Robinson D, Scrimgeour C (1997) Intraspecific transfer of carbon between plants linked by a common mycorrhizal network. Plant Soil 192:153–159

Griffiths RP, Baham JE, Caldwell BA (1994) Soil solution chemistry of ecto-mycorrhizal mats in forest soils. Soil Biol Biochem 26:331–337

Grime JP, Mackey JML, Hillier SH, Read DJ (1987) Floristic diversity in a model system using experimental microcosms. Nature 328:420–422

Hagerman SM, Sakakibara SM, Durall DM (2001) The potential for woody understory plants to provide refuge for ecto-mycorrhizal inoculum at an Interior Douglas-fir forest after clear-cut logging. Can J For Res 31:711–721

Hamel C, Smith DL (1991) Interspecific N-transfer and plant development in a mycorrhizal field-grown mixture. Soil Biol Biochem 23:661–665

Hamel C, Smith DL (1992) Mycorrhizal-mediated ^{15}N transfer from soybean to corn in field-grown intercrops: effect of component crop spatial relationships. Soil Biol Biochem 24:499–501

Hamel C, Barrantes-Cartín, Furlan V, Smith DL (1991a) Endomycorrhizal fungi in nitrogen transfer from soybean to maize. Plant Soil 138:33–40

Hamel C, Furlan V, Smith DL (1991b) N_2-fixation and transfer in a field grown mycorrhizal corn and soybean intercrop. Plant Soil 133:177–185

Hamel C, Nesser C, Barrantes-Cartín, Smith DL (1991 c) Endomycorrhizal fungal species mediate ^{15}N transfer from soybean to maize in non-fumigated soil. Plant Soil 138:41–47

Hamel C, Furlan V, Smith DL (1992) Mycorrhizal effects on interspecific plant competition and nitrogen transfer in legume-grass mixtures. Crop Sci 32:991–996

Hampp R, Schaeffer C (1995) Mycorrhiza-carbohydrate and energy metabolism. In: Varma A, Hock B (eds) Mycorrhiza structure, function, molecular biology and biotechnology. Springer, Berlin Heidelberg New York, pp 267–296

Hampp R, Wiese J, Mikolajewski S, Nehls U (1999) Biochemical and molecular aspects of C/N interaction in ecto-mycorrhizal plants: an update. Plant Soil 215:103–113

Harley JL, Smith SE (1983) Mycorrhizal symbiosis. Academic Press, London

Hartnett DC, Hetrick BAD, Wilson GWT, Gibson DJ (1993) Mycorrhizal influence on intra- and interspecific neighbour interactions among co-occurring prairie grasses. J Ecol 81:787–795

Hatano R, Jwanga K, Okajima H, Sakuma T (1988) Relationship between the distribution of soil macropores and root elongation. Soil Sci Plant Nutr 34:535–546

Haystead A, Malajczuk N, Grove TS (1988) Underground transfer of nitrogen between pasture plants infected with vesicular-arbuscular mycorrhizal fungi. New Phytol 108:417–423

Heap AJ, Newman EI (1980) Links between roots by hyphae of vesicular-arbuscular mycorrhizas. New Phytol 85:169–171

Helgason T, Daniell TJ, Husband R, Fitter AH, Young J (1998) Ploughing up the wood-wide web. Nature 6692:431

Hildebrand EE (1994) The heterogeneous distribution of mobile ions in the rhizosphere of acid forest soils: facts, causes, and consequences. J Environ Sci Health A29:1973–1992

Hirrel MC, Gerdemann JW (1979) Enhanced carbon transfer between onions infected with a vesicular-arbuscular mycorrhizal fungus. New Phytol 83:731–738

Hodge A (1996) Impact of elevated CO_2 on mycorrhizal associations and implications for plant growth. Biol Fertil Soils 23:388–398

Högberg P (1989) Growth and nitrogen inflow rates in mycorrhizal and non-mycorrhizal seedlings of *Pinus sylvestris*. For Ecol Manage 28:7–17

Horton TR, Bruns TD (1998) Multiple-host fungi are the most frequent and abundant ecto-mycorrhizal types in a mixed stand of Douglas-fir (*Pseudotsuga menziesii*) and bishop pine (*Pinus muricata*). New Phytol 139:331–339

Horton TR, Bruns TD, Parker VT (1999) Ecto-mycorrhizal fungi associated with *Arctostaphylos* contribute to *Pseudotsuga menziesii* establishment. Can J Bot 77:93–102

Ikram A, Jensen ES, Jakobsen I (1994) No significant transfer of N and P from *Pueraria phaseoloides* to *Hevea brasiliensis* via hyphal links of arbuscular mycorrhiza. Soil Biol Biochem 26:1541–1547

Ingestad T, Ågren GI (1992) Theories and methods on plant nutrition and growth. Physiol Plant 84:177–184

Javelle A, Chalot M, Botton B (1999) Ammonium transport by the ecto-mycorrhizal fungi *Paxillus involutus* and ecto-mycorrhiza. Dynamics of Physiological Processes in Woody Roots. 2nd International Symposium Nancy, France. Programme and abstracts, p 120

Jennings DH (1995) The physiology of fungal nutrition. Cambridge Univ Press, Cambridge

Jentschke G, Brandes B, Kuhn AJ, Schröder J, Becker S, Godbold DL (2000) The mycorrhizal fungus *Paxillus involutus* transports magnesium to Norway spruce seedlings. Plant Soil 220:243–246

Johansen A, Jensen ES (1996) Transfer of N and P from intact or decomposing roots of pea to barley interconnected by an arbuscular mycorrhizal fungus. Soil Biol Biochem 28:73–81

Jones DL (1998) Organic acids in the rhizosphere – a critical review. Plant Soil 205:25–44

Jones MD, Hutchinson TC (1988) Nickel toxicity in mycorrhizal birch seedlings infected with *Lactarius rufus* or *Scleroderma flavidum*. Effects on growth, photosynthesis, respiration and transpiration. New Phytol 108:451–459

Jones MD, Durall DM, Tinker PB (1990) Phosphorus relationships and production of extramatrical hyphae by two types of willow ecto-mycorrhizas at different soil phosphorus levels. New Phytol 115:259–267

Jones MD, Durall DM, Tinker PB (1991) Fluxes of carbon and phosphorus between symbionts in willow ecto-mycorrhizas and their changes with time. New Phytol 119:99–106

Jones MD, Durall DM, Harniman SMK, Classen DC, Simard SW (1997) Ecto-mycorrhizal diversity on *Betula papyrifera* and *Pseudotsuga menziesii* seedlings grown in the greenhouse or outplanted in single-species and mixed plots in southern British Columbia. Can J For Res 28:1872–1889

Jones MD, Durall DM, Tinker PB (1998) A comparison of arbuscular and ecto-mycorrhizal *Eucalyptus coccifera*: growth response, phosphorus uptake efficiency and external hyphal production. New Phytol 140:125–134

Jordy MN, Azémar-Lorentz S, Brun A, Botton B, Pargney JC (1998) Cytolocalization of glycogen, starch, and other insoluble polysaccharides during ontogeny of *Paxillus involutus-Betula pendula* ecto-mycorrhizas. New Phytol 140:331–341

Jongmans A, Van Breemen N, Lunström U, Van Hees PW, Finlay R, Srinivasan M, Unestam T, Giesler R, Melkerud P, Olsson M (1997) Rock-eating fungi. Nature 389:682–683

Jumpponen A (1999) Spatial distribution of discrete RAPD phenotypes of a root endophytic fungus, *Phialocephala fortinii*, at a primary successional site on a glacier forefront. New Phytol 141:333–344

Karabaghli-Degron C, Scotta B, Bonnet M, Gay G, LeTacon F (1998) The auxin transport inhibitor 2,3,5-triiodobenzoic acid (TIBA) inhibits the stimulation of in vitro lateral root formation and the colonization of the tap-root cortex of Norway spruce (*Picea abies*) seedlings by the ecto-mycorrhizal fungus *Laccaria bicolor*. New Phytol 140:723–733

Kårén O, Nylund J-E (1997) Effects of ammonium sulphate on the community structure and biomass of ecto-mycorrhizal fungi in a Norway spruce stand in southwestern Sweden. Can J Bot 75:1628–1642

Kaye JP, Hart SC (1997) Competition for nitrogen between plants and soil microorganisms. Tree 12:139–143

Kielland K (1994) Amino acid absorption by arctic plants: implications for plant nutrition and nutrient cycling. Ecology 75:2373–2383

Kirk GJD (1999) A model of phosphate solubilization by organic anion excretion from plant roots. Eur J Soil Sci 50:369–378

Kloepper JW (1993) Plant growth-promoting rhizobacteria as biological control agents. In: Metting FB (ed) Soil microbial ecology – applications in agricultural and environmental management. Dekker, New York, pp255–274

Koide R, Elliott G (1989) Cost, benefit and efficiency of the vesicular-arbuscular mycorrhizal symbiosis. Funct Ecol 3:252–255

Kranabetter JM, Hayden S, Wright EF (1999) A comparison of ecto-mycorrhiza communities from three conifer species planted on forest gap edges. Can J Bot 77:1193–1198

Kretzer AM, Bidartondo MI, Grubisha LC, Spatafora JW, Szaro TM, Bruns TD (2000) Regional specialization of *Sarcodes sanguinea* (Ericaceae) on a single fungal symbiont from the *Rhizopogon ellenae* (Rhizopogonaceae) species complex. Am J Bot 87:1778–1782

Kytöviita MM, Sarjala T (1997) Effects of defoliation and symbiosis on polyamine levels in pine and birch. Mycorrhiza 7:107–111

Lapeyrie F, Picatto C, Gerard J, Dexheimer J (1990) T.E.M. study of intracellular and extracellular calcium oxalate accumulation by ecto-mycorrhizal fungi in pure culture or in association with *Eucalyptus* seedlings. Symbiosis 9:163–166

Leake JR (1994) The biology of mycoheterotrophic (saprophytic) plants. New Phytol 127:171–216

Leake JR, Donnelly DP, Saunders EM, Read DJ, Boddy L (1999) Rates and quantities of carbon flux to ecto-mycorrhizal mycelium following ^{14}C pulse labelling of tree seedlings. Dynamics of Physiological Processes in Woody Roots. 2nd International Symposium, 26–30 Sept 1999, Nancy, France. Programme and abstracts, p30

Ledgard SF (1991) Transfer of fixed nitrogen from white clover to associated grasses in swards grazed by dairy cows, estimated using ^{15}N methods. Plant Soil 131:215–223

Leprince F, Quiquampoix H (1996) Extracellular enzyme activity in soil – effect of pH and ionic strength on the interaction with montmorillonite of two acid phosphatases secreted by the ecto-mycorrhizal fungus *Hebeloma cylindrosporum*. Eur J Soil Sci 47:511–522

Lewis DH (1976) Interchange of metabolites in biotrophic symbioses between angiosperms and fungi. In: Sutherland N (ed) Botany. Perspectives in experimental biology, vol 2. Pergamon Press, Oxford, pp 207–219

Lewis DH, Harley JL (1965) Carbohydrate physiology of mycorrhizal roots of beech. III. Movement of sugars between host and fungus. New Phytol 64:256–269

Leyval C, Berthelin J (1993) Rhizodeposition and net release of soluble organic compounds by pine and beech seedlings inoculated with rhizobacteria and ecto-mycorrhizal fungi. Biol Fertil Soils 15:259–267

Li CY, Massicote HB, Moore LVH (1992) Nitrogen-fixing *Bacillus* sp. associated with Douglas-fir tuberculate ecto-mycorrhiza. Plant Soil 140:35–40

Liu G, Chambers SM, Cairney JWG (1998) Molecular diversity of ericoid mycorrhizal endophytes isolated from *Woollsia pungens*. New Phytol 140:145–154

Marschner H (1995) Mineral nutrition of higher plants, 2nd edn. Academic Press, London

Marschner H (1996) Mineral nutrient acquisition in nonmycorrhizal and mycorrhizal plants. Phyton 36:61–68

Mårtensson AM, Rydberg I, Vestberg M (1998) Potential to improve transfer of N in intercropped systems by optimising host-endophyte combinations. Plant Soil 205:57–66

Martin F, Ramstedt M, Soderhall K (1987) Carbon and nitrogen metabolism in ecto-mycorrhizal fungi and ecto-mycorrhizas. Biochimie 69:569–581

Martin F, Lapeyrie F, Tagu D (1997) Altered gene expression during ecto-mycorrhizal development. Mycota 5:223–242

Martin RC, Voldeng HD, Smith DL (1991) Nitrogen transfer from nodulating soybean [*Glycine max* (L.) Merr.] to corn (*Zea mays* L.) and non-nodulating soybean in intercrops: direct ^{15}N labelling methods. New Phytol 117:233–241

Martin RC, Eagelsham RJ, Voldeng HD, Smith SL (1995) Factors affecting nitrogen benefit from soybean (*Glycine max* (L.) Merr. CV Lee) to interplanted corn (*Zea mays* L. CV Co-op S259). Environ Exp Bot 35:497–505

Martins MA (1993) The role of the external mycelium of arbuscular mycorrhizal fungi in the carbon transfer process between plants. Mycol Res 97:807–810

Massicotte HB, Molina R, Luoma DL, Smith JE (1994) Biology of the ecto-mycorrhizal genus, *Rhizopogon*. II. Patterns of host-fungus specificity following spore inoculation of diverse hosts grown in monoculture and dual culture. New Phytol 126:677–690

Massicotte HB, Molina R, Tackaberry LE, Smith JE, Amaranthus MP (1999a) Diversity and host specificity of ecto-mycorrhizal fungi retrieved from three adjacent forest sites by five host species. Can J Bot 77:1053–1076

Massicotte HB, Melville LH, Peterson RL, Unestam T (1999b) Comparative studies of ecto-mycorrhiza formation in *Alnus glutinosa* and *Pinus resinosa* with *Paxillus involutus*. Mycorrhiza 8:229–240

McGonigle TP, Fitter AH (1990) Ecological specificity of vesicular-arbuscular mycorrhizal associations. Mycol Res 94:120–122

McKendrick SL, Leake JR, Taylor DL, Read DJ (2000a) Symbiotic germination and development of myco-heterotrophic plants in nature: ontogeny of *Corallorhiza trifida* and characterization of its mycorrhizal fungi. New Phytol 145:523–537

McKendrick SL, Leake JR, Read DJ (2000b) Symbiotic germination and development of myco-heterotrophic plants in nature: transfer of carbon from ecto-mycorrhizal *Salix repens* and *Betula pendula* to the orchid *Corallorhiza trifida* through shared hyphal connections. New Phytol 145:539–548

Melin E, Nilsson H (1955) Ca45 used as an indicator of transport of cations to pine seedlings by means of mycorrhizal mycelium. Svensk Bot Tidskr 49:119–122

Michelsen A, Quarmby C, Sleep D (1998) Vascular plant ^{15}N natural abundance in heath and forest tundra ecosystems is closely correlated with presence and type of mycorrhizal fungi in roots. Oecologia 115:406–418

Miller SL, Allen EB (1992) Mycorrhiza, nutrient translocation, and interactions between plants. In: Allen MF (ed) Mycorrhizal functioning: an integrative plant-fungal process. Chapman and Hall, New York, pp 301–332

Miller SL, Parsons WFJ, Knight DH (1989) Small scale hydro-excavation of soil monoliths from a lodgepole pine forest. Bull Ecol Soc Am 70:205–206

Miller SL, Koo CD, Molina R (1991) Characterisation of red alder ecto-mycorrhiza: a preface to monitoring belowground ecological responses. Can J Bot 69:516–531

Molina R, Trappe JM (1994) Biology of the ectomycorrhial genus *Rhizopogon*. I. Host associations, specificity and pure culture syntheses. New Phytol 126:653–675

Molina R, Massicotte H, Trappe JM (1992) Specificity phenomena in mycorrhizal symbiosis: community-ecological consequences and practical implications. In: Allen MF (ed) Mycorrhizal functioning: an integrative plant-fungal process. Chapman and Hall, New York, pp 357–423

Molina R, Smith JE, McKay D, Melville LH (1997) Biology of the ecto-mycorrhizal genus, *Rhizopogon*. III. Influence of co-cultured conifer species on mycorrhizal specificity with the arbutoid hosts *Arctostaphylos uva-ursi* and *Arbutus menziesii*. New Phytol 137:519–528

Molina R, Trappe JM, Grubisha LC, Spatafora JW (1999) Rhizopogon. In: Cairney JWG, Chambers SM (eds) Ecto-mycorrhizal fungi: key genera in profile. Springer, Berlin Heidelberg New York, pp 129–161

Moora M, Zobel M (1996) Effect of arbuscular mycorrhiza on inter- and intraspecific competition of two grassland species. Oecologia 108:79–84

Näsholm T, Ekblad A, Nordin A, Giesler R, Hogberg M. Hogberg P (1998) Boreal forest plants take up organic nitrogen. Nature 392:914–916

Nehls U, Berguiristain T, Ditengou F, Lapeyrie F, Martin F (1998) The expression of a symbiosis-regulated gene in eucalypt roots is regulated by auxins and hypaphorine, the tryptophan betaine of the ecto-mycorrhizal basidiomycete *Pisolithus tinctorius*. Planta 207:296–302

Newberry DM, Alexander IJ, Rother JA (2000) Does proximity to conspecific adults influence the establishment of ecto-mycorrhizal trees in rain forest? New Phytol 147:401–409

Newman EI (1988) Mycorrhizal links between plants: their functioning and ecological significance. Adv Ecol Res 18:243–270

Newman EI, Eason WR (1993) Rates of phosphorus transfer within and between ryegrass (*Lolium perenne*) plants. Funct Ecol 7:242–248

Newman EI, Ritz K (1986) Evidence on the pathways of phosphorus transfer between vesicular-arbuscular mycorrhizal plants. New Phytol 104:77–87

Newman EI, Eason WR, Eissenstat DM, Ramos MIRF (1992) Interactions between plants: the role of mycorrhiza. Mycorrhiza 1:47–53

Newman EI, Devoy CLN, Easen NJ, Fowles KJ (1994) Plant species that can be linked by VA mycorrhizal fungi. New Phytol 126:691–693

Norby RJ, O'Neill EG, Hood WG, Luxmoore RJ (1987) Carbon allocation, root exudation and mycorrhizal colonization of *Pinus echinata* seedlings grown under CO_2 enrichment. Tree Physiol 3:203–210

Nurmiaho-Lassila EL, Timonen S, Haahtela K, Sen R (1997) Bacterial colonization patterns of intact *Pinus sylvestris* mycorrhizospheres in dry pine forest soil: an electron microscopy study. Can J Micro 43:1017–1035

Nye PH, Tinker PB (1977) Solute movement in the soil-root system. Blackwell Scientific, Oxford

Olsson PA, Gray SN (1998) Patterns and dynamics of 32P-phosphate and labelled 2-aminoisobutyric acid (14C-AIB) translocation in intact basidiomycete mycelia. FEMS Microbiol Ecol 26:109–120

Olsson PA, Wallander H (1998) Interactions between ecto-mycorrhizal fungi and the bacterial community in soils amended with various primary minerals. FEMS Microbiol Ecol 27:195–205

Olsson PA, Chalot M, Bååth E, Finlay RD, Söderström B (1996) Ecto-mycorrhizal mycelia reduce bacterial activity in a sandy soil FEMS Microb Ecol 21:77–86

Paul EA, Clark FE (1989) Soil microbiology and biochemistry. Academic Press, San Diego

Pedersen CR, Sylvia DM, Shilling DG (1999) *Pisolithus arhizus* ecto-mycorrhiza affects plant competition for phosphorus between *Pinus elliottii* and *Panicum chamaelonche*. Mycorrhiza 9:199–204

Perotto S, Actis-Perino E, Perugini J, Bonfante P (1996) Molecular diversity of fungi from ericoid mycorrhizal roots. Mol Ecol 5:123–131

Perry DA (1998) A moveable feast: the evolution of resource sharing in plant-fungus communities. Tree 13:432–434

Perry DA (1999) Reply from D.A. Perry. Tree 14:70–71

Perry DA, Margolis H, Choquette C, Molina R, Trappe JM (1989a) Ecto-mycorrhizal mediation of competition between coniferous tree species. New Phytol 112:501–511

Perry DA, Amaranthus MP, Borchers J, Borchers S, Brainerd R (1989b) Bootstrapping in ecosystems. BioScience 39:230–237

Perry DA, Bell T, Amaranthus MP (1992) Mycorrhizal fungi in mixed-species forests and other tales of positive feedback, redundancy and stability. In: Cannell MGR, Malcolm DC, Robertson PA (eds) The ecology of mixed-species stands of trees. British Ecological Society, Spec Publ 11. Blackwell, Oxford, pp151–179

Persson H (1978) Root dynamics in a young Scots pine stand in central Sweden. Oikos 30:508–519

Priha O, Lehto T, Smolander A (1999) Mycorrhizas and C and N transformations in the rhizopheres of *Pinus sylvestris*, *Picea abies* and *Betula pendula* seedlings. Plant Soil 206:191–204

Pritsch K, Much JC, Buscot F (1997) Morphological and anatomical characterisation of black alder *Alnus glutinosa* (L.) Gaertn. ecto-mycorrhizas. Mycorrhiza 7:201–216

Qualls RG, Haines BL, Swank WT (1991) Fluxes of dissolved organic nutrients and humic substances in a deciduous forest. Ecology 72:254–266

Quoreshi AM, Timmer VR (1998) Exponential fertilization increases nutrient uptake and ectomycorrhizal development of black spruce seedlings. Can J For Res 28:674–682

Read DJ (1994) Plant-microbe mutualisms and community structure. In: Schulze E-D, Mooney HA (eds) Biodiversity and ecosystem function. Springer, Berlin Heidelberg New York, pp 181–209

Read DJ, Boyd R (1986) Water relations of mycorrhizal fungi and their host plants. In: Ayres PG, Boddy L (eds) Water, fungi and plants. Cambridge Univ Press, Cambridge, pp 287–303

Read DJ, Francis R, Finlay RD (1985) Mycorrhizal mycelia and nutrient cycling in plant communities. In: Fitter AH, Atkinson D, Read DJ, Usher MB (eds) Ecological interactions in soil. Blackwell Scientific, Oxford, pp 193–217

Reid CPP, Kidd FA, Ekwebelam SA (1983) Nitrogen nutrition, photosynthesis and carbon allocation in ecto-mycorrhizal pine. Plant Soil 71:415–432
Rejon A, Garcia-Romera I, Ocampo JA, Bethlenfalvay GJ (1997) Mycorrhizal fungi influence competition in a wheat-ryegrass association treated with the herbicide diclofop. Appl Soil Ecol 7:51–57
Rennenberg H (1999) The significance of ecto-mycorrhizal fungi for sulfur nutrition of trees. Plant Soil 215:115–122
Rieger A, Guttenberger M, Hampp R (1992) Soluble carbohydrates in mycorrhized and non-mycorrhized fine roots of spruce seedlings. Z Naturforsch 47 c:201–204
Ritz K, Newman EI (1984) Movement of ^{32}P between intact grassland plants of the same age. Oikos 43:138–142
Ritz K, Newman EI (1985) Evidence for rapid cycling of phosphorus from dying roots to living plants. Oikos 45:174–180
Ritz K, Newman EI (1986) Nutrient transport between ryegrass plants differing in nutrient status. Oecologia 70:128–131
Robinson D, Fitter AH (1999) The magnitude and control of carbon transfer between plants linked by a common mycorrhizal network. J Exp Bot 50:9–13
Rossow LJ, Bryant JP, Kielland K (1997) Effects of above-ground browsing by mammals on mycorrhizal infection in an early successional taiga ecosystem. Oecologia 110:94–98
Rouhier H; Read DJ (1998) Plant and fungal responses to elevated atmospheric carbon in mycorrhizal seedlings of Pinus sylvestris. Environ Exp Bot 40:237–246
Rygiewicz PT, Anderson CP (1994) Mycorrhiza alter quality and quantity of carbon allocated below-ground. Nature 369:58–60
Salzer P, Hager A (1991) Sucrose utilization of the ecto-mycorrhizal fungi *Amanita muscaria* and *Hebeloma crustuliniforme* depends on the cell wall-bound invertase activity of their host *Picea abies*. Bot Acta 104:439–445
Salzer P, Hager A (1993) Characterization of wall-bound invertase isoforms of *Picea abies* cells and regulation by ecto-mycorrhizal fungi. Phys Plant 88:52–59
Sanders IR, Fitter AH (1992) The ecology and functioning of vesicular arbuscular mycorrhizas in co-existing grassland species. 1. Seasonal patterns of mycorrhizal occurrence and morphology. New Phytol 120:517–524
Schack-Kirchner H, Wilpert KV, Hildebrand EE (2000) The spatial distribution of soil hyphae in structured spruce-forest soils. Plant Soil 224:195–205.
Schaeffer C, Wallenda T, Guttnberger M, Hampp R (1995) Acid invertase in mycorrhizal and non-mycorrhizal roots of Norway spruce (*Picea abies* [L.] Karst.) seedlings. New Phytol 129:417–424
Schelkle M, Ursic M, Faruhar M, Peterson RL (1996) The use of laser scanning confocal microscopy to characterize mycorrhizas of *Pinus strobus* L. and to localize associated bacteria. Mycorrhiza 6:431–440
Scheromm P, Plassard C, Salsac L (1990) Nitrate nutrition of maritime pine (*Pinus pinaster* Soland *in* Ait.) ecto-mycorrhizal with *Hebeloma cylindrosporum* Romagn. New Phytol 114:93–98
Schmidt O, Curry JP (1999) Effects of earthworms on biomass production, nitrogen allocation and nitrogen transfer in wheat-clover intercropping model systems. Plant Soil 214:187–198
Seegmüller S, Schulte M, Herschbach C, Rennenberg H (1996) Interactive effects of mycorrhization and elevated atmospheric CO_2 on sulphur nutrition of young pedunculate oak (*Quercus robur* L.) trees. Plant Cell Environ 19:418–426
Setälä H (1995) Growth of birch and pine seedlings in relation to grazing by soil fauna on ecto-mycorrhizal fungi. Ecology 76:1844–1851
Shishido M, Massicotte HB, Chanway CP (1996a) Effect of plant growth promoting *Bacillus* strains on pine and spruce seedlings growth and mycorrhizal infection. Ann Bot 77:433–441

Shishido M, Petersen, DJ, Massicotte HB, Chanway CP (1996b) Pine and spruce seedling growth and mycorrhizal infection after inoculation with plant growth promoting *Pseudomonas* strains. FEMS Microbiol Ecol 21:109–119

Simard SW (1995) Interspecific carbon transfer in ecto-mycorrhizal tree species mixtures. PhD dissertation. Oregon State University, Corvallis, Oregon

Simard SW (1999) Below-ground connections among trees: implications for enhanced silviculture and restoration. In: Egan B (ed) Helping the land heal: ecological restoration in British Columbia. BC Environmental Network Educational Foundation, Victoria, BC, pp179–184

Simard SW, Perry DA, Jones MD, Myrold DD, Durall DM, Molina R (1997a) Net transfer of carbon between tree species with shared ecto-mycorrhizal fungi. Nature 388:579–582

Simard SW, Molina R, Smith JE, Perry DA, Jones MD (1997b) Shared compatibility of ecto-mycorrhiza on *Pseudotsuga menziesii* and *Betula papyrifera* seedlings grown in mixture in soils from southern British Columbia. Can J For Res 27:331–342

Simard SW, Perry DA, Smith JE, Molina R (1997c) Effects of soil trenching on occurrence of ecto-mycorrhiza on *Pseudotsuga menziesii* seedlings grown in mature forests of *Betula papyrifera* and *Pseudotsuga menziesii*. New Phytol 136:327–340

Simard SW, Jones MD, Durall DM, Perry DA, Myrold DD, Molina R (1997d) Reciprocal transfer of carbon isotopes between ecto-mycorrhizal *Betula papyrifera* and *Pseudotsuga menziesii*. New Phytol 137:529–542

Simard SW, Durall DM, Jones MD (1997e) Carbon allocation and carbon transfer between *Betula papyrifera* and *Pseudotsuga menziesii* seedlings using a 13C pulse-labeling method. Plant Soil 191:41–55

Smith DC, Muscatine L, Lewis DH (1969) Carbohydrate movement from autotrophs to heterotrophs in parasitic and mutualistic symbiosis. Biol Rev 44:17–90

Smith FA, Smith, SE (1988) Solute transfer at the interface: ecological implications. In: Mejstrik V (ed) Proceedings of the 2nd European Symposium on Mycorrhiza. August 1988, Prague, Czechoslovakia. p 100

Smith SE, Read DJ (1997) Mycorrhizal symbiosis, 2nd edn. Academic Press, San Diego

Smith SE, Smith FA (1990) Structure and function of the interface in biotrophic symbioses as they relate to nutrient transport. New Phytol 114:1–38

Smith JE, Molina R, Perry DA (1995) Occurrence of ecto-mycorrhizas on ericaceous and coniferous seedlings grown in soils from the Oregon Coast Range. New Phytol 129:73–81

Söderström B, Finlay RD, Read DJ (1988) The structure and function of the vegetative mycelium of ecto-mycorrhizal plants. IV. Qualitative analysis of carbohydrate contents of mycelium interconnecting host plants. New Phytol 109:163–166

Ta TC, Faris MA (1987) Species variation in the fixation and transfer of nitrogen from legumes to associated grasses. Plant Soil 98:265–274

Taylor DL, Bruns TD (1997) Independent, specialized invasions of ecto-mycorrhizal mutualism by two nonphotosynthetic orchids. Proc Natl Acad Sci USA 94:4510–4515

Taylor DL, Bruns TD (1999) Population, habitat and genetic correlates of mycorrhizal specialization in the 'cheating' orchids *Corallorhiza maculata* and *C. mertensiana*. Mol Ecol 8:1719–1732

Tibbett M, Grantham K, Sanders FE, Cairney JWG (1998) Induction of cold active acid phosphomonoesterase activity at low temperature in psychrotrophic ecto-mycorrhizal *Hebeloma* spp. Myco Res 102:1533–1539

Timonen S, Finlay RD, Olsson S, Söderström B (1996) Dynamics of phosphorus translocation in intact ecto-mycorrhizal systems: non-destructive monitoring using a β-scanner. FEMS Microbiol Ecol 19:171–180

Timonen S, Jorgensen KS, Haahtela K, Sen R (1998) Bacterial community structure at defined location of *Pinus sylvestris-Suillus bovinus* and *Pinus sylvestris-Paxillus invo-*

lutus mycorhizospheres in dry pine forest humus and nursery peat. Can J Microbiol 44:499–513

Tinker PB, Durall DM, Jones MD (1994) Carbon use efficiency in mycorrhizas: theory and sample calculations. New Phytol 128:115–122

Torti SD, Coley PD (1999) Tropical monodominance: a preliminary test of the ecto-mycorrhizal hypothesis. Biotropica 31:220–228

Trappe JM, Fogel R (1977) Ecosystematic functions of mycorrhiza. Colorado State Univ Range Sci Dep Sci Ser 26:205–214

Turnbull MH, Schmidt S, Erskine PD, Richards S, Stewart GR (1996) Root adaptation and nitrogen source acquisition in natural ecosystems. Tree Physiol 16:941–948

Van Breeman N, Finlay R, Lundström U, Jongmans AG, Giesler R, Olsson M (2000) Mycorrhizal weathering: a true case of mineral plant nutrition? Biogeochemistry 49:53–67

Van Tichelen KK, Colpaert JV (2000) Kinetics of phosphate absorption by mycorrhizal and non-mycorrhizal Scots pine seedlings. Physiol Plant 110:96–103

Van Tichelen KK, Vanstraelen T, Colparet JV (1999) Nutrient uptake by intact mycorrhizal *Pinus sylvestris* seedlings: a diagnostic tool to detect copper toxicity. Tree Physiol 19:189–196

Varga AM (1998) Characterisation and seasonal ecology of ecto-mycorrhiza associated with Sitka alder and lodgepole pine from naturally regenerating young and mature forests in the Sub-Boreal Spruce zone of British Columbia. MSc Thesis. University of Northern British Columbia, Prince George, BC

Vogt KA, Grier CC, Meier CE, Edmonds RL (1982) Mycorrhizal role in net primary production and nutrient cycling in *Abies amabilis* ecosystems in western Washington. Ecology 63:370–380

Wallander H (2000) Uptake of P from apatite by *Pinus sylvestris* seedlings colonised by different ecto-mycorrhizal fungi. Plant Soil 218:249–256

Wallander H, Wickman T (1999) Biotite and microcline as potassium sources in ecto-mycorrhizal and non-mycorrhizal *Pinus sylvestris* seedlings. Mycorrhiza 9:25–32

Wallander H, Wickman T, Jacks G (1997a) Apatite as a P source in mycorrhizal and non-mycorrhizal *Pinus sylvestris* seedlings. Plant Soil 196:123–131

Wallander H, Arnebrant K, Östrand F, Kårén O (1997b) Uptake of ^{15}N-labelled alanine, ammonium and nitrate in *Pinus sylvestris* L. ecto-mycorrhiza growing in forest soil treated with nitrogen, sulphur or lime. Plant Soil 195:329–338

Wallenda T, Read DJ (1999) Kinetics of amino acid uptake by ecto-mycorrhizal roots. Plant Cell Environ 22:179–187

Wallenda T, Schaeffer C, Einig W, Wingler A, Hampp R, Seith B, George E, Marschner H (1996) Effects of varied soil nitrogen supply on Norway spruce (*Picea abies* [L.] Karst.). Plant Soil 186:361–369.

Walter LEF, Hartnett DC, Hetrick AD, Schwab AP (1996) Interspecific nutrient transfer in a tallgrass prairie plant community. Am J Bot 83:180–184

Warner A, Mosse B (1983) Spread of vesicular-arbuscular mycorrhizal fungi between separate root systems. Trans Br Mycol Soc 80:353–354

Waters JR, Borowicz VA (1994) Effect of clipping, benomyl, and genet on ^{14}C transfer between mycorrhizal plants. Oikos 71:246–252

Watkins NK, Fitter AH, Graves JD, Robinson D (1996) Carbon transfer between C_3 and C_4 plants linked by a common mycorrhizal network, quantified using stable carbon isotopes. Soil Biol Biochem 28:471–477

Watteau F, Berthelin J (1994) Microbial dissolution of iron and aluminium from soil minerals: efficiency and specificity of hydroxamate siderophores compared to aliphatic acids. Eur J Soil Biol 30:1–9

Wilkinson DM (1998) The evolutionary ecology of mycorrhizal networks. Oikos 82: 407–410.

Wilkinson DM (1999) Mycorrhizal networks are best explained by a plurality of mechanisms. A comment on Fitter et al. 1998. Funct Ecol 13:435–436

Wilpert KV, Schack-Kirchner H, Hoch R, Günther S, Hildebrand EE (1996) Bodenchemische und physikalische Faktoren des Myzelwachstums von Mykorrhizapilzen: Bodenstruktur, Gashaushalt und Hyphenverteilung In: von Wilpert K (ed) Verteilung und Aktivität von Mykorrhizen in Abhängigkeit von der Nährelement-, Wasser- und Sauerstoffverfügbarkeit. Forschungsberichte. ISSN 0948–535X. FZKA-PEF 146, pp 1–97

Zelmer CD, Currah RS (1995) Evidence for a fungal liaison between *Corallorhiza trifida* (Orchidaceae) and *Pinus contorta* (Pinaceae). Can J Bot 73:862–866

Zhou M, Sharik TL (1997) Ecto-mycorrhizal associations of northern red oak (*Quercus rubra*) seedlings along an environmental gradient. Can J For Res 27:1705–1713

Zhu Y-G, Laidlaw AS, Christie P, Hammond MER (2000) The specificity of arbuscular mycorrhizal fungi in a perennial ryegrass-white clover pasture. Agric Ecosys Environ 77:211–218

Zobel M, Moora M (1995) Interspecific competition and arbuscular mycorrhiza: importance for the coexistence of two calcareous grassland species. Folia Geobot Phytotax 30:223–230

Zobel M, Moora M, Haukioja E (1997) Plant coexistence in the interactive environment: arbuscular mycorrhiza should not be out of mind. Oikos 78:202–208

3 Function and Diversity of Arbuscular Mycorrhizae in Carbon and Mineral Nutrition

I. Jakobsen, S.E. Smith, F.A. Smith

Contents

3.1	Summary	75
3.1	Introduction	76
3.2	Carbon Nutrition and Growth of the Fungus	76
3.3	Mineral Nutrition of the Fungus	79
3.4	Nutrition of the Plant: The Autotrophic Symbiont	82
3.5	Mycorrhizal Responsiveness	84
3.6	From Individual Symbioses to Plant Communities	87
References		88

3.1 Summary

The function of arbuscular mycorrhizae (AM) is discussed in terms of reciprocal nutrient exchange within single pairs of symbionts. Emphasis is on carbon and phosphate nutrition and on the importance of diversity of the organisms involved. Growth of the obligate biotrophic AM fungi in roots relies on carbon transferred across interfaces between the plant and the fungus. Carbon use by the fungus is discussed in qualitative and quantitative terms. Attention is paid to the plant species-dependent variability in the formation and probably also function of root internal fungal structures where the *Paris* type of AM may be most important in the ecological perspective. Different fungi vary greatly in the amount of phosphorus transported to the plant. Plant identity is an important determinant of the amount of phosphate transferred from a fungus. Changes in time and space in expression of fungal and plant P transporter genes are discussed with the aim of providing a better understanding of the co-ordination of mechanisms leading to a net transfer of P from fungus to plant in the symbiosis. Future studies with single pairs of symbionts should focus on those fungi which are dominant root colonisers in the field. Careful isolation and controlled experiments with these fungi will contribute to resolving frequent discrepancies between studies in pots and in the field.

3.2 Introduction

Arbuscular mycorrhizal fungi (AMF) are important players in the development of plant communities and appear to have maintained this role during more than 400 million years (Remy et al. 1994). Reciprocal exchange of nutrients between the symbionts is the central process in mycorrhizal functioning, except for myco-heterotrophs. The association is usually facultative for individual plants, at least when growing in soils rich in nutrients, whereas it is always obligatory for the biotrophic AMF. This means that the fitness of AMF entirely depends on their direct supply of organic carbon (C) from the plant, whereas the plant may or may not benefit nutritionally from the association.

This chapter provides a basic framework: the functioning of single pairs of symbionts in terms of nutrient exchange processes and their regulation. Modified functioning caused by diverse AMF isolates and diverse plant species will be considered. Even when the physiology of individual symbioses is well understood during non-reproductive growth in pots, this knowledge may have limited value in an ecological context. Potential confounding factors in translation of data from pot experiments to field situations are briefly considered. We aim to present the most recent advances and readers are referred to Smith and Read (1997) for a comprehensive review. The discussion will in particular refer to symbiotic modifications of plant C relationships and fungal uptake of mineral nutrients and their transfer to the plant. The role of the external AM fungal mycelium in resource allocation in the soil ecosystem is discussed by Olsson et al. (Chap. 4, this Vol.). The ecophysiology of ecto-mycorrhizal fungi is reviewed by Simard et al. (Chap. 2, this Vol.). The different categories of mycorrhizal associations that occur in nature are summarised by Read (Chap. 1, this Vol.; Fig. 1.1).

3.3 Carbon Nutrition and Growth of the Fungus

The key features of arbuscular mycorrhizae are the root, the internal fungal structures and the external mycelium in the soil. Fungal development is driven primarily by plant-derived C and in vitro ^{13}C-NMR studies have shown that fungal structures inside colonised roots, but not external hyphae, take up exogenously supplied hexoses. Lipids constitute by far the largest C pool in AMF and are apparently synthesised in the internal structures before being translocated to the external mycelium (Bago et al. 2000).

The same fungus may develop quite differently in different plant species, but analysis of developmental differences and their effects on nutrient exchange and growth is only just beginning. In Arum types (defined by Gallaud 1905), relatively fast-growing hyphae extend through the intercellular spaces of the root cortex, developing short side branches which penetrate the walls of the cortical cells and branch repeatedly to form relatively large arbuscules, which are completely surrounded by the plant plasma membrane. Arbuscules are relatively short-lived structures with large surface areas, and experimental evidence is consistent with an important role in transfer of phosphate (P) and other nutrients such as Zn (Bürkert and Robson 1994) from fungus to plant. Transfer of sugars to the fungus may be active, by proton co-transport, in which case a proton gradient generated by an H^+-ATPase in the fungal membrane would be necessary. Hypotheses involving hexose uptake by arbuscules have been put forward, based on unpublished molecular evidence (Blee and Anderson 1998). However, a fungal H^+-ATPase has been detected in the intercellular hyphae, but not in membranes of the arbuscule branches (Gianinazzi-Pearson et al. 1991) – a distribution which might exclude the latter as a site for sugar transfer. The issue might be resolved if uptake were passive, mediated by specific sugar carrier(s) in the fungal membrane, with the concentration gradient between the fungus and the cortical apoplast maintained by rapid conversion of sugars in the fungus (see Smith et al. 1969). The process is in fact selective: AMF absorb glucose and to a lesser extent fructose, in preference to sucrose, and conversion of these to trehalose and glycogen and perhaps also lipid within the arbuscules would maintain the necessary concentration difference (Bago et al. 2000). The same mechanisms could of course operate in intercellular hyphae or coils and investigation of the expression and localisation of the necessary transporters in plant and fungus is needed.

From an ecological perspective it may be important that in most plant families *Paris* types (Gallaud 1905) are more common than *Arum* types (see Smith and Smith 1997). Intercellular hyphae are lacking in *Paris* types and the fungi spread directly from cortical cell to cortical cell, forming well developed intracellular coils from which small, branched arbuscules may develop. These coils are, like the arbuscules, surrounded by the plant plasma membrane and remain within the apoplastic compartment of the root. Coils, like intercellular hyphae, are relatively long-lived structures and have largely been ignored as potential nutrient transfer sites, even though a coil may re-present as great a surface area of intracellular interface as an arbuscule (Dickson and Kolesik 1999). *Paris* type mycorrhizae are generally believed to develop more slowly than *Arum* types, but more data are required. There are variations in detail in both classes, e.g. in location of arbuscules within the cortex, and arbuscular mycorrhizae of some plants are intermediate between the two classes (see Smith and Smith 1997). More work is again needed, including conclusive identification of the interface(s) involved in C transfer in both *Arum* and *Paris* types of mycorrhizae.

Knowledge of seasonal and ontogenetic variations in the relative occurrence of fungal structures is crucial in ecological studies as it may provide clues to what the fungi are doing at different times in relation to nutrient transfer and plant growth. Arbuscules, used as markers for active fungus involved in P transfer to the plant, usually appear early in development or at stages of active root growth when nutrient requirements and rates of nutrient uptake are high; they may be absent at other times (e.g. Allen 1983; Mullen and Schmidt 1993). Functions of the intercellular hyphae and coils have not been clarified. Storage vesicles, where formed, appear at later stages of colonisation and are important contributors to C drain from the plant. Most data have been obtained by non-vital staining of all fungal structures inside the root, so that amounts of living fungus present at any one time are not known. Yet this information is critical to the development of models to assess 'mycorrhizal benefit' in terms of the balance between increased nutrient acquisition and C drain. One possibility is to use the fatty acid 16:1ω5, which is present in some AMF in much higher quantities than in other organisms (Olsson et al. 1995).

The mycelium of an AM fungus (constituting up to 80–90% of the fungal biomass – Olsson et al. 1999) may derive its C from several individuals of one or more plant species in a community, but colonisation cannot be solely related to provision of C to the mycelium. Heterotrophic AM plants, at least, derive their C along with nutrients such as P from the fungus (for a discussion on heterotrophic plants, see Taylor et al., Chap. 15, this Vol.). Quantitative estimates of the amount of organic C required by the fungus are difficult to obtain. Values of 10–20% of total photosynthate are generally accepted, but there will be large variations depending on individual plant-fungus combinations, amount of fungal biomass and environmental conditions (see Jakobsen 1999 for references). Most of the C allocated to the fungus is respired, but about 25% of the remaining extraradical C can in fact be found in the external mycelium (Jakobsen and Rosendahl 1990). This mycelium grows in a patchy and fluctuating soil environment, completely dependent on a continuing supply of organic C from the root in order to forage for mineral nutrients and to compete with other soil organisms for them. As Lewis (1973) pointed out, it is the ability of the mycelium to access resources otherwise unavailable to the plants that has allowed the fungi to evolve to mutualism.

Transfer of C from plant to fungus in an established mycorrhiza may be regulated by factors associated with either symbiont. Limitation by photosynthate availability is suggested by decreased mycorrhiza formation in response to defoliation (Allsopp 1998) and also to low light levels under certain conditions (Son and Smith 1988), but results are conflicting. Elevated levels of atmospheric CO_2, which increase photosynthesis as long as nutrients, light or other factors are not limiting, can increase root colonisation and production of external hyphae (Sanders et al. 1998 and references therein; Rillig et al., Chap. 6, this Vol.). However, time-course studies with several species of mono-

and dicotyledons showed that elevated CO_2 had no effect on colonisation by *Glomus mosseae,* when effects on plant size were accounted for (Staddon et al. 1999). This is in accord with an absence of effects of elevated CO_2 on root growth, root colonisation or growth of mycelium in soil in experiments where mineral nutrient levels were not limiting (Gavito et al. 2000; Rillig et al., this Vol.). When below-ground allocation is affected by the atmospheric CO_2 level, it appears that the proportion of organic C available to the fungus does not increase. The factors regulating this proportion are still unknown.

As with internal colonisation, AMF differ in pattern of growth in soil (Abbott et al. 1992) and biomass of spores produced (Sieverding et al. 1989) so that interfungal variation in C demand on the plant is likely. The influence of three fungi on below-ground C allocation in cucumber has been shown to be rather similar, although root colonisation and length of external hyphae differed (Pearson and Jakobsen 1993a). *Scutellospora calospora* apparently produced the lowest fungal biomass and associated plants acquired the least amount of P, but allocation of fixed C to below-ground respiration was highest with this fungus. Interfungal variation in C use is also suggested by a study with ten AMF, where the most aggressive species produced growth depressions in wheat both at low and high P levels (Graham and Abbott 2000). These growth depressions were associated with reduced sucrose concentrations in the roots, a possible consequence of high transfer to, and conversion in, the fungal partner. The frequently observed negative effect of high plant P status on colonisation may well be related to reduced availability of C to the fungus, as the P status influences both the relative below-ground C allocation (Marschner 1995) and membrane leakiness (Graham et al 1981). There is considerable variation between plant species in the sensitivity of colonisation to environmental variables. Wheat may be virtually non-colonised if soil P is high (e.g. Baon et al. 1992), whereas clover or onion may be mycorrhizal over a much wider range of P supply (Oliver et al. 1983; Son and Smith 1988). In addition, there are differences in response of colonisation by individual AMF to increasing soil P (Thomson et al. 1986). Such variations are also correlated with differences in plant responsiveness to colonisation (see below), but this is not always the case and data for wild, rather than cultivated plants are needed. The relationship between extent of colonisation and 'benefit' to the plant, whether positive or negative, has been recently discussed by Gange and Ayres (1999). Their model requires further elaboration, to take into account the variations summarised above.

3.4 Mineral Nutrition of the Fungus

It is presumed that the fungi obtain all their mineral nutrients from the soil and that their growth in soil is limited by availability of mineral nutrients under extreme conditions only. Compartmented systems that separate roots from external mycelium have been used either in conjunction with isotopes or by measurement of depletion to show AM fungal uptake and transfer to the plant of P, N, K, Zn and Cu (see Marschner 1995 for references). The external mycelium gets direct access to nutrients outside the zones of nutrient depletion that develop close to the root and to nutrients in inaccessible microsites. Hyphae may also proliferate rapidly and extensively in enriched patches (Olsson et al., Chap. 4; Leake et al., Chap. 14, this Vol.), and compete effectively with soil saprophytes during periods of rapid mineralisation of P and more especially N. The uptake by AMF can have dramatic effects on growth of the associated plant when root uptake of a nutrient exceeds its rate of replacement from the bulk soil (i.e. when there is a nutrient depletion zone close to the root); P is by far the most important nutrient in this respect (see Sect. 4). Due to their much smaller diameter, the solution at the surface of the hyphae will be less depleted for P than the solution at the root surface and uptake kinetics will be much more important than they are for roots (Barber 1984). This will enable continuing P uptake via proton co-transport through high affinity P transporters, such as the one (GvPT) which has been shown to be expressed in the external mycelium of *Glomus versiforme* (Harrison and van Buuren 1995). A high effectiveness in hyphal uptake of P at low concentrations in the soil is also suggested by estimates of the uptake kinetic parameters V_{max} and K_m; the maximal uptake estimates (V_{max}), in particular, are much higher for hyphae than for roots (Schweiger and Jakobsen 1999).

AMF remove nutrients from fertile soils in root-free compartments with remarkable effectiveness (Table 3.1) and hyphal uptake may be just as efficient as the uptake of roots and hyphae together (Pearson and Jakobsen 1993b). It is generally accepted that the capacity for nutrient uptake by the hyphae is much more important for P than for the more mobile N species (nitrate and ammonium), but under certain conditions, such as low soil moisture, N transport through soil is limited by diffusion instead of mass flow. In any case, the relative importance relates to the plant, rather than the fungal requirements. Most experiments on hyphal nutrient uptake have been carried out in soils that are relatively high in nutrients, such as agricultural soils and soils amended with nutrients. Levels of nutrients are generally much lower in natural soils than the initial values of the experiments reported in Table 3.1. Under these conditions, it is particularly important that highly effective nutrient scavenging systems are present or able to develop rapidly to capture nutrient flushes and to ensure the maximum nutrient availability for both symbionts when the requirement of the plant is high, e.g. during early growth and the reproductive phase.

Inorganic nutrients in the soil solution constitute the primary source for the fungi, but the fungi also appear to have some influence on the availability of nutrients from organic sources. Extracellular phosphatase activity has been detected in excised mycelium of two AMF (Joner and Johansen 2000) and *Glo-*

Table 3.1. Depletion of inorganic nutrients by mycelia of AM fungi in root-free soil compartments

Nutrient	Fungus	Sampling site (mm from root compartment)	Nutrient level (mg kg^{-1} soil)		Depletion by mycelium (%)	Hyphal length (m g^{-1} soil)	Reference
			− Hyphae	+ Hyphae			
P	G. mosseae[a]	20	80	18	78	17	Li et al. (1991)
	G. mosseae	50	60	30	50	2–5	George et al. (1992)
	G. invermaium	5–47	16.2	13.3	18	27	Joner and Jakobsen (1994)
	G. caledonium	5–47	16.2	12.9	20	16	Joner and Jakobsen (1994)
	G. invermaium	5	8.5	4.3	51	17	Joner et al. (1995)
	G. caledonium	5	8.5	5.7	33	17	Joner et al. (1995)
NH$_4$-N	G. intraradices[b]	50	13	1.5	88	13	Johansen et al. (1992)[d]
	G. intraradices[b]	50	42	8	81	21	Johansen et al. (1994)
	G. mosseae[c]	50	9	2	78	2–6	George et al. (1992)
NO$_3$-N	G. intraradices[b]	50	5	0.5	90	13	Johansen et al. (1992)
	G. intraradices[b]	50	16	3	81	21	Johansen et al. (1994)
	G. mosseae[c]	50	85	25	71	2–6	George et al. (1992)

[a] Extracted by 0.5 M NaHCO$_3$.
[b] Extracted by 2 M KCl.
[c] Extracted by 0.01 M CaCl$_2$.
[d] The nitrification inhibitor *N serve* added to soil.

mus intraradices has been shown to take up P from AMP in monoxenic cultures (Joner et al. 2000). Uptake of glycine and glutamate by excised hyphae of several AMF isolates and by *G. intraradices* in monoxenic cultures has also been demonstrated (Hawkins et al. 2000). This uptake did not contribute significantly to N uptake by the roots, but may be important for fungal nutrition.

Nutrient uptake by the external hyphae can vary greatly between fungi. In some cases such variation is primarily caused by different growth patterns of the mycelium in the soil (Jakobsen et al. 1992a), but the physical presence of hyphae does not necessarily imply effective uptake of nutrients (Pearson and Jakobsen 1993b). In a recent study with *Medicago truncatula*, both *Scutellospora calospora* and *Glomus caledonium* colonised an artificial 'nutrient patch' labelled with ^{33}P, but the former fungus was much less effective at transferring the labelled P to the plant. Instead it transferred P obtained closer to the root surface (Smith et al. 2000a). The causality behind the observed interspecific variation in nutrient uptake by AMF is not clear, although the observation of P accumulation in the "inefficient" *S. calospora* (Jakobsen et al. 1992b) suggests that transfer across the symbiotic interface may be an important rate-limiting step. Generalisations concerning the effectiveness of nutrient uptake by a fungus should be avoided, because the plant species plays an important role for the outcome of an experiment. A well-established mycelium of one fungus may transport P at high effectiveness in symbiosis with one plant but not with another (Ravnskov and Jakobsen 1995). Such functional diversity may be of considerable importance at the level of plant communities (see Sect. 3.7).

3.5 Nutrition of the Plant: the Autotrophic Symbiont

Many experimental studies have shown that colonisation by AMF increases P uptake and plant growth compared with non-mycorrhizal controls, though this is not always the case (see below). Not much is known about the co-ordination of plant and fungal nutrient transport mechanisms leading to net transfer of P from fungus to plant (see Harrison 1999; Smith et al. 1999). Up- or down-regulation of tissue expression of fungal and plant P transporter genes have provided some helpful insights. The fungal P transporter expressed in external mycelium is not expressed within the root (Harrison and van Buuren 1995), a pattern which might prevent the fungus in scavenging P from the interfacial apoplast and promoting the net efflux necessary to support the overall transfer from fungus to plant. Changes in expression of plant membrane transporters also occur following colonisation by AMF and/or alterations in external P supply (H. Liu et al. 1998; C. Liu et al. 1998; Rosewarne et al. 1999; Burleigh and Jakobsen, in prep.), indicating possible adjustments in the kinetics of P uptake of the root cells involved. In non-mycorrhizal tomato, transporter genes are expressed in epidermal cells, at or close behind the root apex, appropriate for P absorption from the undepleted soil into which the root tip would have just grown (C. Liu et al. 1998). Transporter expression seems to be lower in older regions of the root, perhaps related to the likely depletion of P, and is also decreased in P-sufficient plants, potentially limiting accumulation of P to toxic concentrations (C. Liu et al. 1998; H. Liu et al 1998; Rosewarne et al. 1999). In P-stressed tomato, mycorrhizal colonisation causes little alteration in overall expression of the P transporter *LePt1*, but the site of expression changes from the root epidermis to cortical cells containing arbuscules where P would be delivered by the fungus (Rosewarne et al. 1999). This is an indication that the fungus may 'take over' the main P supply to the plant (see also Pearson and Jakobsen 1993b). However, there is new information indicating firstly, not all plant species respond in the same way and secondly, different fungal species evoke different responses in the plants, with respect to expression of P-sensitive genes and also P nutrition (Burleigh and Jakobsen, in prep.). These results are relevant to understanding variations in fungal effectiveness and in predicting likely outcomes of colonisation by mixtures of AMF in field soil.

Mycorrhizal plants often accumulate higher concentrations of P in their shoots than non-colonised plants, indicating that factors other than P supply then limit growth (see Stribley et al. 1980). This 'luxury' storage has been interpreted as inefficiency of nutrient use, but in the context of the whole life cycle could be a significant insurance against future (e.g. seasonal) deficits or a means to ensure high quality seed reserves (e.g. see Koide 1991; Merryweather and Fitter 1996). There is much less information about mycorrhizal effects on acquisition of other potentially scarce or immobile nutrients, although a limited amount of work has been done with Zn and Cu. Acquisition of N by external mycelium rarely leads to increased concentrations in the plant, possibly reflecting a high demand for this nutrient (see Smith and Read 1997 for references). Read (2000) has provocatively suggested that 'it is increasingly accepted that nutrient acquisition by most plants growing in natural ecosystems is mediated by mycorrhiza-forming symbiotic fungi.' However, in very few studies is it clear what proportion of acquisition, even of immobile soil nutrients such as P or Zn, actually occurs via the fungi as opposed to the roots, especially when the whole life-span of the plant is taken into account. It is certainly unwise to attempt to generalise the role of an AM fungus to uptake of the whole suite of mineral nutrients, for the simple reason that increased growth caused by the improved uptake of the growth-limiting nutrient (e.g. P) will provide a larger surface area of roots for nutrient uptake generally.

Rates of transfer of organic C across the intraradical interfaces from plant to fungus (see above) lie in the same range as those for P, but this does not imply any direct or indirect coupling (see Smith and Read 1997, for calculations and discussion). Release from P stress as a result of increased P uptake via the fungal symbiont increases the rate of photosynthesis and hence the supply of sugars to both symbionts. Photosynthesis may also be increased by the sink effect of the fungal symbiont as shown by comparing with non-mycorrhizal plants having N and P concentrations no lower than those of mycorrhizal plants (Wright et al. 1998). Ongoing C drain to the fungus may reduce total plant dry matter accumulation and increase shoot fresh weight: dry weight ratio (e.g. Son and Smith 1988). However, the production of hyphae is much more economical in terms of organic C than the production of an equivalent length of root, so that plants may also adjust below-ground C allocation and 'manage' with a smaller, mycorrhizal root system.

There is little compelling evidence to support the idea that AMF hyphal linkages allow net transport of organic C, or indeed any other nutrient, from one autotrophic plant to another in physiologically and hence ecologically relevant quantities (see Newman 1988; Robinson and Fitter 1999). The occurrence of inter-plant carbon and nutrient transport is further discussed by Simard et al. (Chap. 2, this Vol.). However, differential support of parts of the mycelial network by 'donor plants', may have significant effects on community interactions. The situation for achlorophyllous plants (mycoheterotrophs) associated with AMF is quite different, as they must be dependent on supplies of photosynthate from an autotrophic third party, as well as mineral nutrients from the soil, delivered via their fungal symbiont. The mechanisms have not been investigated, but it is interesting that these plants all seem to be *Paris* types, some of which may lack arbuscules entirely (Leake 1994; Imhof 1999), indicating that arbuscules may not be essential for mineral nutrient or organic C transfer to these specialised plants. Further information about the ecology of mycoheterotrophic plants is given by Taylor et al. (Chap. 15, this Vol.).

3.6 Mycorrhizal Responsiveness

Different plant species growing under standard soil conditions show large differences in responsiveness to colonisation. In simple terms, a plant that only achieves low uptake of the growth-limiting nutrient (P) from soil in the absence of mycorrhizal colonisation should show high mycorrhizal responsiveness, provided that other extraneous factors do not prevent this (see Koide 1991; Koide et al. 2000). Much experimental work has been carried out with crop species and the lack of positive growth or yield responses, particularly at high fertiliser applications, has reinforced the view that the symbiosis in these plants is facultative. However, some plants grow very poorly unless they are mycorrhizal even in soils high in nutrients: two very different examples are *Elais* (oil palm: Blal et al. 1990) and *Centaurium* (Grime et al. 1987). In fact, a continuum of responsiveness exists, from (1) species which appear to be obligately mycorrhizal in order to survive, through (2) those whose performance is increased in the mycorrhizal state, to (3) mycorrhiza-forming plants, which are responsive under certain environmental conditions only, and then (4) plants which do not form any mycorrhizal associations. A plant community could include members of any of these groups and of course, a single AM fungus might colonise members of all except the last.

However defined, at the level of individual plants, as opposed to communities, mycorrhizal responsiveness is based to a large extent on costs and benefits, i.e. the 'trade-off' between increased uptake of scarce nutrients and C drain to the fungus. A difference between plant species in ability to change root:shoot ratios in response to colonisation provides further potential mechanisms underlying differences in responsiveness. Table 3.2, adapted from Smith (2000) and Smith et al. (2000b), is an attempt to summarise the most important factors that can lead to high mycorrhizal responsiveness.

Low responsiveness would be produced by the converse of the features listed (i.e. for 'high' read 'low' etc). The first three columns include both architectural and physiological features. They are mostly discussed elsewhere in this chapter and it must again be emphasised that these features will depend on the identity of the fungal and plant partners involved in an individual AM

Table 3.2. Factors that may increase mycorrhizal responsiveness of autotrophic plants. Factors related to fungus, interface and plant will be influenced by the identity of the organisms and may be dependent on their age. Factors in italics are 'physiological', relating directly to resource acquisition and transfer

Fungus	Interface(s)	Root	Plant ecosystem
External hyphae: Fast colonisation High growth rate High extension into soil *High nutrient influx capacity* *High nutrient translocation* Internal hyphae: High growth rate Fast nutrient delivery to interface(s)	Fast development Large area of contact High longevity *High organic C flux from roots* *High nutrient flux to roots*	Short length (low root/shoot ratio) Little branching Large diameter Few or short root hairs Selectively flexible root/shoot ratio[a] Inability to modify rhizosphere[b] *Low nutrient influx capacity* *Fast organic C delivery to interface(s)*	Low soil nutrient availability High light intensity Other 'suitable' soil conditions[c] Interplant connections[d] Interactions with rhizosphere 'helper' bacteria Prevention of infection by root pathogens and other harmful organisms Low plant density (low competition)

[a] Inflexible in response to low soil nutrient levels, not to mycorrhizal colonisation.
[b] Rhizosphere modification to increase nutrient availability.
[c] Good soil aeration, low salinity, etc.
[d] Increased benefits to receiver plants.

symbiosis. There are interactions between architectural and physiological factors, such as those between the area of a surface involved in membrane transport (including interfaces) and 'transport activity'. The latter involves properties that determine the solute flux, e.g. numbers of transport proteins per unit surface area of membrane, combined with the transport parameters, K_m and V_{max}. The properties will be influenced by the age of the symbiotic partners. For example, mycorrhizal responsiveness is often highest during early phases of plant growth, after the initial lag during which colonisation is established. This must involve a combination of time-dependent plant and fungal properties, including those of interfaces. Table 3.2 also lists some of the most important properties of plant ecosystems that can greatly modify mycorrhizal responsiveness from values obtained when plants are grown individually in pots. These are discussed elsewhere in this book. Plants with effective methods of dealing with problems of low nutrient supply other than by development of mycorrhizae are expected to be less responsive to mycorrhizal colonisation than less efficient plants. The strategies are again both architectural and physiological and include rapid root growth into unexploited regions of soil, production of long, dense root hairs and exudation of ions (protons, organic anions) that release P or other nutrients from insoluble sources and hence increase the soil solution concentration (see Marschner 1995; Smith et al. 2000b). Alterations in nutrient uptake processes may also contribute, so that when P supply is very low, activity of the high affinity P transporter(s) in the roots is high – increasing the scavenging capacity (V_{max}) (Clarkson 1985 and see above). This control is partly exerted at the level of expression of the genes encoding the transporters, in response to signals from the shoot (see Schachtman et al. 1998).

Smith et al. (2000b) have suggested that interactions between inherent growth rates of plants and their mycorrhizal responsiveness are likely to be complex, on the basis of growth analysis below-ground. It was suggested that the extent to which mycorrhizal colonisation increases growth rate will depend on (1) the maximum growth rate achieved by a (non-mycorrhizal) plant under low nutrient conditions; (2) the extent to which mycorrhizal colonisation increases supply of additional nutrients to the plant; (3) changes in nutrient use efficiency; and (4) differences in root structural parameters and root:shoot ratios in the absence and presence of colonisation.

Johnson et al. (1997) surveyed the functioning of mycorrhizal symbioses, with emphasis on autotrophic plants, along a mutualism–parasitism continuum. They concluded that 'negative responsiveness', i.e. associations where an AMF parasitises a plant and produces growth depressions, might be induced environmentally, developmentally at different stages in the plant's life cycle, or even constitutively under all conditions. In this last case the fungus would be a 'cheater' (sensu Soberon and Martinez del Rio 1985; see also Smith and Smith 1996) by draining the plant of resources while providing minimal nutritional benefits. Cheating has evolved in a range of symbioses, but there are no known examples of AMF that constitutively cheat *all* their potential host plants. However, examples where different AMF produces large differences in responsiveness in a single plant are common in the literature (Burleigh and Jakobsen, in prep.), with potential ecological consequences covered elsewhere in this book (e.g. Chaps. 9, 10).

3.7 From Individual Symbioses to Plant Communities

As increasing emphasis is placed on the ecological role of AMF it will become more and more critical to assess mycorrhiza functioning in terms of the structural variations in development of arbuscular mycorrhizae in different plant species, colonisation of roots and soil by different fungi, and turnover of both intraradical and external mycelium. These often subtle differences have the potential to be of major importance in influencing biomass production, competition and vegetation dynamics in plant communities and are also critical to the analysis of effects of the underground fungal community on plant species diversity (Wardle 1999). The picture is complicated by ecological factors summarised in the last column of Table 3.2 that we cannot consider here, including effects of other micro-organisms – beneficial and otherwise – in the mycorrhizosphere (e.g. see Chaps. 13, 14 by Gange and Brown and Leake et al., respectively, this Vol.). Density of plants is receiving increasing attention both in the laboratory and field, because mycorrhizae increase relative competition intensity (Facelli et al. 1999 and references therein), associated with overlap of nutrient depletion zones around roots and hyphae and decreased mycorrhizal responsiveness in plant populations.

In experimental studies of interspecific interactions, AMF colonisation can increase the competitive ability of species with high mycorrhizal responsiveness, with consequent effects (negative and neutral, as well as positive) on plant diversity in ecosystems (e.g. Grime et al. 1987; van der Heijden et al. 1998; Hartnett and Wilson 1999). The suggestion that increased competitive ability involves significant net receipt of photosynthate via hyphal links as well as increased nutrient acquisition from soil is, however, still speculation. The observed positive relationship between mycorrhizal responsiveness and net receipt of photosynthate and phosphate (Fig. 10.3) points to the role of mycorrhizal fungi in mediating competitive relationships between plants. Bergelson and Crawley (1988) noted that mycorrhizal effects on species diversity will depend on the characteristics of dominant versus suppressed species (for a discussion see Sect. 10.3.1 in van der Heijden, Chap. 10, this Vol.). In this respect, the physiological and architectural factors summarised above are likely to be very important.

In conclusion, the dangers of extrapolating from individual AM symbioses to associations in the field are well known (e.g. Fitter 1985). The relevance increases at the expense of precision when moving from individual symbiosis to complex systems as observed in the field (Fig. 1.2 in Read, Chap. 1, this Vol.). Previous rather simple perceptions of the role and function of AMF in ecosystems clearly need to be revised, and there appears to be a great potential for detailed studies of nutrient transport in mycorrhizae formed by symbionts isolated from natural ecosystems in order to provide the background for an adequate understanding of the complex effects on plant community structure and diversity. A combination of physiology and molecular biology has the potential to unravel very interesting questions relating to variations in responsiveness and effectiveness in different plant-fungus combinations, to enable accurate identification of the species of fungi colonising the roots in natural ecosystems and hence to contribute to the analysis of interactions in ecological situations.

References

Abbott LK, Robson AD, Jasper D, Gazey C (1992) What is the role of VA mycorrhizal hyphae in soil? In: Read DJ, Lewis DH, Fitter AH, Alexander IJ (eds) Mycorrhizas in ecosystems. CAB International, Wallington, pp 37–41

Allen MF (1983) Formation of vesicular-arbuscular mycorrhizae in *Atriplex gardneri* (Chenopodiaceae): seasonal response in a cold desert. Mycologia 75:773–776

Allsopp N (1998) Effect of defoliation on the arbuscular mycorrhizas of three perennial pasture and rangeland grasses. Plant Soil 202:117–124

Bago B, Pfeffer PE, Shachar-Hill Y (2000) Carbon metabolism and transport in arbuscular mycorrhizas. Plant Physiol 124:949–957

Baon JB, Smith SE, Alston AM, Wheeler RD (1992) Phosphorus efficiency of three cereals as related to indigenous mycorrhizal infection. Aust J Agric Res 43:479–491

Barber SA (1984) Soil Nutrient Bioavailability – a mechanistic approach. Wiley-Interscience, New York

Bergelson JM, Crawley MJ (1988) Mycorrhizal infection and plant species diversity. Nature 334:202

Blal B, Morel C, Gianinazzi-Pearson V, Fardeau JC, Gianinazzi S (1990) Influence of vesicular-arbuscular mycorrhizae on phosphate fertilizer efficiency in two tropical acid soils planted with micropropagated oil palm (*Elaeis guineensis* jacq.). Biol Fert Soils 9:43–48

Blee KA, Anderson AJ (1998) Regulation of arbuscule formation by carbon in the plant. Plant J 16:523–530

Bürkert B, Robson A (1994) ^{65}Zn uptake in subterranean clover (*Trifolium subterraneum* L) by 3 vesicular-arbuscular mycorrhizal fungi in a root-free sandy soil. Soil Biol Biochem 26:1117–1124

Clarkson DT (1985) Factors affecting mineral nutrient acquisition by plants. Annu Rev Plant Physiol 36:77–115

Dickson S, Kolesik P (1999) Visualisation of mycorrhizal fungal structures and quantification of their surface area and volume using laser scanning confocal microscopy. Mycorrhiza 9:205–213

Facelli E, Facelli J, McLaughlin MJ, Smith SE (1999) Interactive effects of arbuscular mycorrhizal symbiosis, intraspecific competition and resource availability using *Trifolium subterraneum* L. cv. Mt Barker. New Phytol 141:535–547

Fitter AH (1985) Functioning of vesicular-arbuscular mycorrhizas under field conditions. New Phytol 99:257-265
Gallaud I (1905) Etudes sur les mycorrhizes endotrophs. Rev Gen Bot 17:5-48, 66-83, 123-135, 223-239, 313-325, 425-433, 479-500
Gange AC, Ayres RL (1999) On the relation between arbuscular mycorrhizal colonisation and plant 'benefit'. Oikos 87:615-621
Gavito ME, Curtis PS, Mikkelsen TN, Jakobsen I (2000) Atmospheric CO_2 and mycorrhiza effects on biomass allocation and nutrient uptake of nodulated pea (*Pisum sativum*) plants. J Exp Bot 51:1931-1938
George E, Haussler KU, Vetterlein D, Gorgus E, Marschner H (1992) Water and nutrient translocation by hyphae of *Glomus mosseae*. Can J Bot 70:2130-2137
Gianinazzi-Pearson V, Smith SE, Gianinazzi S, Smith FA (1991) Enzymatic studies on the metabolism of vesicular-arbuscular mycorrhizas. V. Is H^+-ATPase a component of ATP hydrolysing enzyme activities in plant-fungus interfaces? New Phytol 117:61-74
Graham JH, Abbott LK (2000) Wheat responses to aggressive and non-aggressive arbuscular mycorrhizal fungi. Plant Soil 220:207-218
Graham JH, Leonard RT, Menge JA (1981) Membrane-mediated decrease in root exudation responsible for inhibition of vesicular-arbuscular mycorrhiza formation. Plant Physiol 68:548-552
Grime JP, Mackey JML, Hillier SH, Read DJ (1987) Floristic diversity in a model system using experimental microcosms. Nature 328:420-422
Harrison MJ (1999) Molecular and cellular aspects of the arbuscular mycorrhizal symbiosis. Annu Rev Plant Physiol Plant Mol Biol 50:361-389
Harrison MJ, van Buuren ML (1995) A phosphate transporter from the mycorrhizal fungus *Glomus versiforme*. Nature 378:626-632
Hartnett DC, Wilson GWT (1999) Mycorrhizae influence plant community structure and diversity in a tallgrass prairie. Ecology 80:1187-1195
Hawkins H-J, Johansen A, George E (2000) Uptake, uptake mechanisms and transport of organic and inorganic nitrogen by arbuscular mycorrhizal fungi. Plant Soil 226:275-285
Imhof S (1999) Root morphology, anatomy and mycotrophy of the achlorophyllous *Voyria aphylla* (Jaq.) Pers. (Gentianaceae). Mycorrhiza 9:33-39
Jakobsen I (1999) Transport of phosphorus and carbon in arbuscular mycorrhizas. In: Varma A, Hock B (eds) Mycorrhiza: structure, function, molecular biology and biotechnology, 2nd edn. Springer, Berlin Heidelberg New York, pp 305-332
Jakobsen I, Rosendahl L (1990) Carbon flow into soil and external hyphae from roots of mycorrhizal cucumber plants. New Phytol 115:77-83
Jakobsen I, Abbott LK, Robson AD (1992a) External hyphae of vesicular-arbuscular mycorrhizal fungi associated with *Trifolium subterraneum*. 1: Spread of hyphae and phosphorus inflow into roots. New Phytol 120:371-380
Jakobsen I, Abbott LK, Robson AD (1992b) External hyphae of vesicular-arbuscular mycorrhizal fungi associated with *Trifolium subterraneum*. 2: Hyphal transport of ^{32}P over defined distances. New Phytol 120:509-516
Johansen A, Jakobsen I, Jensen ES (1992) Hyphal transport of ^{15}N-labelled nitrogen by a vesicular-arbuscular mycorrhizal fungus and its effect on depletion of inorganic soil N. New Phytol 122:281-288.
Johansen A, Jakobsen I, Jensen ES (1994) Hyphal N-transport by a vesicular-arbuscular mycorrhizal fungus associated with cucumber grown at three nitrogen levels. Plant Soil 160:1-9
Johnson NC, Graham JH, Smith FA (1997) Functioning of mycorrhizal associations along the mutualism-parasitism continuum. New Phytol 135:575-586
Joner E, Jakobsen I (1994) Contribution by two arbuscular mycorrhizal fungi to P uptake by cucumber (*Cucumis sativus* L.) from ^{32}P-labelled organic matter during mineralization in soil. Plant Soil 163:203-209

Joner E, Magid J, Gahoonia TS, Jakobsen I (1995) Phosphorus depletion and activity of phosphatases in the rhizosphere of mycorrhizal and non-mycorrhizal cucumber (*Cucumis sativus* L.). Soil Biol Biochem 27:1145–1151

Joner EJ, Johansen A (2000) Phosphatase activity of external hyphae of two arbuscular mycorrhizal fungi. Mycol Res 104:81–86

Joner EJ, Ravnskov S, Jakobsen I (2000) Arbuscular mycorrhizal phosphate transport under monoxenic conditions using radio-labelled inorganic and organic phosphate. Biotechnol Lett 22:1705–1708

Koide RT (1991) Nutrient supply, nutrient demand and plant response to mycorrhiza l infection. New Phytol 117:365–386

Koide RT, Goff MD, Dickie IA (2000) Component growth efficiencies of mycorrhizal and nonmycorrhizal plants. New Phytol 148:163–168

Leake JR (1994) The biology of myco-heterotrophic ('saprophytic') plants. Tansley review no 69. New Phytol 127:171–216

Lewis DH (1973) Concepts in fungal nutrition and the origin of biotrophy. Biol Rev 48:261–278

Li X-L, George E, Marschner H (1991) Phosphorus depletion and pH decrease at the root-soil and hyphae-soil interfaces of VA mycorrhizal white clover fertilised with ammonium. New Phytol 119: 397–404

Liu C, Muchhal US, Uthappa M, Kononowicz AK, Ragothama KG (1998) Tomato phosphate transporter genes are differentially regulated in plant tissues by phosphorus. Plant Physiol 116:91–99

Liu H, Trieu AT, Blaylock LA, Harrison MJ (1998) Cloning and characterisation of two phosphate transporters from *Medicago truncatula* roots: regulation in response to phosphate and response to colonisation by arbuscular mycorrhizal (AM) fungi. Mol Plant Microbe Interact 11:14–22

Marschner H (1995) Mineral nutrition of higher plants. Academic Press, London

Merryweather J, Fitter A (1996) Phosphorus nutrition of an obligately mycorrhizal plant treated with the fungicide benomyl in the field. New Phytol 132:307–311

Mullen RB, Schmidt SK (1993) Mycorrhizal infection, phosphorus uptake and phenology in *Ranunculus adoneus*: implications for the functioning of mycorrhizas in alpine systems. Oecologia 94:229–234

Newman EI (1988) Mycorrhizal links between plants: their functioning and ecological significance. Adv Ecol Res 18:243–270

Oliver AJ, Smith SE, Nicholas DJD, Wallace W, Smith FA (1983) Activity of nitrate reductase in *Trifolium subterraneum*: effects of mycorrhizal infection and phosphate nutrition. New Phytol 94:63–79

Olsson PA, Bååth E, Jakobsen I, Söderström B (1995) The use of phospholipid and neutral lipid fatty acids to estimate biomass of arbuscular mycorrhizal fungi in soil. Mycol Res 99:623–629

Olsson PA, Thingstrup I, Jakobsen I, Bååth E (1999) Estimation of the biomass of arbuscular mycorrhizal fungi in a linseed field. Soil Biol Biochem 31:1879–1887

Pearson JN, Jakobsen I (1993a) Exchange of carbon and phosphorus in symbioses between cucumber and three VA mycorrhizal fungi. New Phytol 124:481–488

Pearson JN, Jakobsen I (1993b) The relative contribution of hyphae and roots to phosphorus uptake by arbuscular mycorrhizal plants measured by dual labelling with ^{32}P and ^{33}P. New Phytol 124:489–494

Ravnskov S, Jakobsen I (1995) Functional compatibility in arbuscular mycorrhizas measured as hyphal P transport to the plant. New Phytol 129:611–618

Read DJ (2000) Links between genetic and functional diversity – a bridge too far? New Phytol 145:363–365

Remy W, Taylor TN, Hass H, Kerp H (1994) Four hundred-million-year-old vesicular arbuscular mycorrhizas. Proc Natl Acad Sci USA 91:11841–11843

Robinson D, Fitter AH (1999) The magnitude and control of carbon transfer between plants linked by a common mycorrhizal network. J Exp Bot 50:9-13

Rosewarne GM, Barker SJ, Smith SE, Smith FA, Schachtman DP (1999) A *Lycopersicon esculentum* phosphate transporter (*LePT1*) involved in phosphorus uptake from a vesicular-arbuscular mycorrhizal fungus. New Phytol 144:507-516

Sanders IR, Streitwolf-Engel R, Heijden MGA van der, Boller T, Wiemken A (1998) Increased allocation to external hyphae of arbuscular mycorrhizal fungi under CO_2 enrichment. Oecologia 117:496-503

Schachtman DP, Reid RJ, Ayling SM (1998) Phosphorus uptake by plants: from soil to cell. Plant Physiol 116: 447-453

Schweiger PF, Jakobsen I (1999) The role of mycorrhizas in plant P nutrition: fungal uptake kinetics and genotype variation. In: Gissel-Nielsen G, Jensen A (eds) Plant nutrition - molecular biology and genetics. Kluwer, Dordrecht, pp 277-289

Sieverding E, Toro S, Mosquera O (1989) Biomass production and nutrient concentrations in spores of VA mycorrhizal fungi. Soil Biol Biochem 21:60-72

Smith DC, Muscatine L, Lewis DH (1969) Carbohydrate movement from autotrophs to heterotrophs in parasitic and mutualistic symbioses. Biol Rev 44:17-90

Smith FA (2000) Measuring the influence of mycorrhizas. New Phytol 148:4-6Smith FA, Smith SE (1996) Mutualism and parasitism: diversity in function and structure in the "arbuscular" (VA) mycorrhizal symbiosis. Adv Bot Res 22:1-43

Smith FA, Smith SE (1997) Structural diversity in (vesicular)-arbuscular mycorrhizal fungi. New Phytol 137:373-388

Smith FA, Jakobsen I, Smith SE (2000a) Spatial differences in acquisition of soil phosphate between the arbuscular mycorrhizal fungi S *utellospora calospora* and *Glomus caledonium* in symbiosis with *Medicago truncatula*. New Phytol 147:357-366

Smith FA, Timonen S, Smith SE (2000b) Mycorrhizas. In: Blom CWPM, Visser EJW (eds) Root ecology. Springer, Berlin Heidelberg New York

Smith SE, Read DJ (1997) Mycorrhizal symbiosis, 2nd edn. Academic Press, London

Smith SE, Rosewarne G, Ayling SM, Dickson S, Schachtman DP, Barker SJ, Reid RJ, Smith FA (1999) Phosphate transfer between vesicular-arbuscular mycorrhizal symbionts: insights from confocal microscopy, microphysiology and molecular studies. In: Lynch JP, Deikman J (eds) Phosphorus in plant biology: regulatory roles in molecular, cellular, organismic and ecosystem processes. American Society of Plant Physiologists, Rockville, pp 111-123

Soberon MJ, Marinez del Rio C (1985) Cheating and taking advantage in mutualistic symbioses. In: Boucher D (ed) The biology of mutualism. Croom Helm, London, pp 192-216

Son CL, Smith SE (1988) Mycorrhizal growth responses: interactions between photon irradiance and phosphorus nutrition. New Phytol 108:305-314

Staddon PL, Graves JD, Fitter AH (1999) Effect of enhanced atmospheric CO_2 on mycorrhizal colonisation and phosphorus inflow in 10 herbaceous species of contrasting growth strategies. Funct Ecol 13:190-199

Stribley DP, Tinker PB, Rayner JH (1980) Relation of internal phosphorus concentration and plant weight in plants infected by vesicular-arbuscular mycorrhizas. New Phytol 86:261-266

Thomson BD, Robson AD, Abbott LK (1986) Effects of phosphorus on the formation of mycorrhizas by *Gigaspora calospora* and *Glomus fasciculatum* in relation to root carbohydrates. New Phytol 103:751-765

van der Heijden MGA, Klironomos JN, Ursic M, Moutoglis P, Streitwolf-Engel R, Boller T, Wiemken A, Sanders IR (1998) Mycorrhizal fungal diversity determines plant biodiversity, ecosystem variability and productivity. Nature 396:69-72

Wardle DA (1999) Is 'sampling effect' a problem for experiments investigating biodiversity – ecosystem function relationships? Oikos 87:403–407

Wright DP, Scholes JD, Read DJ (1998) Effects of VA mycorrhizal colonisation on photosynthesis and biomass production of *Trifolium repens* L. Plant Cell Environ 21:209–216

4 Foraging and Resource Allocation Strategies of Mycorrhizal Fungi in a Patchy Environment

PÅL AXEL OLSSON, IVER JAKOBSEN, HÅKAN WALLANDER

Contents

4.1	Summary	93
4.2	Introduction	94
4.3	The Mycorrhizal Mycelium	98
4.3.1	Structures and Growth Rates	98
4.3.2	Density of the Mycelium Front	99
4.3.3	Energy Storage	100
4.4	Foraging of the Ecto-mycorrhizal Mycelium	100
4.4.1	Response to Inorganic Nutrients	100
4.4.2	Response to Organic Nutrient Sources	101
4.4.3	Response to Plant Roots	102
4.5	Foraging of the Arbuscular Mycorrhizal Mycelium	102
4.5.1	Response to Inorganic Nutrients	102
4.5.2	Response to Organic Material and Compounds	103
4.5.3	Response to Plant Roots	104
4.6	Differences in Foraging Strategies Between Arbuscular Mychorhizal and Ecto-mycorrhizal Fungi	104
4.7	Models and Theories on the Foraging of Mycorrhizal Fungi	105
4.8	Conclusions and Future Perspectives	109
References		110

4.1 Summary

Foraging for nutrients and carbon are essential components of the mycorrhizal symbiosis. Foraging strategies of mycorrhizal fungi have received little attention compared to the interaction with the plant. Proliferation of hyphae, resource allocation (carbon and nutrients) within a mycelium and spatial distribution of resource capturing structures (internal mycelium for carbon and external mycelium for nutrients) can be considered as foraging strategies. The arbuscular mycorrhizal fungi (AMF) form a uniformly distributed mycelium in soil, but hyphal proliferation occurs in response to several types of organic material and near potential host roots. The ecto-mycorrhizal fungi (EMF) normally form denser hyphal fronts than AMF, and they respond to both organic material and inorganic nutrients by increased growth. This is especially evident for the EMF that form extensive mycelia connected by differentiated hyphal strands, so-called rhizomorphs. We hypothesise that the growth strategy of the AM fungal mycelium reflects an evolution towards optimal search for potential new host roots. The growth strategy of EMF instead seems to reflect evolution towards optimised nutrient capture in competition with other mycelia. Foraging behaviour of mycorrhizal fungi is discussed and we suggest two conceptual models for resource allocation in the mycorrhizal mycelium. These models consider both the internal and the external mycelium and the trade-offs between different foraging strategies of mycorrhizal fungi. From the experimental data available, it is clear that mycorrhizal fungi forage. It needs to be investigated whether observed foraging strategies are optimal for the mycelium as one individual.

4.2 Introduction

Studies of mycorrhizal fungi have most often focused on their symbiosis with the plants while the external mycelium remains relatively unexplored. Even less attention has been paid to the ecology of individual mycelia of mycorrhizal fungi involving both the external and internal phase and the partitioning of carbon (C) resources between these two phases. Progress may benefit from ecological knowledge already generated for other fungi and organisms. We utilise this knowledge as a platform to formulate novel hypotheses and experimental approaches to study the ecology of the fungal mycelia of different types of mycorrhizae. We focus on two groups of mycorrhizal fungi, which have some characteristics in common, but which are also very different: the arbuscular mycorrhizal fungi (AMF) and the ecto-mycorrhizal fungi (EMF). Detailed information about arbuscular mycorrhizal symbiosis and about the ecto-mycorrhizal symbiosis are given by Jakobsen et al. (Chap. 3, this Vol.) and Simard et al. (Chap. 2, this Vol.) respectively.

Mycelial fungi are complex organisms with efficient nutrient and C translocation from source to sink regions (Olsson 1995). The mycelia differentiate for example through the formation of rhizomorphs (Jennings and Lysek 1996) and possess mobility in the sense of resource and biomass allocation to different parts of the mycelium (Fig. 14.3 in Leake et al., Chap. 14, this Vol.; Boddy 1999). Furthermore, the mycelia have possible communication signals through stimuli-sensitive action potential-like activities (Olsson and

Foraging and Resource Allocation Strategies of Mycorrhizal Fungi 95

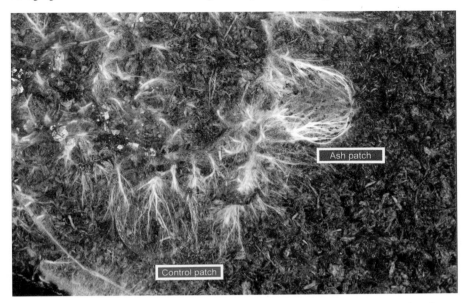

Fig. 4.1. External mycelium on peat growing from ecto-mycorrhizal root tips in an experimental system. The fungus forms a dense mycelium in a patch of wood ash (diameter of the patch is 1.5 cm). The same fungus had previously been found to colonise wood ash granules that had been applied to a Norway spruce forest in south-west Sweden (Mahmood et al. 2001). Molecular analysis suggests that this fungus belongs to the genus *Piloderma*. (Photograph by Shahid Mahmood)

Hansson 1995). A communicating network is established through the ability of individual hyphae to merge together through anastomosis of hyphae, which allows the flow of resources between source and sink regions in all directions (Rayner 1991). A special feature of the fungal mycelium is the "trophic growth" (Jennings and Lysek 1996) where the hyphal tip represents the growing unit and at the same time releases enzymes important for nutrient uptake behind the hyphal tip. Growth of mycelia in soil is probably limited by energy C which, in the case of mycorrhizal fungi, is in the form of plant carbohydrates. The mycorrhizal fungi take up this C directly across their interfaces with the plant and the net transfer will be influenced by the plant and other competing mycorrhizal fungi. Structures of AMF in different plant roots are connected by the mycelium network and AMF move C according to their own C demands, while true exchange with the plants seems less important (Robinson and Fitter 1999). The ecological significance of this interplant C transport is further discussed by Simard et al. (Chap. 2, this Vol.). Performance of individual mycelia without host plants has been studied more frequently in EMF than in AMF since many EMF can be cultivated without their host. Such studies with AMF are restricted to the initial germ tube formation and growth stages based on stored carbohydrates and lipids in the spore (Bago et al. 1999).

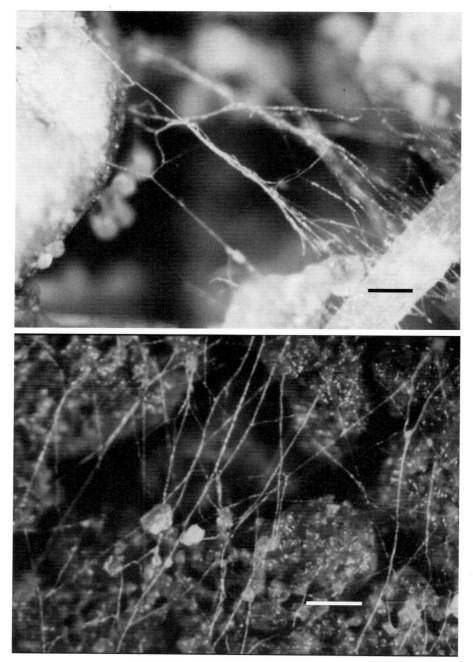

Fig. 4.2.A. AM fungal runner hyphae of *Glomus caledonium* radiating out from the host root of *Trifolium repens* (photograph by Iver Jakobsen). **B** AM fungal hyphae of *G. intraradices* in soil from an experimental system (see Larsen et al. 1998), observed when a plastic lid covering the soil had been removed (photograph by John Larsen). *Bars* 300 μm

Mycelia should be studied in their natural substrate, which is soil in the case of external mycelium of mycorrhizal fungi. This is because mycelial growth in artificial media (e.g. agar plates) is greatly influenced by the substrate. The growth of several saprotrophic fungi can be studied visually on the soil surface (Rayner 1991) and similar studies are possible for EMF in growth systems with a soil layer in microcosms (see Fig. 4.1 and Fig. 14.3 in Leake et al., Chap. 14, this Vol.; see also Finlay and Read 1986a). Both extension rates, hyphal density and fractal dimensions of mycelia can be determined in such systems (Ritz and Crawford 1990; Arnebrant 1994; Baar et al. 1997; Boddy 1999). Such studies are more difficult with AMF, which form no similar surface mycelium, but microscopy has been used to study AMF mycelia on soil surfaces and in soil pores in microcosms (Friese and Allen 1991; Kling and Jakobsen 1998; see Fig. 4.2). The commonly used way to extract mycelia from soil will inevitably destroy the spatial organisation of the mycelium. Some techniques and approaches are now available, in addition to the use of microscopy, to study hyphae, spores and root colonisation. These include ergosterol as signature for fungal membranes (Nylund and Wallander 1992), lipid signatures analysed by gas chromatography (Olsson et al. 1995; Olsson 1999), detection of hydrophobic proteins (glomalin) released from AMF hyphae as a tool for the estimation of mycelium extension rate (Wright and Upadhyaya 1999), stable isotopes that can be used to trace C allocation to different fungal molecules by use of combined gas chromatography and mass spectrometry (Tunlid et al. 1987; Watkins et al. 1996; Högberg et al. 1999) and monoaxenic cultures of the mycorrhizal fungi (Bécard and Fortin 1988). Fatty acid signatures can be used to estimate overall fungal biomass in soil and roots, but can also provide information on C resource allocation to storage compartments and vegetative growth (Olsson et al. 1997; Olsson 1999). These techniques will enable us to reveal and describe trade-offs between maintenance of mycelia in resource poor patches, vegetative growth into new patches and reproduction (in the case of AMF).

Effects of resource availability on external growth of mycorrhizal mycelium can be studied under three different circumstances: (1) application of the resource to the whole system, including both roots and mycelium. (2) Application of the resource to the external mycelium only. (3) Application of the resource to only a fraction of the external mycelium. In the first case, the effects of the plant on the mycelium cannot be distinguished from direct effects on the mycelium. In the third case, changes in mycelia growth pattern are direct effects of the mycelium sensing and responding to a patchy environment.

The aim of this review is to describe growth responses and foraging strategies identified in mycorrhizal fungi with the focus on growth in patchy environments. Roots can be considered as C-rich patches for mycorrhizal fungi and the soil/root matrix is a patchy environment. Foraging of mycorrhizal fungi will be considered as the searching, branching and proliferation within their habitat (Hutchings and De Kroon 1994; De Kroon and Hutchings 1995). Approaches to such studies will be suggested where potential host plants are regarded as patches for proliferation of mycelium. Different organic and inorganic nutrient sources also have a typical patchy distribution in the soil. Possible strategies for optimal foraging will be discussed.

4.3 The Mycorrhizal Mycelium

4.3.1 Structures and Growth Rates

Ecto-mycorrhizal fungi form extensive mycelia that radiate from EM root tips and explore the soil for resources. The proportion of fungal biomass present as external mycelium in EMF has been estimated to be between 60–80% among several species in a laboratory system (Colpaert et al. 1992). Uptake of nutrients occurs at the mycelia fronts mainly, and nutrients are transported within the mycelia between different resource patches (Finlay and Read 1986b). EMF can vary their mode of growth depending on their ecological needs. Hydrophilic hyphae are formed in order to exploit water-soluble nutrient resources in the soil. Hydrophobic hyphae are formed when nutrients need to be transported (Unestam and Sun 1995). The mycelium morphology differs between EM fungal species and some general groups can be recognised (Ogawa 1985). One group of EMF forms large hydrophobic mycelia, which may differentiate into mycelial strands behind the exploring hyphal front. Some of them, such as *Hydnellum* and *Hysterangium*, form dense fungal mats in restricted areas (Griffiths et al. 1996), whereas others form more dispersed mycelia (e.g. *Suillus*, *Paxillus*, *Amanita* and *Boletus*; Dahlberg and Stenlid 1990; Read 1992). Specialised, nutrient-absorbing, hydrophilic hyphae are sometimes formed in response to wet surfaces in the soil (Unestam and Sun 1995). A second group of EMF forms more profuse mycelia that are hydrophilic altogether and some fungi in this group do not form strands (e.g. *Laccaria laccata*), whereas others do (e.g. *Hebeloma* and *Thelephora*). A third group of EMF forms smooth mantles with little external mycelia extending into the soil (e.g. some species of *Lactarius*; Brand and Agerer 1986). It seems likely that a significant proportion of the nutrient uptake in these fungi occurs directly through the fungal mantles surrounding the root tips. In the present review, we will mainly consider EMF that produce mycelial strands (Fig. 4.1), which allow the formation of a large mycelium and for resource transfer in the strands. Energy storage in EMF occurs in the hyphal mantles around the root tip (Read 1991). Sclerotia in *Paxillus involutus* and *Cenococcum graniforme* may also serve as storage in a dormant stage. The EM fungal mycelium grows at a uniform rate which has been estimated to be 2–4 mm/day under laboratory conditions (Read 1992); similar extension rates have been estimated under field conditions during the warmest season (Coutts and Nicoll 1990). Estimates of genet size of *Suillus pungens* in the field indicate a growth rate of 0.5 m/year (Bonello et al. 1998).

The principal AM fungal hyphae are the thick runner hyphae (Fig. 4.2A), the diameter of which decreases with increased order of branching (Friese and Allen 1991). The extension rates of the runner hyphae were between 1–3 mm/day for three different fungi in microcosms under laboratory conditions (Jakobsen et al. 1992) and the extension rate for AMF was 1.4 mm/day in sand dunes in September in southern Sweden (Olsson and Wilhelmsson 2000). The runner hyphae may colonise new roots and thereby establish hyphal bridges between individual roots of the same or different plant species. Runner hyphae may also grow along the already colonised roots and establish new entry points for colonisation (Friese and Allen 1991). Finely branched networks are formed as absorbing structures along the runner hyphae (Bago et al. 1998a) and as pre-infection structures for colonisation of new roots (Friese and Allen 1991; Giovannetti et al. 1993b). Spores are normally formed terminally on absorbing hyphae (Bago et al. 1998b). The ability of AM fungal hyphae to anastomose has convincingly been demonstrated for *Glomus* by Giovannetti et al. (1999); contacts of hyphae from spores of the same isolate resulted in the formation of anastomoses in 34–90% of the cases. In contrast, anastomoses were absent in two species from Gigasporineae and no anastomoses were observed at the interspecific or intergeneric levels.

4.3.2 Density of the Mycelium Front

The characteristics of the mycelium front differ strikingly between AMF and EMF. The EMF mycelium has a dense front, which can easily be observed on soil surfaces (Fig. 4.1), while the mycelium of AMF can only be seen under the microscope (Fig. 4.2B). The differences in mycelium density between AMF and EMF can easily be observed in axenic cultures on Petri plates. The length of external mycelium per unit length of colonised root of *Eucalyptus coccifera* was 2.5–6 times higher in mycorrhizae with EMF than in mycorrhizae with AMF (Jones et al. 1998).

The mycelium length density in soil of AMF, growing with a productive host plant, is typically around 15 m cm^{-3} soil (Table 3.1 in Jakobsen et al., Chap. 3, this Vol.). If these hyphae grow in a parallel mode through a soil volume, each cm^2 of a cross section will be passed by 1500 hyphae. This corresponds to a mean interhyphal distance of around 250 μm, or 50 times the average diameter of AM fungal hyphae (see Fig. 4.2.B). The hyphal densities of AMF in field soil can be as high as 100 m g^{-1} soil (Miller et al. 1995), but such estimates reflect a mixture of several individual mycelia. Hyphal densities of 200 m g^{-1} dry soil (2000 m cm^{-3} fresh soil) have been measured in mycelial fronts of the EMF *Suillus bovinus* and *Paxillus involutus* colonising forest soil in microcosms (Read and Boyd 1986; Read 1992). Assuming a uniform and parallel distribution as in the example for AMF, the mean distance between two parallel hyphae would be only 20 µm. These fungi may, however, belong to the upper range in variation of hyphal densities and many other EMF appear to produce lower amounts of external mycelium.

4.3.3 Energy Storage

The amount of energy stored in fungal mycelia may reflect their C foraging success. AMF spores are lipid-filled (Cooper and Lösel 1978; Olsson and Johansen 2000) and the number of spores in soil correlates well to AMF neutral lipids (Olsson et al. 1997). Hyphae instead correlate well to the signature phospholipid fatty acid 16:1ω5 (Olsson 1999). Consequently, a sporulating mycelium in soil will have a high ratio of neutral lipids to phospholipids while this ratio will be low in a mycelium forming large amounts of thin absorbing hyphae (Olsson 1999). The neutral lipids are the main storage compounds of AMF and this is a major difference to EMF where trehalose, mannitol and glycogen seem to be the most important energy storage compounds (Lewis and Harley 1965; Lewis 1991). Saprophytic fungi may form high amounts of storage lipids, at least on agar media (Larsen et al. 1998), while ecto-mycorrhizal basidiomycetes such as *Suillus* species and *Paxillus involutus* do not accumulate storage lipids (P.A. Olsson and H. Wallander, unpubl.).

4.4 Foraging of the Ecto-mycorrhizal Mycelium

4.4.1 Response to Inorganic Nutrients

Ecto-mycorrhizal fungi are particularly dominant in ecosystems where plants are limited by N (Read 1991), and many reports have demonstrated the beneficial influence of EMF on N uptake to their hosts. Still, external EMF mycelia often do not respond, in terms of biomass production, to patches of inorganic N (Read 1991; Ek 1997). Respiration by EMF was, however, increased in N-enriched patches (Ek 1997). Contrary to these studies, Brandes et al. (1998) found up to 400 % increase in production of external EM fungal mycelium of *P. involutus* in patches rich in NH_4Cl-N and KH_2PO_4-P. This inconsistency may be related to differences in the general growth medium, which was natural soils in the studies by Read (1992) and Ek (1997) and nutrient-free quartz sand in the study by Brandes et al. (1998). Consequently, mycelia of EMF in natural soils are in some cases not N or P limited. Moreover, elevated levels of N from fertilisers or anthropogenic deposition to the whole ecosystem inhibit growth of external EM fungal mycelia and fruit bodies (Fig. 7.2 in Erland and Taylor, Chap. 7, this Vol.; Wallander and Nylund 1992; Arnebrant 1994; Wallander 1995; Wiklund et al. 1995). This inhibition of fungal growth probably results from a reduced pool of available C in the plant, or directly in the mycelium, since nitrogen assimilation will lead to a large consumption of carbohydrates for amino acid synthesis. This would leave little carbohydrates for vegetative growth of the mycorrhizal mycelia (Wallander 1995; Wallenda and Kottke 1998).

Growth of external mycelium of EMF appears to respond to variation in pH. Erland et al. (1990) demonstrated that some EMF produce a much denser mycelium in areas where the substrate of the mycorrhizae was amended with CaO (increasing the pH from 3.8 to 7.3) while other species grew more sparsely at the higher pH. Similarly, Mahmood et al. (2001) found intensive colonisation of patches with $Ca(CO_3)_2$-rich wood ash by certain EMF (Fig. 4.1), while other species avoided the ash. It is not clear from that study whether the fungus reacts to the pH increase directly, to the $Ca(CO_3)_2$ or other elements in the ash, such as P. The EMF mycelium growing on ash particles exuded large amounts of oxalic acid, which converts into calcium oxalate crystals (Mahmood et al. 2001). Similar crystals have been demonstrated on mycorrhizal roots in the field (Cromack et al. 1979). This exudation of oxalic acid has been suggested to be one way to forage for P and base cations (Griffiths et al. 1994; Paris et al. 1995; Wallander et al. 1997; Wallander and Wickman 1999; Wallander 2000a,b). Hyphae have been observed in pores of hornblende and feldspar minerals in E horizon of forest soil and it was suggested that these pores were formed by EMF exuding organic acids while foraging for mineral nutrients (Jongmans et al. 1997). The response of EMF to a range of other environmental factors are further discussed by Erland and Taylor (Chap. 7, this Vol.).

4.4.2 Response to Organic Nutrient Sources

Addition of organic matter such as fresh forest litter in patches (Unestam 1991; Bending and Read 1995a,b) and senescent mosses (Carleton and Read 1991) generally resulted in proliferation of EM fungal mycelium in the organic matter. Such patches with dense mycelium were sinks for ^{14}C supplied to the host and ^{32}P supplied to mycelium outside the patch (Fig. 14.3 in Leake et al., Chap. 14, this Vol.; Finlay and Read 1986a,b; Carleton and Read 1991). Exploitation of organic matter is metabolically expensive since C and inorganic P are needed for the production of enzymes and ATP, respectively (Cairney and Burke 1996). Increased activity of several enzymes involved in nutrient mobilisation was indeed observed in these patches and significant amounts of N, P and K were removed from the patches by the EM fungal mycelia (Bending and Read 1995a,b). Effective substrate exploitation will depend on the exudation of extracellular enzymes from hyphal tips (Wessel

1993), which therefore need to be abundant. It is interesting to note that EMF, such as *Thelephora terrestris,* which have no large capacity to produce nutrient mobilising enzymes (Bending and Read 1995a; Colpaert and Van Tichelen 1996) do not proliferate in response to litter amendments (Bending and Read 1995b). Moreover, EMF with a limited ability to degrade organic nitrogen substances, so called non-protein fungi, increase in abundance when nitrogen deposition is high (Fig. 7.2 in Erland and Taylor, Chap. 7, this Vol.). Colpaert and Van Tichelen (1996) compared colonisation of fresh beech litter by EMF and a saprotrophic fungus (*Lepista nuda*). They found that the saprotrophic fungus was superior in removing nutrients from the beech litter and induced much higher enzymatic activity. It could be that EMF preferentially forage in substrates already decomposed to a certain extent. Despite this, it has been suggested that uptake of organic nutrients by EMF is important for ecto-mycorrhizal trees and it leads to a short-circuiting of the nitrogen cycle (Aerts, Chap. 5; Fig. 14.1 in Leake et al., Chap. 14, this Vol.).

4.4.3 Response to Plant Roots

External EM fungal mycelia have a chemotrophic response to substances exuded by host roots (Horan and Chilvers 1990). This suggests that active foraging for roots occurs. Growth of the external mycelium radiating out from mycorrhizal root tips depends on the C status of the host (e.g. Wallander and Nylund 1992; Ekblad et al. 1995). Ecto-mycorrhizal fungi appear to differ in their ability to obtain C from the host and it has been suggested that species colonising root tips closer to the stem will have access to larger supply of C by being closer to the source (Bruns 1995). The quality of different root tips as C resources is probably highly different, even within a root system.

4.5 Foraging of the Arbuscular Mycorrhizal Mycelium

4.5.1 Response to Inorganic Nutrients

Growth of the external mycelium of AMF is not stimulated by application of inorganic P to root-free compartments (Li et al. 1991b; Olsson and Wilhelmsson 2000). In spite of their central role in P nutrition of the symbiosis, the AMF do not seem to actively forage for P (in the sense of hyphal proliferation). It appears that high soil P may even inhibit both spore germination and early hyphal growth (Miranda and Harris 1994). Although there is no clear indication of active foraging for mineral nutrients, the depletion of P in soils colonised by AM hyphae (Li et al. 1991a; Joner et al. 1995) indicates the effectiveness of the AM hyphae in P uptake from soil (Table 3.1 in Jakobsen et al., Chap. 3, this Vol.). This high effectiveness is observed also with a patchy distribution of the available P (Cui and Caldwell 1996). A similar P uptake to the plant via the AM fungal mycelium was observed when the P was applied in a few patches as when applied uniformly to root-free soil. In contrast, ammonium-N supplied to one of two lateral root-free compartments gave increased hyphal length in *Glomus intraradices* by 20–50 % (Johansen et al. 1994). Substrate pH could also influence the foraging of AM fungal mycelia, but only the effects on colonisation (Abbott and Robson 1985) and spore germination have been studied (Siqueira et al. 1984). Arbuscular mycorrhizal fungi may also deplete the soil for inorganic N (Table 3.1 in Jakobsen et al., Chap. 3, this Vol.; Johansen et al. 1992) even in soil compartments with roots (Johansen 1999). Still, the depletion of inorganic nutrients is not complete at the hyphal front and the AMF do not seem to directly utilise less available fractions of soil P (Bolan 1991; Joner et al. 2000). These observations indicate that mycelium proliferation occurs not only to capture of mineral nutrients.

4.5.2 Response to Organic Material and Compounds

Proliferation of the external AM fungal mycelium in environments rich in organic matter is well-documented. This was first reported as "haustoria-like" hyphal branches attached to soil particles in the rhizosphere of three-year old apple seedlings (Mosse 1959). Accordingly, St. John et al. (1983) observed that organic particles in sand were colonised by AM fungal hyphae. Later experiments showed that AM fungal mycelium proliferated in root-free soil provided with wheat straw (Joner and Jakobsen 1995), dry yeast (Larsen and Jakobsen 1996), wheat bran (Green et al. 1999) and albumin (Ravnskov et al. 1999). It led to an increase in total length of external hyphae (Joner and Jakobsen 1995; Ravnskov et al. 1999). The stimulation of hyphal growth by the organic substrates seems to be local, as a marked stimulatory effect of dry yeast applied to a sand-filled compartment was not observed in an identical unamended compartment in the same growth unit (Ravnskov et al. 1999). Growth increases can be higher than 100 % and have been documented by means of hyphal length counts, gravimetric measurements and an AM fungal signature phospholipid fatty acid. The trade-off between investment in vegetative hyphae and storage energy of the AM fungal mycelium also appears to be affected by organic material as indicated by a decrease in the neutral lipid: phospholipid ratio, suggesting a specific proliferation of vegetative hyphae and less relative allocation to storage after addition of organic material (Green et al. 1999; Ravnskov et al. 1999). It further implies that the organic substrates did not represent a C source for the hyphae and root-external mycelium of AMF. This could be expected since AMF do not take up carbohydrates via the external mycelium (Pfeffer et al. 1999). The mechanism for the growth stimulation is still not known, but the need for certain amino acids by AMF (Hepper and Jakobsen 1983) could play a role. Negative effects on growth of AM fungal mycelium have been observed for certain organic substrates such as starch and cellulose (Ravnskov et al. 1999).

4.5.3 Response to Plant Roots

The host roots evidently affect the external AM fungal mycelium through the nutrient exchange at the interface, but interactions at other levels are also interesting for the ecology of the fungi. The growth of external mycelium (Olsson et al. 1997) and sporulation during early colonisation (Gazey et al. 1992) is rather closely related to colonised root length and the external mycelium appears to constitute around 90 % of the total mycelium in established mycorrhizae (Olsson et al. 1997, 1999). The AM fungal runner hyphae grow distantly from the point of colonisation (Fig. 4.2) and root-free soil compartments may have a higher mycelium biomass than soil with high root densities (Olsson et al. 1997). The runner hyphae will sooner or later meet new roots and there is increasing evidence for a recognition of potential host plants (Giovannetti et al. 1993a, 1994; Giovannetti and Sbrana 1998). The interception of a host root induces hyphal branching and proliferation elicited by root signals (Giovannetti et al. 1993b. The production of spores is probably the most important fitness determinant of AMF, at least in disturbed habitats. Host roots stimulate germination of spores and if no host root is reached, the growth of initial hyphae may be arrested and accompanied by reallocation of C resources into the spore (Logi et al. 1998). In that way, the spore may germinate again with preserved energy resources (Tommerup 1984).

4.6 Differences in Foraging Strategies Between Arbuscular Mycorrhizal and Ecto-mycorrhizal Fungi

Mycelia of AMF and EMF clearly differ and the differences can be viewed from two aspects:
1. Differences in Morphology and Growth (see Fig. 4.1). The EMF form dense mycelia adjacent to the root, as the hyphal mantle and as the external mycelium (Fig. 4.1). They may also form dense mycelia distant from the roots and these are then connected to the host root by rhizomorphs. Growth of the AM fungal mycelium is in the form of runner hyphae associated with lateral absorbing hyphal networks (Fig. 4.2). This growth has a uniform pattern and can occur over at least 5–10 cm (e.g. Li et al. 1991b; Jakobsen et al. 1992; Olsson et al. 1995; Olsson and Wilhelmsson 2000). AM fungal spores are not wind-dispersed like most EMF and spread of AMF is

therefore more restricted. Instead the underground storage spores of AMF are able to enter a dormant stage in a relatively short time.
2. Differences in Nutrient Capturing Mechanisms. The foraging strategies of mycorrhizal fungi include not only the mobility of biomass and energy C between different parts of a mycelium. Optimal foraging of the fungi could also be linked to the energy spent on activities such as nutrient uptake (Harrison 1999), production and release of nutrient-hydrolysing enzymes and release of organic acids (Dighton 1995). Uptake of soluble nutrients is so far the major nutrient capture mechanism detected in AM fungal mycelia. EM fungal mycelia seem to have the widest and most active set of enzymes for foraging on complex organic material and also seem to be more active in exudation of organic acids. This higher activity in EM fungal mycelia is reflected in a high degree of interactions with saprotrophic micro-organisms (Leake et al., Chap. 14, this Vol.; Olsson et al. 1996b; Olsson and Wallander 1998) which is less pronounced in AMF (Olsson et al. 1996a). All these strategies are associated with energy expenses and should therefore be considered when identifying factors that are limiting the growth of fungal mycelium (Sinsabaugh 1994; Dighton 1995).

We suggest that growth strategies of AMF with fast growing runner hyphae mainly reflect evolution towards optimal search for an alternative host. This will increase the probability of survival when one host is not active anymore or when the C flow from a host declines. EM fungal mycelium may have evolved to optimise the search for new root tips to colonise, but the growth strategies also seem to reflect evolution towards optimal nutrient uptake from the soil and translocation into the host roots. In that way, the EM fungal mycelium can increase the quality of the root tip as an energy source by influencing the C allocation within the plant. These differences between AMF and EMF might have evolved from differences in size and persistence of energy supply by the host species. The energy supply would be smaller and more temporal from the arbuscular mycorrhizal herbs and grasses than from ecto-mycorrhizal trees.

4.7 Models and Theories on the Foraging of Mycorrhizal Fungi

Evolution gives rise to foraging strategies in which organisms optimise their search for limiting resources. Optimal foraging theory has been developed to predict foraging under specified conditions (Begon et al. 1990). Foraging models consider which resource to consume and when to leave a patch of resources (Stephens and Krebs 1986). If an organism manages to do the right thing at the right time, then it will be favoured by natural selection. The time spent in resource capture is the critical currency in the case of animals

(Charnov 1976) while energy in the form of carbohydrate is most probably the growth-limiting factor for mycorrhizal fungi. This energy is supplied by the plants, which differ from other resources by providing a potential continuous flow of energy by photosynthesis and translocation of carbohydrates to the root. The balance between energy spent and energy gained by the fungus in the root patch is critical for the success of mycorrhizal fungi. Optimal foraging will consequently depend on the plant or root tip which is colonised by the fungus, just as animals choose the items that form its diet (MacArthur and Pianka 1966). In the soil the mycelium meets roots of varying potential as C sources. Some species may, for example, be non-hosts (Giovannetti and Sbrana 1998) and others may host other types of mycorrhizal fungi. The variation in mycorrhizal colonisation due to plant species and environmental conditions (Smith and Read 1997) may also indicate that variations in quality as patches for C foraging occur. Colonisation strategies that maximise the overall energy intake will be favoured by natural selection.

During the life-cycle of an organism, compromises have to be made where allocation of resources to one activity or structure is at the expense of others. This is called a physiological trade-off. Since different traits are probably required for optimal nutrient uptake, search and colonisation of new plants or root tips, it is likely that there is a trade-off between the traits connected to these two resource capture processes. Since fungi are usually considered to be C limited, the energy capture is of particular interest and the placement of potential energy capturing structures in a patchy environment is critical (compare De Kroon and Hutchings 1995). The energy intake may, however, also depend on the nutrient capture since the C-flow from the plant may depend on the amount of inorganic nutrient delivered to the plant (Wallander 1995). Application of foraging models also requires defined fitness measures. These are usually a function of several life history traits. For most fungi, the number of spores can be a relevant fitness parameter, but vegetative hyphal growth is also important as a way of propagation and reflects access to potential new resources. Spores of AMF are easily extracted from soil and can be counted, weighed or analysed for energy storage. We believe that the relative energy or biomass accumulation of AMF is the best measure to reflect its fitness in established natural vegetation. For further discussion of fitness traits for AMF see Chapter 8 by Clapp et al., Chapter 11 by Bever et al. and Chapter 16 by Sanders (this Vol.). For EMF, sporulation usually does not occur in experimental systems, instead we may speculate that for such systems a high relative C supply to a fungus reflects a high fitness

We suggest two conceptual models that can be used to describe foraging strategies of mycorrhizal mycelia (Figs. 4.3, 4.4). These models are based on known properties of the mycorrhizal mycelium. These models enable us to identify trade-offs between investment into different patches as well as between colonisation and vegetative growth. The models also give the possibility to compare strategies and life-history trade-offs of different fungal

Foraging and Resource Allocation Strategies of Mycorrhizal Fungi 107

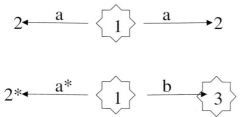

Fig. 4.3. Model 1. A two-dimensional model describing the foraging of mycorrhizal fungi. *1* The fungal investment in a host root. *2* The mycelium grows into a uniform substrate in mainly two directions (*above*) or it grows into substrates of different qualities (*below*). These substrates can be either similar to the dominating substrate (*2*) or a patch enriched by some compound or containing a new host root (*3*). The numbers *1–3* thus denote the energy or resources spent by a fungus within a certain type of patch. The letters *a* and *b* represent the flow of resources from the initially colonised host into new patches. The flow into the dominating substrate (*a*) may be influenced by the flow to the enriched patch (*b*) and is denoted (*a**)

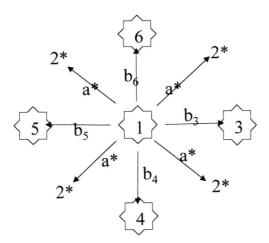

Fig. 4.4. Model 2. A multidimensional model describing a more realistic case of foraging by mycorrhizal fungi. The mycelium grows in all directions and colonises a variety of patches. Allocation of resources to enriched patches (*3–6*) as well as to the uniform substrate between patches (*2**) can be estimated in experiments based on this model. The flow of resources towards enriched patches (b_3–b_6) and towards the uniform substrate (*a* or *a**) can be measured

species. The numbers (1 to 6) in the models denote the different types of patches. The letters (a and b) indicate flow of energy. Where a is the energy flow to an established patch and b is the energy flow to a new patch. This can be measured as C flow subsequent to labelling by tracers. In Fig. 4.3, we present a two-dimensional model where the mycelium spreads uniformly from the host plant (1) into a homogenous substrate (2). Patches enriched by different compounds or new roots can be introduced into the system (3) and the effects on the whole system can be measured when the mycelium has reached the patch. The new patch may also influence the growth in the original patch, which therefore will be denoted as 2*. The timing of the measurements is especially crucial if the patch is a new root. In that case, the cost for the fungus of the initial colonisation can be changed into a gain of energy from the newly colonised root after a certain time. The total energy supply may also influence the choice of further foraging (Stephens and Krebs 1986). Basic assumptions for model 1 are as follows:

1. If b>a, resource allocation occurs to the patch 3. This will lead to 3>2 in terms of fungal resources.
2. If b<a there is an inhibition or avoidance of the patch.
3. If a≠a* there is an effect of the patch on the mycelium also outside the patch.
4. If a*+b>a+a there is an increased gain of resources by the fungus. This is due to resource acquisition from the new patch or due to enhanced gain from the original host plant (1)

Assumption 3 depends on both the quality and quantity of the patch reached by the mycelium. In a study where organic material stimulated proliferation of AM fungal mycelium when applied in one of two sand compartments, there was no change of fungal growth in the other compartment (Ravnskov et al. 1999). This means that 2=2* and 3>2 which indicate that a=a* and b>a for that situation, although the flow was not specifically measured. More complex situations can be analysed in a multidimensional model (Fig. 4.4). This model enables us to test the effects on the system when the mycelium reaches patches of different qualities at the same time. The numbers represent different patches where 1 is a host root, 2 is the general substrate and 3–6 represent different patches which can be roots of the same or different plant species as well as different substrates. The patches are located in a more uniform substrate (2) as in model 1.

We exemplify some further assumptions for model 2:
1. If b_3, b_4, b_5 or b_6 ≠ a there is a patch response.
2. If 2*>2 there is a secondary effect arising from energy gain from any of the patches. This is a likely situation if the patch represents a new plant that supplies additional amounts of resources to the fungi.

Hypotheses based on these assumptions can be formulated in relation to different patch types. It can be tested if the fungi have strategies for optimising energy gain in a complex situation with different hosts. Such an experiment could be performed in two steps:
1. Establish a rank of different plant species according to the gain a fungus receives from them in pure stands.
2. Use the different plant species in a model system according to Fig. 4.4. Test fungal strategies when less beneficial plants are introduced into mixtures

with beneficial plants. Optimal foraging and the preferential colonisation of beneficial host plants could be tested in such a system. The potential effects of the necessity of AMF to forage for resource patches (new roots) is also presented as a possible mechanism preventing the evolution of specificity in the symbiosis but this clearly needs to be investigated with more optimality models (Sanders, Chap. 16, this Vol.).

4.8 Conclusions and Future Perspectives

This chapter reviews the foraging strategies of mycorrhizal fungi. Both mycelia of AMF and of EMF respond to external factors in a way that is not mediated by the plant. The responses are selective which lead us to conclude that active foraging occurs by mycorrhizal mycelia. The foraging strategies of AMF and EMF are clearly different and also within EMF different strategies have evolved. The observed responses of mycorrhizal mycelia to different resources (as discussed in Sects. 4.4, 4.5) provide a framework that can be used to test hypotheses in future studies about foraging strategies in mycorrhizal fungi. By applying the conceptual models suggested here it would be possible to investigate foraging of mycorrhizal fungi and also to ask the question: Is there optimal foraging in mycorrhizal fungi?

We suggest, based on the suggested models, four areas of future research:
1. Studies of the resource allocation within a mycelium to nutrient patches of different quality and quantity. Over which distances does resource allocation occur in different fungi?
2. Studies of the C capture of mycorrhizal fungi within ecosystems. This research should include determination of fungal preferences for different hosts in microcosm plant communities as well as measurements of the C flow through the symbiotic interface in different plants and under different environmental conditions.
3. One critical parameter in foraging theory is when to leave a patch? This is summarised in the marginal value theorem (Charnov 1976; see also Begon et al. 1990). Does a mycorrhizal fungus leave a patch when it is not providing energy anymore? This could be tested in a system where several plants are colonised by a single mycelium and the shoot of one of them is aborted or shaded.
4. Differences in foraging strategies between AM and EMF. Special interest could be paid to responses to different mineral nutrients and organic materials. EM and AMF isolated from the same ecosystem could be used in experimental systems with natural substrates. Compartmented systems could be used where the proliferation and interactions of AM and EM hyphae could be studied in root-free soil.

References

Abbott LK, Robson AD (1985) The effect of soil pH on the formation of VA mycorrhizas by two species of *Glomus*. Aust J Soil Res 23:253–261

Arnebrant K (1994) Nitrogen amendments reduce the growth of extramatrical ecto-mycorrhizal mycelium. Mycorrhiza 5:7–15

Baar J, Comini B, Oude Elfererink M, Kuyper TW (1997) Performance of four ecto-mycorrhizal fungi on organic and inorganic nitrogen sources. Mycol Res 101:523–529

Bago B, Azcon-Aguilar C, Goulet A, Piché Y (1998a) Branched absorbing structures (BAS): a feature of the extraradical mycelium of symbiotic arbuscular mycorrhizal fungi. New Phytol 139:375–388

Bago B, Azcón-Aguilar C, Piché Y (1998b) Architecture and developmental dynamics of the external mycelium of the arbuscular mycorrhizal fungus *Glomus intraradices* grown under monoaxenic conditions. Mycologia 90:52–62

Bago B, Pfeffer PE, Douds Jr DD, Brouillette J, Bécard G, Sachar-Hill Y (1999) Carbon metabolism in spores of the arbuscular mycorrhizal fungus *Glomus intraradices* as revealed by nuclear magnetic resonance spectroscopy. Plant Physiol 121:263–271

Bécard G, Fortin JA (1988) Early events of vesicular-arbuscular mycorrhiza formation on Ri T-DNA transformed roots. New Phytol 108:211–218

Begon M, Harper JL, Townsend CR (1990) Ecology: individuals, populations, and communities, 2nd edn. Blackwell, Boston

Bending GD, Read DJ (1995a) The structure and function of the vegetative mycelium of ecto-mycorrhizal plants. V. The foraging behaviour of ecto-mycorrhizal mycelium and the translocation of nutrients from exploited organic matter. New Phytol 130:401–409

Bending GD, Read DJ (1995b) The structure and function of the vegetative mycelium of ecto-mycorrhizal plants. VI. Activities of nutrient mobilizing enzymes in birch litter colonized by *Paxillus involutus* (Fr.) Fr. New Phytol 130:411–417

Boddy L (1999) Saprotrophic cord-forming fungi: meeting the challenge of heterogeneous environments. Mycologia 91:13–32

Bolan L (1991) A critical review on the role of mycorrhizal fungi in the uptake of phosphorus by plants. Plant Soil 134:189–207

Bonello P, Bruns TD, Gardes M (1998) Genetic structure of a natural population of the ecto-mycorrhizal fungus *Suillus pungens*. New Phytol 138:533–542

Brand F, Agerer R (1986) Studies on ecto-mycorrhizae VIII Mycorrhizae formed by *Lactarius subdulcis*, *L. vellerus* and *Laccaria amethystina* in beech. Z Mykol 52:287–320

Brandes B, Godbold DL, Kuhn AJ, Jentschke G (1998) Nitrogen and phosphorus acquisition by the mycelium of the ecto-mycorrhizal fungus *Paxillus involutus* and its effect on host nutrition. New Phytol 140:735–743

Bruns TD (1995) Thoughts on the processes that maintain local species diversity of Ecto-mycorrhizal fungi. Plant Soil 170:63–73

Carleton TJ, Read DJ (1991) Ecto-mycorrhizas and nutrient transfer in conifer – feather moss ecosystems. Can J Bot 69:778–785

Cairney JWG, Burke RM (1996) Physiological heterogeneity within fungal mycelia: an important concept for a functional understanding of the ecto-mycorrhizal symbiosis. New Phytol 134:685–695

Charnov EL (1976) Optimal foraging: attack strategy of a mantid. Am Nat 110:141–151

Colpaert JV, Van Tichelen KK (1996) Decomposition, nitrogen and phosphorus mineralization from beech leaf litter colonized by ecto-mycorrhizal or litter-decomposing basidiomycetes. New Phytol 134:123–132

Colpaert JV, Van Assche JA, Luijtens K (1992) The growth of the extramatrical mycelium of ecto-mycorrhizal fungi and the growth response of *Pinus sylvestris* l. New Phytol 120:127–135

Cooper KM, Lösel DM (1978) Lipid physiology of vesicular-arbuscular mycorrhiza. New Phytol 80:143–151

Coutts MP, Nicoll BC (1990) Growth and survival of shoots, roots and mycorrhizal mycelium in clonal Sitka spruce during the first growing season after planting. Can J For Res 20:861–868

Cromack K, Sollins P, Graustein WC, Speidel K, Todd AW, Spycher G, Li CY, Todd RL (1979) Calcium oxalate accumulation and soil weathering in mats of hypogeous fungus, *Hysterangium crassum*. Soil Biol Biochem 11:463–468

Cui M, Caldwell MM (1996) Facilitation of plant phosphate acquisition by arbuscular mycorrhizas from enriched soil patches. II. Hyphae exploiting root-free soil. New Phytol 133:461–467

Dahlberg A, Stenlid J (1990) Population structure and dynamics in *Suillus bovinus* as indicated by spatial distribution of fungal clones. New Phytol 115:487–493

De Kroon H, Hutchings MJ (1995) Morphological plasticity in clonal plants: the foraging concept reconsidered. J Ecol 83:143–152

Dighton J (1995) Nutrient cycling in different terrestrial ecosystems in relation to fungi. Can J Bot 73 [Suppl 1]:S1349–S1360

Ek H (1997) The influence of nitrogen fertilization on the carbon economy of *Paxillus involutus* in ecto-mycorrhizal association with *Betula pendula*. New Phytol 135:133–142

Ekblad A, Wallander H, Carlsson R, Huss-Danell K (1995) Fungal biomass in roots and extramatrical mycelium in relation to macronutrients and plant biomass of ectomycorrhizal *Pinus sylvestris* and *Alnus incana*. New Phytol 131:443–451

Erland S, Söderström B, Andersson S (1990) Effects of liming on ecto-mycorrhizal fungi infecting *Pinus sylvestris* L. 2. Growth rates in pure culture at different pH values compared to growth rates in symbiosis with the host plant. New Phytol 115:683–688

Finlay RD, Read DJ (1986a) The structure and function of the vegetative mycelium of ecto-mycorrhizal plants I. Translocation of ^{14}C-labelled carbon between plants interconnected by a common mycelium. New Phytol 103:143–156

Finlay RD, Read DJ (1986b) The structure and function of the vegetative mycelium of ecto-mycorrhizal plants II. The uptake and distribution of phosphorus by mycelial strands and interconnecting host plants. New Phytol 103:157–165

Friese CF, Allen MF (1991) The spread of VA mycorrhizal fungal hyphae in the soil: inoculum types and external hyphal architecture. Mycologia 83:409–418

Gazey C, Abbott LK, Robson AD (1992) The rate of development of mycorrhizas affects the onset of sporulation and production of external hyphae by two species of *Acaulospora*. Mycol Res 96:643–650

Giovannetti M, Sbrana C (1998) Meeting a non-host: the behaviour of AM fungi. Mycorrhiza 8:123–130

Giovannetti M, Avio L, Sbrana C, Citernesi S (1993a) Factors affecting appressorium development in the vesicular-arbuscular mycorrhizal fungus *Glomus mosseae* (Nicol. & Gerd.) Gerd. & Trappe. New Phytol 123:115–122

Giovannetti M, Sbrana C, Avio L, Citernesi AS, Logi C (1993b) Differential hyphal morphogenesis in arbuscular mycorrhizal fungi during pre-infection stages. New Phytol 125:587–593

Giovannetti M, Sbrana C, Logi C (1994) Early processes involved in host recognition by arbuscular mycorrhizal fungi. New Phytol 127:703–709

Giovannetti M, Azzolini D, Citernesi AS (1999) Anastomosis formation and nuclear and protoplasmatic exchange in arbuscular mycorrhizal fungi. Appl Environ Microbiol 65:5571–5575

Green H, Larsen J, Olsson PA, Funck Jensen D, Jakobsen I (1999) Suppression of the biocontrol agent *Tricoderma harzianum* by external mycelium of the arbuscular mycorrhizal fungus *Glomus intraradices*. Appl Environ Microbiol 65:1428–1434

Griffiths RP, Baham JE, Caldwell BA (1994) Soil solution chemistry of ecto-mycorrhizal mats in forest soil. Soil Biol Biochem 26:331–337

Griffiths RP, Bradshaw GA, Marks B, Lienkaemper GW (1996) Spatial distribution of ecto-mycorrhizal mats in coniferous forests of the Pacific Northwest, USA. Plant Soil 180:147–158

Harrison MJ (1999) Molecular and cellular aspects of the arbuscular mycorrhizal symbiosis. Annu Rev Plant Physiol Plant Mol Biol 50:361–389

Hepper CM, Jakobsen I (1983) Hyphal growth from spores of the mycorrhizal fungus *Glomus caledonium*: effects of amino acids. Soil Biol Biochem 1:55–58

Högberg P, Plamboeck AH, Taylor AFS, Fransson P (1999) Natural ^{13}C abundance reveals trophic status of fungi and host-origin in mycorrhizal fungi in mixed forests. Proc Natl Acad Sci USA 96:8534–8539

Horan DP, Chilvers GA (1990) Chemotropism – the key to ecto-mycorrhizal formation? New Phytol 116:297–301

Hutchings MJ, De Kroon H (1994) Foraging in plants: the role of morphological plasticity in resource acquisition. Adv Ecol Res 25:159–238

Hutchinson LJ (1999) *Lactarius*. In: Cairney JWG, Chambers SM (eds) Ecto-mycorrhizal Fungi, key genera in profile. Springer, Berlin Heidelberg New York

Jakobsen I, Abbott LK, Robson AD (1992) External hyphae of vesicular-arbuscular mycorrhizal fungi associated with *Trifolium subterraneum* L. I. Spread of hyphae. New Phytol 120:371–380

Jennings DH, Lysek G (1996) Fungal biology: understanding the fungal lifestyle. BIOS Scientific Publishers, Oxford

Johansen A (1999) Depletion of soil mineral N by roots of *Cucumis sativus* L. colonized or not by arbuscular mycorrhizal fungi. Plant Soil 209:119–127

Johansen A, Jakobsen I, Jensen ES (1992) Hyphal transport of ^{15}N labelled nitrogen by a VA mycorrhizal fungus and its effect on depletion of inorganic soil N. New Phytol 122:281–288

Johansen A, Jakobsen I, Jensen ES (1994) Hyphal N transport by a vesicular-arbuscular mycorrhizal fungus associated with cucumber grown at three nitrogen levels. Plant Soil 160:1–9

Joner EJ, Jakobsen I (1995) Growth and extracellular phosphatase activity of arbuscular mycorrhizal hyphae as influenced by soil organic matter. Soil Biol Biochem 27:1153–1159

Joner EJ, Magid J, Gahoonia TS, Jakobsen I (1995) P depletion and activity of phosphatases in the rhizosphere of mycorrhizal and non-mycorrhizal cucumber (*Cucumis sativus* L.). Soil Biol Biochem 27:1145–1151

Joner EJ, Van Aarle IM, Vosatka M (2000) Phosphatase activity of extra-radical arbuscular mycorrhizal hyphae: a review. Plant Soil 226:199–210

Jones MD, Durall DM, Tinker PB (1998) A comparison of arbuscular and ecto-mycorrhizal *Eucalyptus coccifera*: growth response, phosphorus uptake efficiency and external hyphal production. New Phytol 140:125–134

Jongmans AG, Van Bremen N, Lundström U, Van Hess PAW, Finlay RD, Srinivasan M, Unestam T, Giesler R, Melkerud R, Olsson M (1997) Rock eating fungi. Nature 389:682–683

Kling M, Jakobsen I (1998) Arbuscular mycorrhiza in soil quality assessment. Ambio 27:29–34

Larsen J, Jakobsen I (1996) Interactions between a mycophagous Collembola, dry yeast and the external mycelium of an arbuscular mycorrhizal fungus. Mycorrhiza 6:259–264

Larsen J, Olsson PA, Jakobsen I (1998) The use of fatty acid signatures to study mycelial interactions between the arbuscular mycorrhizal fungus *Glomus intraradices* and the saprophytic fungus *Fusarium culmorum* in root-free soil. Mycol Res 102:1491–1496

Lewis DH (1991) Fungi and sugars – a suite of interactions. Mycol Res 95:897–904

Lewis DH, Harley JL (1965) Carbohydrate physiology of mycorrhizal roots of beech. II. Utilization of exogenous sugars by uninfected and mycorrhizal roots. New Phytol 64:224–237

Logi C, Sbrana C, Giovannetti M (1998) Cellular events involved in survival of individual arbuscular mycorrhizal symbionts growing in the absence of the host. Appl Environ Microbiol 64:3473–3479

Li X-L, George E, Marschner H (1991a) Phosphorus depletion and pH decrease at the root-soil and hyphae-soil interfaces of VA-mycorrhizal white clover fertilized with ammonium. New Phytol 119:397–404

Li X-L, Marschner H, George E (1991b) Acquisition of phosphorus and copper by VA mycorrhizal hyphae and root-to-shoot transport in white clover. Plant Soil 136:49–57

Mac Arthur RH, Pianka ER (1966) On optimal use of a patchy environment. Am Nat 100:603–609

Mahmood S, Finlay RD, Erland S, Wallander H (2001) Solubilisation and colonisation of wood ash by ecto-mycorrhizal fungi isolated from a wood ash fertilised spruce forest. FEMS Microbiol Ecol 35:151–161

Miller RM, Reinhardt DR, Jastrow JD (1995) External hyphal production of vesicular-arbuscular mycorrhizal fungi in pasture and tallgrass prairie communities. Oecologia 103:17–23

Miranda JCC, Harris PJ (1994) Effects of soil phosphorus on spore germination and hyphal growth of arbuscular mycorrhizal fungi. New Phytol 128:103–108

Mosse B (1959) Observations on the extra-matrical mycelium of a vesicular-arbuscular endophyte. Trans Br Mycol Soc 42:439–448

Nylund J-E, Wallander H (1992) Ergosterol analysis as a means of quantifying mycorrhizal biomass. In: Norris JR, Read DJ, Varma AK (eds) Methods in microbiology, vol 24. Academic Press, London, pp 77–88

Ogawa M (1985) Ecological characters of ecto-mycorrhizal fungi and their mycorrhizae. JARQ 18:305–314

Olsson PA (1999) Signature fatty acids provide tools for determination of distribution and interactions of mycorrhizal fungi in soil. FEMS Microbiol Ecol 29:303–310

Olsson PA, Johansen A (2000) Lipid and fatty acid composition of hyphae and spores of arbuscular mycorrhizal fungi at different growth stages. Mycol Res 104:429–434

Olsson PA, Wallander H (1998) Interactions between ecto-mycorrhizal fungi and the bacterial community in soils amended with various primary minerals. FEMS Microbiol Ecol 27:195–205

Olsson PA, Wilhelmsson P (2000) The growth of external AM fungal mycelium in sand dunes and in experimental systems. Plant Soil 226:161–169

Olsson PA, Bååth E, Jakobsen I, Söderström B (1995) The use of phospholipid and neutral lipid fatty acids to estimate biomass of arbuscular mycorrhizal fungi in soil. Mycol Res 99:623–629

Olsson PA, Bååth E, Jakobsen I and Söderström B (1996a) Soil bacteria respond to presence of roots but not to arbuscular mycorrhizal mycelium. Soil Biol Biochem 28:463–470

Olsson PA, Chalot M, Bååth E, Finlay RD, Söderström B (1996b) Ecto-mycorrhizal mycelia reduce bacterial activity in a sandy soil. FEMS Microbiol Ecol 21:77–86

Olsson PA, Bååth E, Jakobsen I (1997) Phosphorus effects on the mycelium and storage structures of an arbuscular mycorrhizal fungus as studied in the soil and roots by analysis of fatty acid signatures. Appl Environ Microbiol 63:3531–3538

Olsson PA, Thingstrup I, Jakobsen I, Bååth E (1999) Estimation of the biomass of arbuscular mycorrhizal fungi in a linseed field. Soil Biol Biochem 31:1879–1887

Olsson S (1995) Mycelial density profiles of fungi on heterogeneous media and their interpretation in terms of nutrient reallocation patterns. Mycol Res 99:143–153

Olsson S, Hansson BS (1995) Action potential-like activity found in fungal mycelia is sensitive to stimulation. Naturwissenschaften 82:30–31

Paris F, Bonnaud P, Ranger J, Robert M, Lapeyrie F (1995) Weathering of ammonium or calcium-saturated 2:1 phyllosilicates by ecto-mycorrhizal fungi in vitro. Soil Biol Biochem 27:1237–1244

Pfeffer PE, Douds Jr DD, Bécard G, Shachar-Hill Y (1999) Carbon uptake and the metabolism and transport of lipids in an arbuscular mycorrhiza. Plant Physiol 120:587–598

Ravnskov S, Larsen J, Olsson PA, Jakobsen I (1999) Effects of various organic compounds on growth and P uptake of an arbuscular mycorrhizal fungus. New Phytol 141:517–524

Rayner ADM (1991) The challenge of the individualistic mycelium. Mycologia 83:48–71

Read DJ (1991) Mycorrhizas in ecosystems. Experientia 47:376–391

Read DJ (1992) The mycorrhizal mycelium. In: Allen MF (ed) Mycorrhizal functioning. Chapman and Hall, New York

Read DJ, Boyd R (1986) Water relations of mycorrhizal fungi and their host plants. In: Ayres P, Boddy L (eds) Water, fungi and plants. Cambridge Univ Press, Cambridge, pp 287–303

Ritz K, Crawford J (1990) Quantification of the fractal nature of colonies of *Trichoderma viride*. Mycol Res 94:1138–1152

Robinson D, Fitter AH (1999) The magnitude and control of carbon transfer between plants linked by a common mycorrhizal network. J Exp Bot 50:9–13

Sinsabaugh RL (1994) Enzymatic analysis of microbial pattern and process. Biol Fertil Soils 17:69–74

Siqueira JO, Hubbell DH, Mahmud AW (1984) Effect of liming on spore germination, germ tube growth and root colonization by vesicular-arbuscular mycorrhizal fungi. Plant Soil 76:115–124

Smith SE, Read DJ (1997) Mycorrhizal symbiosis. Academic Press, San Diego

Stephens DW, Krebs JR (1986) Foraging theory. Princeton Univ Press, Princeton

St John TV, Coleman DC, Reid CPP (1983) Association of vesicular-arbuscular mycorrhizal hyphae with soil organic particles. Ecology 64:957–959

Tommerup IC (1984) Persistence of infectivity by germinated spores of vesicular-arbuscular mycorrhizal fungi in soil. Trans Br Mycol Soc 82:275–282

Tunlid A, Ek H, Westerdahl G, Odham G (1987) Determination of ^{13}C-enrichment in bacterial fatty acids using chemical ionization mass spectrometry with negative ion detection. J Microbiol Methods 7:77–89

Unestam T (1991) Water repellency, mat formation, and leaf-stimulated growth of some ecto-mycorrhizal fungi. Mycorrhiza 1:13–20

Unestam T, Sun Y-P (1995) Extramatrical structures of hydrophobic and hydrophilic ecto-mycorrhizal fungi. Mycorrhiza 5:301–311

Wallander H (1995) A new hypothesis to explain allocation of dry matter between mycorrhizal fungi and pine seedlings in relation to nutrient supply. Plant Soil 168/169:243–248

Wallander H (2000a) Uptake of P from apatite by *Pinus sylvestris* seedlings colonised by different ecto-mycorrhizal fungi. Plant Soil 218:249–256

Wallander H (2000b) Use of strontium isotopes and foliar K content to estimate weathering of biotite induced by pine seedlings colonised by ecto-mycorrhizal fungi from two different soils. Plant Soil 222:215–229

Wallander H, Nylund J-E (1992) Effects of excess nitrogen and phosphorous starvation on extramatrical mycelium in Scots pine seedlings. New Phytol 120:495–503

Wallander H, Wickman T (1999) Biotite and microcline as a K source in mycorrhizal and non-mycorrhizal *Pinus sylvestris* seedlings. Mycorrhiza 9:25-32

Wallander H, Wickman T, Jacks G (1997) Apatit as a P source in mycorrhizal and non-mycorrhizal *Pinus sylvestris* seedlings. Plant Soil 196:123-131

Wallenda T, Kottke I (1998) Nitrogen deposition and ecto-mycorrhizas. New Phytol 139:169-187

Watkins NK, Fitter AH, Graves JD, Robinson D (1996) Carbon transfer between C3 and C4 plants linked by a common mycorrhizal network, quantified using stable carbon isotopes. Soil Biol Biochem 28:471-477

Wessel JH (1993) Wall growth, protein extraction and morphogenesis in fungi. Tansley review no 45. New Phytol 123:397-413

Wiklund K, Nilsson L-O, Jacobsson S (1995) Effects of irrigation, fertilization, and artificial drought on basidioma production in a Norway spruce stand. Can J Bot 73:200-208

Wright SF; Upadhyaya A (1999) Quantification of arbuscular mycorrhizal fungi activity by the glomalin concentration on hyphal tips. Mycorrhiza 8:283-285

5 The Role of Various Types of Mycorrhizal Fungi in Nutrient Cycling and Plant Competition

RIEN AERTS

Contents

5.1	Summary	117
5.2	Introduction	118
5.3	Soil Nutrient Sources	120
5.4	Different Mycorrhizal Types Can Tap Different Soil Nutrient Sources	122
5.5	Vascular Plant ^{15}N Natural Abundance as an Indicator for the Type of Mycorrhiza-Mediated Plant Nutrition?	124
5.6	A Conceptual Model for Mycorrhizal Impacts on Plant Competition and Coexistence	125
5.7	Effects of Increased Nitrogen Deposition	127
5.8	Plant-Soil Feedbacks: Mineralization or 'Organicisation'?	129
5.9	Future Challenges	130
References		131

5.1 Summary

More than 90 % of terrestrial plant species associate with mycorrhizal fungi. These fungi play an important role in the mineral nutrition of plants. This chapter focuses on the ability of various mycorrhiza-types: arbuscular mycorrhizal fungi (AMF), ecto-mycorrhizal fungi (EMF) and ericoid mycorrhizal fungi to utilise different soil nutrient sources (EMF and ericoid mycorrhizal fungi: organic N; AMF: inorganic P) and the consequences of this ability for plant competition and plant-soil feed-backs. A conceptual model is presented which shows that this differentiation in the use of various inorganic and organic nutrient sources may create positive feed-backs between plant species dominance, litter chemistry, litter decomposition, and the dominant mycorrhiza type. However, the mycorrhizal impact, especially of ericoid mycorrhizal fungi and ecto-mycorrhizal fungi, on this triad of plant nutrition, plant competition, and ecosystem functioning can be strongly reduced under high levels of atmospheric nitrogen input. As increased N input leads to a relative P shortage in ecosystems, it can be expected that arbuscular mycorrhizal fungal species become progressively more dominant. However, the effects of increased N supply on AMF colonisation are very variable. Studies show that positive, negative or no effects are found in almost equal proportions.

A review of current experimental evidence shows that although alternative N cycling routes may be important for plant nutrition, nutrient cycling, and ecosystem functioning, the many questions which are still open do not allow any firm conclusions to be drawn yet. Thus, one of the major challenges for mycorrhizal ecology in the coming years is to investigate this conceptual model and test its relevance under field conditions. This is a complicated task, because it requires detailed knowledge of the identity of various mycorrhizal taxa (especially in the arbuscular mycorrhizal fungi), their specific function in plant nutrition, and knowledge of the dynamics of organic compounds in the soil and their actual uptake by mycorrhizal plants. Due to recent advances in molecular ecology and in stable isotope applications for ecology, it should now be possible to unravel this mechanism at field scale under realistic conditions. Confirmation of our conceptual model would prove that mycorrhizal fungi play a crucial role in the belowground controls on aboveground ecosystem functioning.

5.2 Introduction

The species composition and structure of plant communities is largely determined by the relation between abiotic conditions (notably nutrient availability) and competition between plants (Aerts 1999). Most plant individuals experience nutrient availability through their fungal partner and thus it might be expected that there is a relation between the abundance and diversity of mycorrhizal fungi, plant nutrition, plant competition, and thus biodiversity and ecosystem functioning. Until recently, the impact of mycorrhizal networks relative to other mechanisms which determine botanical biodiversity remained largely unclear and little attention has been paid to the effects of microbe-plant interactions, particularly the mycorrhizal symbiosis, on ecosystem nutrient cycling and plant species composition (Ozinga et al. 1997). This is surprising, as more than 90% of the terrestrial plant species associate with mycorrhizal fungi (Trappe 1987).

The various mycorrhiza types, as described in Chapter 1 (Fig. 1.1), play different roles in the nutrition of the plant with which they associate. The most abundant group, the arbuscular mycorrhizal fungi (AMF), is especially efficient in the uptake of inorganic phosphorus (e.g. see Jakobsen et al., Chap. 3, this Vol.). The less common ecto-mycorrhizal fungi (EMF) and ericoid mycorrhizal fungi are especially abundant on soils with low concentrations of inorganic nitrogen (Read 1991). It has been shown that these fungi are potentially capable of taking up organic nitrogen compounds that can be assimilated by the host plant (Read 1996; Simard et al., Sect. 2.3.2; Leake et al., Sect.. 14.3, this Vol.). The ability of mycorrhizal plants to use this 'short-cut' of the N cycle may be of great adaptive significance in nitrogen-poor habitats, because it gives plants access to a nitrogen source of which other species are deprived (Chapin 1995). In temperate ecosystems the ability to take up organic N sources is mainly restricted to ericoid mycorrhizal and ecto-mycorrhizal plants. Little evidence exists that species with AMF or non-mycorrhizal plants have access to organic N (Aerts and Chapin 2000). In addition to their capacity to take up organic N sources, ericoid mycorrhizal fungi show phospho-diesterase activity and are thereby able to take up organic P sources as well (Leake and Miles 1996). Thus, colonisation by different types of mycorrhizal fungi may lead to differentiation in the ability to take up various forms of limiting soil resources such as organic and inorganic nitrogen or phosphorus. These mycorrhiza-mediated nutrient uptake patterns may be important mechanisms to promote plant coexistence and thus botanical biodiversity in ecosystems differing in the primary limiting element for plant growth.

These patterns of coexistence may further be promoted by plant-mediated feed-backs on litter chemistry and soil nutrient sources through the process of litter decomposition and nutrient mineralisation. Ericoid mycorrhizal plants usually produce relatively low amounts of litter. Litter from both ericoid and ecto-mycorrhizal plants usually contains higher concentrations of secondary compounds, such as lignin and polyphenols, than litter from AMF plants and non-mycorrhizal plants (Aerts and Heil 1993; Northup et al. 1995; Aerts and Chapin 2000). Due to the high concentrations of secondary plant compounds in the litter of ericoid and ecto-mycorrhizal plants, N and P mineralisation may remain low and so does the availability of inorganic N and P in the soil (Van Vuuren et al. 1993; Northup et al. 1995), despite the presence of high levels of organic N and P sources, to which only the EM and ECM species have access.

Mycorrhiza-mediated plant nutrition, plant competition, and soil nutrient cycling can be depicted in a conceptual model (Fig. 5.1). This figure shows a typical example of the situation in north-west European heathlands. These relatively species-poor ecosystems are dominated by plants associated with ericoid mycorrhizal fungi (*Erica tetralix, Calluna vulgaris*) which compete with grass species associated with arbuscular mycorrhizal fungi (*Molinia caerulea, Deschampsia flexuosa*). According to this model, the differential use of soil nutrient sources by different mycorrhiza types and the impact of the litter decomposition characteristics of the plant species associated with the different mycorrhiza types promotes plant coexistence and ecosystem stabil-

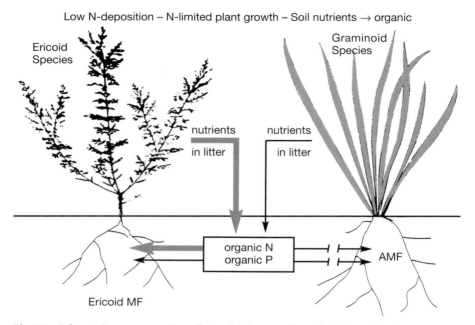

Fig. 5.1. Schematic representation of N and P flows and uptake in heathland ecosystems dominated by species differing in the type of mycorrhizal colonisation (with ericoid mycorrhizal fungi, *Ericoid MF*) or arbuscular mycorrhizal fungi (*AMF*). The thickness of the *arrows* indicates the magnitude of the flows

ity. However, so far there is only a very limited amount of evidence for this hypothetical model.

In this chapter I will explore the actual availability of inorganic and organic nutrient sources in the soil, to what extent different mycorrhiza types are capable of tapping different nutrient sources, how this affects inter-specific competition between plants or plant coexistence, how changes in external nutrient inputs may change patterns of coexistence and, finally, how the litter decomposition characteristics of plant species may contribute to the stability of this mycorrhiza-driven pattern of ecosystem nutrient cycling. As nitrogen is in most terrestrial ecosystems the primary limiting element for plant growth (Aerts and Chapin 2000), the focus of this chapter will be on nitrogen. However, when relevant, phosphorus nutrition will also be discussed.

5.3 Soil Nutrient Sources

Nitrogen is the primary limiting element for plant growth in most terrestrial ecosystems (Aerts and Chapin 2000). In many ecosystems, the bulk of nitrogen is present in soils. For example, in arctic, boreal and temperate ecosystems (including tundra, heathland, coniferous forest, deciduous forest, mire, meadow, bog, grassland and desert, ranging in latitude from 75–40°N) about 87–99 % of the total amount of N in the ecosystem, is present in the soil. The total nitrogen content of soils ranges from <0.2 mg N g^{-1} in subsoils to >25 mg N g^{-1} in peat soil (Van Cleve and Alexander 1981).

Nitrogen is present in multiple forms in the soil: as ammonium, nitrate or as organic nitrogen. The inorganic N forms originate from mineralisation and nitrification processes in the soil, or from external inputs. However, for these inorganic N compounds internal cycling is quantitatively far more important than external inputs (Morris 1991; Koerselman and Verhoeven 1992). Plant litter and its breakdown products are the main organic nitrogen sources that are made available to most terrestrial plant communities by internal cycling. Annual litter production ranges from about 900 kg dry mass ha^{-1} $year^{-1}$ in arctic tundra (Van Cleve and Alexander 1981) to about 14,400 kg ha^{-1} $year^{-1}$ in tropical rain forest (Kumar and Deepu 1992). The amount of nitrogen contained in that litter varies between 10 and 170 kg N ha^{-1} $year^{-1}$ (Van Cleve and Alexander 1981). Many soil animals, bacteria, and fungi are involved in the breakdown of plant litter and the subsequent release of mineral nutrients, which are then available again for uptake by plants and/or micro-organisms. Due to exudation and their death, these organisms also contribute to the organic matter content of the soil (Dighton 1991). This makes the soil a highly heterogeneous medium in which various forms of organic nitrogen are present. These forms range from simple low molecular weight compounds, such as urea, amides and amino acids, to the more complex peptides and proteins, and finally to complex forms such as cell wall components of plants, fungi and soil organisms, or N forms complexed with humic substances. The amount in which these organic N forms are present is dependent on the abundance of organic matter in the soil. Moreover, there appears to be a large temporal variation in the concentration of various organic N forms in the soil (Abuarghub and Read 1988a). Although the amount of data is still very scarce, the currently available data show that in many soils the organic N forms occur in much higher concentrations than ammonium and nitrate (Aerts and Chapin 2000; see also Fig. 14.1 in Leake et al., Chap. 14, this Vol.). For example, Chapin et al. (1993) reported that, in most and wet Arctic soils, the concentrations of water-extractable free amino acids range from 0.14 to 0.57 µmol N g^{-1} (DW) of soil, whereas the concentrations of inorganic N are between 0.04 to 0.08 µmol N g^{-1} (DW) of soil. Similarly, Northup et al. (1995) found that the ratio of dissolved organic N to mineral during nitrogen release from decomposing litter of *Pinus muricata* can be as high as 10. Thus, the capacity of plant species to assimilate these compounds may certainly give them an advantage in competition for nutrients.

5.4 Different Mycorrhizal Types Can Tap Different Soil Nutrient Sources

In most plant nutrition and plant competition models it is assumed that plants depend for their nutrition on the inorganic forms of N and P in the soil. However, as we have seen, nitrogen in organic form is far more abundant in most ecosystems than the inorganic forms. For phosphorus, uptake occurs mainly in inorganic form, and it is well known that plants with AMF have a much higher capacity for inorganic P uptake than non-mycorrhizal plants (Smith and Read 1997). However, ericoid mycorrhizal fungi show phosphodiesterase activity and are, thereby, able to take up organic P sources as well (Leake and Miles 1996). The quantitative importance of this uptake mechanism is not clear.

In recent years, it has been shown that both non-mycorrhizal and mycorrhizal plant species are capable of taking up (simple) organic N compounds (e.g. Chapin et al. 1993; Kielland 1994; Näsholm et al. 1998). Actually, most of this work is building forth on the pioneering work of Read and co-workers who showed in the 1980s that very many plant species with ericoid and ecto-mycorrhizae are capable of taking up various simple and complex organic N compounds (Read 1991). This is an important adaptive trait for plant species on acidic organic soils, such as in heathlands, where most of the N is occluded in organic soil N fractions. The distribution of the various mycorrhiza types on a global scale is clearly related to the distribution of soil types (Read 1991). Ericoid mycorrhizae are usually found in highly organic soils from high latitudes where plant growth is N-limited. Ecto-mycorrhizae dominate in N-limited forest ecosystems with surface litter accumulation at intermediate altitudes and latitudes. Arbuscular mycorrhizae dominate in mainly P-limited herbaceous and woody vegetation on mineral soils at lower latitudes. Non-mycorrhizal species are most abundant in disturbed sites with high nutrient availability and in some high latitude arctic soils (Michelsen et al. 1996, 1998).

The potential importance of the alternative N cycling route is supported by the observation that measured mineralisation rates in highly organic soils very often account for only one-third of the annual N requirement of the vegetation (Aerts and Chapin 2000) and even values of only one tenth have been reported (Jonasson et al. 1999). This strongly suggests that the uptake of organic N compounds may explain this discrepancy. However, currently there are many open ends in this hypothesised N cycling route. The most important remaining questions concern the following issues:

1. So far, it has only been shown that many plant species are capable of assimilating organic N compounds. However, the total amount of organic N taken up by plants has not yet been quantified or considered in terms of seasonal dynamics of total N uptake by the vegetation.

2. In most studies in which it was shown that plants were potentially capable of assimilating organic N forms, the investigators used excised roots or free-living mycorrhizal fungi. These studies thus do not prove the actual occurrence of the described phenomenon under realistic, natural field conditions.
3. Until recently, it has been extremely difficult to identify the mycorrhizal taxa which are specifically associated with certain plant species. This is a serious problem, as recent experimental evidence by Van der Heijden et al. (1998a,b) has shown that the role of mycorrhizal fungi in structuring plant communities is, at least for the arbuscular mycorrhizal fungi, very taxon-specific (see van der Heijden, Chap. 10, this Vol.). New molecular techniques may help in the identification of these taxa and, thus, contribute to elucidating their role in the uptake of various nutrient sources (see Clapp et al., Chap. 8, this Vol.).
4. In many studies the uptake of organic N compounds was not measured adequately. To do so, the use of double-labelled (^{13}C, ^{15}N) organic N compounds is required. Thus, in these studies the possibility that the organic N compounds were mineralised before actual plant uptake occurred, cannot be ruled out.
5. In most organic N uptake studies amino acids were used as model components. However, other more complex organic N compounds can be taken up as well. By reviewing published literature, Vermeer (1995) has shown that both non-mycorrhizal and mycorrhizal plants can take up various organic N compounds ranging from urea, amines and amides, peptides and proteins to even more complex compounds. However, the uptake capacity of non-mycorrhizal plants is much lower and is mainly restricted to the structurally more simple compounds.
6. Most studies have concentrated on organic N uptake. However, in many ecosystems with organic acidic soils, inorganic P supply may, when accounting for the different relative N and P requirements of plant species, even be in shorter supply than inorganic N supply. This is the case in Dutch heathlands where plant growth is P-limited (Aerts and Heil 1993). Thus, when plant species are capable of taking up organic P compounds, this would certainly give them a competitive advantage under these circumstances.
7. It has been shown that there may be severe competition for soil nutrients between plants and soil microbes (Kaye and Hart 1997). The strength of the microbial sink for nutrients, which may be very high (Jonasson et al. 1999), is hardly ever quantified and may also involve microbial immobilisation of nutrients in organic form (Johnson et al. 1998).

Thus, although alternative N cycling routes may be important for plant nutrition, nutrient cycling, and ecosystem functioning, the many questions that are still open do not allow any firm conclusions to be drawn yet.

5.5 Vascular Plant 15N Natural Abundance as an Indicator for the Type of Mycorrhiza-Mediated Plant Nutrition?

Due to the above-mentioned problems, a simple external indicator of the type of nitrogen nutrition of plants would be very welcome. It has been suggested that analysis of the natural abundance of the stable nitrogen isotope ^{15}N, the so-called ∂^{15}N value, can give insight into various aspects of the N cycle (Michelsen et al. 1996, 1998; Högberg 1997). The ∂^{15}N of a sample is calculated as:

$$\partial^{15}N_{sample} = 1000 * (R_{sample} - R_{standard}) / R_{standard}$$

in which R_{sample} and $R_{standard}$ are the ^{15}N/^{14}N ratios of sample and standard, respectively. By definition, the standard has a ∂^{15}N value of 0‰. Samples with negative ∂ values are 'depleted' in the heavier isotope relative to the standard; those which are positive are 'enriched'.

Organic soil N compounds have a lower ∂^{15}N than ammonium. If there is no fungal discrimination against the ^{15}N isotope during uptake or assimilation of nitrogen by mycorrhizal plants, then the ∂^{15}N of the host plant can be considered as a 'signature' representing the type of N nutrition: a lower ∂^{15}N would be an indicator for the fact that ECM or EM plants depend strongly on organic N nutrition. It has been shown by Michelsen et al. (1996) that the ∂^{15}N of ecto-mycorrhizal or ericoid mycorrhizal plant species collected from a heath and a fellfield in northern Sweden was much lower than that of non- or arbuscular mycorrhizal plant species. In a larger study which included plants from northern Sweden, Greenland and Siberia, Michelsen et al. (1998) also found that the vascular plant ^{15}N natural abundance in heath and forest tundra ecosystems is closely related to the presence and type of mycorrhizal fungi in roots (Table 5.1). The lower ∂^{15}N values in leaves of ecto-mycorrhizal and ericoid mycorrhizal species as compared with non-mycorrhizal species are a strong indication that the mycorrhizal species depend to a large extent

Table 5.1. Mean ∂^{15}N (±SE) in leaves of plant species without mycorrhiza (NON), with ecto-mycorrhizal fungi (EMF) or with ericoid mycorrhizal fungi at four heath and forest tundra sites in arctic and subarctic ecosystems. Number of species of each group between parentheses. Data from Michelsen et al. (1998)

Site	Soil	NON	EMF	Ericoid MF
N Sweden (heath tundra)	+0.2±0.2 (7)	−1.6±1.1 (6)	−5.0±0.8 (6)	−8.0±0.3 (5)
N Sweden (forest tundra)	+0.2±0.4 (7)	−1.5±0.4 (13)	−2.1±0.4 (3)	−2.7±0.3 (5)
E Siberia (forest tundra)	−1.5±0.4 (6)	+1.9±0.7 (3)	−5.9±1.4 (3)	−4.3±1.0 (3)
NE Greenland (heath tundra)	−0.9±0.1 (6)	+0.8±0.7 (5)	−5.2±0.6 (5)	−7.2±0.4 (4)

on organic N sources for their nitrogen nutrition. However, this interpretation of the $\partial^{15}N$ patterns is only correct when there is indeed no fungal discrimination against ^{15}N during uptake or assimilation. Recent data by Hobbie et al. (1999) suggest, however, that net fractionation occurs during mycorrhizal transfer of nitrogen to vegetation. Thus, the usefulness of isotopic signatures to assess the type of nitrogen nutrition of plants is still questionable and requires further testing under controlled environmental conditions.

5.6 A Conceptual Model for Mycorrhizal Impacts on Plant Competition and Coexistence

In most competition models, nutrient competition is implicitly considered to be competition for inorganic forms of nutrients (nitrate, ammonium, phosphate). This idea has been the basis for the competition experiments with heathland species that we and others have performed in the past (e.g. Aerts and Berendse 1988; Aerts et al. 1990, 1991; Aerts 1993; Hartley and Amos 1999). As we have seen, it has become clear that the uptake of organic nitrogen compounds by both mycorrhizal and non-mycorrhizal plants is an important pathway in the terrestrial nitrogen cycle. Colonisation by different types of mycorrhizal fungi may, therefore, lead to differentiation in the ability to take up various forms of limiting soil resources such as (in)organic nitrogen or phosphorus. These mycorrhiza-mediated nutrient uptake patterns may be important mechanisms in competition between plants and may affect plant coexistence.

A possible mechanism for this type of interaction is depicted in Fig. 5.2. This analysis is based on Tilman's (1988) R* model of competition. In this model, plants are assumed to compete for two limiting resources. The response of the competing species is presented by drawing their zero growth isoclines as a function of the availability of both limiting resources. According to Tilman's analysis, the species which can reduce the resource availability to the lowest level and still maintain growth (lowest R* value) wins in competition. Figure 5.2 presents a theoretical analysis of competition for N and P between two heathland species, *Calluna vulgaris* (a plant colonised by ericoid mycorrhizal fungi) and *Molinia caerulea* (a graminoid species with AMF). In the absence of mycorrhizal colonisation, *Molinia* wins in competition from *Calluna* because it has a faster uptake of both N and P and, thus, has lower R* values for both resources. This is due to both the higher uptake kinetics of the roots and a higher root allocation (Aerts and van der Peijl 1993). However, when both species are mycorrhizal, the situation is different. Due to its AMF colonisation *Molinia* increases its P uptake and thus assumes a lower R* value for P. *Calluna*, on the other hand, due to its ericoid mycorrhizal fungal colonisation, can take up organic N compounds and, thus, reduces its R* value for N.

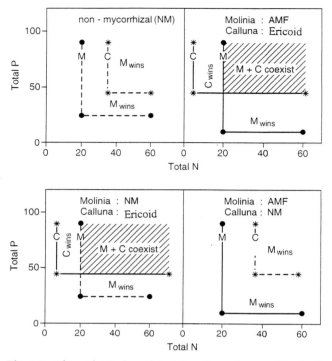

Fig. 5.2. A hypothetical model of the effect of mycorrhizal colonisation on species coexistence in heathlands, based on Tilman's (1988) R* model. Depicted are zero-growth isoclines (R* values) in relation to N and P availability. The species which can grow on the lowest resource concentration (R*) is competitively superior to the other species. In the non-mycorrhizal situation *Molinia caerulea* (M) out-competes *Calluna vulgaris* (C) due to its higher capacity to take up both inorganic N and P. NM = non mycorrhizal; AMF = with arbuscular mycorrhizal fungi; Ericoid = with ericoid mycorriczal fungi. In the mycorrhizal situation coexistence is possible due to *Calluna*'s capacity to take up organic N and *Molinia*'s capacity to take up extra inorganic P. Competitive exclusion depends on the ratio between total N and total P. (Adapted from Van der Heijden, Chap. 10, this Vol.)

As a result, several outcomes of inter-specific competition are possible, including coexistence. The actual outcome depends on the absolute amounts and ratios of the N and P supply. Thus, this hypothetical model shows that the mycorrhizal type determines the competitive balance between plant species and may even lead to coexistence. The mechanism is therefore, due to the use of different soil nutrient sources. If true, this implies that the current nutrient competition models have to be revised. Unfortunately, there is currently no experimental evidence to support this hypothesis. The same basic models are adapted in Chapter 10 by van der Heijden (this Vol.) to explain how AMF promote plant diversity and plant species coexistence (Figs. 10.4, 10.5).

5.7 Effects of Increased Nitrogen Deposition

This fascinating mechanism of species coexistence as a result of differential capacity to use various soil nutrient sources may be disrupted due to increased levels of atmospheric N deposition, as reported for The Netherlands (Aerts and Heil 1993). This results in both higher availability of inorganic N and in an increase of the ratio between inorganic and organic N in the soil. Aerts and Bobbink (1999) summarised published data from the literature on the effect of increased N supply on mycorrhizal colonisation of different plant species (Table 5.2). These data clearly show that the effects of N deposition on mycorrhizal colonisation are dependent on the identity of the mycorrhizal associate. In ecto-mycorrhizal fungal species the amount of colonisation has been shown to decrease with increasing N supply, whereas in AMF species and in ericoid mycorrhizal species the results are very variable. However, it should be noted that these data only refer to the degree of mycorrhizal colonisation. As Rillig et al. (Chap. 6, this Vol.) point out, the effects of increased N deposition on ECM are complex and can also result in no effect or a decrease in colonisation. This is not necessarily equivalent to mycorrhizal function. Currently, it is not clear if mycorrhizal function is reduced to the same extent as mycorrhizal colonisation. Furthermore, the species composition of ecto-mycorrhizal communities changes greatly when nitrogen deposition increases (Rillig et al., Chap. 6, this Vol; Fig. 7.2 in Erland and Taylor, Chap- 7, this Vol.). Another important point is that the nitrogen effects depend on the nitrogen form: in almost all cases where both the effects of ammonium and nitrate were studied separately the most pronounced effects were found when N was supplied as ammonium. As ammonium is the dominant N form in atmospheric N deposition (Aerts and Heil 1993), this implies that atmospheric N deposition generally has negative

Table 5.2. Summary of nitrogen addition experiments in the field and under controlled conditions on mycorrhizal colonisation of different plant species (adapted from Aerts and Bobbink 1999). *n* Number of experiments

Type of mycorrhizal colonisation	n	Effect of N supply
Arbuscular mycorrhizal fungi (AMF)	15	6 Experiments: negative effect 5 Experiments: positive effect 4 Experiments: no effect
Ecto-mycorrhizal fungi (ECM)	7	6 Experiments: negative effect[a] 1 Experiment: no clear effect
Ericoid mycorrhizal fungi (EM)	4	2 Experiments: negative effect 1 Experiment: positive effect 1 Experiment: no effect

[a] More negative effects of NH_4^+ compared with NO_3^-

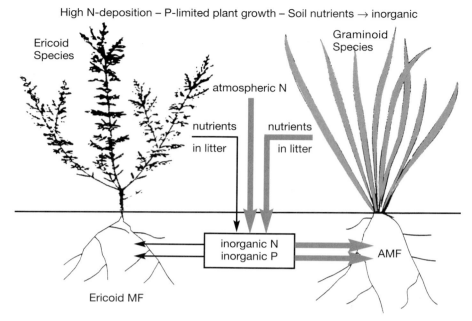

Fig. 5.3. Nitrogen and phosphorus flows and uptake in heathland ecosystems, dominated by species differing in the type of mycorrhizal colonisation, at high levels of atmospheric N deposition. Note that flows of nitrogen in organic form are now of minor importance (cf. Fig. 5.1)

effects on ecto-mycorrhizal fungi and to a lesser extent on AMF and ericoid mycorrhizal fungi.

Reduction of the colonisation of ecto-mycorrhizal and ericoid mycorrhizal plants deprives these species of their relative advantage in nitrogen-poor soils, and may increase the competitive ability of the grasses with AMF, because of their high capacity to use inorganic N and P (Fig. 5.3). An indication for the decreased importance of organic N uptake under high levels of atmospheric N deposition is the observation of less negative $\partial^{15}N$ values of plants growing under such a deposition regime (Hobbie et al. 1999).

In addition to changes in the type of N nutrition, the high atmospheric N deposition leads to a shift from N-limited to P-limited plant growth, a situation to which AMF species are better adapted. It should be noted, however that the dominance of perennials in nutrient-poor environments is not only determined by an efficient nutrient acquisition mechanism, but also by high nutrient retention (Aerts 1999).

5.8 Plant-Soil Feedbacks: Mineralisation or 'Organicisation'?

In most terrestrial ecosystems, plants depend far more on the internal recirculation of nutrients for their mineral nutrition than on external inputs (Aerts and Chapin 2000). Thus, litter decomposition and nutrient mineralisation are key processes in the nutrient cycles of terrestrial ecosystems. However, as we have seen, in particular the ecto-mycorrhizal and ericoid mycorrhizal plants are capable of assimilating organic N compounds (e.g. see Fig. 14.1). This implies that these species do not fully depend on the process of nutrient mineralisation. Instead, it could be argued that they depend on 'organicisation' of plant litter, which is defined here as the process of the transformation of plant litter to organic substances which can be assimilated by (usually) mycorrhizal plants. If we consider plant nutrition from this point of view, then some fundamental problems arise: first of all, contrary to the process of mineralisation there is no well-defined end-product of 'organicisation': mycorrhizal plants are capable of assimilating a wide variety of, sometimes unknown, organic N compounds. These compounds should all be analysed to estimate soil nutrient availability. Apart from the analytical problems with these types of analyses, especially in highly organic soils, it can safely be assumed that the uptake kinetics of all these compounds show substantial differences. This implies that the relative abundance of these N-containing compounds strongly determines the actual N assimilation rate by ecto-mycorrhizal or ericoid mycorrhizal plants.

In addition, the secondary compounds occurring in decomposing plant litter are likely to cause strong interference with the uptake of mineral nutrients. Plant litter initially consists of soluble cell contents and solid cell wall components. Woody litter especially, is rich in lignin, cellulose and hemicellulose. The cell content contains a wide array of compounds, from inorganic ammonium, amino acids, peptides to proteins, but also secondary metabolites may be present in plant litter, like tannins and phenolics. These high concentrations of phenolic compounds may lead to precipitation of proteins prior to protein hydrolysis, which reduces nutrient resorption from senescing tissues (Chapin and Kedrowski 1983) and thus leads to higher nutrient concentrations in fresh plant litter. Tannins may play a key role in the mineral nutrition of plants from infertile sites. By fractionation of soil, Northup et al. (1995) revealed that most (over 60%) dissolved organic nitrogen (DON) consists of humic substances such as protein-tannin complexes, rather than of free amino acids or proteins. Tannins form strong complexes with proteins that are sparingly soluble and recalcitrant to decomposition and mineralisation. Northup et al. (1995) found a highly significant negative correlation between release rates of both organic and mineral forms of nitrogen and litter polyphenol concentrations. They suggested that plants may regulate the availability of organic N by the polyphenol-content in their litter. If some plants are able to take up these protein-tannin complexes by means of mycorrhizal associations then they have a strong competitive advantage compared with species which are not able to do so. This might explain the evolution of tannin-rich plant communities on strongly acidic and infertile soils throughout the world as an adaptation to N-limitation in these soils. Another advantage of this process would be the reduction of N loss, since these forms of N, unlike nitrate, strongly adsorb to soil surfaces, rarely leach beyond the rooting zone and are not subject to gaseous loss by denitrification. Neither are these forms readily available to most soil organisms, like ammonium, or easily converted to a form that might be lost from the ecosystem. However, this suggestion by Northup et al. (1995) requires investigation into two basic questions: Does the individual plant, which produces these compounds, actually acquire more N from these complexes than competing organisms, and does the benefit of increased N-absorption exceed the cost of production of these compounds (Chapin 1995)?

Some secondary compounds may be present in very high quantities in the soil, but these compounds do not all originate from vascular plant litter. Fungi form a significant proportion of the soil biomass in heathland and boreal forest ecosystems. The cell walls of fungi comprise 10%, sometimes up to 40%, of chitin which contains 69 mg N g^{-1}. Therefore, the F and H layers of heathlands and boreal forest ecosystems might be particularly rich in these polymers, and chitin and hyphal wall hexosamines can be major potential sources of N in these soils (Kerley and Read 1995).

From these observations, it can be concluded that research on plant-mediated feed-backs on soil nutrient cycling requires a new approach: apart from the standard decomposition and mineralisation studies, more focus should be on the dynamics of organic compounds in the soil and their actual uptake by mycorrhizal plants.

5.9 Future Challenges

The basic hypothesis treated in this chapter is that the differential use of soil nutrient sources by different mycorrhiza types and the impact of the litter decomposition characteristics of the plant species associated with the different mycorrhiza types promotes plant coexistence and ecosystem stability. As we have seen, so far there is only a very limited amount of evidence for this hypothetical model. Thus, one of the major challenges for mycorrhizal ecology in the coming years is to investigate this hypothesis and test its relevance under field conditions. This is a complicated task, because it requires detailed knowledge of the identity of various mycorrhizal taxa (not only in the AMF group, but also in the other groups), their function in plant nutrition, and knowledge of the dynamics of organic compounds in the soil and their actual uptake by mycorrhizal plants. However, due to recent advances in molecular ecology (see Clapp et al., Chap. 8, this Vol.) and in stable isotope applications for ecology it should now be possible to unravel this mechanism at field scale under realistic conditions. Confirmation of our hypothesis would prove that mycorrhizal fungi play a crucial role in the triad plant nutrition, plant competition, and ecosystem functioning. If we are able to prove this, then it would have great consequences for the way these aspects of plant ecology have to be studied. This would imply that there are strong belowground controls on aboveground ecosystem functioning.

References

Abuarghub SM; Read DJ (1988a) The biology of mycorrhiza in the Ericaceae. XI. The distribution of nitrogen in soil of a typical upland *Callunetum* with special reference to the 'free' amino acids. New Phytol 108:425–431

Abuarghub SM, Read DJ (1988b) The biology of mycorrhiza in the Ericaceae. XII. Quantitative analysis of individual 'free' amino acids in relation to time and depth in the soil profile. New Phytol 108:433–441

Aerts R (1993) Competition between dominant plant species in heathlands. In: Aerts R, Heil GW (eds) Heathlands, patterns and processes in a changing environment. Kluwer, Dordrecht, pp 125–151

Aerts R (1999) Interspecific competition in natural plant communities: mechanisms, trade-offs, and plant-soil feedbacks. J Exp Bot 50:29–37

Aerts R, Berendse F (1988) The effect of increased nutrient availability on vegetation dynamics in wet heathlands. Vegetatio 76:63–69

Aerts R, Bobbink R (1999) The impact of atmospheric nitrogen deposition on vegetation processes in terrestrial, non-forest ecosystems. In: Langan SJ (ed) The impact of nitrogen deposition on natural and semi-natural ecosystems. Kluwer, Dordrecht, pp 85–122

Aerts R, Chapin FS (2000) The mineral nutrition of wild plants revisited: a re-evaluation of processes and patterns. Adv Ecol Res 30:1–67

Aerts R, Heil GW (eds) (1993) Heathlands, patterns and processes in a changing environment. Kluwer, Dordrecht

Aerts R, van der Peijl MJ (1993) A simple model to explain the dominance of low-productive perennials in nutrient-poor habitats. Oikos 66:144–147

Aerts R, Berendse F, De Caluwe H, Schmitz M (1990) Competition in heathland along an experimental gradient of nutrient availability. Oikos 57:310–318

Aerts R, Boot RGA, van der Aart PJM (1991) The relation between above- and belowground biomass allocation patterns and competitive ability. Oecologia 87:551–559

Caporn SJM, Song W, Read DJ, Lee JA (1995) The effect of repeated nitrogen fertilization on mycorrhizal colonization in heather [*Calluna vulgaris* (L.) Hull]. New Phytol 129:605–609

Chapin FS (1995) New cog in the nitrogen cycle. Nature 377:199–200

Chapin FS, Kedrowski RA (1983) Seasonal changes in nitrogen and phosphorus fractions and autumn retranslocation in evergreen and deciduous taiga trees. Ecology 64:376–391

Chapin FS, Moilanen L, Kielland K (1993) Preferential use of organic nitrogen for growth by a non-mycorrhizal arctic sedge. Nature 361:150–153

Dighton J (1991) Acquisition of nutrients from organic resources by mycorrhizal autotrophic plants. Experientia 47:362-369

Hartley SE, Amos L (1999) Competitive interactions between *Nardus stricta* L. and *Calluna vulgaris* (L.) Hull: the effect of fertilizer and defoliation on above- and belowground performance. J Ecol 87:330-340

Hobbie EA, Macko SE, Shugart HH (1999) Interpretation of nitrogen isotope signatures using the NIFTE model. Oecologia 120:405-415

Högberg P (1997) ^{15}N natural abundance in soil-plant systems. New Phytol 137:179-203

Johnson D, Leake JR, Lee JA, Campbell CD (1998) Changes in soil microbial biomass and microbial activities in response to 7 years simulated pollutant nitrogen deposition on a heathland and two grasslands. Environ Pollut 103:239-250

Jonasson S, Michelsen A, Schmidt IK (1999) Coupling of nutrient cycling and carbon dynamics in the Arctic, integration of soil microbial and plant processes. Appl Soil Ecol 11:135-146

Kaye JP, Hart SC (1997) Competition for nitrogen between plants and soil microorganisms. Trends Ecol Evol 12:139-143

Kerley SJ, Read DJ (1995) The biology of mycorrhiza in the Ericaceae, XV Chitin degradation by *Hymenoscyphus ericae* and transfer of chitin-nitrogen to the host plant. New Phytol 131:369-375

Kielland K (1994) Amino acid absorption by arctic plants: implications for plant nutrition and nitrogen cycling. Ecology 75:2373-2383

Koerselman W, Verhoeven JTA (1992) Nutrient dynamics in mires of various trophic status: nutrient inputs and outputs and the internal nutrient cycle. In: Verhoeven JTA (ed) Fens and bogs in the Netherlands: vegetation, history, nutrient dynamics and conservation. Kluwer, Dordrecht, pp 397-432

Kumar BM, Deepu JK (1992) Litter production and decomposition dynamics in moist deciduous forests of the Western Ghats in Peninsular India. For Ecol Manage 50:181-201

Leake JR, Miles W (1996) Phosphodiesters as mycorrhizal P sources. I. Phosphodiesterase production and the utilization of DNA as a phosphorus source by the ericoid mycorrhizal fungus *Hymenoscyphus ericae*. New Phytol 132:435-443

Michelsen AS, Schmidt IK, Jonasson S, Quarmby C, Sleep D (1996) Leaf ^{15}N abundance of subarctic plants provides field evidence that ericoid, ectomycorrhizal and non- and arbuscular mycorrhizal species access different sources of soil nitrogen. Oecologia 105:53-63

Michelsen AS, Quarmby C, Sleep D, Jonasson S (1998) Vascular plant ^{15}N natural abundance in heath and forest tundra ecosystems is closely correlated with presence and type of mycorrhizal fungi in roots. Oecologia 115:406-418

Morris JT (1991) Effects of nitrogen loading on wetland ecosystems with particular reference to atmospheric deposition. Annu Rev Ecol Syst 22:257-279

Näsholm T, Ekblad A, Nordin A, Gieslr R, Högberg M, Högberg P (1998) Boreal forest plants take up organic nitrogen. Nature 392:914-916

Northup RR, Yu Z, Dahlgren RA, Vogt KA (1995) Polyphenol control of nitrogen release from pine litter. Nature 377:227-229

Ozinga WA, van Andel J, McDonnell-Alexander MP (1997) Nutritional soil heterogeneity and mycorrhiza as determinants of plant species diversity. Acta Bot Neerl 46:237-254

Read DJ (1991) Mycorrhizas in ecosystems. Experientia 47:376-391

Read DJ (1996) The structure and function of the Ericoid Mycorrhizal root. Ann Bot 77:365-374

Smith SE, Read DJ (1997) Mycorrhizal symbiosis, 2nd edn. Academic Press, San Diego

Tilman D (1988) Plant strategies and the dynamics and structure of plant communities. Princeton Univ Press, Princeton

Trappe JM (1987) Phylogenetic and ecologic aspects of mycotrophy in the angiosperms from an evolutionary standpoint. In: Safir GR (ed) Ecophysiology of VA Mycorrhizal plants. CRC Press, Boca Raton, pp 5–25

Van Cleve K, Alexander V (1981) Nitrogen cycling in tundra and boreal ecosystems. In: Clark FE, Rosswall T (eds) Terrestrial nitrogen cycles. Ecol Bull 33:375–404

Van der Heijden MGA, Klironomos JN, Ursic M, Moutoglis P, Steitwolf-Engel R, Boller T, Wiemken A, Sanders IR (1998a) Mycorrhizal fungal diversity determines plant biodiversity, ecosystem variability and productivity. Nature 396:69–72

Van der Heijden MGA, Boller T, Wiemken A, Sanders IR (1998b) Different arbuscular mycorrhizal fungal species are potential determinants of plant community structure. Ecology 79:2082–2091

Van Vuuren MMI, Berendse F, De Visser W (1993) Species and site differences in the decomposition of litters and roots from wet heathlands. Can J Bot 71:167–173

Vermeer C (1995) Uptake of organic nitrogen by plants: implications for plant nutrition and the nitrogen cycle. MSc Thesis, Utrecht University

6 Global Change and Mycorrhizal Fungi

MATTHIAS C. RILLIG, KATHLEEN K. TRESEDER, MICHAEL F. ALLEN

Contents

6.1	Summary	135
6.2	Introduction	136
6.2.1	What Is Global Change?	136
6.2.2	Rationale for Considering Mycorrhizae in Global Change Biology	138
6.3	Global Change: Multiple Factors	139
6.3.1	Elevated Atmospheric CO_2	139
6.3.2	Nitrogen Deposition	143
6.3.3	Altered Precipitation and Temperature	144
6.3.4	Ozone	147
6.3.5	UV Radiation	148
6.4	Global Change Factor Interactions	149
6.5	Long-Term Versus Short-Term Responses	150
6.6	Global Change and Effects on Symbionts	151
6.7	Conclusions	152
References		153

6.1 Summary

Mycorrhizae, due to their key position at the plant-soil interface, are important to consider in the study of ecosystem impacts of global changes. Human-induced changes in the earth's environment are clearly multi-factorial. Examples of important factors are: elevated concentrations of atmospheric gases (for example carbon dioxide or ozone), increased input of nutrients into ecosystems by atmospheric deposition (for example nitrogen), climate change (including altered precipitation and temperature regimes), invasive species, and increased UV-radiation. All of these components of future or present global changes can have positive or negative impacts on mycorrhizal associations. However, a more fundamental distinction has to be made between these factors, paying tribute to the fact that in the mycorrhizal symbiosis we are dealing with two classes of organisms with partially independent biology. There are those factors that directly affect only the host plant (e.g., carbon fixation), and that only have indirect effects on mycorrhizal fungi (mycobionts) via altered carbon allocation from the host. Examples include atmospheric changes, against which soil serves largely as a buffer. Other factors can (in addition) directly affect the mycobionts, for example warming or altered precipitation. This distinction is crucial for a mechanistic understanding of the impact of global change factors, and for experimental approaches. Global change factors rarely occur in isolation. The complexity of regional combinations of global change factors further highlights the need for mechanistic studies, since direct experimental exploration of a large number of scenarios would be virtually impossible. Finally, processes and patterns at larger temporal and spatial scales have to be considered in an assessment of global change impacts on mycorrhiza. Most experiments only allow access to short-term responses, while longer-term responses are really relevant. Possible approaches include the use of natural experiments, for example, CO_2 springs. Large-scale processes such as shifts in the global distribution of plant communities (or their regional extinction) due to climate change would affect mycorrhiza, for example, alter the current distribution of mycorrhizal types on the globe. With potential impacts on host biodiversity, mycobiont species diversity may also be impacted at regional scales. In addition, changes in the mycorrhizal fungal community that are independent of changes in the plant community may be one of the least-understood, but potentially most important, mycorrhizal responses to global change.

6.2 Introduction

6.2.1 What is Global Change?

Global change encompasses the full range of natural and human-induced changes in the earth's environment. In this review, we will focus on the anthropogenic component of global change in relation to mycorrhiza. While climate change, in particular global warming, has received a lot of media attention, it is now clear that anthropogenic influences on a global scale are not just restricted to changes in climate. Global changes, those that have already occurred or are expected in the future, are without a doubt multi-factorial (Vitousek 1994).

Fossil fuel combustion and deforestation contribute to the accumulation of carbon dioxide in the atmosphere. Elevated atmospheric carbon dioxide is probably one of the best-documented (e.g., Keeling et al. 1995) and most intensely researched factors of global change in general (e.g., Bazzaz 1990; Field et al. 1992; Amthor 1995; Pritchard et al. 1999), and in particular with respect to global change effects on mycorrhiza. Atmospheric CO_2 concentrations have increased from 280 µmol mol^{-1} before the Industrial Revolution to around 360 µmol mol^{-1} today, with a growth rate of ca. 1.8 µmol mol^{-1} year^{-1}. Some of the main effects on plants include increased water-use efficiency and increased photosynthesis, often leading to increased growth and increased belowground carbon allocation. Other changes in atmospheric composition include pollutants, for example, increased concentrations of tropospheric ozone. Tropospheric O_3 is a major air pollutant that has adverse effects on the growth and health of plants (e.g., Heagle 1989; Sandermann 1996). It is equally clear that the global nitrogen cycle has been profoundly altered by human activity, as illustrated by the fact that human-linked nitrogen fixation exceeds naturally occurring fixation (Vitousek 1994). While many atmospheric changes are global in nature due to efficient atmospheric mixing, nitrogen deposition varies in intensity at a regional scale; however, areas of enrichment are distributed around the globe. Another climate change factor, precipitation, will also exhibit regional scale variations, with increases, decreases, no changes, and changes in timing possible. The occurrence of global warming, the increase in near surface temperature of the earth due to the enhanced greenhouse effect, is being increasingly supported (Santer et al. 1996), and may already be detectable as a biological signal (Hughes 2000). Predicted continual reductions in the earth's protective stratospheric ozone layer due to human activity will lead to increases in ultraviolet-B radiation (UV-B: 280–320 nm). UV-B exerts strong effects on living organisms since this radiation can be absorbed by macromolecules such as proteins and nucleic acids, and deleterious effects on plants and ecological interactions are known (Caldwell et al. 1989; Caldwell and Flint 1994; Lumsden 1997; Rousseaux et al. 1998). One of the most significant factors of global change is the invasion of native ecosystems by exotic species (D'Antonio and Vitousek 1992; Dukes and Mooney 1999), made possible by the breakdown of historic long-distance dispersal barriers due to human activity. From the mycorrhizal perspective this is of interest, because new host species are presented to (presumably native) mycorrhizal fungal communities. Information on invasive fungal species seems to be restricted to numerous examples of pathogen introductions (Agrios 1997), but there appears to be little evidence for invasion and effects of exotic mycorrhizal species – mostly due to the fact that resident mycorrhizal fungal communities are rarely known in the first place (e.g., see Clapp et al., Chap. 8, this Vol., and Hart and Klironomos, Chap. 9.4, this Vol.).

These current and future global changes do not occur in isolation, but in combination. Since some of the factors (e.g., precipitation, nitrogen deposition) occur at regional levels, there is a huge variety of potential global change forces to be considered – a major challenge to global change biology.

6.2.2 Rationale for Considering Mycorrhiza in Global Change Biology

Soil has been described as the "chief organizing center for ecosystem function" (Coleman et al. 1992), and the role of soil biota and processes as modifiers of ecosystem or plant responses to global change is becoming increasingly recognized. Rhizosphere-associated microorganisms are known to provide strong feedback to plant growth (e.g., Bever 1994; Bever et al., Chap. 11, this Vol.), and any change in the functioning of this group of organisms due to global change factors is of importance, since this may alter the response of plants to these disturbances.

Mycorrhiza, by virtue of their ubiquity, their key position at the root-soil interface, and because of their influence on plant physiology, plant communities, and ecosystems, warrant a detailed consideration in global change biology (O'Neill et al. 1991). Arbuscular mycorrhizal fungi (AMF), in particular, are completely dependent on the host plant for carbon acquisition; thus any global change factors affecting the plant are likely to affect these fungi as well. The role of mycorrhiza can be viewed as that of a modifier of plant, plant community, and ecosystem responses to global change factors. Mycorrhiza become particularly important, if there is a significant statistical interaction term between the presence of the symbiosis and the global change factor, i.e., if the effect of the global change factor on plants, communities or an ecosystem depends on the presence and functioning of mycorrhizal symbionts. Although the presence of mycorrhiza may not often be limiting in natural ecosystems, this can be the case in disturbed ecosystems (and, in fact, habitat fragmentation and land use change is another factor of global change; Vitousek 1994). Far more importantly, however, an interaction of global change factors with mycorrhizal presence points out that there is a potentially important mechanism in operation that needs to be appreciated in order to understand the effects of global change. In view of the complex interplay of global change factors that could lead to various different combinations at the regional level, an understanding of mechanisms should be a priority over documentation of effects.

As has been pointed out previously (Rillig and Allen 1999), it is vitally important to recognize that mycorrhiza cannot just be experimentally "substituted" with a fertilization treatment, since the effects of mycorrhiza on plants and ecosystems are far more multifaceted than merely nutritional (Newsham et al. 1995). In this context, it is noteworthy that one of the main functions of mycorrhizal fungi (and fungi in general) at the ecosystem level is their contribution to the formation and maintenance of soil structure (Tisdall and Oades 1982). Any global change effects on the extraradical mycelium of mycorrhizal fungi can, therefore, secondarily impact soil structure (Young et al. 1998; Rillig et al. 1999b). Soil structure is crucial for facilitating water infiltration, soil-borne aspects of biogeochemical cycling processes, success of sustainable agriculture, and for providing resistance against erosional loss of soil (Oades 1984; Elliott and Coleman 1988; Hartge and Stewart 1995; Jastrow and Miller 1997). Potential effects of mycorrhizal fungal hyphae on such a basic ecosystem property alone call for an inclusion of mycorrhiza in global change research.

6.3 Global Change: Multiple Factors

Global change is a complex phenomenon comprising a variety of factors, the most prominent of which we review in this section with respect to their effects on mycorrhiza. An overview of known and unknown responses of global change factors on mycorrhiza is given in Table 6.1. New manifestations of global changes may become important in the future (e.g., urbanization effects, other kinds of global pollutions), and others, such as habitat fragmentation, may additionally develop into worthwhile areas of study in mycorrhizal ecology.

6.3.1 Elevated Atmospheric CO2

Elevated CO_2 is perhaps one of the best-researched global change factors with respect to mycorrhizal effects. This issue is a critical one, as several studies have indicated that plant- and ecosystem-level responses to elevated CO_2 can be mediated through the presence of mycorrhizal fungi (i.e., evidence of a significant mycorrhizal-by-CO_2 interaction). For example, Seegmuller and Rennenberg (1994) reported that the tree height of young *Quercus robur* trees increased 26 % with a doubling of CO_2 when individuals were non-mycorrhizal, but rose 54 % when colonized by *Laccaria laccata*, an ecto-mycorrhizal fungus (EMF). Likewise, in *Pinus sylvestris* seedlings, %N and %P concentrations in both roots and shoots declined significantly under elevated CO_2 with the presence of *Paxillus involutus*, but had smaller or opposite responses in the absence of this EMF (Rouhier and Read 1998a). In addition, root biomass, shoot-to-root ratio, and short-term C dynamics of *Pinus sylvestris* seedlings responded differently to elevated CO_2 when colonized by the EMF *Suillus bovinus* versus *Laccaria bicolor* (Gorissen and Kuyper 2000).

Arbuscular mycorrhizal fungi (AMF) appear to influence CO_2 effects as well. Under elevated CO_2, root biomass of a nitrogen-fixing plant (*Robinia pseudoacacia*) increased 114 % in non-mycorrhizal individuals but had a much smaller increase (31 %) when AMF species of the genus *Glomus* were

Table 6.1. Response matrix[a] of three different mycorrhizal fungal types (arbuscular mycorrhizal fungi (AMF), ecto-mycorrhizal fungi (EMF) and ericoid mycorrhizae to various global change factors

	Parameter	AMF	EMF	Ericoid mycorrhizae	Plant growth
Elevated CO_2	% Infection	+/0[b]	+/0[c]	?	+[b,e]
	Hyphal length	+/0[d]	+/0[e]	?	
Nitrogen deposition	% Infection	+/0/−[f]	+/0/−[g]	0/−[h] (few studies)	+/0/−[f,g,h,i]
	Hyphal length	+/0/−[i]	+/0/−[j]	?	
	Fruit bodies		−[k]		
Altered precipitation		+/0[m]	+/0[n]	+/−[o]	
Warming		+/0[p]	0/−[q]	−[r]	
Ozone	% Infection	0/− [s] (few studies)	0/− (few studies)[t]	?	0/−[s,t,u]
	Hyphal length	?	− (few studies)[u]	?	
UV radiation	% Infection	0/− (few studies)[v]	0 (few studies)[w]	?	0/−[v,w]

[a] Responses are noted as positive (+), negative (−), no response found (0), and unknown (?). When too few studies were available to make strong conclusions this is noted.
[b] Reviewed in Diaz (1996); Hodge (1996); Staddon and Fitter (1998); Cairney and Meharg (1999).
[c] Reviewed in Diaz (1996); Hodge (1996); Cairney and Meharg (1999).
[d] Reviewed in Staddon and Fitter (1998); Treseder and Allen (2000)
[e] Reviewed in Treseder and Allen (2000).
[f] Reviewed in Cairney and Meharg (1999).
[g] Reviewed in Jansen and Dighton (1990); Wallenda and Kottke (1998); Cairney and Meharg (1999).
[h] Decrease: Stribley and Read (1976); Yesmin et al. (1996); No response: Johansson (1992); Lee et al. (1992); Caporn et al. (1995).
[i] Reviewed in Treseder and Allen (2000).
[j] Reviewed in Wallenda and Kottke (1998); Treseder and Allen (2000).
[k] Reviewed in Arnolds (1988, 1991).
[m] Edwards (1985); Allen et al. (1987, 1995); Miller and Bever (1999).
[n] Nilsen et al. (1998).
[o] Read (1983); Thormann et al. (1999).
[p] Allen et al. (1995); Kandeler et al. (1998); Pattinson et al. (1999).
[q] Allen et al. (1995); Jonasson and Shaver (1999).
[r] Read (1983); Jonasson and Shaver (1999).
[s] Decrease: McCool and Menge (1984); No response: Duckmanton and Widden (1994).
[t] Reviewed in Cairney and Meharg (1999).
[u] McQuattie and Schier (1992).
[v] Decrease: Van de Staaij et al. (2001); No response: Klironomos and Allen (1995).
[w] Newsham et al. (1999).

present. In the same experiment, nitrogen fixation increased 212 and 90 % in non-mycorrhizal and mycorrhizal seedlings, respectively (Olesniewicz and Thomas 1999). In a study with *Plantago lanceolata* and the AMF species *Glomus mosseae,* CO_2 enrichment increased net photosynthesis and root biomass, and this effect was greater in mycorrhizal plants (Staddon et al. 1999b). Although some studies have found no significant mycorrhizal-by-CO_2-treatment interaction (Jongen et al. 1996; Rouhier and Read 1998b), the remainder suggests that both AMF and EMF may provide one or several mechanisms for CO_2 effects on plants and ecosystems. These potential mediating factors are not necessarily independent and include expanded capacity for nutrient foraging by plants, alterations in carbon allocation within plant structures or between plants and their symbiotic root microbes, changes in soil structure, and increases in soil C storage.

What is the primary mechanism? For plants, mycorrhizal fungi are (among other things) a means of nutrient acquisition. Arbuscular mycorrhizal fungi are known to increase plant uptake of inorganic phosphorus (Jakobsen et al., Chap. 3, this Vol.; Smith and Read 1997) and nitrogen (e.g., Ames et al. 1983; Tobar et al. 1994), and ecto-mycorrhizal fungi provide access to organic nutrient pools in the soil (Simard et al., Chap. 2, this Vol.; Smith and Read 1997). These fungi require a carbon source and can use a significant amount of the plant's net photosynthate (generally about 10–20 %; Allen 1991). Since photosynthetic capacity, water use efficiency, and growth generally increase in plants exposed to elevated CO_2 (e.g., Bazzaz 1990), the amount of carbon available for investment in mycorrhizal fungi should increase as well. In addition, as carbon acquisition rises, so should demand for soil nutrients like nitrogen and phosphorus (although this requirement may be amended somewhat through changes in nutrient use efficiency). Therefore, assuming plants allocate resources efficiently according to economic theories (Bloom et al. 1985; Read 1991), investment in mycorrhizal fungi could increase under elevated CO_2.

Researchers have used several approaches to assess the effects of CO_2 enrichment on mycorrhizal dynamics. The most common unit of measure for the presence of AMF is percent root length infected by mycorrhizal structures (arbuscules, vesicles, or internal hyphae), and for EMF, percent root tips infected. Increases in percent infection of both AMF and EMF with high CO_2 have been reported in several pot and field experiments, with a varying degree of response among ecosystems, host plants, fungal types, harvest date, and nutrient availability. Notably, many additional studies have documented no change, while very few have found decreases (reviewed in Diaz 1996; Hodge 1996; Staddon and Fitter 1998; see also Table 6.1).

Despite its widespread use, percent root infection is a somewhat limited index of mycorrhizal growth or nutrient transfer to the plant (Rillig and Allen 1999). For example, since root biomass often increases under elevated CO_2, total root length infected (and foraging capacity) per plant may increase even if percent infection does not (Diaz 1996; Staddon and Fitter 1998; Staddon et al. 1999a). In addition, the intensity of infection within a given root section may vary. Rillig et al. (1998) recorded infection intensity as the number of intraradical hyphae intersecting a microscope cross hair within a cross section of root. They found that although percent root infection by hyphae in *Bromus hordeaceus* did not change with CO_2 level, infection intensity increased in two size classes of roots. In addition, the allocation of AMF biomass between intraradical and extraradical structures can be altered under elevated CO_2. For instance, Staddon et al. (1999b) noted a reduction in the ratio of AMF hyphal length to total root length infected by AMF in *Plantago lanceolata* and *Trifolium repens* growing under elevated CO_2. This ratio also ranged nearly two-fold among CO_2 and N treatments in *Gutierrezia sarothrae* (Rillig and Allen 1998), and was nearly three times greater under ambient versus elevated CO_2 in a serpentine grassland (Rillig et al. 1999a). Measurements of percent infection alone may underestimate CO_2 effects on mycorrhizal growth.

Alternately, determination of external hyphal biomass in the soil under ambient versus elevated CO_2 provides additional information that can be scaled up to assess ecosystem-level factors (Rillig and Allen 1999). Hyphal length of AMF and EMF often increases under CO_2 enrichment, sometimes four- or fivefold, and this effect has been documented in individual greenhouse-grown plants as well as natural ecosystems (Treseder and Allen 2000). Hyphal length can also respond more strongly to CO_2 than do roots, resulting in a shift in C allocation toward the symbiont (Sanders et al. 1998). As with percent infection, the magnitude of response varies among studies and may be related to a number of factors including plant and fungal species composition. Reports of non-significant CO_2 effects on hyphal length are also common, but we are unaware of any instances of a significant decrease (Treseder and Allen 2000). At the plant level, increases in hyphal length could imply greater nutrient uptake by plants, higher influx of carbon to the fungi, and alterations in carbon allocation among plant components.

In addition, greater total hyphal length in soil under elevated CO_2 can affect a number of ecosystem characteristics, including physiochemical properties of soil. For instance, AMF hyphae contain glomalin, a glycoprotein that is likely to be long-lived in the soil (Wright and Upadhyaya 1996; Wright et al. 1996; Wright and Upadhyaya 1998). Glomalin is thought to bind water-stable soil macroaggregates (large conglomerations of smaller soil particles), which play a significant role in reducing water- and wind erosion (Angers and Mehuys 1993; Lyles et al. 1983). Rillig et al. (1999b) found that glomalin content and macroaggregate abundance increased significantly under CO_2 enrichment in two Californian ecosystems (grassland and chaparral). These CO_2 responses are accompanied by higher rates of carbon input into macroaggregates in the chaparral system (K.K. Treseder, unpubl. data). This latter result indicates that soil carbon storage may be influenced via mycorrhizal fungi under elevated CO_2. Since soils provide a substrate for numerous microbes responsible for nutrient transformations, these changes in soil quality can feed back to affect a number of ecosystem-level processes such as carbon flux, nutrient immobilization, and plant productivity.

6.3.2 Nitrogen Deposition

While elevated atmospheric CO_2 occurs at a global scale due to atmospheric mixing, increased anthropogenic nitrogen deposition is a more localized phenomenon centered primarily in ecosystems that surround urban or agricultural areas. Nevertheless, anthropogenic sources of nitrogen, including N fertilization and N fixation during fossil fuel combustion, are now approaching a worldwide distribution and are common in tropical as well as temperate systems (Matson et al. 1999). Effects of N deposition on natural ecosystems have therefore received much recent attention (e.g., Vitousek 1994; Vitousek et al. 1997; Nadelhoffer et al. 1999). Since mycorrhizal fungi are involved in nitrogen uptake by plants (see Aerts, Chap. 5, this Vol.), their responses to increased N availability have likewise been the subject of many reviews (Arnolds 1988, 1991; Jansen and Dighton 1990; Colpaert and van Tichelen 1996; Wallenda and Kottke 1998; Cairney and Meharg 1999; Erland and Taylor, Chap. 7, this Vol.). It has been suggested that the ratio of carbohydrate to nitrogen supply affects the formation of EMF fruiting bodies, external hyphae, and root tips, and this may serve as a mechanism underlying potential N effects in this fungal group (Wallander 1995; Wallenda and Kottke 1998, and citations therein). If this ratio declines under N deposition, EMF biomass may drop as well.

To date, empirical studies have indicated that increases in nitrogen availability (through deposition or fertilization) have inconsistent effects on dynamics of ericoid, AMF, and EMF in natural ecosystems. The production of EMF fruit bodies has recently declined in Europe (Arnolds 1988, 1991), presumably due in part to nitrogen deposition (see also Fig. 7.2 in Chap. 7, this Vol.). However, influences on belowground processes are less clear. Studies of ericoid mycorrhiza have reported a reduction (Stribley and Read 1976; Yesmin et al. 1996) or no response (Johansson 1992; Lee et al. 1992; Caporn et al. 1995) in percent infection or total mycorrhizal biomass with N additions. For both AMF and EMF, greater nitrogen availability (as ammonium or nitrate) can increase, decrease, or evoke no response in root infection or production of external hyphae in long-term field studies (reviewed in Jansen and Dighton 1990; Wallenda and Kottke 1998; Cairney and Meharg 1999; Treseder and Allen 2000). Many controlled greenhouse studies indicate that EMF growth increases as N availability rises from deficient to optimal conditions, then declines at higher levels (reviewed in Jansen and Dighton 1990; Wallenda and Kottke 1998). Therefore, the initial (pre-deposition) nitrogen status of ecosystems may determine the direction of response. In many cases, this condition is difficult to ascertain without long-term experiments. It appears that effects of nitrogen deposition on mycorrhizal growth should vary among localities and may be difficult to predict.

Perhaps the most consistent response of mycorrhizal fungi to N availability is a shift in community composition (Wallenda and Kottke 1998; Cairney and Meharg 1999; Treseder and Allen 2000). Ecto-mycorrhizal fungi vary in their growth response to nitrogen (Wallander and Nylund 1992; Arnebrant 1994; Wallander et al. 1999), as may AMF. Accordingly, several studies have detailed changes caused by N deposition or fertilization in the assemblage of mushrooms (e.g., Menge and Grand 1978; Arnolds 1988, 1991; Termorshuizen 1993) and root tip morphotypes (Taylor and Alexander 1989; Arnebrant and Soderstrom 1992; Karen and Nylund 1997) of EMF. For instance, Erland and Taylor observed that the proportion of EMF species that are able to use proteins as organic nitrogen decrease when soil nitrogen levels increase (Fig. 7.2 in Chap. 7, this Vol.). Assessments of AMF communities have noted shifts toward *Glomus aggregatum*, *Glomus leptotichum*, and *Glomus geosporum* in coastal sage scrub (Egerton-Warburton and Allen 2000), and toward *Gigaspora gigantea* and *Glomus mosseae* in a tallgrass prairie (Eom et al. 1999) under N deposition or fertilization. Johnson (1993) reported increases in abundance of *Gigaspora gigantea*, *Gigaspora margarita*, *Scutellospora calospora*, and *Glomus occultum* after fertilization with a suite of nutrients including nitrogen and phosphorus. Alterations in the community composition of mycorrhizal fungi appear to be a likely outcome of widespread N deposition. Since the identity and diversity of mycorrhizal species can influence several plant- and ecosystem-level parameters including productivity, soil P content, and plant species composition (van der Heijden et al. 1998; Daniell et al. 1999), these shifts may have significant feedbacks at several scales. Moreover, due to N deposition, ecosystems may change from being N-limited to P-limited. It has been hypothesized that this could lead to a shift of the dominance of different mycorrhizal types (Aerts, Chap. 5, this Vol.).

6.3.3 Altered Temperature and Precipitation

Changes in temperature and water relations, which are tightly interrelated, could provide very strong effects on mycorrhiza. Temperature can have both direct and indirect effects on mycorrhiza. Direct, in that the symbiosis responds to changes in air and soil temperature (see discussion in Sect. 6.6; Fig. 6.1). Indirect, in that nutrient immobilization/mineralization processes are altered, changing the availability of nutrients (e.g., Fitter et al. 1999) and thereby affecting the mycorrhizal association. Water availability can be influenced by various factors. First, precipitation can change. Second, plant water use efficiency can change. Soil moisture can clearly also respond to increasing soil temperatures.

Global Change and Mycorrhizal Fungi

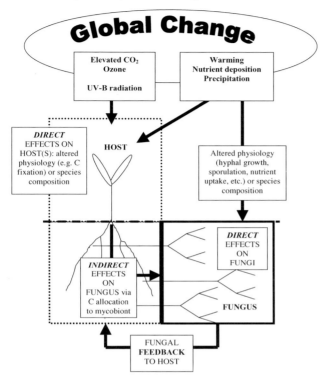

Fig. 6.1. Diagrammatic representation of the effects of different global change factors on the mycorrhizal symbiosis. Some factors only affect the plant directly, with indirect effects on the mycobiont (fungus), via altered carbon allocation from the host to the fungus. Other global change factors can affect the fungus in addition directly. Altered symbiotic functioning of the mycobionts can feed back on host growth. See text for details

With changing climate, it is important to realize that events at larger spatial scales can affect mycorrhiza. For example, plant communities, and therefore mycorrhizal hosts, may shift in distribution, or even cease to exist regionally. One example is the tropical cloud forest that is being strongly affected by climate change, as warming has raised the average altitude of the cloud bank (Pounds et al. 1999). In these forests, there is a high (but mostly undescribed) diversity of ecto-mycorrhizal plants (largely Pinaceae and Fagaceae) and fungi (Allen et al. 1995). As the cloud forests migrate upwards they may cease to exist, and with them possibly the mycobiont community.

6.3.3.1 Mycorrhiza and Temperature

Greenhouse gases have their most dramatic effect on re-radiation of solar energy, and night-time re-radiation into space is decreased as cloud cover increases (Dai et al. 1999). The most dramatic change in temperature is a likely increase in night-time minimum temperatures. Because vegetation takes the bulk of the direct sunlight heat loading, when vegetation is present, the soil temperatures reflect night-time values more than daytime highs. Thus, a major change is a likely increase in soil temperatures where most nutrient cycling dynamics occur and where mycorrhiza function. The data base on soil temperatures and AMF is quite limited (Monz et al. 1994). We would hypothesize that few mycorrhiza are likely to shift, either compositionally or functionally *directly*, in response to temperature shifts in the range caused by global change, except in high latitude areas. However, the indirect consequences of temperature shifts, caused by changes in mineralization and nitrification activity could dramatically alter both the composition and types of mycorrhiza, and could alter the relationship between non-mycotrophic species and mycorrhizal plants.

The biggest changes in temperature are predicted near the poles (Rowntree et al. 1997). In taiga and polar ecosystems, the dominant mycorrhizal types are ecto-mycorrhiza and ericoid mycorrhiza (Read 1991), with only scattered arbuscular mycorrhiza. These mycorrhiza substitute for direct decomposition and uptake by directly transporting organic N from organic matter to the host. If the temperatures were to increase 4–8 °C, as projected by some models for higher latitudes (Barber et al. 2000), mineralization and nitrification would be increased. With AM plants dominant in areas of high N mineralization, this could lead to an increased presence of AM (Read 1991). At the most extreme high latitudes, the temperature shift is expected to be 4–8 °C. In summer, temperatures hover around freezing, sometimes above, often below. Currently in areas like the Antarctic Dry Valleys, there are only algae and few or no mycorrhiza (Virginia and Wall 1999). A small increase would make a growing season possible in these regions, even if short, with possibilities for an expansion of AM distributions. Thus, clearly, temperature effects could lead to shifts in the distribution of AMF around the globe.

6.3.3.2 Mycorrhiza and Altered Water Availability

Arbuscular mycorrhizal fungi affect plant water relations in two distinct ways: by transporting water to the host (increased nutrient uptake during drought also occurs along with the water transport; Faber et al. 1991) and by altering the host stomata or root physiology. Root and leaf physiology are also altered by mycorrhiza. The formation of arbuscular mycorrhizae is also associated with changing root and stomatal hormone levels that facilitate open stomata, and increasing overall rates of water flux (Allen et al. 1982; Augé 2000). Importantly, AMF appear to have little or no effect on water-use efficiency (WUE) (Allen et al. 1981; Di and Allen 1991). It appears that AMF (and possibly EMF) increase total water throughput, thereby allowing additional

CO_2 to be fixed. As drought is initiated, the fungi can prolong the acquisition of water (Allen and Allen 1986), presumably by accessing pockets not accessible to larger roots and probably by maintaining the contact between soil particles and roots as root shrinkage begins.

Water availability likely influences the presence of AMF, but there appear to be few relevant studies. In an irrigation and rainout shelter study, Allen et al. (1989) reported that AMF root colonization increased with drought and decreased water irrigation. In a drought-adapted region, high soil moisture reduced AMF colonization, possibly by nematode or parasite activity (Allen et al. 1987). Water also influences the availability of nutrients which, in turn, alters AMF functioning. In an elaborate set of studies, Mullen and Schmidt (1993) and Mullen et al. (1998) demonstrated in the alpine that as snow melted, plants took up NO_3^-. As the soils began to dry out, AMF rapidly formed and P was taken up. Late in the growing season, during extreme drought, AMF disappeared and nutrient uptake ceased. Allen (1983) also found that arbuscular formation in *Atriplex gardneri* was limited to a very short period in spring when soil moisture was high. Miller and Bever (1999) observed changes in AMF species composition with water depth with the same plant community. This suggests that AMF species are not physiologically equivalent in their tolerance to water availability. For an individual ecosystem, it thus appears likely that altered water availability will affect functioning and/or species composition of mycorrhiza.

6.3.4 Ozone

Ozone is produced when strong sunlight or hot weather catalyzes the transformation of the ozone precursors nitrogen oxides (NO_x) and volatile organic compounds (VOC) to ozone (O_3). As industrial emissions and automobile exhaust are major emitters of NO_x and VOC, low-altitude ozone production is primarily centered on urban areas and the ecosystems downwind from them. Effects of ozone on plant- and ecosystem-processes in these regions have received relatively little attention compared to elevated CO_2 and N deposition. Evidence to date suggests that ozone appears to partially inhibit photosynthesis in plants, either due to a decrease in stomatal conductance accompanied by a decline in internal CO_2 concentration, or due to a reduction in capacity to fix CO_2 within the chloroplast (Heath 1998). A decreased growth rate or biomass may follow.

A few studies have investigated the possibility that AMF and EMF can influence plant responses to O_3. For example, in non-mycorrhizal seedlings of *Pinus sylvestris,* ozone exposure increased the incidence of the root pathogen *Heterobasidion annosum*. However, inoculation by the EMF species *Hebeloma crustuliniforme* prevented this ozone effect (Bonello et al. 1993). However, in *Pinus ponderosa* seedlings inoculated with the same EMF species, Andersen and Rygiewicz (1995) noted no significant mycorrhizal-by-ozone effect on respiration, assimilation, or allocation of carbon. As for AMF, the tomato variety "Heinz 1350" had significant reductions in biomass when exposed to 15 pphm ozone if also colonized by *Glomus* or *Gigaspora* (McCool and Menge 1984). Uncolonized individuals had no negative response at that ozone level. Similarly, a concentration of 50 ppb ozone reduced biomass in *Trifolium subterraneum* individuals infected with *Glomus margarita*, but not in non-mycorrhizal plants (Miller et al. 1997).

How are mycorrhiza affected by ozone? Since photosynthetic rate is often reduced, carbohydrate supplies to the fungi may become limited, decreasing root infection or production of external hyphae (Stroo et al. 1988; Jansen and Dighton 1990). Empirical evidence seems to support this hypothesis. In a majority of studies, ozone significantly reduces percent infection of EMF and may damage external hyphae (reviewed in Cairney and Meharg 1999). Kasurinen et al. (1999) noted an initial increase in ecto-mycorrhizal root tips of *Pinus sylvestris* saplings after one year of ozone exposure, but this stimulation had disappeared by the following year. Studies of ozone effects on AMF are sparse. McCool and Menge (1984) reported that ozone exposure reduced percent infection in tomato. Duckmanton and Widden (1994) found no significant effect on overall colonization, but the proportional frequency of arbuscules (versus vesicles, hyphal coils, and internal hyphae) was decreased. It is possible that even though colonization levels tend to drop under ozone exposure, mycorrhizal fungi may still form a substantial enough carbohydrate sink to reduce growth of their host plants. This stress could be responsible for the interactive role of mycorrhiza in exacerbating negative effects of ozone on plant growth, and potentially, ecosystem productivity.

6.3.5 UV Radiation

Depletions in the earth's stratospheric ozone layer are exposing an increasing number of ecosystems to elevated UV-B radiation (Gurney et al. 1993). UV-B can damage the DNA, cell membranes, and photosynthetic mechanisms (e.g., photosystem II, RuBP carboxylase, and chlorophyll content) of plants (reviewed in Caldwell et al. 1989; Rozema et al. 1997). Reductions in photosynthetic capacity may alter the carbon economy of plants and reduce allocation of carbohydrates to mycorrhiza. However, we are aware of few studies that have examined the influence of UV-B radiation on mycorrhizal fungi. Klironomos and Allen (1995) exposed seedlings of *Acer saccharum* to elevated UV-A or UV-B radiation for eight weeks, harvested the plants, and measured AMF infection. While overall percent infection did not change significantly with UV radiation, UV-B light reduced the incidence of arbuscules and increased the abundance of vesicles. The authors suggest that since arbuscules can be sites of nutrient exchange between the fungi and its host plant

(Bonfante-Fasolo 1986), and vesicles are resting structures (Brundrett 1991), these two shifts indicate a decrease in the amount of carbohydrate available to the mycorrhiza. Van de Staaij et al. (2001) found a decrease in percent root colonization with AMF in dune grassland plants after five years of exposure to UV-B. Percent root colonization was reduced by approximately 20%, and the authors also documented a decrease in the amount of arbuscular colonization. In another study, saplings of *Quercus robur* were exposed to elevated UV-A or UV-B light for 2 years with no effect on the frequency of ecto-mycorrhizae (Newsham et al. 1999). Clearly, further study of a potential mycorrhizal response, and its implications for plant and ecosystem processes, is warranted.

6.4 Global Change Factor Interactions

Since the global change factors discussed in the previous section do not occur independently of one another, we must consider how these perturbations interact to affect mycorrhizal dynamics. Of all these factors, elevated CO_2 is the most uniformly distributed and will likely affect more ecosystems than do the others. Globally, its effects on mycorrhizal dynamics may predominate. A subset of habitats (near populated areas) will receive N deposition and/or ozone exposure in addition to CO_2 enrichment. Overall, interactions between elevated CO_2 and the other factors may require the most attention, followed by interactions between N deposition and ozone.

Studies of interactions between elevated CO_2 and nitrogen availability indicate that nitrogen additions may reduce CO_2 effects on mycorrhizal growth in some systems (Treseder and Allen 2000). For example, Klironomos et al. (1997) reported that on *Populus tremuloides* seedlings, AMF hyphal lengths increased with CO_2 in the low N, but not the high N, treatment. Similar results were found with AMF on *Gutierrezia sarothrae* (Rillig and Allen 1998). In *Pinus ponderosa* inoculated with *Pisolithus tinctorus* (an EMF), elevated CO_2 decreased percent infection under high and low N availability, but not at medium levels of N (Walker et al. 1995). Lack of significant interactions on percent infection or hyphal length has also been reported, with both EMF (Tingey et al. 1995; Walker et al. 1997) and AMF (Lussenhop et al. 1998).

Information regarding interactions between ozone and other global change factors is limited. In a *Pinus sylvestris* stand in eastern Finland, elevated CO_2 reduced a slight stimulatory effect of ozone on the number of total mycorrhizal root tips (Kasurinen et al. 1999), while Gorissen et al. (1991) reported no significant interaction between ozone and ammonium sulfate treatments on ecto-mycorrhizal frequency in *Pseudotsuga menziesi*. Finally, in *Pinus sylvestris* saplings, exposure to either ammonia gas or ozone each reduced percent colonization by EMF. If the two treatments were combined, however, infection levels were close to controls. The same study indicated no significant CO_2 interaction with ozone or ammonia on mycorrhizal infection. Additional studies may provide a better understanding of the mechanisms underlying these interactions (or lack of them).

6.5 Long-Term Versus Short Term Responses

While long-term responses to global changes are really of interest, only short-term results are available in many cases, often due to cost and time considerations. This represents a challenge in global change biology, and mycorrhiza studies are no exception. Extrapolating short-term responses to predict longer-term trends is problematic (Luo and Reynolds 1999), because initial responses to a step-change of an experimentally imposed global change factor may not be indicative of a longer-term equilibrium.

A modeling approach to predict long-term responses from initial ones (Luo and Reynolds 1999), as has been used for ecosystem carbon storage, may not be an option in the near future since our knowledge of the inherent mechanisms of the mycorrhizal symbiosis is probably not adequate. In addition, and perhaps more importantly, such a modeling analysis is unlikely to account for potential micro-evolutionary changes in the organisms involved. For plants, population-level studies have been proposed as a way to study possible evolutionary changes to global change (Schmid et al. 1996). However, in mycology, studies of populations are often very difficult (Rayner 1991; Smith et al. 1991; Anderson and Kohn 1998). Additionally, in the case of a symbiosis the genetic variability of host and mycobiont to global change factors would have to be considered. The other option is to use long-term data sets to verify short-term results. Two options present themselves here: long term experiments (e.g., Drake et al. 1996), or the use of natural experiments. While there are probably no natural experiment "analogs" for all global changes, there are for some of them. For example, areas with geothermal or mineral springs that vent CO_2 can be exploited for the study of long-term effects of elevated CO_2 (e.g., Miglietta and Raschi 1993; Newton et al. 1996). Rillig et al. (2000) have shown that long-term exposure of vegetation to elevated CO_2 by a natural CO_2 spring in New Zealand could produce similar AMF responses (increase in percent colonization, hyphal lengths, and glomalin concentration) as were observed in short-term experiments. Native ecosystems of the Tierra del Fuego National Park (southern Argentina), an area of the globe that is frequently under the Antarctic "ozone hole" in early spring, can be used to study effects of elevated UV radiation (e.g., Rousseaux et al. 1998). The study of mycorrhizal responses to global change could profit by taking advantage of these possibilities to compare short-term with potential long-term responses.

6.6 Global Change and Effects on Symbionts

The study of mycorrhizal responses to global change, in addition to the points discussed above, has to face up to an additional challenge: we are dealing not only with two types of organisms (photobiont and mycobiont), living in tight symbiotic association with integrated development (Peterson and Farquhar 1994), but also with partially *independent* biology (Rillig and Allen 1999). In the study of the ecology of mycorrhiza, it is often the integrated aspect of the symbiosis that receives most attention, while the separate biological characteristics are neglected, particularly in the case of arbuscular mycorrhiza (where the fungi are obligate biotrophs).

A classical example for the independence of the ecology of the symbiotic partners is dispersal, which is unlinked for photobiont and fungi. Each forms its own dispersal agents, seeds and spores. This is in contrast to other symbioses, for example some lichen associations, where photo- and mycobiont can disperse together as soredia or isidia (Alexopoulos et al. 1996). Can dispersal of the partners be differentially affected by a global change factor? Possible movement of plant species in response to climate change may be so slow that the fungi have no problem "keeping up" with their hosts. However, exotic plant hosts can be transported quickly over long distances (as seeds) and interact with the fungal community present at the new location in ways that can apparently either increase the success of the invader (Marler et al. 1999) or decrease it (Gange et al. 1999). Hence, in influencing the performance of invasive plant host species, independent dispersal of the symbionts can become very important indeed to global change biology.

Physiologically and architecturally, fungi and the photobionts are obviously quite distinct, as one would expect given that the kingdoms Fungi and Plantae are less likely related than Fungi and Animalia (e.g., Baldauf and Palmer 1993). For example, growth in fungi occurs as tip growth for hyphae, contrasted to meristematic growth for plants. The fungal thallus is made up of hyphae, compared to various tissue types in plants. We, therefore, cannot expect that the fungus and the plant show the same proportional response to any given global change factor. It is equally unlikely that the two partners would respond with the same time constants. This distinction in the biology of the two symbiosis partners becomes particularly important when one considers that some global change factors can *directly* influence the mycobionts, whereas others only influence the mycobionts via their effects on the plant host, as illustrated in Fig. 6.1.

Some global change factors can directly affect only the host plant; the fungus only "perceives" the global change *indirectly* via the host plant. Examples of these are elevated CO_2, UV-B radiation, and ozone, as well as other atmospheric pollutants (Fig. 6.1). This is by virtue of the fact that the fungi are largely isolated from atmospheric gas concentrations and radiation by being imbedded in a medium with its own microclimate: soil. Studies of direct effects of these factors on the fungus are hence generally considered of limited to no importance, and the examination of symbiotic responses is the key.

All global change factors that can directly affect soil and soil biota can directly affect mycorrhizal fungi, most notably their extraradical mycelium. Direct effects on the mycobionts could lead to altered production of fruiting structures (ecto-mycorrhiza) or spores (arbuscular mycorrhiza), quite possibly as a function of fungal species, thus contributing to shifts in fungal community composition. The physiology of hyphae (e.g., growth rate, nutrient uptake or translocation ability) of different fungal species, or their distribution in soil may be differentially affected, thereby leading to altered symbiotic efficiencies of mycobionts. Warming is clearly a global change factor that can affect the above- and below-ground components of ecosystems, and hence plant shoots, roots and the fungi directly. Symbiotic responses to warming are therefore a (so far unknown) function of independent direct effects on each of the partners of the symbiosis, and their interactions. Increased atmospheric nitrogen (or other nutrient, e.g., phosphorus) deposition can also directly affect the host and its roots, and the fungi. Change in precipitation (and accompanying change in soil moisture) is another example of a global change factor that can affect photobiont and mycobiont directly, since it can also act via the soil. In these cases, direct effect studies on the mycobiont(s) are of critical importance for a mechanistic understanding of symbiotic responses. However, these are often limited by the difficulty to examine the fungal response in isolation from the plant's response, for example in the case of arbuscular mycorrhizal fungi.

6.7 Conclusions

The study of global change effects on mycorrhiza represents major challenges. We are dealing with a multitude of factors, and their interactions, calling for a more mechanistic understanding of the symbioses. Global change factors present themselves in two different modes of action: they either affect the mycobionts directly, or only indirectly via the host plant (Fig. 6.1). In the former, it is important to increasingly study direct effects on the mycobiont alone in order to come to a mechanistic understanding of effects. Additionally, it is necessary to often infer long-term responses with only short-term experimental results at hand, where these even exist for a particular factor/type of mycorrhiza combination (Table 6.1). Wherever possible, mycorrhiza research should take advantage of natural experiments that permit the study of longer-term responses to validate short-term results.

Besides mechanistic studies focusing on host-mycobiont(s) interactions, it is also important to keep in mind that mycorrhizal fungi cannot be viewed in isolation in the soil ecosystem. Global change factors that influence other soil biota (for example fungal grazers, mycorrhization helper organisms, and/or other root-inhabiting microbes) or physical factors (such as soil structure) can have potentially large, but as yet unexplored *indirect* effects on the community composition of mycorrhizal fungi and mycorrhizal functioning. At the same time, large-scale processes such as movement of entire plant communities (or their regional extinction) due to climate change could certainly affect mycorrhiza, for example, alter the current distribution of mycorrhizal types on the globe or lead to regionally or globally extinguished mycorrhizal fungal biodiversity.

References

Agrios GN (1997) Plant pathology. 4th edn. Academic Press, San Diego
Alexopoulos CJ, Mims CW, Blackwell M (1996) Introductory Mycology. 4th edn. Wiley, New York
Allen EB, Allen MF (1986) Water relations of xeric grasses in the field: Interactions of mycorrhiza and competition. New Phytol 104:559–571
Allen EB, Allen MF, Helm DJ, Trappe JM, Molina R, Rincon E (1995) Patterns and regulation of arbuscular and ectomycorrhizal plant and fungal diversity. Plant Soil 170:47–62
Allen MF (1983) Formation of vesicular-arbuscular mycorrhizae in *Atriplex gardneri* (Chenopodiaceae): seasonal response in a cold desert. Mycologia 75:773–776
Allen MF (1991) The ecology of mycorrhiza. Cambridge Univ Press, Cambridge
Allen MF, Smith WK, Moore TS Jr, Christensen M (1981) Comparative water relations and photosynthesis of mycorrhizal and nonmycorrhizal *Bouteloua gracilis* H.B.K. Lag ex Steud. New Phytol 88:683–693
Allen MF, Moore TS Jr, Christensen M (1982) Phytohormone changes in *Bouteloua gracilis* infected by vesicular-arbuscular mycorrhiza II. Altered levels of gibberellin-like substances and abscisic acid in the host plant. Can J Bot 60:468–471
Allen MF, Allen EB, West NB (1987) Influence of parasitic and mutualistic fungi on *Artemisia tridentata* during high precipitation years. Bull Torrey Bot Club 114:272–279
Allen MF, Richards JH, Busso CA (1989) Influence of clipping and water status on vesicular-arbuscular mycorrhiza of two semiarid tussock grasses. Biol Fertil Soils 8:285–289
Allen MF, Morris SJ, Edwards F, Allen EB (1995) Microbe-plant interactions in Mediterranean-type habitats: shifts in fungal symbiotic and saprophytic functioning in response to global change. In: Moreno JM, Oechel WC (eds) Global change and Mediterranean-type ecosystems. Ecological studies 117. Springer, Berlin Heidelberg New York, pp 287–305
Ames RN, Reid CPP, Porter LK, Cambardella C (1983) Hyphal uptake and transport of nitrogen from two ^{15}N-labeled sources by *Glomus mosseae*, a vesicular arbuscular mycorrhizal fungus. New Phytol 95:381–396
Amthor JS (1995) Terrestrial higher-plant responses to increasing atmospheric [CO_2] in relation to the global carbon cycle. Global Change Biol 1:243–274
Andersen CP, Rygiewicz PT (1995) Allocation of carbon in mycorrhizal *Pinus ponderosa* seedlings exposed to ozone. New Phytol 131:471–480

Anderson JB, Kohn LM (1998) Genotyping, gene genealogies and genomics bring fungal population genetics above ground. Trends Ecol Evol 13:444–449

Angers DA, Mehuys GR (1993) Aggregate stability to water. In: Carter MR (ed) Soil sampling and methods of analysis. Lewis Publishers, Boca Raton, pp 651–658

Arnebrant K (1994) Nitrogen amendments reduce the growth of extramatrical ectomycorrhizal mycelium. Mycorrhiza 5:7–15

Arnebrant K, Soderstrom B (1992) Effects of different fertilizer treatments on ectomycorrhizal colonization potential in two Scots Pine forests in Sweden. For Ecol Manage 53:77–89

Arnolds E (1988) The changing macromycete flora in the Netherlands. Trans Br Mycol Soc 90: 391–406

Arnolds E (1991) Decline of ectomycorrhizal fungi in Europe. Agricult Ecosyst Environ 35:209–244

Augé R (2001) Water relations, drought and vesicular-arbuscular mycorrhizal symbiosis. Mycorrhiza 11:3–42

Baldauf SL, Palmer JD (1993) Animals and fungi are each other's closest relatives: congruent evidence from multiple proteins. Proc Natl Acad Sci USA 90:11558–11562

Barber VA, Juday GP, Finney BP (2000) Reduced growth of Alaskan white spruce in the twentieth century from temperature-induced drought stress. Nature 405:668–673

Bazzaz FA (1990) The response of natural ecosystems to the rising global CO_2 levels. Annu Rev Ecol Syst 21:167–196

Bever JD (1994) Feedback between plants and their soil communities in an old field community. Ecology 75:1965–1977

Bloom AJ, Chapin FS III, , Mooney HA (1985) Resource limitation in plants – an economic analogy. Annu Rev Ecol Syst 16:363–393

Bonello P, Heller W, Sandermann H (1993) Ozone effects on root disease susceptibility and defense responses in mycorrhizal and nonmycorrhizal seedlings of Scots Pine (*Pinus sylvestris* L). New Phytol 124:653–663

Bonfante-Fasolo P (1986) Anatomy and morphology of VA mycorrhiza. In: Powell C, Bagyaraj D (eds) VA Mycorrhiza. CRC Press, Boca Raton, pp 2–33

Brundrett M (1991) Mycorrhizas in natural ecosystems. Adv Ecol Res 21:171–313

Cairney JWG, Meharg AA (1999) Influences of anthropogenic pollution on mycorrhizal fungal communities. Environ Pollut 106:169–182

Caldwell MM, Flint SD (1994) Stratospheric ozone reduction, solar UV-B radiation and terrestrial ecosystems. Climatic Change 28:375–394

Caldwell MM, Teramura AH, Tevini M (1989) The changing solar ultraviolet climate and the ecological consequences for higher plants. Trends Ecol Evol 4:363–367

Caporn SJM, Song W, Read DJ, Lee JA (1995) The effect of repeated nitrogen fertilization on mycorrhizal infection in heather [*Calluna vulgaris* (L.) Hull]. New Phytol 129:605–609

Coleman DC, Odum EP, Crossley DA (1992) Soil biology, soil ecology and global change. Biol Fertil Soils 14:104–111

Colpaert JV, van Tichelen KK (1996) Mycorrhizas and environmental stress. In: Frankland JC, Magan N, Gadd GM (eds) Fungi and environmental change. Cambridge Univ Press, Cambridge, pp 109–128

Dai A, Trenberth KE, Karl TR (1999) Effects of clouds, soil moisture, precipitation, and water vapor on diurnal temperature range. J Climate 12:2451–2473

Daniell TJ, Hodge A, Young JPW, Fitter A (1999) How many fungi does it take to change a plant community? Trends Plant Sci 4:81–82

D'Antonio CM, Vitousek PM (1992) Biological invasions by exotic grasses, the grass/fire cycle, and global change. Annu Rev Ecol Syst 23:63–87

Di JJ, Allen EB (1991) Physiological responses of 6 wheatgrass cultivars to mycorrhiza. J Range Manage 44:336–341

Diaz S (1996) Effects of elevated [CO_2] at the community-level mediated by root symbionts. Plant Soil 187:309–320

Drake BG, Peresta G, Beugeling E, Matamala R (1996) Long term elevated CO_2 exposure in a Chesapeake Bay wetland: ecosystem gas exchange, primary production, and tissue nitrogen. In: Koch GW, Mooney HA (eds) Carbon dioxide and terrestrial ecosystems. Academic Press, San Diego, pp 197–214

Duckmanton L, Widden P (1994) Effect of ozone on the development of vesicular-arbuscular mycorrhiza in sugar maple saplings. Mycologia 86:181–186

Dukes JS, Mooney HA (1999) Does global change increase the success of biological invaders? Trends Ecol Evol 14:135–139

Edwards FE (1985) Indicators of low intensity ecosystem perturbations: shifts in herbaceous plants and arbuscular mycorrhizal fungi in two semi-arid ecosystems. MS Thesis, San Diego State University

Egerton-Warburton LM, Allen EB (2000) Shifts in arbuscular mycorrhizal communities along an anthropogenic nitrogen deposition gradient. Ecol Appl 10:484–496

Elliott ET, Coleman DC (1988) Let the soil work for us. Ecol Bull 39:23–32

Eom A-H, Hartnett DC, Wilson GWT, Figge DAH (1999) The effect of fire, mowing and fertilizer amendment on arbuscular mycorrhizas in tallgrass prairie. Am Midland Nat 142:55–70

Faber BA, Zasoski RJ, Munns DN, Shackel K (1991) A method for measuring hyphal nutrient and water uptake in mycorrhizal plants. Can J Bot 69:87–94

Field CB, Chapin FS, Matson PA, Mooney HA (1992) Reponses of terrestrial ecosystems to the changing atmosphere: a resource-based approach. Annu Rev Ecol Syst 23:201–236

Fitter AH, Self GK, Brown TK, Bogie DS, Graves JD, Benham D, Ineson P (1999) Root production and turnover in an upland grassland subjected to artificial soil warming respond to radiation flux and nutrients, not temperature. Oecologia 120:575–581

Gange AC, Lindsay DE, Ellis LS (1999) Can arbuscular mycorrhizal fungi be used to control the undesirable grass Port annua on golf courses? J Appl Ecol 36:909–919

Gorissen A, Joosten NN, Jansen AE (1991) Effects of ozone and ammonium sulfate on carbon partitioning to mycorrhizal roots of juvenile Douglas fir. New Phytol 119:243–250

Gorissen A, Kuyper TW (2000) Fungal species-specific responses of ectomycorrhizal Scots pine (*Pinus sylvestris*) to elevated [CO_2]. New Phytol 146:163–168

Gurney RJ, Foster JL, Parkinson CL (1993) Atlas of satellite observations related to global change. Cambridge Univ Press, Cambridge

Hartge KH, Stewart BA (1995) Soil structure: its development and function. Advances in soil sciences. CRC/Lewis Publishers, Boca Raton

Heagle AS (1989) Ozone and crop yield. Annu Rev Phytopathol 27:397–423

Heath RL (1998) Oxidant induced alteration of carbohydrate production and allocation in plants. In: Bytnerowicz A, Arbaugh MJ, Schilling SL (eds) Proceedings of the international symposium on air pollution and climate change effects on forest ecosystems. Albany, California, Pacific Southwest Research Station, Forest Service, USDA, Riverside, CA, pp 11–18

Hodge A (1996) Impact of elevated CO_2 on mycorrhizal associations and implications for plant-growth. Biol Fertil Soil 23:388–398

Hughes L (2000) Biological consequences of global warming: is the signal already apparent. Trends Ecol Evol 15:56–61

Jansen AE, Dighton J (1990) Effects on air pollutants on ectomycorrhiza. A review. Air Pollut Res Rep 30:1–58

Jastrow JD, Miller RM (1997) Soil aggregate stabilization and carbon sequestration: feedbacks through organomineral associations. In: Lal R, Kimble JM, Follett RF, Stewart BA (eds) Soil processes and the carbon cycle. CRC Press, Boca Raton, pp 207–223

Johansson M (1992) Effects of nitrogen application on ericoid mycorrhizas of *Calluna vulgaris* on a Danish heathland. In: Read DJ, Lewis DH, Fitter AH, Alexander IJ (eds) Mycorrhizas in ecosystems. CAB International Wallingford, p 386

Johnson NC (1993) Can fertilization of soil select less mutualistic mycorrhiza. Ecol Appl 3:749–757

Jonasson S, Shaver GR (1999) Within-stand nutrient cycling in arctic and boreal wetlands. Ecology 80:2139–2150

Jongen M, Fay P, Jones MB (1996) Effects of elevated carbon dioxide and arbuscular mycorrhizal infection on *Trifolium repens*. New Phytol 132:413–423

Kandeler E, Tscherko D, Bardgett RD, Hobbs PJ, Kampichler C, Jones TH (1998) The response of soil microorganisms and roots to elevated CO_2 and temperature in a terrestrial model ecosystem. Plant Soil 202:251–262

Karen O, Nylund JE (1997) Effects of ammonium-sulfate on the community structure and biomass of ectomycorrhizal fungi in a Norway Spruce stand in southwestern Sweden. Can J Bot 75:1628–1642

Kasurinen A, Helmisaari HS, Holopainen T (1999) The influence of elevated CO_2 and O_3 on fine roots and mycorrhizas of naturally growing young Scots pine trees during three exposure years. Global Change Biol 5:771–780

Keeling CD, Whorf TP, Wahlen M, van der Plicht J (1995) Interannual extremes in the rate of rise of atmospheric carbon dioxide since 1980. Nature 375:666–670

Klironomos JN, Allen MF (1995) UV-B-mediated changes on belowground communities associated with the roots of *Acer saccharum*. Funct Ecol 9:923–930

Klironomos JN, Rillig MC, Allen MF, Zak DR, Kubiske M, Pregitzer KS (1997) Soil fungal-arthropod responses to *Populus tremuloides* grown under enriched atmospheric CO_2 under field conditions. Global Change Biol 3:473–478

Lee JA, Caporn SJM, Read DJ (1992) Effects of increasing nitrogen deposition and acidification on heathlands. In: Schneider T (ed) Acidification research, evaluation and policy applications. Elsevier, Amsterdam, pp 97–106

Lumsden PJ (ed) (1997) Plants and UV-B: responses to environmental change. Society for Experimental Biology (Great Britain) 64. Cambridge Univ Press, Cambridge

Luo Y, Reynolds JF (1999) Validity of extrapolating field CO_2 experiments to predict carbon sequestration in natural ecosystems. Ecology 80:1568–1583

Lussenhop J, Treonis A, Curtis PS, Teeri JA, Vogel CS (1998) Response of soil biota to elevated atmospheric CO_2 in poplar model systems. Oecologia 113:247–251

Lyles LL, Hagen LJ, Skidmore EL (1983) Soil conservation: Principles of erosion by wind. In: Dregne HE, Willis WW (eds) Dryland agriculture. American Society of Agronomy, Madison WI, pp 177–188

Marler MJC, Zabinski C, Callaway R (1999) Mycorrhiza indirectly enhance competitive effects of an invasive forb on a native bunchgrass. Ecology 80:1180–1186

Matson PA, McDowell WH, Townsend AR, Vitousek PM (1999) The globalization of N deposition: ecosystem consequences in tropical environments. Biogeochemistry 46:67–83

McCool PM, Menge JA (1984) Interaction of ozone and mycorrhizal fungi on tomato as influenced by fungal species and host variety. Soil Biol Biochem 16:425–427

McQuattie CJ, Schier GA (1992) Effect of ozone and aluminum on pitch pine (*Pinus rigida*) seedlings: anatomy of a mycorrhiza. Can J For Res 22:1901–1916

Menge JA, Grand LF (1978) Effect of fertilization on production of epigeous basidiocarps by mycorrhizal fungi in loblolly-pine plantations. Can J Bot 56:2357–2362

Miglietta F, Raschi A (1993) Studying the effects of elevated CO_2 in the open in a naturally enriched environment in Central Italy. Vegetatio 104/105:391–400

Miller SP, Bever JD (1999) Distribution of arbuscular mycorrhizal fungi in stands of the wetland grass *Panicum hemitomon* along a wide hydrologic gradient. Oecologia 119:586–592

Miller JE, Shafer SR, Schoeneberger MM, Pursley WA, Horton SJ, Davey CB (1997) Influence of a mycorrhizal fungus and/or rhizobium on growth and biomass partitioning of subterranean clover exposed to ozone. Water Air Soil Pollut 96:233–248

Monz CA, Hunt HW, Reeves FB, Elliott ET (1994) The response of mycorrhizal colonization to elevated CO_2 and climate change in *Pascopyrum smithii* and *Bouteloua gracilis*. Plant Soil 165:75–80

Mullen RB, Schmidt SK (1993) Mycorrhizal infection, phosphorus uptake, and phenology in *Ranunculus adoneus*: implications for the functioning of mycorrhiza in alpine systems. Oecologia 94:229–234

Mullen RB, Schmidt SK, Jaeger CH (1998) Nitrogen uptake during snowmelt by the snow buttercup, *Ranunculus adoneus*. Arct Alp Res 30:121–125

Nadelhoffer KJ, Emmett BA, Gundersen P, Kjonaas OJ, Koopmans CJ, Schleppi P, Tietemal A, Wright RF (1999) Nitrogen deposition makes a minor contribution to carbon sequestration in temperate forests. Science 398:145–147

Newsham KK, Fitter AH, Watkinson AR (1995) Multi-functionality and biodiversity in arbuscular mycorrhizas. Trends Ecol Evol 10:407–411

Newsham KK, Greenslade PD, McLeod AR (1999) Effects of elevated ultraviolet radiation on *Quercus robur* and its insect and ectomycorrhizal associates. Global Change Biol 5:881–890

Newton PCD, Bell CC, Clark H (1996) Carbon dioxide emissions from mineral springs in Northland and the potential of these sites for studying the effects of elevated carbon dioxide on pastures. N Z J Agric Res 39:33–40

Nilsen P, Borja I, Knutsen H, Brean R (1998) Nitrogen and drought effects on ectomycorrhiza of Norway spruce (*Picea abies* L. (Karst.)). Plant Soil 198:179–184

Oades JM (1984) Soil organic matter and structural stability: mechanisms and implications for management. Plant Soil 76:319–337

Oechel WC, Callaghan T, Gilmanov T, Hiolten JI, Maxwell B, Molau U, Sveinbjoernsson B (1997) Global and regional patterns of climate change: recent predictions for the Arctic. In: Rowntree PR (ed) Global change and Arctic terrestrial ecosystems. Ecological studies. Springer, Berlin Heidelberg New York, pp 82–109

Olesniewicz KS, Thomas RB (1999) Effects of mycorrhizal colonization on biomass production and nitrogen fixation of black locust (*Robinia pseudoacacia*) seedlings grown under elevated atmospheric carbon dioxide. New Phytol 142:133–140

O'Neill EG, O'Neill RV, Norby RJ (1991) Hierarchy theory as a guide to mycorrhizal research on large-scale problems. Environ Pollut 73:271–284

Pattinson GS, Hammill KA, Sutton BG, McGee PA (1999) Simulated fire reduces the density of arbuscular mycorrhizal fungi at the soil surface. Mycol Res 103:491–496

Peterson RL, Farquhar ML (1994) Mycorrhizas – integrated development between roots and fungi. Mycologia 86:311–326

Pounds JA, Fogden MPL, Campbell JH (1999) Biological response to climate change on a tropical mountain. Nature 398:611–615

Pritchard SG, Rogers HH, Prior SA, Peterson CM (1999) Elevated CO_2 and plant structure: a review. Global Change Biol 5:807–837

Rayner ADM (1991) The challenge of the individualistic mycelium. Mycologia 83:48–71

Read DJ (1983) The biology of mycorrhiza in the Ericales. Can J Bot 61:985–1004

Read DJ (1991) Mycorrhizas in ecosystems – Nature's response to the "Law of the minimum". In: Hawksworth DL (ed) Frontiers in mycology. CAB International, Regensburg, pp 101–130

Rillig MC, Allen MF (1998) Arbuscular mycorrhiza of *Gutierrezia sarothrae* and elevated carbon dioxide: evidence for shifts in C allocation to and within the mycobiont. Soil Biol Biochem 30:2001–2008

Rillig MC, Allen MF (1999) What is the role of arbuscular mycorrhizal fungi in plant-to-ecosystem responses to elevated atmospheric CO_2? Mycorrhiza 9:1–8

Rillig MC, Allen MF, Klironomos JN, Field CB (1998) Arbuscular mycorrhizal percent root infection and infection intensity of *Bromus hordeaceus* grown in elevated atmospheric CO_2. Mycologia 90:199–205

Rillig MC, Field CB, Allen MF (1999a) Soil biota responses to long-term atmospheric CO_2 enrichment in two California annual grasslands. Oecologia 119:572–577

Rillig MC, Wright SF, Allen MF, Field CB (1999b) Long-term CO_2 elevation affects soil structure of natural ecosystems. Nature 400:628

Rillig MC, Hernandez GY, Newton PCD (2000) Arbuscular mycorrhiza respond to elevated atmospheric CO_2 after long-term exposure: evidence from a CO_2 spring in New Zealand supports the resource-balance model. Ecol Lett 3:475–478Rouhier H, Read DJ (1998a) Plant and fungal responses to elevated atmospheric carbon dioxide in mycorrhizal seedlings of *Pinus sylvestris*. Environ Exp Bot 40:237–246

Rouhier H, Read DJ (1998b) The role of mycorrhiza in determining the response of *Plantago lanceolata* to CO_2 enrichment. New Phytol 139:367–373

Rousseaux MC, Ballare CL, Scopel AL, Searles PS, Caldwell MM (1998) Solar ultraviolet B radiation affects plant insect interactions in a natural ecosystem of Tierra del Fuego (southern Argentina). Oecologia 116:528–535

Rozema J, van de Staaij J, Bjorn LO, Caldwell M (1997) UV-B as an environmental factor in plant life: stress and regulation. Trends Ecol Evol 12:22–28

Sandermann H (1996) Ozone and plant health. Annu Rev Phytopathol 34:347–366

Sanders IR, Streitwolf-Engel R, van der Heijden MGA, Boiler T, Wiemken A (1998) Increased allocation to external hyphae of arbuscular mycorrhizal fungi under CO_2 enrichment. Oecologia 117:496–503

Santer BD, Wigley TML, Barnett TP, Anayamba E (1996) Detection of climate change and attribution of causes. In: Houghton J, Meira Filho LG, Callander BA, Harris N, Kattenberg A, Maskell K (eds) Climate change 1995: science of climate change. Cambridge Univ Press, Cambridge, pp 407–444

Schmid B, Birrer A, Lavigne C (1996) Genetic variation in the response of plant populations to elevated CO_2 in a nutrient-poor, calcareous grassland. In: Körner C, Bazzaz FA (eds) Carbon dioxide, populations, and communities. Academic Press, San Diego, pp 31–50

Seegmuller S, Rennenberg H (1994) Interactive effects of mycorrhization and elevated carbon dioxide on growth of young pedunculate oak (*Quercus robur* L) trees. Plant Soil 167:325–329

Smith ML, Bruhn JN, Anderson JB (1992) The fungus *Armillaria bulbosa* is among the largest and oldest living organisms. Nature 356:428–431

Smith SE, Read DJ (1997) Mycorrhizal symbiosis. Academic Press, San Diego

Staddon PL, Fitter AH (1998) Does elevated atmospheric carbon dioxide affect arbuscular mycorrhizas? Trends Ecol Evol 13:455–458

Staddon PL, Fitter AH, Graves JD (1999a) Effect of elevated atmospheric CO_2 on mycorrhizal colonization, external mycorrhizal hyphal production and phosphorus inflow in *Plantago lanceolata* and *Trifolium repens* in association with the arbuscular mycorrhizal fungus *Glomus mosseae*. Global Change Biol 5:347–358

Staddon PL, Fitter AH, Robinson D (1999b) Effects of mycorrhizal colonization and elevated atmospheric carbon dioxide on carbon fixation and below-ground carbon partitioning in *Plantago lanceolata*. J Exp Bot 50:853–860

Stribley DP, Read DJ (1976) Biology of mycorrhiza in Ericaceae 6. Effects of mycorrhizal infection and concentration of ammonium nitrogen on growth of cranberry (*Vaccinium macrocarpon* Ait) in sand culture. New Phytol 77:63–72

Stroo HF, Reich PB, Schoettle AW, Amundson RG (1988) Effects of ozone and acid-rain on white-pine (*Pinus-Strobus*) seedlings grown in 5 soils 2. Mycorrhizal infection. Can J Bot 66:1510–1516

Taylor AFS, Alexander IJ (1989) Demography and populations dynamics of ectomycorrhizas of Sitka spruce fertilized with N. Ecosyst Environ 28:493–496

Termorshuizen AJ (1993) The influence of nitrogen fertilizers on ectomycorrhizas and their fungal carpophores in young stands of *Pinus sylvestris*. For Ecol Manage 57:179–189

Thormann MN, Currah RS, Bayley SE (1999) The mycorrhizal status of the dominant vegetation along a peatland gradient in southern boreal Alberta, Canada. Wetlands 19:438–450

Tingey DT, Johnson MG, Phillips DL, Storm MJ (1995) Effects of elevated CO_2 and nitrogen on ponderosa pine fine roots and associated fungal components. J Biogeogr 22:281–287

Tisdall JM, Oades JM (1982) Organic matter and water-stable aggregates in soils. J Soil Sci 33:141–163

Tobar R, Azcon R, Barea JM (1994) Improved nitrogen uptake and transport from ^{15}N-labeled nitrate by external hyphae of arbuscular mycorrhiza under water-stressed conditions. New Phytol 126:119–122

Treseder KK, Allen MF (2000) The potential role of mycorrhizal fungi in soil carbon storage under elevated CO_2 and nitrogen deposition. New Phytol 147:189–200

van der Heijden MGA, Klironomos JN, Ursic M, Moutoglis P, Streitwolf-Engel R, Boller T, Wiemken A, Sanders IR (1998) Mycorrhizal fungal diversity determines plant biodiversity, ecosystem variability and productivity. Nature 396:69–72

van de Staaij J, Rozeman J, van Beem A, Aerts R (2001) Increased solar UV-B radiation may reduce infection by arbuscular mycorrhizal fungi (AMF) in dune grassland plants; evidence from five year field exposure. Plant Ecol 154:169–177

Virginia RA, Wall DH (1999) How soils structure communities in the Antarctic Dry Valleys. BioScience 49:973–983

Vitousek PM (1994) Beyond global warming: ecology and global change. Ecology 75:1861–1876

Vitousek PM, Aber JD, Howarth RW, Likens GE, Matson PA, Schindler DW, Schlesinger WH, Tilman DG (1997) Human alteration of the global nitrogen cycle: sources and consequences. Ecol Appl 7:737–750

Walker RF, Geisinger DR, Johnson DW, Ball JT (1995) Interactive effects of atmospheric CO_2 enrichment and soil N on growth and ectomycorrhizal colonization of ponderosa pine seedlings. For Sci 41:491–500

Walker RF, Geisinger DR, Johnson DW, Ball JT (1997) Elevated atmospheric CO_2 and soil N fertility effects on growth, mycorrhizal colonization, and xylem water potential of juvenile ponderosa pine in a field soil. Plant Soil 195:25–36

Wallander H (1995) A new hypothesis to explain allocation of dry matter between mycorrhizal fungi and pine seedlings in relation to nutrient supply. Plant Soil 169:243–248

Wallander H, Nylund JE (1992) Effects of excess nitrogen and phosphorus starvation on the extramatrical mycelium of ectomycorrhizas of *Pinus sylvestris* L. New Phytol 120:495–503

Wallander H, Arnebrant K, Dahlberg A (1999) Relationships between fungal uptake of ammonium, fungal growth and nitrogen availability in ectomycorrhizal *Pinus sylvestris* seedlings. Mycorrhiza 8:215–223

Wallenda T, Kottke I (1998) Nitrogen deposition and ectomycorrhizas. New Phytol 139:169–187

Wright SF, Upadhyaya A (1996) Extraction of an abundant and unusual protein from soil and comparison with hyphal protein of arbuscular mycorrhizal fungi. Soil Sci 161:575–586

Wright SF, Upadhyaya A (1998) A survey of soils for aggregate stability and glomalin, a glycoprotein produced by hyphae of arbuscular mycorrhizal fungi. Plant Soil 198:97–107

Wright SF, Franke-Snyder M, Morton JB, Upadhyaya A (1996) Time-course study and partial characterization of a protein on hyphae of arbuscular mycorrhizal fungi during active colonization of roots. Plant Soil 181:193–203

Yesmin L, Gammack SM, Cresser MS (1996) Effects of atmospheric nitrogen deposition on ericoid mycorrhizal infection of *Calluna vulgaris* growing in peat. Appl Soil Ecol 4:49–60

Young IM, Blanchart E, Chenu C, Dangerfield M, Fragoso C, Grimaldi M, Ingram J, Monrozier L (1998) The interaction of soil biota and soil structure under global change. Global Change Biol 4:703–712

Section C:

Biodiversity, Plant and Fungal Communities

7 Diversity of Ecto-mycorrhizal Fungal Communities in Relation to the Abiotic Environment

SUSANNE ERLAND, ANDY F.S. TAYLOR

Contents

7.1	Summary	163
7.2	Introduction	164
7.3	Natural Abiotic Factors Influencing Ecto-mycorrhizal Community Structure	172
7.3.1	Edaphic Factors	172
7.3.2	Soil Organic Matter and Spatial Heterogeneity	173
7.3.3	Moisture	175
7.3.4	pH	176
7.3.5	Temperature	177
7.3.6	Wildfire	177
7.4	Effects of Pollution and Forest Management Practices on Ecto-mycorrhizal Fungal Communities	179
7.4.1	Elevated CO_2	179
7.4.2	Ozone	180
7.4.3	Heavy Metals	180
7.4.4	N-Deposition and Fertilisation	181
7.4.5	Acidification	184
7.4.6	Liming	185
7.4.7	Wood Ash Application and Vitality Fertilisation	186
7.5	Conclusions	187
7.6	Future Progress	192
References		193

7.1 Summary

In boreal forest ecosystems, the richness and complexity of ecto-mycorrhizal (EM) fungal communities are in striking contrast to the often species-poor stands of host trees. The factors that influence community development and maintain this high EM fungal diversity are, however, poorly understood. There are very few studies that have examined determinants of EM fungal diversity under natural undisturbed systems, with most studies examining diversity in relation to changes in abiotic factors due to pollution and/or forest management practices. In this chapter, we attempt to compile what little data are available on natural factors and suggest some areas for future studies. The great majority of the chapter is however, concerned with anthropogenic influences upon EM fungal diversity.

Typically, the below ground EM fungal community consists of a few common species, colonising 50–70 % of the available fine roots, and a large number of rarer species. A common feature of EM community response to changes due to anthropogenic factors is a shift in the community structure such that dominance increases and species richness declines. Additionally, new species may appear as important colonisers after a perturbation. However, we suggest that until we have a better understanding of the spatial distribution and have developed sampling strategies to deal with the non-random distribution of EM fungi in soil, results suggesting changes in species richness should be treated with caution. Root tip density can change markedly following a perturbation and since the number of root tips in a sample can significantly affect the number of EM fungal species found, sampling strategies must be able to accommodate this. Soil acidification and N-fertilisation typically result in lower fine root densities and lower below ground species richness. Conversely, liming and elevated CO_2 often result in higher fine root densities and higher below ground species richness. We have included a brief discussion on how some of the structural and ecological features of individual EM fungal species may affect their response to changes in the abiotic environment. The significance of shifts in EM fungal diversity at the ecosystem level remains unclear due to a lack of knowledge on the functional capabilities, under field conditions, of most EM fungal taxa. It is known, however, that considerable interspecific variation exists with regard to a number of physiological attributes. Changes in dominance may therefore have measurable effects upon the nutritional status of the host.

It is suggested that the extramatrical mycelium (EMM) is likely to be the component of the belowground EM community that is most sensitive and responsive to environmental change. There are however, as yet, no reliable methods available to study the EMM of individual EM species under field conditions.

7.2 Introduction

It has been hypothesised that morphological and physiological differences between mycorrhizal types determine the success of the associated plants in different soils and have led to distinctive distribution patterns of the major mycorrhizal symbioses in different biomes (Janos 1980; Read 1991). Key environmental factors that are thought to favour and delimit the development of ecosystems dominated by different mycorrhizal associations have been discussed by Smith and Read (1997; see also Read, Chap. 1, this Vol.). In temperate and boreal forest soils, one of the most abundant and diverse groups of organisms are the obligately symbiotic ecto-mycorrhizal (EM) fungi. The richness and complexity of EM fungal communities is usually in striking contrast to the inherent simplicity of the often species-poor stands of host trees. Even within small areas (<1 ha), a large number of EM fungi can be found, with individual trees supporting many different fungal species simultaneously, even on the same root tip.

In the first real attempt to identify some of the factors that may contribute to the high diversity of EM fungi, Bruns (1995) suggested that resource partitioning, disturbance, competition and interactions with other soil organisms were important. Although several testable hypotheses were presented in relation to these factors, there has been very limited progress in the intervening 5 years in elucidating determinants of diversity. In this respect, mycorrhizal research lags behind that on soil fauna ecology, where considerable experimental and theoretical work has been carried out on both biotic and abiotic factors which influence species richness (see Giller 1996).

Two factors that have undoubtedly hindered progress are the identification and quantification of the EM community below ground. Gardes et al. (1991) first used molecular techniques to identify EM fungi on root tips and this development has led to a rapid increase in the number of studies which have directly examined EM fungal communities on host root systems. However, the great majority of these studies have involved perturbations of systems and an examination of the community response to that disturbance. While these studies may indicate which factors, when altered, influence diversity, they do not necessarily provide evidence for establishing the factors as important determinants of diversity prior to the disturbance. Due to the lack of available literature, this chapter will therefore be primarily concerned with summarising EM fungal community data derived from below ground studies in which the effect of a disturbance has been examined.

Community diversity is usually considered to have two components, number of species or species richness and the relative abundance of species or community evenness (Magurran 1988). As a result of the non-random spatial distribution of fungal species within the soil profile, our ability to detect changes within these two components may differ significantly. We have therefore tried, where possible, to distinguish between these two components of diversity in our discussions.

Ecto-mycorrhizal fungal community diversity and the importance of individual fungal species have traditionally been assessed using sporocarp production (e.g. Termorshuizen 1991; Brunner et al. 1992; Cripps and Miller

1993). Over the last 15 years, however, there has been a monumental increase in our ability to examine EM fungal communities directly on the roots of the host plants. Using a combination of morphological characterisation (Agerer 1986–1998) and molecular identification methodologies (Egger 1995), many significant advances have been made. One of the most important findings from studies using these techniques is that the picture obtained aboveground from the sporocarps is usually a poor indicator of the community structure below ground (Mehmann et al. 1995; Gardes and Bruns 1996; Kårén and Nylund 1996; Kraigher et al. 1996; Dahlberg et al. 1997; Pritsch et al. 1997; T. Jonsson et al. 1999). The reasons for this disparity are numerous and have been discussed previously (Egger 1995), but they are repeated here since many are relevant to this chapter.

The ephemeral nature and the irregularity with which many species produce sporocarps means that unless recording is carried out regularly and frequently (one to two times per week) over an extended period of time (4–5 years), many species capable of producing visible sporocarps will remain undetected (Richardson 1970; Lange 1978). Species which produce hypogeous (belowground) or inconspicuous sporocarps also present considerable sampling problems (Fogel 1976; Adams et al. 1978). Some species which are generally considered to be saprotrophic fungi may in fact be mycorrhizal and will be overlooked during an analysis of EM fungal sporocarps (Erland and Taylor 1999).

The density of root tips in upper organic layers of boreal forests is typically in the region of $2-4\times10^6$ m^2 (Dahlberg et al. 1997). It is, therefore, impossible to analyse more than a small fraction of the total EM fungal community on the root tips and many species must go unrecorded. In addition, the heterogeneity within soil is high (see Sect. 7.3.2), and the distribution of species in both space and time makes representative sampling very difficult. All of these reasons may lead to different estimates of species richness derived from above- and belowground assessments of a single EM fungal community. It is, however, the numerical relationship between sporocarps and the associated mycorrhizal tips that leads to the greatest errors in estimates of species abundance or community evenness. Gardes and Bruns (1996) found that sporocarp production by *Suillus pungens* and *Amanita franchetii* was prolific in a stand of *Pinus muricata* but the relative abundance of mycorrhizae formed by these two species was low. By contrast, mycorrhizal root tips formed by *Russula amoenolens* made up nearly 30 % of the total number of mycorrhizal root tips that had been analysed. However, the numbers of sporocarps produced by this species was small.

The structure of most EM fungal communities, as analysed on the host roots, seems be typical of most other species-rich communities (Putnam 1994). A small number of species, ca. five to ten, are relatively abundant and typically colonise ca. 50–70 % of the root tips. It seems reasonable to assume that these species are those best adapted to the prevailing soil conditions at the time of sampling and that they are of more importance to the nutritional status of the host plant than rarer species. The remaining species, which usually make up the great majority of the species richness, occur with varying degrees of rareness (Visser 1995; Kårén and Nylund 1996; Erland et al. 1999; L. Jonsson et al. 1999; Mahmood et al. 1999; Taylor et al. 2000). However, it is worth recalling that, due to the high root tip density, even species which constitute only 0.1 % of the EM community may still be present on $2-4\times10^3$ mycorrhizae m^{-2} of forest floor.

Interpretation of the functional significance of a change in EM community diversity to the host plants and indeed to the ecosystem as a whole is currently constrained by a lack of knowledge concerning the functional capabilities under field conditions of most EM taxa. However, laboratory experiments have established that there is considerable interspecific variation with regard to a number of physiological attributes including phosphorus uptake (Table 2.1 in Simard et al., Chap. 2, this Vol. Thomson et al. 1994), uptake kinetics of amino acids (Wallenda and Read 1999), efficiency in transferring acquired N to the host (Colpaert et al. 1996) and the ability to access N and P bound in organic matter (Leake and Read 1997). Therefore, shifts in community structure that alter the most abundant species could potentially lead to a change in nutrient acquisition and subsequent transfer to the host plant. In addition, it is likely that EM fungi differ markedly in their carbon use efficiency. Shifts in dominance or species composition which therefore affect the overall carbon use efficiency of the community could, given the magnitude of the carbon flow into EM fungi (10-20 % of NPP; Simard et al., Chap. 2, this Vol.; Smith and Read 1997), have important implications for the carbon balance of the forest ecosystem. There is recent evidence that the possibility of a forest being a net source or sink of carbon may be very finely balanced (Goulden et al. 1998).

This chapter is split into two parts. The first part attempts to bring together a number of studies where diversity within mainly undisturbed systems has been examined. Many of these are now relatively old and have limited resolution of species, but they clearly illustrate some of the edaphic factors that affect EM fungal community composition and which deserve further analysis. The second part is a short synthesis of the numerous anthropogenic influences that have been shown to affect EM fungal diversity (see also Table 7.1 for a summary). Several issues discussed in this part are also considered by Rillig et al. (Chap. 6, this Vol.). In the discussion, we have tried to suggest how the distribution of EM fungi in relation to natural factors may influence their reaction to anthropogenic disturbances. In addition, aspects of EM fungal ecology are examined in relation to how they may influence individual species response to changing abiotic factors

Table 7.1. Summary of known effects of management and pollution upon the EMF-community

Factor	Chapter section	Mycorrhizal tip numbers	Colonisation (%)	Sporocarp production
Elevated CO_2	7.4.1	Increase in fine root production often recorded (Rey and Jarvis 1997; Runion et al. 1997)	No effect recorded	No data
Ozone	7.4.2	Possible decrease (Edwards and Kelly 1992)	No effect (Roth and Fahey 1998)	No data
Heavy metals (HM)	7.4.3	Effects dependent upon metal spp. and conc. (Hartley et al. 1999)	Effects dependent upon metal spp. and conc. (Hartley et al. 1999)	Decrease. Species richness negatively affected (Rühling and Söderström 1990)
N-deposition	7.4.4	Decreases have been reported (Kraigher et al. 1996; Erland et al. 1999)	No change (Wallenda and Kottke 1998; Erland et al, 1999; Taylor et al. 2000)	Initial change in community structure. Reduction in sporocarp production (Baar and ter Braak 1996) but increased production by tolerant spp. may mask decrease in sensitive spp. (Wallenda and Kottke 1998)
N-fertilisation	7.4.5	Short-term decrease after large single N additions (Meyer 1962; Ahlström et al. 1988)	Short-term decrease after large single N additions (Wallenda and Kottke 1998)	Differential response – some spp. increase, e.g. Lactarius rufus, most spp. decline (Wallenda and Kottke 1998)

EM community belowground	Extramatrical mycelium	General comments
Changes in species composition in pot cultures (Godbold et al. 1997; Rey and Jarvis 1997)	Increased production (Godbold et al. 1997; Rouhier and Read 1998, 1999)	Insufficient data, particularly field data
Change in community structure (Edwards and Kelly 1992; Qiu et al. 1993)	Insufficient data	Insufficient data
Increase in tolerant species (Hartley et al. 1999)	Insufficient data	Large inter- and intraspecific differences. Complex interactions between plant/fungus/metal (Leyval et al. 1997). Percent colonisation may decrease, especially if host more tolerant than mycobionts.
Decrease in diversity (Kraigher 1996; Lilleskov and Fahey 1996; Taylor et al. 2000) decrease in protein spp. (Taylor et al. 2000)	Insufficient data	Decrease in diversity, both in terms of spp. richness and evenness. More severe effect above ground. "Specialist species" more adversely affected. (Wallenda and Kottke 1998). Deficiency of other nutrients may lead to higher numbers of EM
Insufficient data, particularly with regard to long term effects. Changes in community structure recorded (Arnebrant and Söderström 1992; Kåren and Nylund 1996)	Possible decrease in EMM of EM fungal spp. after large single N additions (Arnebrant 1994; Wallander and Nylund 1992)	Most studies record decreasing diversity after large single N additions

Table 7.1 (*Continued*)

Factor	Chapter section	Mycorrhizal tip numbers	Colonisation (%)	Sporocarp production
Acidification	7.4.5	Decrease in fine root numbers. (Dighton and Skeffington 1987)	No change (Dighton and Skeffington 1987)	Decline in diversity (Arnolds 1991; Dighton and Skeffington 1987; Agerer et al. 1998). Increased production by acidophilous spp. (Agerer et al. 1998)
Liming	7.4.6	Often large increase in root tips (Erland and Söderström 1991; Persson and Ahlström 1994; T. Jonsson et al. 1999; Bakker et al. 2000)	No change recorded but few data available on immediate effects of liming	Differential response by spp. (Agerer et al. 1998)
Wood ash	7.4.7	Insufficient data	No change (Mahmood 2000)	Insufficient data
Vitality fertilisation	7.4.7	Insufficient data	No change (Kåren and Nylund 1996)	Insufficient data

EM community belowground	Extramatrical mycelium	General comments
Changes in species composition (Roth and Fahey 1998; Qiu et al. 1993). Decreases in spp. with abundant EMM	Decreased production (Dighton and Skeffington 1987)	Increased disturbance due to greater earthworm activity could reduce EMM
Considerable changes in spp. composition often recorded after liming. (Lehto 1984, 1994; Erland and Söderström 1991; Andersson and Söderström 1995; T. Jonsson et al. 1999)	Increase in types with abundant mycelia (Bakker et al. 2000)	There is a great need for more studies into the effects of liming
Some evidence of differential spp. response (Mahmood 2000)	Insufficient data. EM mycelia have been reported to colonise ash granules (Mahmood 2000)	Very few studies
Insufficient data	No data	Very few studies

7.3 Natural Abiotic Factors Influencing Ecto-mycorrhizal Community Structure

7.3.1 Edaphic Factors

Determining the influence of individual edaphic factors upon EM fungal diversity is compounded by the fact that few, if any, factors may change independently from all others. For example, changes in community structure have been attributed to shifts in soil pH (Dighton and Skeffington 1987; Agerer et al. 1998), but there is a close relationship between soil pH and many other soil factors, particularly heavy metal and aluminium availability (Marschner 1995), and these are also likely to be important components of the causal effect. The sections below present a number of soil characteristics that appear to influence EM fungal diversity. Although these are discussed separately, the integrated complex nature of the soil environment means that there may be many amelioratory and exacerbatory interactions between them. However, this is not to say that changes in single factors do not, on their own, affect community structure. Numerous laboratory experiments have shown that growth, colonisation rate, and enzymatic activity of EM fungi can all be significantly altered by single factor changes in experimental conditions (Smith and Read 1997).

The EM symbiosis may be regarded as an adaptation to conditions of low mineral nutrient availability and/or to situations where nutrient inputs are pulsed (Smith and Read 1997). Given this connection between the symbiosis and nutrient availability, it seems likely that the nutrient status of soil will have an important role to play in determining EM fungal community structure. One early opinion concerning the role of EM fungi in plant nutrition was that the fungi increased the uptake of any nutrient in short supply (Hatch 1937). This view is largely unchanged. It is known that EM species vary in their ability to acquire specific nutrients from soil (Thomson et al. 1994; Leake and Read 1997) and this differential efficiency could, in itself, be considered as niche partitioning. The mycelia of several fungi may colonise the same substrate but they may extract different components of the substrate. In theory, the greater the number of nutrients that are in short supply, the greater the number of potential 'nutrient' niches. This idea would form part of the niche differentiation via resource partitioning considered by Bruns (1995) to be involved in the maintenance of high EM fungal diversity. Unfortunately with the exception of nitrogen, there is little information available on the influence of specific nutrients upon EM fungal diversity. The effects of different forms of N-free fertilisers (ash, vitality fertilisation) on EM fungal community structure are presented in Section 7.4.7 and the various quantitative and qualitative impacts of N upon EM fungal communities are discussed in Section 7.4.4.

7.3.2 Soil Organic Matter and Spatial Heterogeneity

Heterogeneity, both spatial and temporal, within soil is high. It has been suggested that this spatial heterogeneity within soil is important for maintaining genetic diversity within plant populations (Mitchell-Olds 1992). This suggestion is based on the assumption that the relative fitness of genotypes changes in different environments. Egger (1995) extended this idea to EM fungi but a test of this hypothesis will be more problematic than when dealing with plants as any test would require defining a meaningful criterion with which to measure the fitness of fungal individuals. Bruns (1995) also suggested heterogeneity to be an important factor in determining the high levels of biodiversity observed within EM fungal communities. However, the emphasis here was on niche differentiation via resource partitioning among species rather than maintaining genetic diversity within populations.

As far as we are aware, no studies have examined the microspatial distribution of individual EM fungal species in relation to the chemico-physical soil environment. Here again, EM research lags behind that on soil fauna where temperature, moisture content and organic matter content in different soil horizons have been shown to be important in determining community structure of microarthropods (Klironomos and Kendrick 1995).

A small number of studies have examined spatial distribution of EM fungi in relation to substrata at a gross scale, e.g. rotting wood versus bulk soil (Kropp 1982). Harvey et al. (1976, 1978) examined the distribution of mycorrhizal tips in a Douglas fir/larch forest and found 5% in the mineral soil, 66% occurred in the humus layer, 21% in decayed wood and 8% in charcoal fragments. Although individual species were not distinguished, this work clearly demonstrates the non-random distribution of mycorrhizal tips in a heterogeneous environment. In a recent study, where EM morphotypes were distinguished on the roots in a Norway spruce stand in north Sweden, Fransson et al. (2000) used a binomial statistical model to show that species differed in their preference for the mineral or organic soil horizons. *Cenococcum geophilum* mycorrhizae were preferentially found in the organic horizons, while mycorrhizae formed by *Tylospora fibrillosa* were associated with the mineral soil. *Piloderma* mycorrhizae, on the other hand, did not show any particular preference for either soil fraction and were distributed evenly within the soil profile. Similarly, in a pot experiment in which *Pinus contorta* seedlings were grown in organic and mineral soil, Danielson and Visser (1989) found that morphotypes formed by E-strain fungi, *Tomentella* spp. and *Suillus* spp., were all more common in the mineral than in the organic soil. Conversely, Stendell et al. (1999) and Taylor and Bruns (1999), using molecular typing, both found *Tomentella sublilicina* to be more common in the organic layers of two pine forests than in the mineral soil.

Much more data are required on the vertical distribution of different mycorrhizal fungal species, as potential niche partitioning in relation to physico-chemical changes with soil depth could be important in maintaining the observed high levels of species richness. However, one important factor which must be considered when determining vertical spatial patterns is that the distribution of root tips is usually (at least when they are well developed) strongly skewed towards the upper organic horizons.

Malajczuk and co-workers have emphasised the close relationship between the occurrence of certain fungi and the accumulation of organic matter (Malajczuk and Hingston 1981; Reddell and Malajczuk 1984; Malajczuk 1987; Hilton et al. 1989). Gardner and Malajczuk (1984) found that the mycorrhizal fungi which colonised transplanted *Eucalyptus* seedlings on rehabilitated bauxite mine sites included *Pisolithus arhizus* and *Scleroderma* spp, which are commonly found on sites which had been recently burned or on sites that were devoid of litter (Jeffries 1999). Species of the genus *Cortinarius* appeared 3 years after planting, around the base of trees and in areas of localised litter accumulation. Gardner and Malajczuk (1984) suggested that EM fungal succession (sensu Mason et al. 1983) is primarily due to changing soil conditions which are related to increased shading by expanding canopies and the accumulation of litter. More recently, Yang et al. (1998) examined the morphotypes occurring on *Larix kaempferi*, which was regenerating on lava flows and found a relationship between accumulating organic matter and mycorrhizal diversity.

However, the proposed link between organic matter accumulation and the development of complex EM fungal communities is not simple. Root tip abundance is also likely to increase with increasing depth of organic matter. It therefore seems likely, given the positive relationship which is often found between the number of species in a sample and the number of root tips in a sample that at least part of the increase in species richness may be explained by the increase in food bases (see Newton 1992).

Recently, Conn and Dighton (2000) and Dighton et al. (2000) have demonstrated that the size and composition of leaf litter patches in the New Jersey pine barrens affected the EM fungal community structure which developed in and under the patches. In addition, they showed that the morphotypes with the highest phosphatase activity were those associated with oak litter which had the highest P immobilisation during decomposition. These simple but elegant studies suggest a strong link between the spatial distribution of morphotypes and their function.

Abuzinadah and Read (1986) suggested that the relative ability of fungi to utilise organic sources of nutrients might significantly influence their distribution within the soil. Fungi with limited ability, termed 'non-protein' fungi, were expected to be more successful in the mineral soil. However, the development of surface organic horizons in boreal forest is often accompanied by considerable transport of organic material, as dissolved organic nutrients and particulate matter, to the underlying mineral soil (Harrison et al. 2000). Under these conditions, the ability to use organic nutrient sources would

Diversity of Ecto-mycorrhizal Fungal Communities

therefore also be of a distinct advantage for fungi growing in the mineral soil.

7.3.3 Soil Moisture

Soil moisture content has been known for a long time to affect EM fungal community composition (Worley and Hacskaylo 1959). One general feature is that where soils are subject to drying out, community diversity is lowered (see Fogel 1980). However, it is not clear whether this is simply due to reduced growth of host roots under these conditions or a direct effect upon the fungi. Another feature is that reduced moisture often leads to an increase in the proportion of tips colonised by *C. geophilum* (Pigott 1982). These field observations are supported by experimental evidence in which *C. geophilum* was found to be less sensitive to water stress than other EM fungi (LoBuglio 1999 and references therein). This fungus with its thick melanised cell walls seems particularly suited to tolerate periodic drying. Glenn et al. (1991) compared the mycorrhizal morphotypes of *Picea rubens* growing in mesic and wetland sites in New Jersey and found *C. geophilum* to be the most commonly encountered morphotype. However, in this study, it was equally common at both sites. There was an inverse relationship between the occurrence of two unknown morphotypes, one of these appeared well adapted to the wet sites. The other (tentatively identified as being formed by a species of *Tuber*) was significantly more common on mesic sites and it was suggested that this was a result of its ability to withstand drought.

Significant effects of excessive moisture have also been recorded. Lorio et al. (1972) found that in a Loblolly pine stand, nodular forms (presumably tuberculate mycorrhizae) were more common during periods of excess soil

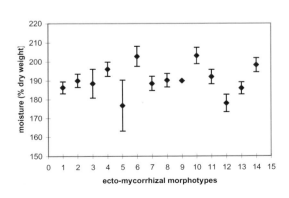

Fig. 7.1. Occurrence of ecto-mycorrhizal morphotypes in relation to average soil moisture content. Soil moisture as percent dry weight. The soil samples (n=19–200, ±SE) were collected from a 37-year-old sitka spruce plantation in which different ecto-mycorrhizal morphotypes were recorded. Type 1 *Tylospora fibrillosa*; 3 *Russula ochroleuca*; 6 *Amanita* sp.; 7 *Cenococcum geophilum*; 8 'Piceirhiza bicolorata'; 11 *Cortinarius* sp.; 12 E-type mycorrhizae; 2, 4, 5, 10, 13, 14 mycobionts unknown; 9 occurred only in a single sample

moisture. They stated that mycorrhizae were very responsive to changes on moisture availability.

In an interesting study on the potential antagonistic interactions between EM fungi and arbuscular mycorrhizal (AM) fungi on the roots of a *Populus* and a *Salix* species, Lodge and Wentworth (1990) found that under moist soil conditions, the EM fungi appeared to displace the AM fungi but not under either drier or wetter conditions. In addition, the negative association appeared to vary among EM morphotypes. This would strongly suggest that soil moisture availability affected the competitive ability of individual EM fungal species relative to the AM fungi. That EM fungi vary in their moisture requirements follows also from Fig. 7.1. This figure shows the average moisture content (percent dry weight) of soil samples of a Sitka spruce plantation in Scotland ($n=19–200\pm SE$) in relation to specific EM morphotypes (Taylor and Alexander, unpubl.). These results suggest that the relative niche width of individual fungi varies with respect to their moisture requirements. A simple analysis such as this gives valuable ecological information concerning the potential niche partitioning among species of EM fungi.

7.3.4 pH

Of all the soil parameters considered here, soil pH is perhaps the most complex, since many other soil characteristics are closely linked to pH. However, in vitro studies clearly indicate that fungi vary with regard to pH optima for growth (Hung and Trappe 1983) and colonisation potential (Erland 1990). In addition, changes in soil pH would be expected to alter the enzymatic capabilities of some fungi, since at least some of the enzymes produced by EM fungi have rather narrow pH optima (Leake and Read 1990). McAfee and Fortin (1987) examined the influence of pH on the competitive ability of *Pisolithus tinctorius* and *Laccaria bicolor* under field conditions and found that the former species was more sensitive to lower pH values than the latter.

In a pioneering study, Kumpfer and Heyser (1986) examined the distribution of mycorrhizal morphotypes in soil adjacent to the base of 400-year-old beech (*Fagus sylvatica* L.) trees. The acidic stem flow created a marked pH gradient with soil becoming less acidic with increasing distance from the tree base. A total of 11 morphotypes were recognised but only one, *Cenococcum geophilum* was found closest to the tree base. A linear correlation was found between the natural log of the percentage of tips colonised by *C. geophilum* and the soil pH, log of the Ca/Al ratio, log Ca and log Pb soil concentrations. There is a great need for more studies like this, which utilise natural gradients under soil conditions.

Lu et al. (1999) examined sporocarp production in *Eucalyptus* plantations in Western Australia and found that diversity was inversely correlated to soil pH. However, an additional complication here is that species richness was, not

Diversity of Ecto-mycorrhizal Fungal Communities

surprisingly, positively related to stand age, which would suggest that soil pH was inversely related to stand age. Separating a cause and effect in this case would be extremely difficult.

7.3.5 Temperature

There are as far as we know no studies that have specifically examined the effects of temperature on EM fungal community structure under field conditions. Such studies could be carried out in conjunction with soil warming experiments. Results from such studies are important since increased soil temperatures may be expected as a consequence of global climate change (see Hantschel et al. 1995). Erland and Finlay (1992) demonstrated in microcosm systems that the rate of colonisation of seedlings by *Piloderma croceum* and *Paxillus involutus* was significantly affected by changes in temperature.

In a series of recent papers, Tibbett et al. (1998a,b, 1999) demonstrated that many EM fungi are still physiologically active at low temperatures (ca. 0–1 °C). If we assume that these species are also capable of growth under these conditions, then they could reasonably be expected to have a competitive advantage over species which are less cold-tolerant, particularly during late autumn and spring when soil temperature is around freezing.

High soil temperatures in exposed sites, for example on mine tailings, have been suggested as a major factor which influences the establishment and subsequent development of EM fungal communities under these conditions (Danielson 1985). Species of the genera *Scleroderma* and *Pisolithus arhizus* are successful colonisers of mine spoils and this ability may be linked to their ability to withstand high temperatures.

7.3.6 Wildfire

Prior to the introduction of fire control methods, wildfire was a major disturbance factor in boreal ecosystems (Zackrisson 1977). Under this disturbance regime, mycorrhizal species that are primarily associated with the organic horizons, would in theory, be more susceptible to the deleterious effects of fires than those found in the mineral layers. However, fire intensity appears to be a major factor in determining the response of the EM fungal community.

Severe fires which kill the host plants and which radically alter the soil environment, perhaps not surprisingly, have been shown to reduce mycorrhizal diversity (Danielson 1984; Torres and Honrubia 1997). However, Horton et al. (1998), using molecular characterisation, demonstrated that although the community structure was radically altered by a stand replacing fire, the dominant components of the pre-burnt EM community were present on establishing seedlings but they were quantitatively reduced.

In a study of the EM fungi on a *Pinus banksiana* Lamb. chronosequence created by severe wildfires, Visser (1995) demonstrated, using morphotypes, that there was a progressive development of an increasingly complex EM community with increasing time. Rank abundance plots (Tokeshi 1993) of the community data from 6 years after wildfire were best described by a geometric series while those of the oldest stands, having the more complex communities, conformed to a lognormal series. This change in pattern is consistent with that of plant succession following disturbance (Bazzaz 1975; Tokeshi 1993). However, the process of EM fungal community development cannot strictly be defined as succession since Visser (1995) found that most of the pioneer fungi which were present in the youngest stands, the so called multistage fungi, were still present in the oldest stands.

Several recent studies have highlighted the importance of resistant propagules of EM fungi (Baar et al. 1999) and arbuscular mycorrhizal fungi (Horton et al. 1998) in the colonisation of tree seedlings establishment in areas affected by stand-replacing fires. Many of the EM fungi which appeared as dominant on the post-fire seedlings were those which either formed only a small proportion of the mycorrhizae in the forest before the fire or those thought to be present only as resistant propagules within the soil. When competition from other fungi was removed, either by the fire or when soil from the pre-burnt forest was removed and used in a seedling bioassay (Taylor and Bruns 1999), these fungi were able to proliferate and become the dominant components of the EM fungal community. The propagules are thought to be spores, mycelial fragments and perhaps sclerotia, which due to their position within the soil profile survive the effects of the fire.

In contrast to stand-replacement fires, low intensity fires, where many larger hosts survive, seem to have a limited long-term effect upon mycorrhizal diversity and community composition (L. Jonsson et al. 1999). However, Stendell et al. (1999) demonstrated that a low intensity fire (ground fire) could in the short term have a very marked effect upon community structure where the dominant species within an EM fungal community are primarily associated with the organic layers.

Although based on an analysis of sporocarp production, the study by Johnson (1995) deserves special mention here as it appears to illustrate an evolutionary adaptation to wildfire by mycorrhizal fungi. The study found enhanced production of hypogeous sporocarps following wildfire by species within the family Mesophelliaceae. Increased mycophagy by mammals was then suggested to facilitate rapid dispersal and establishment of the fungi early on in the post-fire succession. An analysis of the mycorrhizae in this system would clearly be of great interest in order to verify this hypothesis.

In addition to the obvious effects on the vegetation above ground, wildfires can dramatically alter surface soil characteristic. Nutrients bound within organic residues are released creating a nutrient pulse, microbial activity may increase and in general, the rate of nutrient cycling may increase (Tamm 1991;

Wardle et al. 1997). Mycorrhizal species that are able to respond rapidly and exploit these conditions would, at least in the short term, have a competitive advantage over other species. The multi-stage fungi mentioned above in the study by Visser (1995) may, at least in part, be fungi capable of this rapid response.

7.4 Pollution and Forest Management

This section provides a short synthesis of the numerous anthropogenic influences that have been shown to affect EM fungal diversity. The major findings of this section are summarised in Table 7.1.

7.4.1 Elevated CO_2

Increased atmospheric CO_2 concentration, altered air temperature and precipitation are factors associated with climate change (Rygievicz et al. 1997). Many studies have shown that increased atmospheric CO_2 concentrations enhance growth and productivity of several plant species including trees (see Rey and Jarvis 1997). Mycorrhizal fungi, which are among the first of the soil biota to receive carbon from plants, are also likely to be influenced by the increased CO_2 levels (Table 7.1; see also Rillig et al., Chap 6, this Vol.; Rygiewicz et al. 1997). Norby et al. (1987) and O'Neill et al. (1987) found a differential response of different EM-species to elevated CO_2. *Cenococcum geophilum* (*C. graniforme*) varied from abundant under ambient CO_2 to under represented or absent under elevated CO_2 conditions. Elevated CO_2 levels have been shown to increase production of extraradical mycelium by *Paxillus involutus* growing on *Pinus sylvestris* (Rouhier and Read 1998) and on *Betula pendula* (Rouhier and Read 1999). Interestingly, neither of these two studies recorded any increase in mycorrhizal tip numbers, which suggests that there was an increase in carbon transfer through individual root tips in order to support the increased mycelial growth.

Runion et al. (1997) exposed *Pinus palustris* Mill. seedlings to two levels of CO_2 in open top chambers for 20 months. They found that increased fine-root length and EM fungal colonisation under elevated CO_2 resulted in almost double the numbers of EM per seedling. Godbold et al. (1997) investigated the effects over 27–35 weeks of ambient (375 ppm) and elevated (700 ppm) CO_2 on *Betula papyrifera* Marsh, *Pinus strobus* L. and *Tsuga canadensis* L. Carr saplings. They found a significant increase in EM colonisation in *B. papyrifera* and *P. strobus*. They identified 12 EM morphotypes on these two species, in both ambient and elevated CO_2. There was, however, a distinct shift in the EM community on *B. papyrifera* at elevated CO_2. EM types with emanating hyphae and rhizomorphs increased. Rey and Jarvis (1997) examined the growth response and mycorrhization of young birch trees after 4.5 years of elevated CO_2 exposure. At the end of the experiment, fine root density was almost doubled in elevated CO_2 plants and there was a clear difference in EM morphotype composition on the roots. Under ambient CO_2, the dominant fungal species were *Leccinum* sp. although others, such as *Paxillus involutus* and *Laccaria* sp., were present. In elevated CO_2, no single species was dominant, with *Laccaria* sp., *Hebeloma sacchariolens* and *Thelephora terrestris* being the most frequent. Mycorrhizae formed by *Leccinum* sp. were also observed but were much less frequent than in the elevated CO_2 chambers. The authors interpret these findings as evidence to support the hypothesis that elevated CO_2 leads to an acceleration in tree ontogeny since the EM fungal species found on the roots under elevated CO_2 do not normally appear on birch until after 6–7 years (Mason et al. 1982; Last et al. 1985).

7.4.2 Ozone

Edwards and Kelly (1992) studied the effects of ozone on mycorrhization of loblolly pine (*Pinus taeda* L). and reported a shift in ecto-mycorrhizal morphotypes as a response to the treatment. Qiu et al. (1993) studied the effects of ozone on mycorrhization of two families of six-months-old loblolly pine seedlings that differed in their sensitivity to ozone. The seedlings were planted in open top chambers for one growth season and they were exposed to four ozone concentrations. Ecto-mycorrhizae were quantified as numbers of morphotypes per centimetre of long root. The total number of morphotypes per centimetre increased with ozone concentration in the sensitive plant family and elevated ozone caused a shift in mycorrhizal morphotypes in both families. Roth and Fahey (1998) studied the effects of acid precipitation and ozone on the ecto-mycorrhizal morphotypes of red spruce saplings. They found no treatment effect on the percentage of roots colonised and ozone alone did not affect the composition of the EM fungal community on the saplings. There was, however, a strong interactive effect of the treatments leading to a change in the EM fungal community in the organic horizon after 1 year of treatment.

7.4.3 Heavy Metals

Sources of heavy metal contamination in the soil are diverse. Common examples include dumping of waste products from the burning of fossil fuels, mining and smelting of metalliferous ores, municipal refuse, fertilisers, pesticides, sewage sludge amendments and waste products from the manufacture of pigments and batteries. Some of the metals are micronutrients (Zn, Cu, Mn, Ni and Co), while others have no known biological function, such as Cd, Pb and Hg (see Leyval et al. 1997). Since mycorrhizal fungi provide a direct link between the soil and roots, much interest has been focused on the ability of EM fungi to prevent toxic levels of metals from entering the host plants (Hartley et al. 1997; Leyval et al. 1997). Much less information is, however, available on the below ground EM fungal communities at metal contaminated sites. A study of occurrence of macromycetes in Sweden along a complex heavy metal pollution gradient revealed that both the number of fruiting species and the number of sporocarps decreased strongly with increasing levels of pollution (Rühling and Söderström 1990). There were, however, some species which tolerated the metal loads well, many of which were EM fungi. Among those were different species of *Amanita, Albatrellus ovinus* and *Leccinum scabrum*. Hartley et al. (1999) investigated the EM fungal colonisation potential in soil contaminated with Cd, Pb, Zn, Sb and Cu by a chemical accident. Although contaminated field soil significantly inhibited shoot and root growth, total root tip density was not affected. The soil had a toxic effect on EM fungi associated with Scots pine seedlings and caused shifts in EM fungal species composition on the seedlings. Baxter et al. (1999) compared ecto-mycorrhizal community structure on mature oak in urban and rural stands. The urban stands had higher heavy metal content but also higher N deposition and earthworm counts than the rural stands. Baxter et al. (1999) also compared the EM fungal colonisation potential in these soils by planting *Quercus rubra* seedlings in soil cores in the glasshouse. Twenty-six EM morphotypes were distinguished on mature oak in rural sites compared to 16 in urban sites. The seedlings growing in rural soils were colonised by nine morphotypes compared to seven in the urban soils.

7.4.4 N Deposition and Fertilisation

Many ecosystems, where the dominant plants form ecto-mycorrhizae, are characterised by limited availability of mineral N and accumulations of organically bound nutrients (Read 1991). The boreal coniferous forest is perhaps the quintessential example of such a system, which is essentially adapted to minimise N loss. There is concern as to what may happen to the stability of this system when N becomes increasingly available in an inorganic form as a result of atmospheric pollution or from fertiliser additions.

Over the last few decades, there has been an apparent reduction in sporocarp production in European forests (Arnolds 1991) and it has been suggested that this may be related to increases in pollutant N deposition (Baar and ter Braak 1996). In an excellent review on ecto-mycorrhizae and N deposition, Wallenda and Kottke (1998) summarised the available data and concluded that there seemed little doubt that EM fungal species richness, as measured by sporocarp production, was negatively affected by increasing N deposition. They also concluded that 'specialist' species (especially the symbionts of conifers) were more adversely affected than 'generalist' species, which are able to form mycorrhizae with a wide range of host plants. This differential response by individual species was exemplified by *Paxillus involutus* and *Lactarius rufus*, two nitrophilous EM fungal species, which have been shown on a number of occasions to dramatically increase fruit body production following the addition of N fertilisers.

According to available information, the effect of N additions on the mycorrhizal community below ground may be less dramatic than that recorded from sporocarp studies (Wallenda and Kottke 1998). However, alterations in community structures on the roots have been reported (Arnebrant and Söderström 1992; Kårén and Nylund 1996). Using a combination of morphotyping and molecular typing, Kraigher et al. (1996) examined the EM belowground communities at two Norway spruce sites that were subject to different levels of N and S deposition from a thermal power plant in Slovenia. They found a decrease in species richness on the roots at the more polluted site and also recorded a decrease in fine root density at this site. There were 17 EM types on 28,443 root tips in the polluted site compared to 24 different types on 38,502 root tips in the less polluted site. The dominant types that were identified to fungal species in the polluted site were *Paxillus involutus* and *Xerocomus badius*, whereas *Hydnum rufescens*, *Amphinema byssoides* and *Cenococcum geophilum* were among the dominant types on the roots in the less polluted site.

Erland et al. (1999) used molecular typing to study the belowground EM fungal communities in two *Picea abies* stands in southern Sweden, with different yearly N-deposition loads (14–15 and 24–29 kg N ha^{-1} year^{-1} respectively). With a similar sampling effort, 13 taxa were found in the less polluted site compared to seven in the site with higher N deposition. There were significantly higher numbers of fine roots per metre root length in the less polluted site. Additional differences between the sites to N deposition (site history and location in relation to other spruce forests) were suggested to have contributed to the differences in species numbers. Decreases in root tip density are also commonly recorded after N fertilisation (e.g. Meyer 1962; Ahlström et al. 1988).

Taylor et al. (2000) examined the hypothesis that pollutant N deposition contributes to a reduction in diversity of EM fungal communities. They studied both below ground communities using morphotyping and above ground sporocarp production in beech and spruce forests along a N–S transect in Europe, ranging from polluted sites in central Europe to relatively unpolluted sites in Scandinavia (Fig. 7.2).

No apparent negative effects on EM fungal diversity (species richness and community evenness) were found in the beech forests on either morphotypes or sporocarps. There was, in fact, a positive correlation between extractable

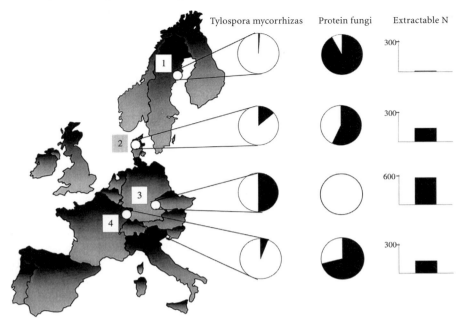

Fig. 7.2. Summary of an analysis of the ecto-mycorrhizal communities at four Norway spruce [*Picea abies* (L.) Karst.] forests along a north-south European transect. *Pie charts* show the proportion of root tips at each site colonised by *Tylospora fibrillosa* and the proportion of species isolates from each site that could utilise protein as a source of nitrogen (protein fungi, sensu Abuzinadh and Read 1986). None of the isolates from site 3 could utilise protein as an N source. The *bar charts* show KCl-extractable inorganic N (µg N g⁻¹ LOI) in the LFH soil horizons (note that the scale at site 3 is twice that at the other three sites). Sites: *1* Åheden – northern Sweden; *2* Klosterhede – western Jutland, Denmark; *3* Waldstein – Fichtelgebirge, Germany; *4* Aubure – Vosges Mountains, NE France. (Data from Taylor et al. 2000; Persson et al. 2000)

mineral N in the soil and morphotype richness. Diversity of EM-morphotypes and sporocarps within the spruce forests appeared to be more sensitive to increasing N availability. Although colonisation was over 90% at all sites, there was a decline in diversity of morphotypes from north to south (from the less to the more polluted sites). Two species, *Tylospora fibrillosa* and *Lactarius rufus*, dominated the EM fungal community at the most polluted site, forming 75% of the mycorrhizae examined. *Tylospora* mycorrhizae showed a clear trend along the transect; increasing in abundance in relation to increases in available mineral N (Fig. 7.2). The opposite pattern was apparent for mycorrhizae formed by members of the family Cortinariaceae. The decline in species richness and increase in dominance along the transect was accompanied by changes in the N nutrition of the fungi present. The proportion of species within communities that could utilise protein as a source of N declined with increasing mineral N availability (Fig. 7.2). Taylor et al. (2000)

followed Dighton and Mason (1985) and used the terminology S (stress-tolerant) and R (ruderal) strategists to categorise species that decreased or increased respectively relative to increasing mineral N availability. The R strategists are described as fast growing fungi of low host specificity while the S strategists are successful in environments where the mineralisation processes are extremely slow for climatic and biotic reasons.

Lilleskov and Fahey (1996) also reported a reduction in EM fungal diversity along an N gradient created by emissions from a fertiliser factory in Alaska. This study, based on both sporocarp studies and mycorrhizal morphotyping, showed that the genus *Cortinarius* was particularly sensitive to increases in N deposition.

In addition to changes in community structure with N additions, increases in the numbers and/or the proportion of uncolonised fine roots have also been recorded as a short-term effect after the application of a single large dose of N fertiliser (see Wallenda and Kottke 1998). Reductions in colonisation which may be attributable to long-term (i.e. chronic) N deposition are more difficult to quantify but Wollecke et al. (1999) reported a significant difference in colonisation level between two pine stands subjected to 10–20 and >35 kg N ha^{-1} year^{-1} respectively. On average, colonisation was only 72 % in the more polluted stand, compared to >90 % in the less polluted stand.

It is still unclear if increases in N availability affect the fungi directly or whether changes in community structure are mediated via the host plant. It has been shown that increases in mineral N availability can negatively effect the growth of the extramatrical mycelium (EMM) of some EM fungi (Arnebrant 1994; Wallander and Nylund 1992). Any species which rely upon the EMM to colonise new root tips and which are sensitive to increasing N, would therefore be at a competitive disadvantage under conditions of high N availability. Moreover, it has been hypothesized that nitrogen deposition could alter plant community composition by changing the community of mycorrhizal fungal symbionts (Aerts, Chap. 5, this Vol.; Rillig et al., Chap. 6, this Vol.).

7.4.5 Acidification

The slow acidification of soil is a natural process under most coniferous forests. This process has, however, been greatly accelerated by anthropogenic influences such as deposition of acidifying airborne pollutants throughout Europe. Several authors have attributed a decline in diversity of ecto-mycorrhizal sporocarps with increased soil acidification (Dighton and Skeffington 1987; Arnolds 1991). Agerer et al. (1998) found a reduction in species richness of ecto-mycorrhizal sporocarps in Norway spruce plots treated with acid irrigation at the Höglwald experimental forest in southern Germany. Below ground, there was also a reduction in the number of morphotypes recorded in samples and an increase in dominance (Taylor et al., unpubl.). It was proposed that this decrease in species richness was a result of a significant increase in the proportion of root tips colonised by *Russula ochroleuca* coupled to a reduction in root tip numbers following acid irrigation.

Meier et al. (1989) found that of five EM morphotypes observed on red spruce, four were not affected by short-term acid rain, whereas one type, *Cenococcum geophilum*, increased in numbers. Roth and Fahey (1998) also found five EM morphotypes on red spruce saplings and a shift in relative abundance of these types after exposure to acid precipitation. Qiu et al. (1993) studied the effects of acid precipitation on mycorrhizal morphotypes on loblolly pine seedlings from two families that were differently sensitive to ozone. In the ozone-sensitive family, the number of EM morphotypes decreased with increasing rainfall acidity whereas the number of EM morphotypes was unaffected in the other plant family.

The EM symbiosis, at least in much of the boreal and temperate regions, is adapted to acidic soil conditions (Read 1991). For example, many of the enzyme systems have rather low pH optima (Leake and Read 1997). Therefore, although it seems clear that a severe reduction in soil pH would negatively affect diversity, the effects of a moderate reductions in pH are likely to be less drastic than those associated with large increases in soil pH that occur as a result of some forest practices (e.g. liming).

7.4.6 Liming

Liming of forest soils has been proposed as a measure to counteract the continuing acidification of forest soil due to deposition of airborne pollutants. Opinion on the suitability of liming is, however, divided (Binkley and Högberg 1997). Shifts in belowground EM fungal communities following the addition of lime have been reported in several studies (Lehto 1984, 1994; Erland and Söderström 1991; Andersson and Söderström 1995; T. Jonsson et al. 1999). Taylor et al. (unpubl. data) observed an increase in belowground EM morphotype diversity on *Picea abies* in limed plots at the Höglwald in Germany. This increase was also accompanied by an increase in fine root density. A large increase in the numbers of fine root after liming has previously been reported by Persson and Ahlström (1994). Increased ramification of the roots and higher numbers of fine roots per root length in limed plots were also found on *Pinus sylvestris* by Erland and Söderström (1991) and on *Picea abies* by T. Jonsson et al. (1999). Bakker et al. (2000) studied the effects of liming on the ectomycorrhizal status of oak at a number of sites. They found a slight but significant increase in total numbers of mycorrhizal root tips at all ten sites studied. Morphotyping of the mycorrhizae revealed a general decrease in smooth mycorrhizae in favour of types with abundant extramatrical mycelium as a result of liming. Erland and Söderström (1991) found a pink mycorrhizal morphotype (*Pinirhiza rosea*) to be twice as abundant on *P. sylvestris* seedlings grown in limed plots (pH 5.2) for 4 months compared to seedlings growing in control plots (pH 3.8). T. Jonsson et al. (1999), using molecular typing, did not detect any significant effect on single EM fungal species on *P. abies* as a result of the lime treatment. However, Taylor et al. (unpubl.) found a very significant increase in the abundance of *Piceirhiza nigra*, *Amphinema byssoides* and *Tuber* (cf) *puberulum* mycorrhizae at both the Höglwald site in Germany and at a site in Southern Sweden. Both of these sites were Norway spruce forests. The mycorrhizal activity measurements by Qian et al. (1998), mentioned in the previous section, also showed the highest activities in limed plots in *Piceirhiza nigra* and *Tuber* (cf) *puberulum/P.abies* mycorrhizae. Due to a lack of sufficient data, it is not possible, at present, to predict how liming may affect EM fungal diversity either in the short term or more importantly in the long term. Further studies covering a wider range of trees and soil conditions are required.

7.4.7 Wood Ash Application and Vitality Fertilisation

There is an increasing interest in the use of CO_2 neutral biofuels of forest origin in the Nordic countries. Repeated outtake of forest residues (trees and twigs were removed after clear-cutting and three subsequent thinnings) has been shown to decrease the thickness of the humus layer and reduce both the density and the total number of fine roots in this horizon (Mahmood et al. 1999). Wood ash has been proposed as a fertiliser to replace some of the nutrients that are removed during residue harvesting. Wood ash is highly alkaline (pH 10–13) and contains the macro nutrients Ca, P, K and Mg and the micronutrients B, Cu and Zn. Exact contents vary depending on the source of the burnt material and a major concern is the level of heavy metals in the ash. Presently, this practise is at an experimental stage, however, in Sweden, 5000 ha of forest are currently treated with ash each year.

There are at present few studies that have examined the effects of ash additions upon EM fungi. Mahmood (2000) used molecular typing to study the effects of granulated wood ash (3 and 6 ton/ha) on the belowground EM fungal community in a Swedish spruce forest 7 years after the treatment. No differences were detected in soil pH or in numbers of fine roots per metre root length. A small shift in the EM fungal community could, however, be attributed to the ash treatment. Mycelia from 3 of the 20 EM fungal taxa detected in the study were also found to colonise ash granules buried in the treated plots. These taxa were not identified to fungal species. Two of them had a tendency to increase with increased ash-application and colonised 20 and 7 % respectively of the analysed roots in the highest ash amendment compared to 4 and 1 % in the control plots. These 'ash' fungi were further studied in the laboratory and they were shown to have a higher ability to solubilise hardened wood ash (HWA) compared to other EM fungi from the same site. Under symbiotic conditions, one of the 'ash' isolates actively colonised ash patches in the substrate with dense mycelium while *Piloderma croceum*, a common EM fungus of boreal forests, avoided the ash patches. The competitive outcome in colonisation of a bait seedling between *P. croceum* and the "ash" isolate was also significantly affected in the favour of the ash fungus when HWA was mixed into the peat substrate (Mahmood 2000).

Erland and Söderström (1991) examined short-term effects of non-granulated wood ash on EM fungal colonisation in *Pinus sylvestris* plots treated with 7.5 tons ha^{-1} wood ash. Twelve months after ash application, 1-year-old *P. sylvestris* seedlings were planted in the plots and left to grow for 4 months (June–September). The mycorrhizal distribution on the roots was estimated using rough morphotyping. Six morphotypes could be distinguished and there were no significant differences in abundance of any of these morphotypes compared to the control plants despite the fact that humus in treated plots was pH 6.4 compared to pH 3.8 in the untreated plots. There was a significant reduction of numbers of root tips per metre root length in the treated plots but not in the total number of tips per plant. In general, the levels of nutrients supplied to the forest with the addition of wood ash seem unlikely to have a major impact upon the EM fungal community. However, the shifts in soil pH which sometimes accompany the addition of wood ash could be expected to have measurable effects. Again, more studies are required.

N-free fertilisers have been suggested as a method for replacing basecations leached from forest soils as a result of N deposition (Evers and Huettl 1991). Kårén and Nylund (1996) studied the effects of an N-free fertiliser (P:K:Ca:Mg:S, 48:43:218:46:75 kg ha^{-1}) on the EM fungal community structure in a Norway spruce forest by morphotyping mycorrhizae. They concluded that moderate levels of the N-free fertiliser are unlikely to drastically affect the abundance of the dominant EM-types.

7.5 Conclusions

In parallel with most other diverse groups of soil organisms, the factors which influence the development and maintenance of high EM fungal species richness continue to be a matter of much conjecture and debate (see Bruns 1995; Giller 1996). Within this chapter, we have presented a number of abiotic factors which have been shown, to varying degrees, to influence EM fungal community diversity. However, whether or not many of these factors actually play a role in determining the initial community diversity is generally difficult to interpret. For example, extremes in soil pH values appear to restrict the development of diverse communities, but is a more moderate soil pH value by itself a determinant of diversity?

Ecto-mycorrhizal fungi appear to be sensitive indicators of forest ecosystem response to environmental perturbations, such as soil acidification and nitrogen deposition (Dighton and Jansen 1991; Meier 1991; Meyer et al. 1988; Nitare 1988). The predictive power of this observation is however, somewhat constrained as it is largely based upon investigations which have utilised changes in the sporocarp production to monitor effects upon EM fungal communities (Brandrud 1987; Termorshuizen and Schaffers 1987; Termorshuizen 1993). As Arnolds (1991) succinctly states "the absence of sporocarps should not be interpreted as necessarily indicating the absence of mycorrhizae". However, a singularly important point is highlighted by these studies that of differential response by individual species to the same perturbation. A consequence of this is that any analysis of an EM fungal community must include estimates of individual species abundance since estimates of total abundance or productivity will fail to detect decreases in more sensitive species which are masked by increases in more tolerant species (Meier 1991). With the exception of large one-off fertiliser additions (e.g. Haug and Feger 1990,1991), decreases in the percentage of root tips colonised are rarely noted in response to changes in abiotic factors. In situations where changes in EM fungal communities occur that are not a result of changes in root tip numbers, then the potential effects upon ecosystem functioning must be expressed through qualitative changes in the composition of the EM fungal community.

Are we currently in a position to make predictions concerning how EM fungal community structure changes in response to perturbations? In addition, can we predict what effect(s) these qualitative changes may have upon ecosystem processes?

In the introduction to this chapter, we briefly described the 'typical' structure of an EM fungal community on the roots of a host plant as having a small number of species colonising the majority of the roots with a large number of rarer species. A rank abundance plot (Tokeshi 1993) of such a community may best be described by a lognormal series or by the 'broken stick' model (Magurran 1988). This has already been mentioned (Sect. 7.3.2) in connection with the work of Visser (1995), who found that EM data from *P. banksiana* stands, 6 years after wildfire, were best described by a geometric series, while that of the oldest stands, having the more complex structure, conformed to a lognormal series. A feature of communities that develop according to the lognormal series is a decrease in species dominance (increasing evenness) and an increase in species richness. In contrast to this, most of the studies included within this chapter have reported decreasing species richness and increasing dominance by one or a small number of species in relation to perturbations. Such communities can be best described by a geometric series. A similar shift to a geometric series was reported by Taylor et al. (2000) in an examination of EM fungal communities on spruce along a gradient of increasing soil mineral N availability in Europe (see Sect. 7.4.4). A feature which was highlighted in this study and which would seem also to apply to several other studies is that changes in community structure may involve an increase in less specialised (R-strategists) species over more specialised, stress-tolerant (S-strategists) species. This shift in species composition could have a significant effect upon nutrient cycling within ecosystems as many of the species which appear to be sensitive to perturbations are those with considerable capacity to utilise organic sources of nutrients (Leake and Read 1997). Further analyses of community changes coupled to examinations of the physiological attributes of the species involved will further elucidate this hypothesis.

Earlier we made the point that diversity is usually considered as having two components, species richness and evenness, and that our ability to detect changes within these two components may differ. At present, our knowledge concerning the relationship between species richness estimates and sample size, species spatial distribution and community evenness is virtually nonexistent. What we do know is that, in common with analyses of most organism-rich communities, sample size (number of root tips sampled) can greatly influence the number of species recorded (Taylor et al. 2000). In a first attempt to examine species-area relationships in EM fungi, Newton and Haigh (1998) found a positive linear relationship between the number of species of EM fungi associated with different host genera and the total area occupied by each tree genus in Britain. A detailed analysis of the same relationship on a small scale (within a single stand) is urgently required as no data exist for minimal sampling areas (root tip numbers) for EM communities. Without this knowledge, it is practically impossible to determine the accuracy of estimates of community structure. This is particularly true for estimates of changes in species richness. In communities where rank-abundance plots reveal the presence of a long tail of rare species, the sampling strategy must be sufficient to gain a reliable estimate of the occurrence of rare species, otherwise false claims of changes in species richness could easily occur. Similarly, changes in species dominance will by definition mean that some species will become rarer; a sampling strategy must accommodate this as with increased sampling of root tips, the supposed extinct species may be found. The technique of rarefaction (Magurran 1988) can help to overcome problems of comparing communities from which different numbers of roots have been sampled (Taylor et al. 2000).

Perturbation events, unless catastrophic, seldom result in the total destruction of existing communities (Bazzaz 1975). It is more usual that the composition of the community changes in favour of species better adapted to the new conditions. The resilience or stability of an EM fungal community in the face of a disturbance will to some extent depend upon the initial composition of the community. When all species are equally sensitive then, unless replacement species immigrate into the site, the decline in the community is likely to be rapid. Replacement species are likely to be those best adapted for dispersal (early-stage fungi, R-strategists, sensu Deacon and Fleming 1992). Where there exists a store of resistant propagules within the soil of species which are better adapted to the new conditions, then these may become dominant without the need for immigration (see Taylor and Bruns 1999). Alternatively, if a proportion of the existing active species is more tolerant, then these are likely to increase to replace the more sensitive species

A number of the anthropogenic factors that have been included in this chapter may be viewed as essentially accelerated natural processes and the response of EM fungal communities to these factors in some ways reflects this. For example, it is well established that the application of a single large dose of N-fertiliser will invariably have a measurable impact upon the EM fungal community (Wallenda and Kottke 1998). However, the deposition of nitrogen onto forest ecosystems occurs naturally at a low background level (1–2 kg N ha^{-1} year^{-1}; Tamm 1991) and when this chronic addition of N is mimicked, even with much higher application rates (e.g. 75 kg N ha^{-1} year^{-1}) in field fertilisation experiments (see Emmett et al. 1995), the response of the EM fungi and indeed of the forest ecosystem as a whole may be limited (Saunders et al. 1996; Fransson et al. 2000). However, this does not mean that EM fungal communities do not respond to chronic atmospheric additions of N, as witnessed by the decline of fungi in The Netherlands (Arnolds 1991). In addition, increasing inputs of atmospheric N and increasing availability of soil mineral N have been suggested as the factors leading to significant shifts in community species structure along N–S transects in Europe (Gulden et al. 1992; Taylor et al. 2000)

Since this chapter is primarily concerned with the impacts of the abiotic environment upon EM fungal diversity, we have so far refrained from discussing the reaction of the host plants to perturbations and how this may mediate the response of the fungal community.

Although it is often mentioned in the literature, there is in fact little evidence to support the idea that EM fungi have the capacity for prolonged growth in the absence of a host plant. Consequently, constant contact with a host root system and the continuous acquisition of new contact points will be major factors determining the survival of an EM mycelium. This necessity has been emphasised on a number of occasions (e.g. Deacon and Fleming 1992; Bruns 1995). Replication of contact points (colonised roots) may be achieved by a number of methods: branching of existing tips results in pre-colonised new tips; primary colonisation of uncolonised root tips, and by secondary colonisation involving the replacement of one fungus by another on the same tip. It is unclear whether this last mechanism involves active displacement or as suggested by Marks and Foster (1967), dual colonisation only occurs after the root has broken free from the mantle envelope. The relative importance of these three mechanisms appears to differ markedly among fungal species.

Some fungi rarely produce branched ecto-mycorrhizae, e.g. *Cenococcum geophilum* (Pigott 1982) and *Piceirhiza gelatinosa* (Haug 1989), and for these fungi, replication of contact points via primary and secondary colonisation will be very important. *Cenococcum geophilum* has been known for a long time as a coloniser of senescing tips (Mikola 1948) and it is frequently found as a secondary colonist on tips previously occupied by other fungi. While the frequency of these observations may, in part, be due to its distinctive appearance, secondary colonisation of tips by *C. geophilum* does seem to be an important part of its ecology. Alexander and Fairley (1983) found that the relative abundance of *C. geophilum* mycorrhizae on spruce seedlings decreased following the addition of ammonium sulphate. While a direct effect on the fungus cannot be ruled out, the additional observation that the treatment increased the longevity of the root tips may suggest that the number of tips susceptible to colonisation by *C. geophilum* had decreased. Some support for this idea may come from the observation that root turnover is higher under harsher environmental conditions and it is exactly under these conditions that *C. geophilum* may become a dominant component of the EM fungal community (LoBuglio 1999).

Changes in root turnover rates may have profound effects upon the mycorrhizal community as a whole. Increased rates of turnover will increase the rate at which fungi must establish new contacts with the root system, which could potentially lead to extinctions of less competitive species. Coupled to the turnover is the actual growth rate of new root tips. Sohn (1981) suggested that a threshold growth limit existed for the extension rate of a root above which mycorrhizal formation was progressively restricted. During flushes of root growth, which often occur in the spring and autumn or after a wetting front passes through the soil, large numbers of non-colonised root tips may be produced and this is likely to contribute to enhanced diversity through disturbance enrichment as suggested by Bruns (1995).

EM fungal species differ markedly in the amount of apparent mycelium which they produce (Agerer 1986–1998). Many species within the family Russulaceae form mycorrhizae with remarkably small amounts of extramatrical mycelium and appear to have little contact with the surrounding soil. It therefore seems likely that primary and secondary colonisation of root tips by these fungi is of limited importance. The production of new tips by branching seems to be the predominant form of replication as many species produce highly branched racemose mycorrhizal complexes (Agerer 1986–1998). Conversely, mycorrhizae identified as being formed by species of the genus *Cortinarius* characteristically have copious quantities of EMM associated with the tips (Agerer 1986–1998) and it seems likely that primary and secondary colonisation is of much greater importance for these fungi.

Perturbations, such as N deposition, which have a detrimental affect upon EMM production would be expected to affect the persistence of these two groups of fungi differently. However, the reaction of the individual species must also be considered. For example, *Paxillus involutus* which produces large amounts of EMM is known as a nitrophilous fungus which, in general, is strongly stimulated by N additions (Wallenda and Kottke 1998).

An increasing number of studies relate to anthropogenic influences upon diversity and some general patterns may be emerging from these studies. But there are still remarkably few studies where 'natural' factors, such as soil spatial heterogeneity, are considered. Until we understand the underlying mechanisms through which EM fungal species interact and which maintain the observed high diversity of EM fungal communities, we will remain ignorant of the fundamental tools with which to enhance diversity in degraded systems.

7.6 Future Progress

In this chapter, we have advocated the use of direct analysis of the mycorrhizae as a means by which we can estimate the importance of individual species with EM fungal communities. However, just as the picture obtained from the sporocarps aboveground may differ significantly from that obtained belowground, so might the importance of species be different if we examine the extra-matrical phase of the mycorrhial symbiosis. We already mentioned that species differ markedly in the amount of EMM they produce. In this respect their importance in nutrient uptake may also be very different. In addition, to facilitate efficient nutrient uptake, the terminal elements of the EMM are in close physical contact with the soil environment (see Raidl 1997). A perturbation that directly alters the chemical and/or physical nature of the soil may therefore be experienced by the EMM before the mycorrhizal root tips. This will be particularly true for associations in which the mycorrhizal tips are hydrophobic (Unestam 1991). Also, since the rate of mycelial turnover is likely to be much more rapid than that of fine roots, changes in EMM may be a more sensitive indicator of EM response to environmental perturbations. However, there are at present no methods available to separate, identify and quantify EM-mycelia colonising a soil profile. The rapid development of molecular tools such as quantitative PCR, DNA microarray chips and the use of monoclonal antibodies to investigate the EMM phase of the EM symbiosis are all likely to significantly advance our current knowledge concerning the factors that affect EM fungal community diversity.

Acknowledgements. Financial support from the Swedish Council for Forest and Agricultural Research (SJFR) and the Knut and Alice Wallenberg foundation is gratefully acknowledged.

References

Abuzinadh RA, Read DJ (1986) The role of proteins in the nitrogen nutrition of ecto-mycorrhizal fungi I. Utilization of peptides and proteins by ecto-mycorrhizal fungi. New Phytol 103:481–493

Adams SN, Cooper JE, Dickson DA, Dickson EL, Seaby DA (1978) Some effects of lime and fertilizer on a Sitka spruce plantation. Forestry 5:57–65

Agerer R (1986–1998) Colour atlas of ecto-mycorrhizae. Einhorn-Verlag, Schwäbisch-Gmünd

Agerer R, Taylor AFS, Treu R (1998) Effects of acid irrigation and liming on the production of fruit bodies by ecto-mycorrhizal fungi. Plant Soil 199:83–89

Ahlström K, Persson H, Börjesson I (1988) Fertilization in a mature Scots pine (*Pinus sylvestris* L.) stand – effects on fine roots. Plant Soil 106:179–190

Alexander IJ, Fairley RI (1983) Effects of N fertilisation on populations of fine roots and mycorrhizas in spruce humus. Plant Soil 71:49–53

Andersson S, Söderström B (1995) Effects of lime ($CaCO_3$) on ecto-mycorrhizal colonisation of *Picea abies* (L.) Karst. seedlings planted in a spruce forest. Scand J For Res 10:149–154

Arnebrant K (1994) Nitrogen amendments reduce the growth of extramatrical ecto-mycorrhizal mycelium. Mycorrhiza 5:7–15

Arnebrant K, Söderström (1992) Effects of different fertilizer treatments on ecto-mycorrhizal colonization potential in two Scots pine forests in Sweden. For Ecol Manage 53:77–89

Arnolds E (1991) Decline of ecto-mycorrhizal fungi in Europe. Agric Ecosyst Environ 35:209–244

Baar J, ter Braak CJF (1996) Ecto-mycorrhizal sporocarp occurrence as affected by manipulation of litter and humus layers in Scots pine stands of different age. Appl Soil Ecol 4:61–73

Baar J, Horton TR, Kretzer AM, Bruns TD (1999) Mycorrhizal colonization of *Pinus muricata* from resistant propagules after a stand replacing wildfire. New Phytol 143:409–418

Bakker MR, Garbaye J, Nys C (2000) Effect of liming on the ecto-mycorrhizal status of oak. For Ecol Manage 126:121–131

Baxter JW, Picket STA, Carreiro MM, Dighton J (1999) Ecto-mycorrhizal diversity and community structure in oak forest stands exposed to contrasting anthropogenic impacts. Can J Bot 77:771–782

Bazzaz FA (1975) Plant species diversity in old-field successional ecosystems in southern Illinois. Ecology 56:485–488

Binkley D, Högberg P (1997) Does atmospheric deposition of nitrogen threaten Swedish forests? For Ecol Manage 92:119–152

Brandrud TE (1987) Mycorrhizal fungi in 30 year old, oligotrophic spruce (*Picea abies*) plantation in SE Norway. A one-year permanent plot study. Agarica 8(16):48–58

Bruns TD (1995) Thoughts on the processes that maintain local species diversity of ecto-mycorrhizal fungi. Plant Soil 170:63–73

Brunner I, Brunner F, Laursen GA (1992) Characterisation and comparison of macrofungal communities in an *Alnus tenuifolia* and an *Alnus crispa* forest in Alaska. Can J Bot 70:1247–1258

Colpaert JV, Van Laere A, Van Assche JA (1996) Carbon and nitrogen allocation in ecto-mycorrhizal and non-mycorrhizal *Pinus sylvestris* L. seedlings. Tree Physiol 16(9):787–793

Conn C, Dighton J (2000) The influence of litter quality on mycorrhizal communities. Soil Biol Biochem 32:489–496

Cripps C, Miller OK Jr (1993) Ecto-mycorrhizal fungi associated with aspen on three sites in the north-central Rocky Mountains. Can J Bot 71:1414–1420

Dahlberg A, Jonsson L, Nylund J-E (1997) Species diversity and distribution of biomass above- and below-ground among ecto-mycorrhizal fungi in an old-growth Norway spruce forest in south Sweden. Can J Bot 75:1323–1335

Danielson RM (1984) Ecto-mycorrhizal association in Jack pine stands in north eastern Alberta. Can J Bot 42(5):932–939

Danielson RM (1985) Mycorrhizae and reclamation of stressed terrestrial environments. In: Robert LT, Klein DA (eds) Soil reclamation processes. Dekker, New York, pp 173–201

Danielson RM, Visser S (1989) Effects of forest soil acidification on ecto-mycorrhizal and vesicular-arbuscular mycorrhizal development. New Phytol 112:41–48

Deacon JW, Fleming LV (1992) Interactions of ecto-mycorrhizal fungi. In: Allen MJ (ed) Mycorrhizal functioning. An integrated plant-fungal process. Chapman and Hall, New York, pp 249–300

Dighton J, Jansen AE (1991) Atmospheric pollutants and ecto-mycorrhizae: more questions than answers? Environ Poll 73:179–204

Dighton J, Mason PA (1985) Mycorrhizal dynamics during forest tree development. In: Moore D, Casselton LA, Wood DA, Frankland JC (eds) Developmental biology of higher Fungi. Cambridge Univ Press, Cambridge, pp 117–139

Dighton J, Skeffington RA (1987) Effects of artificial acid precipitation on the mycorrhizas of Scots pine seedlings. New Phytol 107:191–202

Dighton J, Morale Bonilla AS, Jiminez-Nunez RA, Martinez N (2000) Determinants of leaf litter patchiness in mixed species New Jersey pine barrens forest and its possible influence on soil and soil biota. Biol Fertil Soils 31:288–293

Edwards GS, Kelly JM (1992) Ecto-mycorrhizal colonisation of loblolly-pine seedlings during 3 growing seasons in response to ozone, acid precipitation and soil Mg status. Environ Poll 76:71–77

Egger KN (1995) Molecular analysis of ecto-mycorrhizal fungal communities. Can J Bot 73 [Suppl 1]:S1415–S1422

Emmett BA, Brittain SA, Hughes S, Görres J, Kennedy V, Norris D, Rafarel R, Reynolds B, Stevens PA (1995) Nitrogen additions ($NaNO_3$ and NH_4NO_3) at Aber forest Wales I. Response of throughfall and soil water chemistry. For Ecol Manage 71:45–59

Erland S (1990) Effects of liming on pine ecto-mycorrhiza. Doctoral Thesis, Lund University, ISBN 91-7105-014-0

Erland S, Finlay RD (1992) Effects of temperature and incubation time on the ability of three ecto-mycorrhizal fungi to colonise *Pinus sylvestris* roots. Mycol Res 96:270–272

Erland S, Söderström B (1991) Effects of lime and ash treatments on ecto-mycorrhizal infection of *Pinus sylvestris* L. seedlings planted in a pine forest. Scand J For Res 6:519–526

Erland S, Taylor AFS (1999) Resupinate Ecto-mycorrhizal Fungal Genera. In: Cairney JWG, Chambers SM (eds) Ecto-mycorrhizal fungi: key genera in profile. Springer, Berlin Heidelberg New York, pp 347–363

Erland S, Jonsson T, Mahmood S, Finlay RD (1999) Below-ground ecto-mycorrhizal community structure in two *Picea abies* forests in southern Sweden. Scand J For Res 14:209–217

Evers FH, Huettl RF (1991) A new fertilisation strategy in declining forests. In: Zöttl HW, Huettl RF (eds) Management and nutrition of forests under stress. Kluwer, Dordrecht, pp 495–508

Fogel R (1976) Ecological studies of hypogeous fungi II. Sporocarp phenology in a western Oregon Douglas fir stand. Can J Bot 54:1152–1162

Fogel R (1980) Mycorrhizae and nutrient cycling in natural forest ecosystems. New Phytol 86:199–212

Fransson PMA, Taylor AFS, Finlay RD (2000) Effects of optimal fertilization on belowground ecto-mycorrhizal community structure in a Norway spruce forest. Tree Physiol 20:599–606

Gardes M, Bruns TD (1996) Community structure of ecto-mycorrhizal fungi in a *Pinus muricata* forest: above- and below-ground views. Can J Bot 74:1572–1583

Gardes M, White TJ, Fortin JA, Bruns TD, Taylor JW (1991) Identification of indigenous and introduced symbiotic fungi in ecto-mycorrhizae by amplification of nuclear and mitochondrial ribosomal DNA. Can J Bot 69:180–190

Gardner JH, Malajczuk N (1984) Recolonisation by mycorrhizal fungi of rehabilitated bauxite mine sites in Western Australia. For Ecol Manage 24:27–42

Giller PS (1996) The diversity of soil communities, the 'poor man's tropical rainforest'. Biol Con 5:135–168

Glenn MG, Wagner WS, Webb SL (1991) Mycorrhizal status of mature red spruce (*Picea rubens*) in mesic and wetland sites of northwestern New Jersey. Can J For Res 21:741–749

Godbold DL, Berntson GM, Bazzaz FA (1997) Growth and mycorrhizal colonization of three North American tree species under elevated CO_2. New Phytol 137:433–440

Goulden ML, Wofsy SC, Harden JW, Trumbore SE, Crill PM, Gower ST, Fries T, Daube BC, Fan SM, Sutton DJ, Bazzaz A, Munger JW (1998) Sensitivity of boreal forest carbon balance to soil thaw. Science 279:214–217

Gulden G, Hoiland K, Bendiksen K, Brandrud TE, Foss BS, Jenssen HB (1992) Macromycetes and air pollution: mycocoenological studies in oligotrophic spruce forests in Europe. Bibl Mycol 144. Cramer, Stuttgart

Hantschel RE, Kamp T, Beese F (1995) Increasing the soil temperature to study global warming effects on the soil nitrogen cycle in agroecosystems. J Biogeogr 22(2/3):375–380

Harrison AF, Harkness DD, Rowland AP, Garnett JS, Bacon PJ (2000) Annual carbon and nitrogen fluxes in soils along the European transect, determined using ^{14}C-bomb. In: Schulze E-D (ed) Ecological studies, vol 142. Springer, Berlin Heidelberg New York

Hartley J, Cairney JWG, Meharg AA (1997) Do ecto-mycorrhizal fungi exhibit adaptive tolerance to potentially toxic metals in the environment? Plant Soil 189:303–319

Hartley J, Cairney JWG, Freestone P, Woods C, Meharg AA (1999) The effects of multiple metal contamination on ecto-mycorrhizal Scots pine (*Pinus sylvestris*) seedlings. Environ Pollut 106:413–424

Harvey AE, Larsen MJ, Jurgensen MF (1976) Distribution of ecto-mycorrhizae in a mature Douglas fir larch forest soil in western Montana. For Sci 22(4):393–398

Harvey AE, Larsen MJ, Jurgensen MF (1978) Comparative distribution of ecto-mycorrhizae in soils of three western Montana forest habitat types. For Sci 25(2):350–358

Hatch AB (1937) The physical basis of mycotrophy in the genus *Pinus*. Black Rock For Bull 6:168

Haug I (1989) Intracellular infection in the meristematic region of '*Piceirhiza gelatinosa*' mycorrhizas. New Phytol 111:203–207

Haug I, Feger KH (1990/1991) Effects of fertilisation with $MgSO_4$ and $(NH_4)_2SO_4$ on soil solution chemistry, mycorrhiza and nutrient content of fine roots in a Norway spruce stand. Water Air Soil Pollut 54:453–467

Hilton RN, Malajczuk N, Pearce MH (1989) Larger fungi of the Jarrah forest: an ecological and taxonomic survey. In: Dell B (ed) The Jarrah forest. Kluwer, Dordrecht

Horton TR, Cazares E, Bruns TD (1998) Ecto-mycorrhizal, vesicular-arbuscular and dark septate fungal colonization of bishop pine (*Pinus muricata*) seedlings in the first 5 months of growth after wildfire. Mycorrhiza 8:11–18

Hung LL, Trappe JM (1983) Growth variation between and within species of ecto-mycorrhizal fungi in response to pH in vitro. Mycologia 75:234–241

Janos DP (1980) Mycorrhizae influence tropical succession. Biotropica 12:56–54
Jeffries P (1999) *Scleroderma*. In: Cairney JWG, Chambers SM (eds) Ecto-mycorrhizal fungi: key genera in profile. Springer, Berlin Heidelberg New York, pp 187–200
Johnson CN (1995) Interactions between fire, mycophagous mammals, and dispersal of ecto-mycorrhizal fungi in *Eucalyptus* forests. Oecologia 104:467–475
Jonsson L, Dahlberg A, Nilsson M-C, Zackrisson O, Kårén O (1999) Ecto-mycorrhizal fungal communities in late-successional Swedish boreal forest, and their composition following wildfire. Mol Ecol 8:205–215
Jonsson T, Kokalj S, Finlay RD, Erland S (1999) Ecto-mycorrhizal community structure in a limed spruce forest. Mycol Res 103:501–508
Kårén O, Nylund J-E (1996) Effects of N-free fertilization on ecto-mycorrhiza community structure in Norway spruce stands in southern Sweden. Plant Soil 181:295–305
Klironomos JN, Kendrick B (1995) Relationships among microarthropods, fungi, and their environment. In: Collins HP, Robertson GP, Klug MJ (eds) The significance and regulation of soil biodiversity. Kluwer, Dortrecht, pp 209–223
Kraigher H, Batic F, Agerer R (1996) Types of ecto-mycorrhizae and mycobioindication of forest site pollution. Phyton 36:115–120
Kropp BR (1982) Formation of mycorrhizae on nonmycorrhizal western hemlock outplanted on rotten wood and mineral soil. For Sci 28(4):706–710
Kumpfer W, Heyser W (1986) Effects of stem flow of Beech (*Fagus sylvatica* L.). In: Gianinazzi-Pearson V, Gianinazzi S (eds)Physiological aspects and genetical aspects of mycorrhizae. Proceedings of the 1st European Symposium on Mycorrhizae. Dijon, 1–5 July 1985, INRA, pp 745–750
Lange M (1978) Fungus flora in August. Ten year observations in a Danish beech wood district. Bot Tidsskr 73:21–54
Last FT, Mason PA, Wilson J, Ingleby K, Munrow RC, Fleming LV, Deacon JW (1985) Epidemiology of sheathing (ecto-) mycorrhizas in unsterile soil. A case study of *Betula pendula*. Proc R Soc Edinb 83B:299–315
Leake JR, Read DJ (1990) Proteinase activity in mycorrhizal fungi. I. The effect of extracellular pH on the production and activity of proteinase by ericoid endophytes from soils of contrasting pH. New Phytol 115:243–250
Leake JR, Read DJ (1997) Mycorrhizal fungi in terrestrial ecosystems. In: Wicklow D, Soderström B (eds) The Mycota IV. Experimental and microbial relationships. Springer, Berlin Heidelberg New York, pp 281–301
Lehto T (1984) Kalkituksen vaikutus männyn mykoritsoihin. Folia For 609:1–20
Lehto T (1994) Effects of liming and boron fertilization on mycorrhizas of *Picea abies*. Plant Soil 163:65–68
Leyval C, Turnau K, Haselwandter K (1997) Effect of heavy metal pollution on mycorrhizal colonization and function: physiological, ecological and applied aspects. Mycorrhiza 7:139–153
Lilleskov EA, Fahey TJ (1996) Patterns of ecto-mycorrhizal diversity over an atmospheric nitrogen deposition gradient near Kenai, Alaska. In: Szaro TM, Bruns TD (eds) Abstracts of the 1st International Conference on Mycorrhizae. Univ California, Berkeley, 76
Lo Buglio KF (1999) *Cenococcum*. In: Cairney JWG, Chambers SM (eds) Ecto-mycorrhizal Fungi: key genera in profile. Springer, Berlin Heidelberg New York, pp 287–309
Lodge DJ, Wentworth TR (1990) Negative associations among VA-mycorrhizal fungi and some ecto-mycorrhizal fungi inhabiting the same root system. Oikos 57:347–356
Lorio PL Jr, Howe VK, Martin CN (1972) Loblolly pine rooting varies with microrelief on wet sites. Ecology 53:1134–1140
Lu X, Malajczuk N, Brundrett M, Dell B (1999) Fruiting of putative ecto-mycorrhizal fungi under blue gum (*Eucalyptus globulus*) plantations of different ages in Western Australia. Mycorrhiza 8:255–261

Magurran AE (1988) Ecological diversity and its measurement. Croom Helm, London
Mahmood S (2000) Ecto-mycorrhizal community structure and function in relation to forest residue harvesting and wood ash applications. Doctoral Thesis, Lund University. ISBN 91-7105-136-8
Mahmood S, Finlay RD, Erland S (1999) Effects of repeated harvesting of forest residues on the ecto-mycorrhizal community in a Swedish spruce forest. New Phytol 142:577-585
Malajczuk N (1987) Ecology and management of ecto-mycorrhizal fungi in regenerating forest ecosystems in Australia. In: Sylvia DM, Hung LL, Graham JH (eds) Mycorrhizae in the next decade. 7th NACOM. Gainesville, Florida
Malajczuk N, Hingston FJ (1981) Ecto-mycorrhizae associated with Jarrah. Aust J Bot 29:453-462
Marks GC, Foster RC (1967) Succession of mycorrhizal associations on individual roots of *Radiata* pine. Aust For 31:194-201
Marschner H (1995) Mineral nutrition of higher plants, 2nd edn. Academic Press/Harcourt Brace, London
Mason PA, Last FT, Pelham J, Ingelby K (1982) Ecology of some fungi associated with an ageing stand of birches (*Betula pendula* and *Betula pubescens*). For Ecol Manage 4:19-39
Mason P, Wilson J, Last FT, Walker C (1983) The concept of succession in relation to the spread of sheathing mycorrhizal fungi on inoculated tree seedlings growing in unsterile soils. Plant Soil 71:247-256
McAfee BJ, Fortin JA (1987) The influence of pH on the competitive interactions of ecto-mycorrhizal mycobionts under field conditions. Can J For Res 17:859-864
Mehmann B, Egli S, Braus GH, Brunner I (1995) Coincidence between molecularly and morphologically classified ecto-mycorrhizal morphotypes and fruitbodies in a spruce forest. In: Stocchi V, Bonfante P, Nuti M (eds) Biotechnology of Ecto-mycorrhizae. Plenum Press, New York, pp 229-239
Meier S (1991) Quality versus quantity: optimising evaluation of ecto-mycorrhizae for plants under stress. Environ Pollut 73:205-216
Meier S, Robarge WP, Bruck RI, Grand LF (1989) Effects of simulated rain acidity on ecto-mycorrhizae of red spruce seedlings potted in natural soil. Environ Pollut 59:315-324
Meyer FH (1962) Die Buchen und Fichtenmykorrhiza in verschiedenen Bodentypen, ihre Beeinflussung durch Mineraldünger sowie für die Mykorrhizabildung wichtige Faktoren. Mitteilungen der Bundesforschungsanstalt für Forst- und Holzwirtschaft 54:1-73
Meyer J, Schneider BU, Werk K, Oren R, Schulze E-D (1988) Performance of two *Picea abies* (L.) Karst. stands at different stages of decline. Oecologia 77:7-13
Mikola (1948) On the physiology and ecology of *Cenococcum graniformae* especially as a mycorrhizal fungus on birch. Inst For Fenn Commun 36:1-104
Mitchell-Olds T (1992) Does environmental variation maintain genetic variation? A question of scale. TREE 7(12):397-398
Newton AC (1992) Towards a functional classification of ectomycorrhizal fungi. Mycorrhiza 2:75-79
Newton AC, Haig JM (1998) Diversity of ecto-mycorrhizal fungi in Britain: a test of the species area relationship and the role of host specificity. New Phytol 138:619-627
Nitare J (1988) Changes in the mycoflora – research and species protection. Svensk Bot Tidskr 82:485-489
Norby RJ, O'Neill, EG, Hood WG, Luxmoore RJ (1987) Carbon allocation, root exudation, and mycorrhizal colonisation of *Pinus echinata* seedlings grown under CO_2 enrichment. Tree Physiol 3:203-210

O'Neill EG, Luxmoore RJ, Norby RJ (1987) Increase in mycorrhizal colonisation and seedling growth in *Pinus echinata* and *Quercus alba* in an enriched CO_2 atmosphere. Can J For Res 17:878–883

Persson H, Ahlström K (1994) The effects of alkalizing compounds on fine-root growth in a Norway spruce stand in southwest Sweden. J Environ Sci Health 29:803–820

Persson T, van Oene H, Harrison AF, Karlsson P, Bauer G, Cerny J, Coûteaux M-M, Dambrine E, Högberg P, Kjøller A, Matteucci G, Rudebeck A, Schulze E-D, Paces T (2000) Experimental sites in the NIPHYS/CANIF project. In: Schulze E-D (ed) Carbon and nitrogen cycling in European forest ecosystems. Ecological studies, vol 142. Springer, Berlin Heidelberg New York, pp 14–46

Pigott CD (1982) Survival of mycorrhizas formed by *Cenococcum geophilum* Fr. in dry soils. New Phytol 92:513–517

Pritsch K, Boyle H, Munch JC, Buscot F (1997) Characterization and identification of black alder ecto-mycorrhizas by PCR/RFLP analyses of the rDNA internal transcribed spacer (ITS). New Phytol 137:357–369

Putnam R (1994) Community ecology. Chapman and Hall, London

Qian XM, Kottke I, Oberwinkler F (1998) Influence of liming and acidification on the activity of the mycorrhizal communities in a *Picea abies* (L.) Karst. stand. Plant Soil 199:99–109

Qiu Z, Chepellka AH, Somers GL, Lockaby BG, Meldahl RS (1993) Effects of Ozone and simulated acid precipitation on ecto-mycorrhizal formation on loblolly pine seedlings. Environ Exp Bot 33:423–431

Raidl S (1997) Studien zur Ontigenie an Rhizomorphen von Ektomycorrhizen. Bibl Mykol 169:1–184

Read DJ (1991) Mycorrhizas in ecosystems. Experientia 47:376–391

Reddell P, Malajczuk N (1984) Formation of ecto-mycorrhizae by Jarrah (*Eucalyptus marginata* Donn ex Smith) in litter and soil. Aust J Bot 32:511–520

Rey A, Jarvis PG (1997) Growth response of young birch trees (*Betula pendula* Roth.) after four and a half years of CO_2 exposure. Ann Bot 80:809–816

Richardson MJ (1970) Studies on *Russula emetica* and other agarics in a Scots pine plantation. TBMS 55:217–229

Roth DR, Fahey TJ (1998) The effects of acid precipitation and ozone on the ecto-mycorrhiza of red spruce saplings. Water Air Soil Poll 103:263–276

Rouhier H, Read DJ (1998) Plant and fungal responses to elevated atmospheric carbon dioxide in mycorrhizal seedlings of *Pinus sylvestris*. Environ Exp Bot 40(3):237–246

Rouhier H, Read DJ (1999) Plant and fungal responses to elevated atmospheric CO_2 in mycorrhizal seedlings of *Betula pendula*. Environ Exp Bot 42(3):231–241

Runion GB, Mitchell RJ, Rogers HH, Prior SA, Counts TK (1997) Effects of nitrogen and water limitation and elevated atmospheric CO_2 on ecto-mycorrhiza of longleaf pine. New Phytol 137:681–689

Rühling Å, Söderström B (1990) Changes in fruitbody production of mycorrhizal and litter decomposing macromycetes in heavy metal polluted coniferous forests in north Sweden. Water Air Soil Poll 49:375–387

Rygievicz PT, Johnson MG, Ganio LM, Tingey DT, Storm MJ (1997) Lifetime and temporal occurrence of ecto-mycorrhizae on ponderosa pine (*Pinus ponderosa* Laws.) seedlings grown under varied atmospheric CO_2 and nitrogen levels. Plant Soil 189:275–287

Saunders E, Taylor AFS, Read DJ (1996) Ecto-mycorrhizal community response to simulated pollutant nitrogen deposition in a Sitka spruce stand, North Wales. In: Szaro TM, Bruns TD (eds) Abstracts of the 1st international conference on Mycorrhizae. Univ California, Berkeley 106

Smith SE, Read DJ (1997) Mycorrhizal symbiosis, 2nd edn. Academic Press, London

Sohn RF (1981) *Pisolithus tinctorius* forms long ecto-mycorrhizae and alters root development in seedlings of *Pinus resinosa*. Can J Bot 59:2129–2134

Stendell ER, Horton TR, Bruns TD (1999) Early effects of prescribed fire on the structure of the ecto-mycorrhizal fungus community in a Sierra Nevada ponderosa pine forest. Mycol Res 103:1353–1359

Tamm CO (1991) Nitrogen in terrestrial ecosystems. Springer, Berlin Heidelberg New York

Taylor AFS, Martin F, Read DJ (2000) Fungal diversity in ecto-mycorrhizal communities of Norway spruce (*Picea abies* [L.] Karst.) and Beech (*Fagus sylvatica* L.) along north-south transects in Europe. In: Schulze E-D (ed) Ecological studies, vol 142. Springer, Berlin Heidelberg New York, pp 343–365

Taylor DL, Bruns TD (1999) Community structure of ecto-mycorrhizal fungi in a *Pinus muricata* forest: minimal overlap between the mature forest and resistant propagule communities. Mol Ecol 8:1837–1850

Termorshuizen AJ (1991) Succession of mycorrhizal fungi in stands of *Pinus sylvestris* in the Netherlands. J Veg Sci 2:555–564

Termorshuizen AJ (1993) The influence of nitrogen fertilizers on ecto-mycorrhizas and their fungal carpophores in young stands of *Pinus sylvestris*. For Ecol Manage 57:179–189

Termorshuizen AJ, Schaffers AP (1987) Occurrence of carpophores of ecto-mycorrhizal fungi in selected stands of *Pinus sylvestris* in the Netherlands in relation to stand vitality and air pollution. Plant Soil 104:209–217

Thomson BD, Grove TS, Malajczuk N, Hardy GStJ (1994) The effectiveness of ecto-mycorrhizal fungi in increasing the growth of *Eucalyptus globulus* Labill, in relation to root colonisation and hyphal development in soil. New Phytol 126:517–524

Tibbett M, Sanders FE, Cairney JWG (1998a) The effect of temperature and inorganic phosphorus supply on growth and acid phosphatase production in arctic and temperate strains of ecto-mycorrhizal *Hebeloma* species in axenic culture. Mycol Res 102:129–135

Tibbett M, Grantham K, Sanders FE, Cairney JWG (1998b) Induction of cold active acid phosphomonoesterase activity at low temperature in pychrotrophic ecto-mycorrhizal *Hebeloma* spp. Mycol Res 102:1533–1539

Tibbett M, Sanders FE, Cairney JWG, Leake JR (1999) Temperature regulation of extracellular proteases in ecto-mycorrhizal fungi (*Hebeloma* spp.) grown in axenic culture. Mycol Res 103:707–714

Tokeshi M (1993) Species abundance patterns and community structure. Adv Ecol Res 24:112–186

Torres P, Honrubia M (1997) Changes and effects of natural fire on ecto-mycorrhizal inoculum potential of soil in a *Pinus halepensis* forest. For Ecol Manage 96:189–196

Unestam T (1991) Water repellency, mat formation, and leaf stimulated growth of some ecto-mycorrhizal fungi. Mycorrhiza 1:13–20

Visser (1995) Ecto-mycorrhizal fungal succession in jack pine stands following wild fire. New Phytol 129:389–401

Wallander H, Nylund JE (1992) Effects of excess nitrogen and phosphorus starvation on the extramatrical mycelium of ecto-mycorrhizas of *Pinus sylvestris* L. New Phytol 120:495–503

Wallenda T, Kottke I (1998) Nitrogen deposition and ecto-mycorrhiza. New Phytol 139:169–187

Wallenda T, Read DJ (1999) Kinetics of amino acid uptake by ecto-mycorrhizal roots. Plant Cell Environ 22(2):179–187

Wardle DA, Zackrisson O, Hörnberg G, Gallet C (1997) The influence of island size area upon ecosystem properties. Science 277:1296–1299

Wollecke J, Münzenberger B, Huettl RF (1999) Some effects of N on ecto-mycorrhizal diversity of Scots pine (*Pinus sylvestris* L.) in northeastern Germany. Water Air Soil Pollut 116:135–140

Worley JF, Hacskaylo E (1959) The effects of available soil moisture on the mycorrhizal associations of Virginia pine. For Sci 5:267–268

Yang G, Cha JY, Shibuya M, Yajima T, Takahashi K (1998) The occurrence and diversity of ecto-mycorrhizas of *Larix kaempferi* seedlings on a volcanic mountain in Japan. Mycol Res 102:1503–1508

Zackrisson O (1977) Influence of forest fires on the north Swedish boreal forest. Oikos 29:22–32

8 Genetic Studies of the Structure and Diversity of Arbuscular Mycorrhizal Fungal Communities

Justin P. Clapp, Thorunn Helgason, Tim J. Daniell, J. Peter W. Young

Contents

8.1	Summary	202
8.2	Introduction	202
8.2.1	The Importance of Community Structure in Arbuscular Mycorrhizal Fungi	202
8.2.2	Molecular Identification of Arbuscular Mycorrhizal Fungi	203
8.3	Strategies for Identifying Arbuscular Mycorrhizal Fungi Using Molecular Markers	205
8.3.1	Methods Available	205
8.3.2	Primers Developed for Analysis of Arbuscular Mycorrhizal Fungal Ribosomal Genes: A Short History	206
8.4	Investigations of Arbuscular Mycorrhizal Fungal Communities	208
8.4.1	Primers that Target Single Families or Species	208
8.4.2	Primers Intended to Target all Glomalean Fungi	209
8.4.3	Alternative Approaches to Molecular Characterisation of Arbuscular Mycorrhizal Fungi	212
8.5	Comparison of Molecular and Morphological Data	212
8.5.1	Phylogeny and Taxonomy	213
8.5.2	Field Data	213
8.6	The Genetic Organisation of Arbuscular Mycorrhizal Fungi	215
8.6.1	Heterogeneous Sequences in Spores and Cultures	215
8.6.2	"Contaminant" Sequences in Spores and Cultures	217
8.6.3	How Is the Sequence Heterogeneity Organised?	218
8.6.4	How Is Sequence Heterogeneity Maintained?	219
8.6.5	How Does This Affect Analysis of Field Sequences?	220
8.7	Future Prospects	221
References		221

8.1 Summary

Recent studies have shown that arbuscular mycorrhizal fungi (AMF) are important components of ecosystem diversity. Traditional taxonomy suggests that global diversity is low, yet it is common to find high local diversity in field soils. Molecular methods of identification have, for the first time, allowed a characterisation of the diversity of AMF within actively growing plant roots. In this chapter, we present a summary of the AMF community structure that has been revealed in field-based studies. In particular, there are many AMF genotypes present in field systems that have not been established in culture. The genetic structure of AMF isolates has been shown to be complex. This profoundly affects the interpretation of genetic data obtained from AMF that cannot be cultured. We will review this data, and how it might make an impact on molecular analyses of field samples.

8.2 Introduction

8.2.1 The Importance of Community Structure in Arbuscular Mycorrhizal Fungi

Arbuscular mycorrhizal fungi (Zygomycota: Glomales) are ubiquitous in terrestrial ecosystems although only approximately 150 species have been described. AMF colonise species from around 90% of plant families. An apparently low taxonomic diversity coupled with a broad geographic range has led to the view that AMF are a rather homogeneous group, both functionally and morphologically. It is routinely possible, however, to find 10–30 spore types in the soil at a single site. This high local diversity against a background of low global diversity is paradoxical if all fungi are equally capable of colonising all plants. The apparently high genetic diversity of AMF is similarly paradoxical in relation to the low morphological diversity. Improved methods of isolation from soil, and of microscopy and molecular analysis, have shown that there may be substantial diversity present in field soils, but the relationship between morphological and molecular diversity is still unclear. As no way has yet been found to grow AMF independently of their plant hosts, it has been difficult to arrive at a reliable and objective definition of a species. In addition, traditional ways of naming AMF have relied on pot cultures, which we now know to represent a limited subset of the genetic diversity present in field soils. What is more, only a relatively small number of the 156 described species are routinely maintained in pot culture. The development of alternative, molecular, strategies to delimit our concept of species and to provide tools for the study of AMF communities in situ has therefore become a priority. Nevertheless, until more information is available on the definition of a "species" in AMF, we will continue to use the term in this chapter.

The assumed primary benefit to plants of the mycorrhizal symbiosis is an increased uptake of immobile nutrients, especially phosphorus that are mobilised by the fungus. However, there is increasing evidence that AMF have a range of other effects, for example, protection against pathogens, water stress tolerance (Newsham et al. 1995) and improved soil structure (Wright and Upadyaya 1998). This evidence of multiple functions, host selectivity (or non-random distribution) and higher diversity than is apparent from cultures and morphology leads us to conclude that communities of AMF are much more diverse that previously thought. At the community level, recent experiments have shown that AMF are major players and may influence plant species diversity (van der Heijden et al. 1998; Klironomos et al. 2000; also see Chaps. 9, 10, this Vol., by the same authors) and recruitment of plant species into the population (Kiers et al. 2000). Colonisation of host plants by AMF exhibits significant spatial and temporal heterogeneity (Merryweather and Fitter 1998b; Helgason et al. 1999). High spore diversity has been found in the field (Bever et al. 1996; Eom et al. 2000) and the number of spores produced by each AM fungus may be host-dependent (Sanders and Fitter 1992; Bever et al. 1996). Traditionally, AMF community diversity has been measured using spore counts from field soils, but such counts are a measure of the sporulation activity of the fungi rather than a direct measurement of diversity, and may be highly variable and almost impossible to correlate among habitat types (Morton et al. 1995). Molecular genetic analysis provides a way around this obstacle as it has the potential to identify objectively the taxa present in the roots of plants. In Chapter 11 (this Vol.), Bever et al. discuss the merits of studying AMF communities based on spore counts and present evidence for why spore counts may be a good measure of AMF fitness in the roots of some plant species.

8.2.2 Molecular Identification of Arbuscular Mycorrhizal Fungi

In any study of the functional ecology, biodiversity and phylogenetic relationships of a group of organisms, it is necessary to distinguish individual taxa consistently within the group. This applies equally to AMF communities. Molecular analysis has provided the first opportunity to consistently identify taxa present in the roots of plants. Much of the field data described in this paper shows how important this is, as the dominant taxa in field roots are often those that are neither taken into culture nor found in the soil as spores (Fig. 8.1; Helgason et al. 1998, 1999). As we shall see in this chapter, however, the genetics of AMF have proved to be unexpectedly complex. It is essential, therefore, that this work be carried out in tandem with functional and genetic

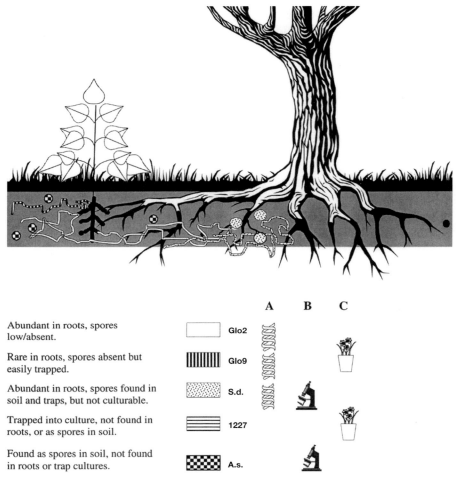

Fig. 8.1. Resolution of methods revealing AMF diversity. Five AMF from Pretty Wood are illustrated: *Glo2* and *Glo9* (sequence types, Helgason et al. 1998), *S.d* (Scut1), 1227 (*Glomus* sp. trap cultured) and *A.s.* (*Acaulospora scrobiculata*, spore type found in soil), Fig. 8.3. The methods that have detected them are shown; *A* sequence type detected in roots, *B* spores found in soil, *C* cultured isolates. It is clear that no single method detects all types present in the soil and roots

studies on cultures in order that we may be able to interpret the data obtained from non-culturable AMF. In this chapter, we review studies that have measured diversity of AMF in field soils.

8.3 Strategies for Identifying Arbuscular Mycorrhizal Fungi Using Molecular Markers

8.3.1 Methods Available

The first application of molecular methods to the study of AMF (Simon et al. 1992) was published just a decade ago, but molecular analysis has already revolutionised the study of AMF. It has opened up possibilities that could only be dreamt of for studying AMF in the laboratory and, especially, in the field. Early molecular studies showed conclusively that the population of spores (upon which most taxonomic criteria are based) in the soil may not reflect root colonisation (Clapp et al. 1995). Molecular methods allow the identification of AMF taxa independently of morphological criteria, and potentially at low levels of colonisation. Analysis of AMF from field samples relies on the Polymerase Chain Reaction (PCR) that allows the DNA from small quantities of starting material to be bulked up into sufficient quantities for analysis. AMF DNA can thus be targeted in a background of DNA of different origin (e.g. within a plant root) or from small amounts of template such as a single spore.

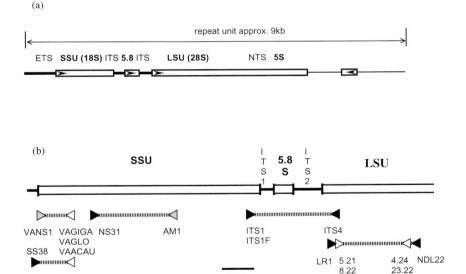

Fig. 8.2.a A complete repeat unit of fungal ribosomal RNA genes, showing the structural RNA genes *white boxes* and the transcribed *thick* and untranscribed *thin spacers*. *Arrows* indicate the direction of transcription. **b** An enlargement showing the positions of the PCR primers and products mentioned in this chapter. *Solid triangles* General primers; *grey triangles* primers for AMF; *white triangles* primers for specific AM taxa

The data from PCR depend upon the design of the primers used to initiate the reaction. The specificity of the primer determines the range of sequences that may be obtained. In theory, it is possible to target a group at any taxonomic level, provided the appropriate sequence can be identified. Studies on AMF have used primers ranging in specificity from those designed to amplify sequences from a single isolate up to those that recognise all AMF. Several general manuals describe the execution and analysis of molecular methods for ecology (e.g. Avise 1994; Carvalho 1998; Innis et al. 1990).

The majority of nucleic acid information derived from the Glomales is from the ribosomal RNA genes. In most organisms, the ribosomal RNA genes are present in multiple copies arranged in tandem arrays, with each repeat unit consisting of a small subunit (SSU or 18S) and a large subunit (LSU or 28S) ribosomal RNA gene separated by an internal transcribed spacer (ITS) that includes the 5.8S rRNA gene (Fig. 8.2). The multiple copies are normally very similar in sequence due to the genetic processes of unequal crossing over and gene conversion. The maintenance of the same sequence in several copies of the same gene is known as concerted evolution. It is presumed that the ribosomal genes of the Glomales are similarly arranged, but an unusual feature that has come to light is the frequent occurrence of multiple rRNA gene sequences, sometimes quite divergent, in a single spore or culture.

8.3.2 Primers Developed for Analysis of Arbuscular Mycorrhizal Fungal Ribosomal Genes: A Short History

In the first molecular analysis of AMF, Simon et al. (1992) isolated DNA from spores of *Glomus intraradices* and *Gigaspora margarita* and sequenced the SSU rRNA gene using PCR fragments generated with universal eukaryotic primers (Fig. 8.2). Using this information to generate a taxon-specific primer (VANS1), it was possible to amplify sequences from several AMF directly from plant roots. A phylogenetic analysis of sequences from spores using VANS1 showed that the molecular phylogeny was congruent with the family classification, and that the sequence divergence was consistent with the fossil record (Simon et al. 1993a). Group-specific primers VAGLO (*Glomus*), VAACAU (*Acaulospora*), VALETC (*G. etunicatum*) and VAGIGA (*Gigaspora* and *Scutellospora*) were subsequently designed by Simon et al. (1993b) and used to amplify unknown taxa from plant roots, revealing the possibilities of such an approach. However, in practice, other workers found it difficult to amplify products directly from roots using VANS1 (Clapp et al. 1995) and it was subsequently found that the VANS1 sequence does not occur, or is only a minor sequence motif, in some taxa (Clapp et al. 1999; Schüßler et al. 2001).

Helgason et al. (1998) designed the AM1 primer to exclude plant sequences and to amplify AMF sequences preferentially. The design of this primer was based on a larger amount of reference sequence information than was available to earlier studies. In combination with the general primer NS31, AM1 amplifies a section of the SSU gene that has a high degree of informative sequence variation (Simon 1996). A drawback is that this primer combination also amplifies sequences from some other fungi, notably pyrenomycetes that frequently occur in samples collected from some habitat types. While the AMF that fall within the three long-accepted families can be detected successfully using this primer, it does not match the SSU sequences of the more divergent clades described by Redecker et al. (2000), which are now designated as the new families Archaeosporaceae and Paraglomaceae (see http://invam.caf.wvu.edu/). Redecker (2000) has described SSU primers that are specific for each of the five Glomalean families.

Other studies have focused on the ITS region of the rRNA gene family. The ITS has been used extensively for molecular taxonomy, and universal primer sequences ITS1 and ITS4 were published by White et al. (1990). These primers were used to obtain PCR-restriction fragment length polymorphism (RFLP) and sequence information from spores collected from the field (Sanders et al. 1995), and to study the relationships among AMF species (Lloyd-MacGilp et al. 1996; Redecker et al. 1997). This is a robust primer combination, and ITS1 and ITS4, with the fungal-specific variant ITS1-F (Gardes and Bruns 1993), have not been superseded. Until the genetic structure of AMF populations is better understood, however, the utility of ITS variation for field identification is limited because of the high level of diversity observed within species and even within spores (see below).

More recently, taxon-specific primers have been designed to amplify fragments (ca. 450 bp) of the large subunit (LSU) rRNA gene (van Tuinen et al. 1998). These have yet to be applied to field samples, but have been successfully used to amplify inoculated taxa in microcosms.

The design of PCR primers for community analysis necessarily evolves as more information becomes available. It is a delicate balance between generality, needed so that important target organisms are not missed (Schüßler et al. 2001), and specificity, to avoid being swamped by non-target organisms (with an estimated 20 % of spores yielding cloned sequences that are considered not to be glomalean, Schüßler et al. 2001). The history of molecular analysis of AMF is one of the evolution of more precise targeting. No single system will satisfy all purposes, and the regular evaluation of a system is necessary as more reference sequence information becomes available and a better understanding of the genetic diversity of AMF is acquired.

8.4 Investigations of Arbuscular Mycorrhizal Fungal Communities

The application of molecular techniques to provide an insight into the diversity of AMF in the environment has been surprisingly limited. The majority of community studies have used microcosms containing particular isolates of described species of fungi.

8.4.1 Primers That Target Single Families or Species

The use of family- or species-specific primer sets has demonstrated the power of molecular approaches for positive identification of fungi. Targets have been the SSU of the ribosomal gene complex (Simon et al. 1992; Di Bonito et al. 1995), the LSU (Van Tuinen et al. 1998) and the ITS (Millner et al. 1998). Specific primers have also been generated from random amplified polymorphic DNA (RAPD) analysis of spores (Wyss and Bonfante 1993; Abbas et al. 1996) or characterisation of highly repeated DNA sequences (Zézé et al. 1996; Longato and Bonfante 1997). The study of Van Tuinen et al. (1998) provides a good example of the use of specific primers. They used nested PCR with species-specific primers in the second amplification to investigate the influence of two host plants (leek and onion) on the community structure of four species of AM fungus. They found that all four species consistently co-colonised the root systems and that *Glomus mosseae* was dominant with both hosts. In addition, there was evidence to suggest that the *Gigaspora* isolates used in the study showed increased colonisation when grown in combination with *Glomus* species.

To our knowledge, only two published studies have used specific primers on samples collected from the field. In the first paper, Clapp et al. (1995) utilised Gigasporaceae-specific primers designed by Simon et al. (1992, 1993b) to investigate the structure of natural AMF communities colonising the roots of bluebell (*Hyacinthoides non-scripta*) in a semi-natural woodland. The VANS1 primer was found to perform poorly and the authors had to resort to using a subtraction technique in order to generate PCR products. The molecular evidence for the presence or absence of each of the three AMF families in roots was compared with counts of spores isolated from the soil immediately surrounding the plants. This comparison showed that spore populations did not accurately reflect the root colonisation patterns of AMF. The study included plants collected from both sycamore and oak-dominated areas and showed that AMF colonisation was affected by the surrounding vegetation.

In the second study, Clapp et al. (1999) concentrated on *Scutellospora*, using the primer combinations SS38/VAGIGA and VANS1/VAGIGA. These primer sets were used to amplify sequences from bluebell roots and field spores. They demonstrated a surprisingly high level of SSU sequence variation with single spores, as discussed earlier. Additionally, it was noted that several of the AMF sequences obtained from spores would not have been amplified using VANS1.

This illustrates the limited value of using specific primers to study community diversity in any situation in which the starting material is not well characterised with regard to its inherent sequence variability. Specific primers are ideally suited for the study of microcosm experiments where the aim is to rapidly test for the presence/absence of fungi added from well-characterised pot cultures, but they are not appropriate for field collected samples unless the aim is to focus on just one component of the community.

8.4.2 Primers Intended to Target all Glomalean Fungi

An allied approach is to use primers designed to amplify glomalean templates but exclude other sequences likely to contaminate DNA extracts. Studies of this kind have all used cloning and PCR-RFLP to separate classes and sequencing to give an insight into the identity of the clone. Interest has again concentrated on the ribosomal gene complex, in particular the SSU and ITS regions. Three primer sets have been used to date. Sanders et al. (1995) generated products from glomalean spores using the ITS1-F/ITS4 primers, a combination that was originally designed to amplify basidiomycete sequences (White et al. 1990) but which amplifies AMF sequences efficiently. PCR-RFLP was used to investigate species diversity in a natural Swiss grassland (Sanders et al. 1995) and revealed genetic diversity both between spores of the same morphotype and, after cloning, within a single spore. This suggested that the diversity of AMF may be higher than previously estimated by morphological methods.

Chelius and Triplett (1999) used the VANS1 primer designed by Simon et al. (1992) with a universal reverse primer to detect the presence of an arbuscular fungus present in an intensively managed turf grass system. This paper is of particular interest as the authors also successfully applied the technique to DNA extracted directly from soil. Surprisingly, they only found a single sequence, which was related to the SSU of *Glomus intraradices*. This lack of diversity might be a reflection of the intensive management regime but runs counter to most other sequence data obtained from AMF. Alternatively, it could be due to the specificity of VANS1, which is known to exclude a number of sequence types even within the traditional three families (Clapp et al. 1999; Schüßler et al. 2001).

The most extensive studies of field communities of AMF yet undertaken have used a primer which was designed to target all AMF sequences then available (Helgason et al. 1998). The authors used PCR-RFLP to separate clones into classes and then sequenced examples of each class to produce a phylogenetic tree grouping sequences by similarity (for an example see Fig. 8.3). Importantly, PCR-RFLP classes corresponded to sequence classes

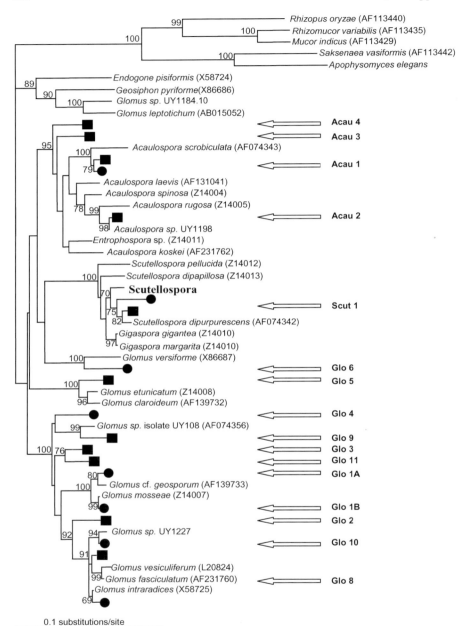

Fig. 8.3. Phylogenetic tree of partial SSU (NS31 to AM1) sequences from AMF, by neighbor-joining method. Named sequences are from laboratory cultures (GenBank accession numbers shown), *squares* are sequences from roots of woodland plants, *circles* are from roots of arable crops (Helgason et al. 1998, 1999; T. Helgason, unpubl.). *Arrows* indicate examples of distinct RFLP groups (Acau 4, etc.)

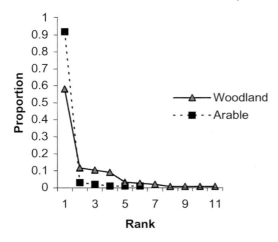

Fig. 8.4. Rank abundance of AMF sequences types in Woodland and Arable soils (data from Helgason et al. 1998). While both are dominated by a single sequence type, there are many more types in the woodland soil, and their abundance is more even

and the classes agreed with an earlier morphological approach which identified fungi in roots at the family level (Merryweather and Fitter 1998a). The York group has applied this method to both woodland and arable systems over an extended period. Initial work by Helgason et al. (1998) compared the diversity of the two systems showing that, despite the proximity of the sites, diversity of AMF sequences is dramatically lower at arable sites (Fig. 8.4). Indeed, only rare sequence groups are shared between the two systems, with the arable system dominated by *Glomus* types not present in the woodland system. This switch may reflect a number of differences between the two habitats including tillage, pH differences and fertilisation. Identical sequences were found across different host plants within both agricultural and woodland communities, suggesting that the broad host specificity attributed to AMF in culture is also realised in the field. At this stage, it is not possible to conclude whether host plant or management regime is the main stimulus for the community change. This result poses many questions about the responses of AMF to their environment as well as to the host plant, and shows again how important it is to develop methods of studying the biology and functional responses of AMF in situ.

The second study (Helgason et al. 1999) investigated the seasonal variation in the diversity of the AMF community colonising bluebell (*Hyacinthoides non-scripta*) in the same woodland. This study confirmed earlier findings that the surrounding vegetation affected the AMF community (Clapp et al. 1995). The temporal sampling highlighted some important seasonal changes in the fungi colonising bluebell roots, with *Glomus* and *Scutellospora* types colonising the roots during the active winter growth stage and *Acaulospora* becoming dominant towards the end of the growth period. Temporal shifts in community structure have also been observed in the arable system; for example, the *Acaulospora* sequence type, while never abundant, was present in early growth stages only (T.J. Daniell, R. Husband, A.H. Fitter and J.P.W. Young, unpubl.).

8.4.3 Alternative Approaches to Molecular Characterisation of Arbuscular Mycorrhizal Fungi

Perhaps because the structure of ribosomal genes proved to be complex, methods that sample the whole genome have been used as an alternative. There are several such 'fingerprinting' methods, which rely on short, sometimes repeated, sequences dispersed throughout the genome. These include RAPD analysis (Wyss and Bonfante 1993; Abbas et al. 1996), M13-primed minisatellite PCR (Zézé et al. 1997), microsatellite-primed PCR (Vandenkoornhuyse and Leyval 1998), amplified fragment length polymorphism (AFLP) (Rosendahl and Taylor 1997) and PCR-generated microsatellite loci (Longato and Bonfante 1997). These studies all revealed more variation than was expected. Three cultures of one isolate showed substantial variation after being cultured in different laboratories for 12 years (Longato and Bonfante 1997). Single spores originating from the same culture were highly heterogeneous when analysed using minisatellites (Zézé et al. 1997) and AFLP (Rosendahl and Taylor 1997).

However, a recent reanalysis (Redecker et al. 1999) of two earlier papers (Hijri et al. 1999; Hosny et al. 1999) showed that many of the sequences obtained from single spores were ascomycete in origin, a phenomenon also observed by Clapp et al. (1999). Obtaining completely sterile spores is difficult and, in any case, bacteria have been found that are intimately associated with the spores. It is estimated that 250,000 live bacteria are present in spores of *Gigaspora margarita*, probably ten times the number of nuclei present (Bianciotto et al. 1996). It seems likely then that much of the variation revealed by these whole-genome approaches may be a result of amplification of DNA from other organisms. This is a particularly important consideration in field studies of AMF. Until there is a way of isolating demonstrably pure AMF DNA we must assume that foreign DNA is also present and use methods that precisely target AMF sequences.

8.5 Comparison of Molecular and Morphological Data

Field populations of AMF have higher molecular genetic diversity than that of the isolates we are able to culture. In theory, molecular analysis has the potential to identify all the individual taxa, unlike trapping or spore sampling which depend on the sporulation of the fungi. As we have seen, however, this relies upon precise targeting of molecular methods. It is important that the molecular data can be related both to the classical taxonomy and to the morphotypes that are identified in the field.

8.5.1 Phylogeny and Taxonomy

For fungi as a whole, ribosomal gene phylogenies are broadly congruent with classical taxonomy based on morphology (Bruns et al. 1992). The three classical families of the Glomales, defined on morphological criteria, are supported by the molecular data (Morton and Benny 1990; Simon et al. 1993a). However, SSU sequence analysis suggests that the genus *Acaulospora* is no more closely allied to the Glomaceae than to the Gigasporaceae, and there is therefore little support for the two sub-orders Glominae and Gigasporinae (Simon 1996, Schüßler 1999). Within families, the relationships among species are unclear. Analysis of partial SSU sequences suggest that *Entrophospora colombiana* is internal to *Acaulospora* and *Gigaspora* internal to *Scutellospora*, so there is little support for the genera being separate in either family. In the former case *E.colombiana* is the only "species" of the genus that can be maintained in culture; the type species for the genus, *E. infrequens,* cannot be cultured.

Two groups of taxa fall outside the three families as currently defined. The recently identified *Acaulospora gerdemannii* and *Glomus occultum* groups appear to be unrelated to the genera to which they are currently assigned, and it has been proposed that these taxa should be in separate families. They are more closely related to *Geosiphon* than they are to other AMF. *Geosiphon*, a non-mycorrhizal fungus that forms a symbiosis with a cyanobacterium (*Nostoc*), was thought to be a basal lineage of the Glomales (Gehrig et al. 1996; Schüßler 1999), but it now appears that it may be less distant than some deeply branching AMF taxa (Redecker et al. 2000; see also Fig. 8.3) which have been placed in the new families Archaeosporaceae and Paraglomaceae (see http://invam.caf.wvu.edu/). Further work is required to resolve the phylogenetic relationships of the taxa within these groups and to other fungi.

All these conclusions have been based on ribosomal genes. The presence of multiple sequences within single spores (discussed in detail below) means that phylogenetic relationships derived from these data within the Glomales could be ambiguous. Sequence analysis of other genes with contrasting rates of evolution may be able to clarify interspecific relationships.

8.5.2 Field Data

Of great importance to those studying the diversity of AMF in the field, is whether the sequences obtained from field roots are a true reflection of the AMF colonisation of the roots. Very little work has been done to establish this, but some results comparing hyphal morphotypes identified by staining and microscopy with sequences obtained from roots have shown that it is possible to correlate these two data types. In a study of seasonal variation in AMF colonising *Hyacinthoides non-scripta* from Pretty Wood (UK), Merryweather and Fitter (1998a,b) were able to identify six AMF morphotypes: three *Glomus*, one *Acaulospora*, one *Scutellospora* and a fine endophyte type. The molecular analysis of Helgason et al. (1999) revealed nine sequence clusters: three *Glomus*, four *Acaulospora*, one *Scutellospora* and another sequence group that we now know to fall within the recently defined deep-branching lineages described by Redecker et al. (2000). The numbers of *Glomus* and *Scutellospora* clusters were in agreement, and although Merryweather and Fitter (1998a) identified only one *Acaulospora* morphotype, they noted morphological variation within it and suggested that it might represent more than one taxon. This observation is in agreement with the finding of multiple *Acaulospora* sequence clusters (Helgason et al. 1999). It was also observed that the relative abundance of the sequence clusters at two sampling times agreed with the colonisation patterns shown in the morphological analysis. *Scutellospora* was more abundant in the early part of the *H. non-scripta* growing season (December) and was gradually replaced by *Glomus* spp. beneath the sycamore (*Acer pseudoplatanus*) canopy and *Acaulospora* spp. beneath the oak (*Quercus petraea*) canopy by July (Merryweather and Fitter 1998a,b; Helgason et al. 1999). The morphological and molecular data for AMF colonisation of bluebell are therefore broadly in agreement, but this study required lengthy and detailed analyses, particularly the extensive numerical analysis of the morphotypes. It has demonstrated that, in principle, molecular and morphological data can be linked, but this time-consuming approach is likely to be impractical for wider field studies. Furthermore, it is likely that the morphology observed in an AMF root is determined in part by the host plant as well as the fungus, so the morphotypes identified by Merryweather and Fitter (1998a) in bluebell may not be readily recognisable in other plant species and extrapolation to other host plants may be problematic if morphological plasticity of the fungus is high. Moreover, the same AMF species may develop quite differently in different plant species (e.g. forming *Paris* types of mycorrhizae in some plant species and *Arum* types in others; e.g. see Jakobsen et al., Chap. 3, this Vol.). In these cases, molecular sequences provide much more reliable evidence that host species share the same AMF taxa so long as it can be shown where the genetic limits between "morphospecies" lie.

In an extension of this study to include five other host plant species at the site (T. Helgason, J.P.W. Young and A.H. Fitter, unpubl.), five additional sequence clusters were found, bringing the total to 13. This matches the numbers of spore types identified from Pretty Wood; eight found in soil samples and five isolates that have been trapped from the soil. Although the numbers of taxa are the same, only three sequence clusters have been matched to spore types. The orders of magnitude appear to be similar, but more work clearly needs to be done. For example, one of the trapped cultures was *Acaulospora trappei*, which fails to amplify with AM1, so the colonisation patterns of this taxon are not known. Only a small number of the host plants have been studied and, given the seasonal patterns already found and the effect of host preference, it seems likely that more sequence types exist in plant roots.

8.6 The Genetic Organisation of Arbuscular Mycorrhizal Fungi

The majority of Zygomycetes can reproduce both asexually and sexually and a vegetative mycelium usually originates from a uninucleate asexual spore. AMF spores have an aseptate, multinucleate mycelium, which originates from an apparently asexual, multinucleate spore. AMF are not typical Zygomycetes and indeed, it is not certain that they are closely related to the other members of this class. This means that there are no readily cultured close relatives of AMF that could serve as models for understanding their biology, so it is not surprising that their genetic systems remain obscure. Tommerup (1988) observed structures that she interpreted as sexual, but no recombination process has been reliably documented. The spores are multinucleate, so there is no uninucleate stage of the life cycle. The number of nuclei found in individual spores varies according to species but ranges between 1000 and 20,000 (Bécard and Pfeffer 1993; Burggraaf and Beringer 1989; Viera and Glenn 1990) with 1000–5000 accepted as most likely.

8.6.1 Heterogeneous Sequences in Spores and Cultures

Sequence diversity within a single spore or a culture was first reported in ITS regions of *Glomus* spp. (Sanders et al. 1995; Lloyd-MacGilp et al. 1996), and later in the ITS regions of *Gigaspora margarita* (Lanfranco et al. 1999) and *Scutellospora castanea* (Hosny et al. 1999). The diversity is not restricted to the ITS, but has also been demonstrated in SSU rDNA sequences (Clapp et al. 1999; Hijri et al. 1999; T. Helgason and J.P.W. Young, unpubl.). Intra-sporal diversity can be extensive: six distinct glomalean sequences were found in a single *Scutellospora* spore and this figure is likely to be conservative since the number was obtained through random sequencing of cloned PCR fragments (Clapp et al. 1999). Most of the sequence variation has been relatively minor (see, for example, the pairs of sequences marked ss in Fig. 8.5), though often as great as the differences between species within a genus (which are sometimes negligible). More dramatically, SSU sequences similar to those found in *Glomus* as well as sequences typical of *Scutellospora* were cloned from single *Scutellospora* spores by Clapp et al. (1999) and Hosny et al. (1999). Similarly, ITS sequences of two radically different types from subcultures of the same *G. fistulosum* isolate that had been grown in different laboratories have been obtained (S. Lloyd-MacGilp and J.P.W. Young, unpubl.; A, B, C in Fig. 8.5). One of these types (Glo-

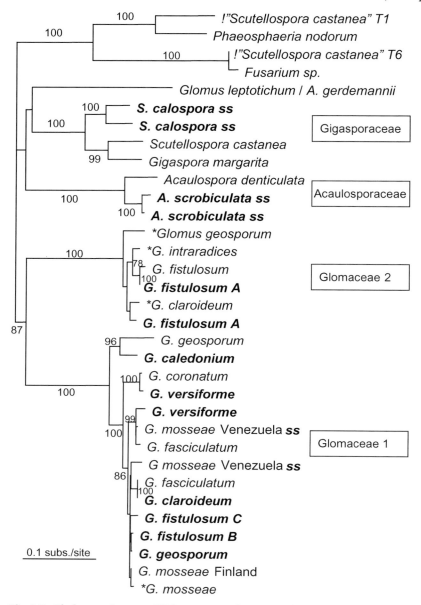

Fig. 8.5. Phylogenetic tree of ITS sequences from AMF, by neighbor-joining method. **ss**: pairs of sequences from single spores; ! ascomycete sequences from *Scutellospora* spores (Hosny et al. 1999); * Millner et al. (1998); others from Lloyd-MacGilp (1996), Redecker et al. (2000 and unpublished GenBank entries). *Bold* sequences are from our laboratory (S. Lloyd-MacGilp and J.P.W. Young, unpubl.)

maceae 1) is very similar to previously published sequences from *G. mosseae* and other *Glomus* species (Sanders et al. 1995; Lloyd-MacGilp et al. 1996), while the other (Glomaceae 2 in Fig. 8.5) includes a sequence recently submitted to GenBank by Redecker (accession no. AJ239126) from the same species, and was also found by Millner et al. (1998) in several other *Glomus* species (* in Fig. 8.2). Millner et al. contrasted these sequences with the ITS sequence that they obtained from *G. mosseae*, which was virtually identical to those reported previously (Sanders et al. 1995; Lloyd-MacGilp et al. 1996). However, ITS sequences determined for some of the same species (*G. geosporum, G. claroideum*; S. Lloyd-MacGilp and J.P.W. Young, unpubl.), in contrast to the results of Millner et al., fell in the Glomaceae 1 clade (Fig. 8.5).

It is clear that many *Glomus* species have two very different classes of ITS sequence; it is likely that many cultures contain both types in proportions that vary among subcultures. This hypothesis is supported by an extensive investigation of *G. coronatum* isolates (Clapp et al. 2001). Five hundred clones from seven *G. coronatum* isolates, and one isolate each of *G. geosporum, G. mosseae* and *G. constrictum*, were screened for variation across 460 bp of the LSU D2 region by PCR-single-strand conformation polymorphism (SSCP) and sequencing. Distinct sequence clusters were present within each isolate and, although a *G. coronatum* "core cluster" could be identified, several sequences from other species occurred in the *G. coronatum* clusters. The precise origin of this variation is unclear; it may represent a sequence continuum between these species, indicating genetic exchange, or possibly a mixed inoculum at the initiation of cultures although no evidence for this could be found. Sequence variation in both type and frequency was also found between subcultures and the parent culture of the *G. coronatum* holotype (BEG 28). Clearly, more work is necessary to delimit the underlying genetic variation present in AMF before sequence variation can be taken to imply species diversity.

8.6.2 "Contaminant" Sequences in Spores and Cultures

Ascomycete sequences have also been found in ostensibly clean AMF spores from the field (Clapp et al. 1999) and from laboratory cultures (Hijri et al. 1999; Hosny et al. 1999; Redecker et al. 1999; ! in Fig. 8.5). It is not clear whether these are arbitrary contaminants or regular associates of AMF. Phylogenetic analysis shows that they appear to be derived from a diverse range of taxa, suggesting that the association may be casual rather than specific. The true extent of this association may be masked by the tendency of many workers not to report such "contaminant" sequences.

Studies of whole genome variation using AFLP markers (Rosendahl and Taylor 1997) and M13 minisatellite-primed PCR (Zézé et al. 1997) both showed genetic diversity among spores from cultures established from single spores. This is consistent with the presence of populations of different nuclei rather than variability generated by recombination. However, any study based on arbitrary markers, which are not specific to glomalean genomes, must be interpreted very cautiously because it is now clear that sequences from other organisms are commonly found in spores and cultures of AMF. It is likely that random amplification methods also amplify fragments from such contaminants, as well as from symbiotic bacteria within the fungal cytoplasm (Bianciotto et al. 1996), and the resulting data are therefore difficult to interpret.

8.6.3 How Is the Sequence Heterogeneity Organised?

Multiple rRNA variants are not unique to AMF, but the high divergence among AMF sequence types recovered from single cultures and even single spores has led to discussion about molecular evolution in AMF (Sanders et al. 1996; Sanders 1999). Since there are between 1000 and 5000 nuclei per spore and about 90 copies of ribosomal RNA genes for each nucleus (Passerieux 1994), there are two possible sources for the observed sequence variation. Ribosomal RNA gene sequences may differ within a single nucleus (intranuclear polymorphism) or nuclei within a single spore may be genetically different (internuclear polymorphism).

So far, there is no conclusive evidence as to whether the genetic variants are located within the same nucleus or in different nuclei. The experiments of Rosendahl and Taylor (1997) and of Zézé et al. (1997) showed segregation of genetic variants within a culture. In the absence of any detectable meiotic process, this is evidence that the variants were not in the same nucleus but, as noted above, these studies may be flawed because the DNA sequences they studied might have been derived from contaminating organisms. Similarly, Hosny et al. (1999) and Hijri et al. (1999) found segregation of rRNA genes, but a re-examination of their data revealed that many of the sequences they were studying matched genes from ascomycetes rather than glomalean fungi, and were probably derived from unnoticed contaminants in the cultures (Redecker et al. 1999; Schüßler 1999). The advantage of studying ribosomal genes rather than "random" DNA is that the origin of each sequence can be traced in this way. In fact, a close examination of the study by Hijri et al. (1999) shows that the sequence variants they call ITS T2 and ITS T4 are both apparently 'genuine' *Scutellospora* sequences, and that they segregated in spores derived from the same pot culture. Out of 24 spores, 8 retained both sequences, 8 had T2 but not T4, 7 had T4 but not T2, and in 1 spore neither sequence could be detected. This is perhaps the best evidence so far that glomalean nuclei within a culture may differ in the sequence variants they carry. The consequences of AMF carrying genetically different nuclei for the evolution of specificity in the symbiosis is also discussed by Sanders in Chapter 16 (this Vol.).

Of course, the existence of differences between nuclei does not mean that there cannot be intranuclear polymorphism as well. In most organisms, the processes of concerted evolution continually homogenise the sequences within the tandem array of ribosomal RNA genes, so that variation within the array is very limited (Li 1997). Sanders et al. (1996) speculated that concerted evolution might be suspended in glomalean fungi if there is no meiotic stage in the life cycle. An alternative explanation might be that there is more than one tandem array of ribosomal RNA genes in the genome. Concerted evolution operates much less effectively between variants that are not adjacent, so genes in different arrays could plausibly diverge. Each array is a potential nucleolar organiser, and may remain spatially discrete in the nucleolus. Trouvelot et al. (1999) used in situ hybridisation with a large subunit rRNA gene probe to visualise the corresponding genes in interphase nuclei from spores of four AMF species. They found that hybridisation was usually concentrated into three or five spots within each nucleus, and their interpretation was that each spot represented a separate array of rRNA genes. In each of the four species (*Scutellospora castanea*, *Glomus mosseae*, *G. intraradices* and *Gigaspora rosea*), nuclei had three or five spots in a ratio that varied from 1:1 to 1:3. One nucleus of *G. intraradices* (out of 25 examined) had seven spots. These results clearly demonstrated variation in the number of nucleoli among the nuclei within an AMF culture, although it is curious that in each of these very divergent taxa the same numerical variants were found. Why were there no nuclei with four spots, for example? It is not necessarily the case that these numerical variants represent genetic variants, because the fusion of nucleolar organisers into a single nucleolus is a dynamic process during the cell cycle (Anastassova-Kristeva 1977); it is possible that three and five nucleoli represent particularly stable configurations for the glomalean genome.

An extension of this in situ hybridisation protocol that included probes specific for different sequence variants would be very valuable. It could demonstrate directly whether nuclei differ in the sequences they carry and, equally importantly, whether different sequences can occur within a single nucleus. A related approach that is also promising in this context is in situ PCR using fluorescent primers (Bago et al. 1998). Whatever method is used, it will, of course, be essential to ensure that only glomalean sequences are recognised.

8.6.4 How is Sequence Heterogeneity Maintained?

If there are different nuclei within a hyphal network or within a single spore, then we have to ask how this diversity can be maintained. Although there is no known stage in the AMF life cycle at which new individuals are derived from single nuclei, variation should nevertheless be lost by random processes. As a hyphal network grows, some branches will, by chance, lack certain nuclear types, and spores formed in that sector will have permanently lost these variants. Why, then, are cultures so often heterogeneous? Either heterogeneity is constantly being generated, or there are mechanisms to prevent its loss. Loss could be prevented if nuclei of different types were somehow physically linked, or if they were mutually interdependent so that a hypha that lacked one type would be unable to grow, or to form spores. However, there is no evidence for any such mechanism, and the segregation of ITS types T2 and T4 in the study by Hijri et al. (1999) suggests that heterogeneity is, in fact, lost at a rapid rate.

Variants could arise by fresh mutation but, while this could explain some of the very slight variants found, much of the diversity that has now been documented involves extensive sequence divergence that must represent millions of years of evolution. It is more plausible that these different sequences have been brought together from different sources. Van Tuinen et al. (1994) demonstrated that hyphae of different mycelia could fuse, while Giovannetti et al. (1999) showed that nuclei could pass through anastomoses formed in this way. However, the latter authors also reported that such anastomoses were only found in *Glomus*, not in *Gigaspora rosea* or *Scutellospora castanea*, and that they were confined to interactions between hyphae of the same species of *Glomus*. Thus, these processes might explain how intraspecific sequence variants come to be in the same mycelium, but there is as yet no direct evidence for a mechanism that would, for example, allow *Glomus* nuclei to enter *Scutellospora* spores, as found by Clapp et al. (1999) and Hosny et al. (1999).

8.6.5 How Does This Affect Analysis of Field Sequences?

If genetic variation arises within individual nuclei and hyphae in the way described above, this clearly poses several questions about the analysis of sequence types obtained from the field. How do we define an individual? Can we relate this with any confidence to species as we currently understand them? Where does an individual fungus begin and end? In most of the studies so far, we have been content to delimit taxa by some sequence grouping criteria, and for the non-culturable taxa this is all that can be done currently. However, gaining a better understanding of the genetic variation within isolates is clearly a priority for future research. Molecular methods that are able to 1) locate genetic variants within a root or spore and 2) relate this to morphology will be an essential part of this process. Accurate molecular methods for quantification of AMF, both in the roots and in the soil, are another area that needs to be addressed.

8.7 Future Prospects

Molecular approaches are clearly capable of yielding much information regarding the structure and diversity of AMF communities. However, if the aim is to study the whole AMF community, the primers used need to be reviewed as our knowledge increases. For example, a new primer for the Glomales is currently being tested at York, as AM1 will not amplify recently described sequences from a number of fungi (*Acaulospora gerdemannii/Glomus leptotichum, Acaulospora trappei, Glomus occultum* and *Glomus brasilianum*) lying outside the traditional three families (Redecker et al. 2000). There is also a need to streamline the methodology to allow higher throughput of samples. Numerous techniques are available including SSCP, which has been used to identify simple mixed communities in microcosms (Simon et al. 1993b) and to assess the magnitude of sequence diversity in AMF cultures (Clapp et al. 2001). It has been tested on environmental samples (Kjøller and Rosendahl 1998). Other possible methodologies to avoid the cloning step include terminal restriction fragment length polymorphism (T-RFLP; e.g. Liu et al. 1997) and temporal temperature gradient electrophoresis (TTGE; e.g. Bosshard et al. 2000). The present emphasis on ribosomal genes has the benefit that it has established a substantial database of sequence information that new studies can build on. However, one locus cannot give a complete picture of population diversity, and there is a need to obtain information from other genes in order to confirm the conclusions and also to establish definitively whether genetic recombination occurs in the Glomales. All these possibilities can only be investigated once the genetics of AMF are more clearly understood. In a recent COST 838 meeting in Faro, Portugal, the conclusion was that we have a large amount of data on the molecular diversity of AMF, but we do not yet know what it means. The molecular analysis of AMF and their communities has only just begun.

References

Abbas JD, Hetrick BAD, Jurgenson JE (1996) Isolate specific detection of mycorrhizal fungi using genome specific primer pairs. Mycologia 88(6):939–946

Anastassova-Kristeva M (1977) The nucleolar cycle in man. J Cell Sci 25:103–10

Avise JC (1994) Molecular markers, natural history and evolution. Chapman and Hall, London

Bago B, Piché Y, Simon L (1998) Fluorescently-primed in situ PCR in arbuscular mycorrhizas. Mycol Res 102(12):1540–1544

Bécard G, Pfeffer PE (1993) Status of nuclear division in arbuscular mycorrhizal fungi during in-vitro development. Protoplasma 174(1/2):62–68

Bever JD, Morton JB, Antonovics J, Schultz PA (1996) Host-dependent sporulation and species diversity of arbuscular mycorrhizal fungi in a mown grassland. J Ecol 84:71–82

Bianciotto V, Bandi C, Minerdi D, Sironi M, Tichy HV, Bonfante P. (1996) An obligately endosymbiotic mycorrhizal fungus itself harbors obligately intracellular bacteria. Appl Environ Microbiol 62(8):3005–3010

Bosshard PP, Santini Y, Gruter D, Stettler R, Bachofen R (2000) Bacterial diversity and community composition in the chemocline of the meromictic alpine Lake Cadagno as revealed by 16S rDNA analysis. FEMS Microbiol Ecol 31(2):173–182

Bruns TD, Vilgalys R, Barns SM, Gonzalez D, Hibbet DS, Lane DJ, Simon L, Stickel S, Szaro TM, Weisberg WG, Sogin ML (1992) Evolutionary relationships within the fungi: analysis of nuclear small subunit rRNA sequences. Mol Phylogenet Evol 1:231– 41

Burggraaf AJP, Beringer JE (1989) Absence of nuclear DNA synthesis in vesicular-arbuscular mycorrhizal fungi during in vitro development. New Phytol 111:25–33

Carvalho GR (ed) (1998) Advances in molecular ecology. NATO science series, vol 306. IOS Press, Amsterdam, The Netherlands

Chelius MK, Triplett EW (1999) Rapid detection of arbuscular mycorrhizae in roots and soil of an intensively managed turfgrass system by PCR amplification of small subunit rDNA. Mycorrhiza 9(1):61–64

Clapp JP, Young JPW, Merryweather JW, Fitter AH (1995) Diversity of fungal symbionts in arbuscular mycorrhizas from a natural community. New Phytol 130:259–265

Clapp JP, Young JPW, Fitter AH (1999) Ribosomal small subunit sequence variation within spores of an arbuscular mycorrhizal fungus, *Scutellospora* sp. Mol Ecol 8:915–921

Clapp JP, Rodriguez A, Dodd JC (2001) Inter- and intra-isolate rRNA large sub-unit variation in spores of *Glomus coronatum*. New Phytol 149:(3)539–554

Di Bonito R, Elliott ML, Jardin EAD (1995) Detection of an arbuscular mycorrhizal fungus in roots of different plant species with the PCR. Appl Environ Microbiol 61(7): 2809–2810

Eom A, Hartnett DC, Wilson GWT (2000) Host plant species effects on arbuscular mycorrhizal fungal communities in tallgrass prairie. Oecologia 122:435–444

Gardes M, Bruns TD (1993) ITS primers with enhanced specificity for basidiomycetes – application to the identification of mycorrhizae and rusts. Mol Ecol 2(2):113–118

Gehrig H, Schüßler A, Kluge M (1996) *Geosiphon pyriforme*, a fungus forming endocytobiosis with *Nostoc* (Cyanobacteria), is an ancestral member of the Glomales: evidence by SSU rRNA analysis. J Mol Evol 43(1):71–81

Giovannetti M, Azzolini D, Citernesi AS (1999) Anastomosis formation and nuclear and protoplasmic exchange in arbuscular mycorrhizal fungi. Appl Environ Microbiol 65:5571–5575

Helgason T, Daniell TJ, Husband R, Fitter AH, Young JPW (1998) Ploughing up the wood-wide web? Nature 394:431

Helgason T, Fitter AH, Young JPW (1999) Molecular diversity of arbuscular mycorrhizal fungi colonising *Hyacinthoides non-scripta* (bluebell) in a seminatural woodland. Mol Ecol 8(4):659–666

Hijri M, Hosny M, van Tuinen, D, Dulieu H (1999) Intraspecific ITS polymorphism in *Scutellospora castanea* (Glomales, Zygomycota) is structured within multinucleate spores. Fungal Genet Biol 26(2):141–151

Hosny M, Hijri M, Passerieux E, Dulieu H (1999) rDNA units are highly polymorphic in *Scutellospora castanea* (Glomales, Zygomycetes). Gene 226(1):61–71

Innis MA, Gelfand DH, Sninsky JJ, White TJ (eds) (1990) PCR protocols: a guide to methods and applications. Academic Press, New York

Kiers ET, Lovelock CE, Krueger EL, Herre EA (2000) Differential effects of tropical arbuscular mycorrhizal fungal inocula on root colonization and tree seedling growth: implications for tropical forest diversity. Ecol Lett 3(2):106–113

Kjøller R, Rosendahl S (1998) Detection of arbuscular mycorrhizal fungi in roots using PCR-SSCP. Abstracts of the 2nd International Conference on Mycorrhiza, 5–10 July 1998, Uppsala, Sweden

Klironomos JN, McCune J, Hart M, Neville J (2000) The influence of arbuscular mycorrhizae on the relationship between plant diversity and productivity. Ecol Lett 3:137–141

Lanfranco L, Delpero M, Bonfante P (1999) Intrasporal variability of ribosomal sequences in the endomycorrhizal fungus *Gigaspora margarita*. Mol Ecol 8(1):37–45

Li WH (1997) Molecular evolution. Sinauer Associates, Sunderland, MA, pp 309–334

Liu WT, Marsh TL, Cheng H, Forney LJ (1997) Characterization of microbial diversity by determining terminal restriction fragment length polymorphisms of genes encoding 16S rRNA. Appl Environ Microbiol 3(11):4516–4522

Lloyd-MacGilp SA, Chambers SM, Dodd JC, Fitter AH, Walker C, Young JPW (1996) Diversity of the ribosomal internal transcribed spacers within and among isolates of *Glomus mosseae* and related fungi. New Phytol 133:103–111

Longato S, Bonfante P (1997) Molecular identification of mycorrhizal fungi by direct amplification of microsatellite regions. Mycol Res 101(4):425–432

Merryweather J, Fitter AH (1998a) The arbuscular mycorrhizal fungi of *Hyacinthoides non-scripta* I. Diversity of fungal taxa. New Phytol 138(1):117–129

Merryweather J, Fitter AH (1998b) The arbuscular mycorrhizal fungi of *Hyacinthoides non-scripta* II. Seasonal and spatial patterns of fungal populations New Phytol 138(1):131–142

Millner PD, Mulbry WW, Reynolds SL, Patterson CA (1998) A taxon-specific oligonucleotide probe for temperate zone soil isolates of *Glomus mosseae*. Mycorrhiza 8(1):19–27

Morton JB, Benny GL (1990) Revised classification of arbuscular mycorrhizal fungi (Zygomycetes): a new order, Glomales, two new suborders, Glomineae and Gigasporineae, and two new families, Acaulosporaceae and Gigasporaceae, with an emendation of Glomaceae. Mycotaxon 37:471–491

Morton JB, Bentivenga SP, Bever JD (1995) Discovery, measurement, and interpretation of diversity in arbuscular endomycorrhizal fungi (Glomales, Zygomycetes). Can J Bot 73:S25–S32

Newsham KK, Fitter AH, Watkinson AR (1995) Multi-functionality and biodiversity in arbuscular mycorrhizas. Trends Ecol Evol 10:407–411

Passerieux E (1994) DEA Thesis, Univ Bourgogne, Dijon, France

Redecker D (2000) Specific PGR primers to identify arbuscular mycorrhizal fungi within colonized roots. Mycorrhiza 10(2):73–80

Redecker D, Thierfelder H, Walker C, Werner D (1997) Restriction analysis of PCR-amplified internal transcribed spacers of ribosomal DNA as a tool for species identification in different genera of the order Glomales. Appl Environ Microbiol 63(5):1756–1761

Redecker D, Hijri M, Dulieu H, Sanders IR (1999) Phylogenetic analysis of a data set of fungal 5.8S rDNA sequences shows that highly divergent copies of internal transcribed spacers reported from *Scutellospora castanea* are of ascomycete origin. Fungal Genet Biol 28(3):238–244

Redecker D, Morton JB, Bruns TD (2000) Ancestral lineages of arbuscular mycorrhizal fungi (Glomales). Mol Phylogenet Evol 14(2):276–284

Rosendahl S, Taylor JW (1997) Development of multiple genetic markers for studies of genetic variation in arbuscular mycorrhizal fungi using AFLP. Mol Ecol 6(9):821–829

Sanders IR (1999) No sex please, we're fungi. Nature 399:737–739

Sanders IR, Fitter AH (1992) Evidence for differential responses between host-fungus combinations of vesicular-arbuscular mycorrhizas from a grassland. Mycol Res 96:415–419

Sanders IR, Alt M, Groppe K, Boller T, Wiemken A (1995) Identification of ribosomal DNA polymorphisms among and within spores of the Glomales: application to studies on the genetic diversity of arbuscular mycorrhizal fungal communities. New Phytol 130:419–427

Sanders IR, Clapp JP, Wiemken A (1996) The genetic diversity of arbuscular mycorrhizal fungi in natural ecosystems – a key to understanding the ecology and functioning of the mycorrhizal symbiosis. New Phytol 133:123–134

Schüßler A (1999) Glomales SSU rRNA gene diversity. New Phytol 144:205–207

Schüßler A, Gehrig H, Schwarzott D, Walker C (2001) Analysis of partial Glomales SSU rRNA gene sequences: implications for primer design and phylogeny. Mycol Res 105:5–15

Simon L (1996) Phylogeny of the Glomales: deciphering the past to understand the present. New Phytol 133:95–101

Simon L, Lalonde M, Bruns TD (1992) Specific amplification of 18S fungal ribosomal genes from VA endomycorrhizal fungi colonising roots. Appl Environ Microbiol 58:291–295

Simon L, Bousquet J, Levesque RC, Lalonde M (1993a) Origin and diversification of endomycorrhizal fungi and coincidence with vascular land plants. Nature 363:67–69

Simon L, Levesque RC, Lalonde M (1993b) Identification of endomycorrhizal fungi colonising roots by fluorescent single-strand conformation polymorphism-polymerase chain reaction. Appl Environ Microbiol 59:4211–4215

Tommerup IC (1988) The vesicular-arbuscular mycorrhizas. Adv Plant Pathol 6:81–91

Trouvelot S, van Tuinen D, Hijri M, Gianinazzi-Pearson V (1999) Visualization of ribosomal DNA loci in spore interphasic nuclei of glomalean fungi by fluorescence in situ hybridization. Mycorrhiza 8(4):203–206

van der Heijden MGA, Klironomos JN, Ursic M, Moutoglis P, Streitwolf-Engel R, Boller T, Wiemken A, Sanders IR (1998) Mycorrhizal fungal diversity determines plant biodiversity ecosystem variability and productivity. Nature 396:69–72

Van Tuinen D, Dulieu H, Zézé A, Gianinazzi-Pearson V (1994) Impact of mycorrhizas on sustainable agriculture and natural ecosystems. Birkhäuser, Basel, pp 13–23

Van Tuinen D, Jacquot E, Zhao B, Gollotte A, Gianinazzi-Pearson V (1998) Characterisation of root colonisation profiles by a microcosm community of arbuscular mycorrhizal fungi using 25S rDNA-targeted nested PCR. Mol Ecol 7:879–887

Vandenkoornhuyse P, Leyval C (1998) SSU rDNA sequencing and PCR-fingerprinting reveal genetic variation within *Glomus mosseae*. Mycologia 90(5):791–797

Viera A, Glenn MG (1990) DNA content of vesicular-arbuscular mycorrhizal fungal spores. Mycologia 82(2):263–267

Wright SF, Upadhyaya A (1998) A survey of soils for aggregate stability and glomalin, a glycoprotein produced by hyphae of arbuscular mycorrhizal fungi. Plant Soil 198:97–107

White, TJ, Bruns T, Lee S, Taylor J (1990) Amplification and direct sequencing of fungal ribosomal genes for phylogenies. In: Innis MA, Gelfand DH, Sninsky JJ, White TJ (eds) PCR protocols: a guide to methods and applications. Academic Press, New York, pp 315–322

Wyss P, Bonfante P (1993) Amplification of genomic DNA of arbuscular mycorrhizal (AM) fungi by PCR using short arbitrary primers. Mycol Res 97(11):1351–1357

Zézé A, Hosny M, Gianinazzi-Pearson V, Dulieu H (1996) Characterisation of a highly repeated DNA sequence (SC1) from the arbuscular mycorrhizal fungus *Scutellospora castanea* and its detection *in planta*. Appl Environ Microbiol 62(7):2443–2448

Zézé A, Sulistyowati E, Ophel-Keller K, Barker S, Smith S (1997) Intersporal genetic variation of *Gigaspora margarita*, a vesicular arbuscular mycorrhizal fungus, revealed by M13 minisatellite-primed PCR. Appl Environ Microbiol 63(2):676–678

9 Diversity of Arbuscular Mycorrhizal Fungi and Ecosystem Functioning

Miranda M. Hart, John N. Klironomos

Contents

9.1	Summary	225
9.2	Introduction	226
9.3	Linking Biodiversity and Ecosystem Function	227
9.4	Arbuscular Mycorrhizal Fungal Diversity and Ecosystem Functioning	229
9.4.1	Arbuscular Mycorrhizal Fungal Networks	229
9.4.2	Functional Specificity	230
9.4.3	Differential Effects	231
9.4.4	Community Effects	231
9.5	Succession of Arbuscular Mycorrhizal Fungi	237
9.6	Conclusions	239
References		239

9.1 Summary

In this chapter, we review recent literature pertaining to the debate linking biodiversity and ecosystem functioning, highlighting the role of arbuscular mycorrhizae. We suggest that the arbuscular mycorrhizal symbiosis should be a vital component of studies designed to elucidate a mechanistic basis linking biodiversity and terrestrial ecosystem function. Most importantly, it is important to consider the role of the mycorrhizal networks, functional specificity among glomalean fungi, and differential effects on host species, all of which can influence plant community dynamics. Finally, we suggest a new direction for mycorrhizal research that includes developing fundamental ecological theory for this group of fungi. Eventually, such theory will allow researchers to include arbuscular mycorrhizal fungi into community and ecosystem models.

9.2 Introduction

The interaction between arbuscular mycorrhizal fungi (AMF) and land plants arguably represents the most widespread symbiosis on earth. These associations have a long history, with fossil and molecular evidence (Simon et al. 1993; Remy et al. 1994) indicating that they evolved during the Devonian period when plants were making the transition onto land. Terrestrial habitats were harsh environments, depauperate of organic matter and available nutrients, and inhospitable to plants with primitive root systems. AMF were likely crucial for the successful invasion of plants onto land. Fungi, with their nutrient scavenging hyphae, were better able to extract nutrients. Today's terrestrial ecosystems are more hospitable to plant invasion, but AMF remain deeply involved in ecosystem processes (in particular the C and P cycles). Although they cannot fix atmospheric CO_2, they stimulate photosynthesis, and thus primary productivity. Although they have limited (if any) saprobic abilities, they alter the nutrient quality of plant tissues, and thus indirectly influence decomposition and the rate of nutrient cycling.

The role of AMF in ecosystems is undeniable, but the significance of AMF *biodiversity* on ecosystem-level processes is less tenable. Since Hutchinson's (1959) famous question, "Why are there so many kinds of animals?", scientists have been studying biodiversity by trying to better understand the distribution and abundance of organisms using evolutionary, population and community perspectives. Only more recently have ecologists started asking, "What is the ecosystem significance of species diversity?" Ecosystem ecologists have become concerned with biodiversity mainly because of the loss of species diversity on a global scale and the uncertainty of the ecosystem consequences of this loss (Jones and Lawton 1995). As a result, recent research is examining whether biodiversity is an important determinant of ecosystem processes. This has required ecologists to think more broadly and integrate organismal interactions (population and community approaches) with material and energy fluxes in ecosystems (ecosystem approach).

The significance of biodiversity on ecosystem function has been studied mainly with plants. Since the Devonian period, terrestrial plants have undergone a significant amount of morphological and functional divergence. Today it is estimated that there are approximately 300,000 plant species. High species diversity is also evident in other groups of organisms (animals, fungi) that ultimately depend on plants. Since high species diversity is often associated with a high variety of forms, each adapted to particular conditions and performing particular functions, it is easy to hypothesize that ecosystem function may be a function of species diversity. Yet with AMF, there is little reason to believe diversity has any bearing on ecosystem processes. The functioning of AMF is closely linked to their mycelial morphologies (Allen 1991), but the morphology of different species is often similar. In fact, based on fossil evidence (Remy et al. 1994), some have likely not changed significantly since early in their evolutionary history. AMF are so similar morphologically that we base species definitions on sub-cellular characteristics of individual spores rather than mycelia (Morton and Benny 1990), and recognize less than 200 species globally. Because of this similarity, along with evidence that most fungal species are not host-specific, there is little reason to believe that species-poor fungal communities will function any differently from species-rich communities. Granted, much of our understanding of AMF biology is based on research using very few fungal isolates (Klironomos and Kendrick 1993), but it suggests that a loss of AMF species diversity should not have significant ecosystem consequences. However, more recent research suggests otherwise. It is becoming clear that AM symbioses are actually multi-functional. Although all species within a fungal community may be able to infect all plants, they may still influence those plants differentially, indicating that AMF diversity may strongly influence ecosystem processes.

In this chapter we will introduce the approaches used to evaluate biodiversity-function relationships in ecology. We will then present recent work on the influence of AMF diversity on ecosystem function. Finally we will discuss practical applications of this, and suggest new avenues for research.

9.3 Linking Biodiversity and Ecosystem Function

"Ecosystem function" encompasses all physical, chemical and biological processes in an ecosystem, but most often refers to ecosystem productivity, stability and resilience. The diversity/stability hypothesis (Elton 1958) predicted diversity and ecosystem stability to have a positive, linear relationship. That is, ecosystem function is impaired with the loss of any species, and is highest with the greatest number of species (Fig. 9.1). More recently, the redundancy hypothesis has stipulated a critical minimum level of diversity for ecosystem functioning, however it stressed that additional species above the threshold were likely to be redundant (Walker 1992). It is described by an asymptotic curve, with the redundant species occurring after the plateau (Fig. 9.1b). Two research groups, in particular, have led the way in attempting to describe the impact of biodiversity on ecosystem function, with very different, yet complimentary approaches.

Tilman and colleagues (University of Minnesota, USA) looked at both production and resilience in field communities with varying levels of plant diversity. Their early diversity experiments examined biodiversity as an indirect result of N-addition (Tilman 1993; Tilman and Downing 1994). They found productivity and resilience to be an asymptotic function of species diversity, as is described by the redundant species hypothesis. Tilman's later experiments eschewed nutrient gradients because of criticism that high nutrient

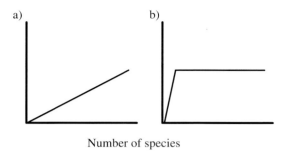

Fig. 9.1. The relationship between species diversity and ecosystem function. a Diversity/stability hypothesis; b redundancy hypothesis

plots are predisposed to low diversity (Givnish 1994; Huston 1997), and used direct manipulation of species numbers to examine diversity/function (Tilman et al. 1996). By randomly selecting species among experimental units, they found productivity and N uptake to increase with species richness, while N leaching decreased. Again, they observed an asymptotic relationship between diversity and function; above ten species, there was no further gain in productivity.

Meanwhile, in the Ecotron (Imperial College at Silwood Park, Berkshire, UK), Naeem and colleagues were addressing the same questions, but on a much different scale (Naeem et al. 1994, 1995). They manipulated plant and animal diversity in controlled environmental chambers, allowing for identical replicates in every aspect except species richness. Unlike Tilman's random diversity treatments, low diversity treatments were "nested" into higher diversity treatments, i.e., low and intermediate diversity treatments were depauperate versions of high diversity treatments. They found high diversity treatments to be more productive, intercept more incipient light, and absorb more CO_2 than low diversity treatments.

While these and other recent studies (Hooper 1998; Hooper and Vitousek 1998; Hector et al. 1999) have correlated ecosystem function with biodiversity, they lack a theoretical framework. Tilman et al. (1997) issued the first theoretical explanation for these results. They presented three mathematical models for the influence of biodiversity on interspecific, competitive interactions within ecosystems. In their first model (competition for a single resource) they showed plant biomass to increase asymptotically with diversity simply because the chance of having a superior competitor increased with increased diversity. Thus the better competitor, through reducing the concentration of the limiting resource, would increase productivity. For two-resource competition, they showed that inherent niche differences between species when competing for different ratios of the same resource resulted in the ecosystem with the most species exploiting the most resources. Thus in a heterogeneous environment, diverse ecosystems would lead to increased productivity. Their final model for a generalized niche described community biomass increasing asymptotically with diversity, depending on spatial heterogeneity relative to niche size, and niche shape. Further theoretical support was provided with mechanistic (Loreau 1998a) and stochastic dynamic (Yachi and Loreau 1999) models. The latter, in particular, demonstrates that increasing diversity species buffers and enhances productivity even in a system with environmental fluctuations. While this might inadvertently suggest resolution to the stability versus redundancy debate – that is, perhaps all systems follow the redundancy hypothesis, and vary only with respect to the threshold of redundancy, further empirical tests are needed.

Both theoretical and experimental diversity research has been necessarily simplified: no experiment can incorporate all trophic levels and their interactions. In fact, one approach is to purposefully annihilate soil microbial populations (Hartnett and Wilson 1999). While it has been well established that increased plant species richness increases ecosystem productivity and stability (Naeem et al. 1994; Tilman and Downing 1994), it is inconclusive to examine plant dynamics in the absence of their fungal symbionts due to the intimacy of the plant-AMF relationship. Because mycotrophic plants differ chemically and physiologically from their non-mycorrhizal counterparts, studies which use normally mycorrhizal plants in the absence of AMF will be biased, and this has been shown experimentally since diversity/productivity relationships are different when investigated with and without AMF (Klironomos et al. 2000).

9.4 Arbuscular Mycorrhizal Fungal Diversity and Ecosystem Function

9.4.1 Arbuscular Mycorrhizal Fungal Networks

One important aspect of diversity and ecosystem functioning to consider is the indirect, community-level effect of the plant-fungus connection. While there is limited specificity among AMF species and their plant hosts (Klironomos 2000), most AMF species can colonize most plants. In turn, most plants host several different AMF species concurrently (Allen 1996). Therefore, not only do mycorrhizal plants acquire more nutrients, they are able to share them via an underground network of hyphal connections linking individuals within and between species (Chap. 2, this Vol.). It has also been shown that interplant carbon transport occurs between different plant species (Fig. 10.3 in Chap. 10, this Vol.). A stable carbon-isotope study by Fitter et al. (1998) found up to 40 % of carbon in the root system of a plant was derived from another plant. However, they found the carbon was sequestered in the fungal structures within the roots, and was not translated into shoot growth of recipient plants. The significance and occurrence of interplant carbon and nutrient transport is, thus, still under debate (Simard et al., Chap. 2, this Vol.).

The existence of such networks confound the issues of niche partitioning, coexistence and competition (Wilkinson 1998), and studies that attempt to resolve the function of biodiversity will be ultimately unsuccessful unless they address resource transfer among con- and inter-specifics.

9.4.2 Functional Specificity

AMF are known mainly for their involvement in P-absorption (e.g. Jakobsen et al., Chap. 3, this Vol.), but they have been associated with other functions as well (Newsham et al. 1995a). In fact, physiological changes in the host plants induced by the AMF association have many functional responses other than increased P-uptake. It has been shown that the AMF relationship increases cytokinin production in host leaves and roots (Allen et al. 1980a). Increased cytokinin levels can lead to increased P-absorption and decreased water resistance. In greenhouse studies on grasses, Allen (1982) found transpiration to increase by 100% in AMF grass relative to non-mycorrhizal controls. Later experiments showed elevated stomatal conductance in AMF treatments (Allen and Boosalis 1983), especially during periods of low soil water potential (Allen and Allen 1986). Further hormonal changes include increased gibberellic acid and decreased abscisic acid production in leaves of plants with AMF associations (Allen et al. 1980b). These changes are associated with chlorophyll synthesis, and ultimately improved fitness of inoculated plants. AMF can also enhance soil aggregate stability, and this has been shown to be a function of hyphal length in soil (Andrade et al. 1998; Jastrow et al. 1998), and the production of the glycoprotein, glomalin (Rillig et al. 1999). Increased resistance to pathogens has also been found for AMF-infected plants (Newsham et al. 1995b; Klironomos 2000). In fact, in situations where there is no net P-transfer from AMF to its host plant, these plants have marked resistance to pathogenic fungi. The experiments described above were executed with different AMF species. That could be one of the reasons why AMF affected such a wide range of functions within plants. This variance in function points also to the existence of functional groups that consist of different AMF species. However, the occurrence of functional groups will be difficult to establish unless different isolates that co-exist in communities are carefully compared for their functional abilities. If this were the case, diversity among AMF species would be important determinants of ecosystem function in so far as all functional groups were represented. From this, it is clear that a supposed lack of host specificity in their ability to infect plant hosts, does not necessarily translate into functional redundancy. These fungi are multi-functional and there may exist tradeoffs among fungi in their abilities to perform these functions.

9.4.3 Differential Effects

In addition to functional differentiation, a single AMF species does not have a uniform effect on all plants. In fact, there is a broad continuum of responses, to the same isolate, ranging from parasitism to mutualism (Johnson et al. 1997). It has been well shown that plants have differential growth in response to the presence of AMF, based on their dependency on the mycorrhizal association (Sanders and Koide 1994), but the identity of the AMF species proves to be a significant factor governing plant response. Sanders and Fitter (1992) provided evidence for a differential response between host/fungus combinations from a grassland. Van der Heijden et al. (1998b) also demonstrated this by subjecting three plant species to four AMF species, alone, and in combination. They found biomass, colonization, and P-concentration among plants to be dependent on the interaction between AMF and plant species. Not all plants benefited to the same extent from AMF, and among those that did, the extent of benefit depended upon the identity of the fungus. Specific AMF had effects on growth but these were not the same for each plant. Interestingly, all isolates in this experiment were from the genus *Glomus*, which suggests even a greater potential for differential response between the other genera of AMF.

9.4.4 Community Effects

The variation within a genus was clearly shown at the community level by Klironomos et al. (2000). They treated various levels of plant diversity to colonization by *Glomus etunicatum*, *Glomus intraradices*, or no AMF. While *G. etunicatum* greatly enhanced plant biomass, *G. intraradices* suppressed biomass, even with respect to the non-AMF treatment. Both fungi were isolated from the same field site.

While it may not be surprising that different species of AMF differ functionally, a large amount of genetic divergence has been recently found within individual fungal isolates (Clapp et al., Chap. 8, this Vol.; Sanders, Chap. 16, this Vol.). This would suggest a potential for significant functional diversity within individual organisms. In fact, it has been shown that the degree of functional variability that can result from a single AMF isolate or individual spore is similar to the degree of inter-specific variability (Fig. 9.2). The variation in the growth response of *Plantago lanceolata* is high in a treatment comprising different genera of AMF, but is surprisingly similar in treatments containing isolates from within and among species of the genus *Glomus*. Although we would also need to investigate intraspecific variation in genera other than *Glomus*, this result indicates that there exists a much greater genetic and functional diversity than is explained by our inadequate under-

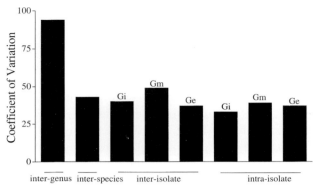

Fig. 9.2. Variation in *Plantago lanceolata* biomass when plants are infected by different (1) genera of AMF, (2) species of *Glomus*, (3) isolates of a single species (*Gi Glomus intraradices*, *Gm Glomus mosseae*, or *Ge Glomus etunicatum*), or (4) different spores from within a single isolate (*Gi Glomus intraradices*, *Gm Glomus mosseae*, or *Ge Glomus etunicatum*). The coefficient of variation shows the amount of variation within each treatment (Zar 1984). Intra-isolate, inter-isolate and inter-species variation is similar and much lower than inter-genus variation. All fungi were isolated from an old-field within the University of Guelph. The "genera of AMF" treatment included *Glomus intraradices, Acaulospora denticulata, Entrophospora colombiana, Scutellospora calospora,* and *Gigaspora gigantea*. The "species of *Glomus*" treatment included *G. intraradices, G. mosseae, G. etunicatum, G. aggregatum* and *G. claroideum*. For all treatments $n=5$. (Klironomos et al., unpubl. data)

standing of Glomalean taxonomy. New molecular techniques should discover the underlying basis for this.

The capacity among plants for differential response to AMF works on many levels, from community assemblages to genotypes within species. A study examining genetic response of *Chamerion angustifolium*, a facultative mycotroph, to AMF infection found 89 % of phenotypic variation among offspring from three full sib-families was explained by differences in AMF colonization (Husband and Klironomos, unpubl.). Progeny from three full-sib families of *Chamerion angustifolium* were inoculated with one of three different AMF (*Gigaspora gigantea, Glomus intraradices,* and *Glomus mosseae*). They found that there was a significant plant family × fungal species interaction for flower production, inflorescence length and shoot height in *C. angustifolium*. This is the first evidence for genetic variation within plant populations for response to AMF. Thus even within a species, or in this case, among siblings, not all plants will benefit equally from colonization by the same AMF.

The fact that AMF differ with respect to function, and because plants differ in their response to AMF, connotes a benefit to systems with high AMF species richness: more AMF species means more functions fulfilled and more opportunities for beneficial relationships to develop. Through these two mechanisms, AMF might play a role in determining plant community composition,

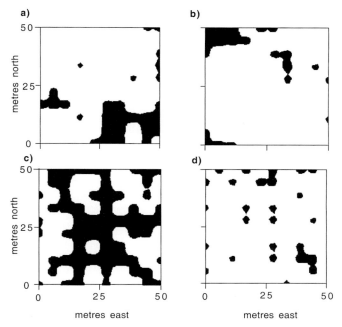

Fig. 9.3. Spatial distribution of four different fungal species within a 50×50 m plot. The plot is located on an old-field within the University of Guelph property. The plot was sampled at every 1-m interval, and the presence/absence of each species was determined after trap-culturing under greenhouse conditions using *Allium porrum* as a bait plant. Trap cultures were inspected after 3 months of growth for the presence of fresh AMF spores, which were identified using a compound microscope. **a** *Acaulospora denticulata* (relative frequency = 0.28), **b** *Scutellospora calospora* (relative frequency = 0.21), **c** *Glomus etunicatum* (relative frequency =0.61), **d** *Glomus intraradices* (relative frequency = 0.26). (Klironomos et al., unpubl. data)

and functioning. Coupled with the patchy distribution of AMF species within a community (Fig. 9.3), there is an enormous capacity for diversity inherent in most natural systems. Results from studies which use no AMF or a limited selection of AMF species, as has been the case throughout the history of mycorrhizal research (Klironomos and Kendrick 1993), may find their results valid only for the artificial conditions of their experiment, and have no bearing on the natural world with its varied and diverse AMF communities. To understand the mechanisms of biodiversity in ecosystems, we are obligated to consider both parts. To this end, only two studies examining diversity/function have considered the impact of AMF.

The first contribution showed that floristic diversity was higher in microcosms containing AMF (Grime et al. 1987). They found AMF increased the biomass of subordinate species relative to dominant species, perhaps through improved nutrient capture, and transfer of assimilate to subordinate plant species from dominants via their mycorrhizal connections. However, in this

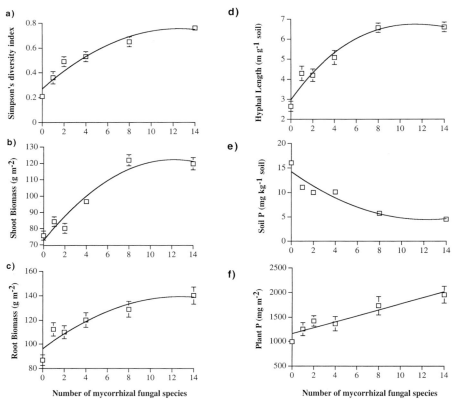

Fig. 9.4. The effect of AMF species richness on **a** plant diversity (fitted curve is $y=0.271+0.077x-0.003x^2$; $r^2=0.63$; $P = 0.0001$), **b** shoot biomass ($y=-0.334x^2+8.129x+72.754$; $r^2=0.69$; $P=0.0001$), **c** root biomass ($y=-0.265x^2+6.772x+96.141$; $r^2=0.55$; $P=0.0001$), **d** length of external mycorrhizal hyphae in soil ($y=0.001x^3-0.046x^2+0.756x+2.979$; $r^2=0.60$; $P=0.0001$), **e** soil phosphorus concentration ($y=0.065x^2-1.593x+14.252$; $r^2=0.67$; $P=0.0001$) and **f** total plant phosphorus content (linear relationship; $y=61.537x+1156.281$; $r^2=0.48$; $P=0.001$). Experiment was performed using macrocosms simulating an old-field ecosystem. *Squares* represent means (±SE). (van der Heijden et al. 1998a)

study, AMF were added by natural inoculum, and species composition of AMF was not qualified.

More than a decade later, van der Heijden et al. (1998a) showed that the diversity and identity of AMF, and not merely the presence or absence of inoculum, was a determinant of plant diversity and biomass production. This was achieved in two separate experiments. The first looked at functional compatibility among AMF and plants, by growing four AMF taxa alone and in combination with microcosms of calcareous grasslands. They discovered different plant species responded differently to AMF, meaning alterations in AMF composition could cause fluctuations in plant community structure and composition. To determine whether AMF species richness would influence plant diversity and ecosystem productivity, they manipulated AMF diversity in macrocosms (Fig. 9.4). To achieve this they added 0, 1, 2, 4, 8, or 14 AMF species to macrocosms containing sterile field soil and a seed mix comprising 15 plant species. All organisms were isolated from an old-field site that had been previously farmed but had been left abandoned for over 30 years. Fungal species were selected at random from a suite of 23 species that had been successfully cultured from that site. Macrocosms were harvested after a single growing season, after which they found that plant diversity and biomass increased with AMF diversity. In addition, diverse AMF communities also produced the most extensive extra-radical mycelium, which may explain why P-concentration in plant tissues increased and soil P decreased with high AMF diversity, indicating a more efficient exploitation of nutrients from soil. It is not certain why plant diversity was stimulated, but the most likely explanation would be that different fungal species associated preferentially with different plant species. Like Tilman and Downing (1994) they found productivity to saturate with increasing diversity (redundancy hypothesis). In this case, plant diversity and productivity reached a plateau at approximately eight AMF species. The ecological mechanisms that can explain how AMF and AMF diversity affect plant diversity and ecosystem functioning are discussed in Chapter 10 by van der Heijden (this Vol.).

Like Tilman and Naeem before, the most significant criticism of van der Heijden et al. (1998a) was the role of "sampling effect" in the interpretation of results. This criticism highlights a problem inherent to all diversity studies. Because manipulating diversity necessarily influences factors which then affect diversity (see Tilman's nutrient gradients), most diversity studies are hobbled by imperfect experimental designs, and hidden treatments. Sampling effect (SE), whereby most diverse treatments have a greater chance of including highly productive species, has been proposed as both a hidden treatment and viable mechanism in diversity experiments (Huston 1997; Hector 1998; Loreau 1998b; Wardle 1999). Tilman et al. (1997) were among the first to describe and support SE as a mechanism through which biodiversity could influence ecosystem function. In one of the only theoretical examinations of the diversity/function relationship, they suggested SE could promote increased biomass because better competitors (which produce more biomass) were most likely to occur in high diversity assemblages.

Those who disagreed saw SE as a flaw in experimental design. Huston (1997) viewed SE as a hidden treatment. He suggested the observed increase in biomass was a result of the indirect effect of high biomass, not high diversity. Since biomass is determined by the size of the largest species, and there is an increased chance of including larger species in high diversity treatments, the effect is a result of plant size rather than species richness (Aarssen 1997; Wardle 1999). Van der Heijden et al. (1999) pointed out SE was a valid mechanism for diversity/function only if it was possible to identify productive/ unproduc-

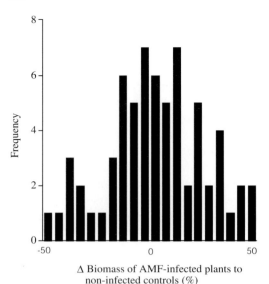

Fig. 9.5. The influence of *Glomus etunicatum* on the biomass of 64 different plant species which co-occur on an old-field within the University of Guelph property. There was no significant difference between the number of plants positively (x=34) or negatively (x=30) influenced by the symbiosis (chi square=0.250, df=1, P=0.617). (Klironomos et al., unpubl. data)

tive species. The premise of SE as a mechanism for increased functioning in diverse treatments is that more diverse treatments contain more highly productive species (Huston 1997; Tilman et al. 1997). However, there is no such thing as a highly productive AMF species since the effect of a single AMF isolate depends on the identity of its host plant (van der Heijden et al. 1999). Therefore, the chance of a treatment including a highly unproductive species is just as high as including a highly productive AMF species. Moreover, a study of 64 plant species grown with a single fungal species from the old-field site showed that AMF are just as likely to produce positive and negative effects, and the final outcome depends on the identity of the plant species (Fig. 9.5). Finally, they argued through SE, it is just as likely to observe a reduction in productivity with increasing AMF diversity. In their study, greater species diversity means greater niche exploitation and resource uptake, and the effects of diversity increase with increasing differences among AMF species. Here, the increased function is due to the added-on benefits of each AMF species.

These studies demonstrate the importance of AMF diversity for plant diversity and productivity, but do not reveal a mechanism (but see Chap. 10, this Vol.). Further, these experiments have dealt with very artificial systems. Studies like these need be replicated in the field for different systems. If AMF truly are critical determinants of community composition and functioning, recent reports of AMF impoverishment resulting from anthropogenic causes may have manifold consequences. For instance, a study examining AMF diversity in the UK found more than 92% of AMF in different, arable sites were represented by a single species (*Glomus mosseae*), one not found in woodland sites, which had much more diverse AMF communities (Chap. 8, this Vol.; Helgason et al. 1998). The authors attributed the disparity to agricultural techniques, such as tilling, fertilization and pesticide use. This study is a nice example as to how molecular tools can be used to answer ecological questions (for further details see Clapp et al., Chap. 8, this Vol.). A related study found fertilization to select for AMF species that are less mutualistic (Johnson 1993). If human activity directly or indirectly influences composition and diversity of AMF communities in already disturbed landscapes, then the sustainability of diversity and function of ecosystems is perilous. Thus, governments and agencies attempting to promote diversity must consider the full role of AMF in so far as ecosystem functioning. And in this respect, restoration ecologists have initiated the development of AMF as a restoration tool, but the role of AMF in this capacity remains largely unexamined and under-utilized.

9.5 Succession of AMF

While existing theoretical framework involving succession, invasion and diversity illustrates plant community dynamics, it tells us little about fungi. What exists of fungal ecology deals almost entirely with saprophytic fungi, or ecto-mycorrhizae (e.g. Erland and Taylor, Chap. 7, this Vol.), and what little we know of AMF succession pertains to plant, not fungal characteristics. For example, it has been demonstrated that in primary succession, there is a progression from non-mycotrophy, through facultative mycotrophy to obligate mycotrophy (Janos 1980a; Allen 1991). Following disturbance, soil resources are not limiting and AMF inoculum potential may be reduced or eliminated. Plants that can quickly establish in the absence of AMF are favored. As resources become limiting and AMF establish, selection is for the better competitors, which are typified by obligate mycotrophs (Janos 1980b).

But what does this tell us about AMF? Is there succession of AMF species corresponding to vegetation succession? A comparison of spore populations along a successional gradient suggests that there is (Johnson et al. 1991). Are there life history strategies for AMF species? The role of AMF in succession has not been tested. It is known that AMF differ in their abilities to absorb nutrients (Jakobsen et al. 1992), colonize roots (Wilson 1984b) and produce spores (Bever et al. 1996), but these traits have not been functionally grouped. One reason is because of the inherent difficulty of working with AMF; we can never completely know the biomass or life span of an individual or even determine what constitutes an individual. Despite these obstacles, there has been some examination of competitive strategies involving AMF. Wilson and Trinick (1983) found *Glomus tenuis* slower growing and less infective, but ultimately a more extensive colonizer than *Glomus monosporus*. Wilson (1984a) described competition among AMF with respect to aggressiveness (competitive strength) and infectivity (colonization). He found the most aggressive species (*Gigaspora decipiens*) was also the least infective, and the most infective (*Glomus fasiculatum*) was the least aggressive. Further, he found spore number correlated with competitive ability: the AMF best able to maintain sporulation in mixed inoculum also produced the fewest spores (*Glomus mosseae*), whereas the AMF which fared most poorly in mixed inoculum produced the highest number of spores in monoculture (*Glomus caledonium*). From these isolated studies, it seems likely that life history strategies can also be applied to AMF on the basis of spore size and production, hyphal growth and colonization rate.

If, in fact, AMF can be separated into distinct life history strategies, we can test to determine whether traditional plant successional and invasion theory apply to AMF. If superior colonizing fungi preferentially associate with superior colonizing plants, then it is likely there is a successional progression of different AMF genotypes through time. Whereas for plants resources are light, nutrients and space, for fungi resources are carbon and space to colonize along a root. Using this framework, we can predict which communities would be resistant to AMF invasion, and which AMF would be most successfully introduced. For successful restoration efforts, we must coordinate the desired plant community with its appropriate AMF successional stage. Failure to do so might result in reversion to an earlier successional plant community, or perhaps proliferation of parasitic, not mutualistic, AMF. While AMF have traditionally been considered to be exclusively mutualistic, it is now thought that the cost of some AMF associations outweigh benefits conferred to the host (Johnson et al. 1997). In these situations, the fungus receives carbon, but instead uses nutrients for its own growth. While any plant/fungus combination can fluctuate within time on the mutualism-parasitism continuum (for instance, the carbon drain on young seedlings often outweighs the nutritive benefits), and some AMF are unilaterally mutualistic regardless of host identity, are there universally parasitic AMF?

It is tempting to incorporate the mutualist-parasite continuum into successional theory. If, for example, more parasitic relationships develop when nutrients are in excess, then early successional, facultative mycorrhizal communities would be more likely to develop unbeneficial AMF associations. There is limited evidence for this. It has been shown that soil fertility correlates with more parasitic AMF (Johnson 1993). This is because nutrient-stressed roots exude more carbon (are better hosts), and thus attract less aggressive (less parasitic) AMF. Johnson found grasses inoculated with AMF from fertile soil were smaller, and produced fewer inflorescences than those inoculated with AMF from nutrient-poor soil, or those with no AMF whatsoever. While there was no difference in the number of vesicles for AMF grown in fertilized/unfertilized soil, the fertilized soils had proportionately fewer arbuscules, indicating the two fungi were storing similar amounts of carbon, but the unfertilized AMF were investing more for it.

It is unlikely, however, that early successional communities are biased for parasitic AM, or selection would favor non-mycotrophic plants. We know this is not true, because obligate mycotrophs are ultimately successful in most communities. More likely, evolution has co-evolved plant-soil-AMF associations, and that detrimental AMF relationships are the exception. But should human activity disrupt this coordination, through fertilization, or introduction of inappropriate AMF or plants, then the emphasis for mutualistic AM might be disrupted. Given the magnitude of human activity in natural systems, this could irrevocably alter the AM association.

9.6 Conclusions

Though it is well established that species richness promotes ecosystem function and integrity, it is clear the diversity/function argument is preoccupied with plant communities at the expense of other essential aspects of ecosystems, such as soil communities. It is now known that AMF can regulate community structure and production through functional differentiation among fungi, and differential effects on plants. Previously, any AMF would suffice, since it was thought AMF were involved solely in P uptake and were not host specific. Not only is this not true, but the varied functions of and responses to AMF are only now being discovered.

Since AMF is developing as a tool in restoration ecology, ecological theory pertaining to the uniqueness of AMF is urgently needed. As it stands, we know little of AMF succession, or consequences of invasion. What is clear is that AMF are no longer of interest only to soil ecologists; an understanding of ecosystems is incomplete without consideration of AMF. Ideas regarding AMF have changed dramatically in the last 10 years, and the number of publications regarding AMF has gone from 84 to over 700 per year in the last four decades (Klironomos and Kendrick 1993). With the incorporation of AMF into mainstream ecology, our understanding of AMF is bound to change even more.

Acknowledgements. The authors wish to thank the Natural Sciences and Engineering Research Council of Canada for financial support.

References

Aarssen LW (1997) High productivity in grassland ecosystems: affected by species diversity or productive species? Oikos 80:183–184

Allen EB, Allen MF (1986) Water relations of xeric grasses in the field: interactions of mycorrhizas and competition. New Phytol 104:559–571

Allen MF (1982) Influence of vesicular-arbuscular mycorrhizae on water movement through *Bouteloua gracilis* (H.B.K.) Lag Ex Steud. New Phytol 91:191–196

Allen MF (1991) The ecology of Mycorrhizae. Cambridge Univ Press, Cambridge

Allen MF (1996) The ecology of arbuscular mycorrhizas: a look back into the 20th century and a peek into the 21st. Mycol Res 100:769–782

Allen MF, Boosalis MG (1983) Effects of two species of VA mycorrhizal fungi on drought tolerance of winter wheat. New Phytol 93:67–76

Allen MF, Moore TS, Christiensen M (1980a) Phytohormone changes in *Bouteloua gracilis* infected by vesicular-arbuscular mycorrhizae: I. Cytokinin increases in the host plant. Can J Bot 58:371–374

Allen MF, Moore TS, Christiensen M (1980b) Phytohormone changes in *Bouteloua gracilis* infected by vesicular-arbuscular mycorrhizae II. Altered levels of gibberellin-like substances and abscisic acid in the host plant. Can J Bot 60:468–471

Andrade G, Mihara KL, Linderman RG, Bethlenfalvay GL (1998) Soil aggregation status and rhizobacteria in the mycorrhizosphere. Plant Soil 202:89–96

Bever JD, Morton JB, Antonovics J, Schultz PA (1996) Host-dependent sporulation and species diversity of arbuscular mycorrhizal fungi in a mown grassland. J Ecol 84:71–82

Elton CS (1958) The ecology of invasions by animals and plants. Methuen, London

Fitter AH, Graves JD, Watkins NK, Robinson D, Scrimgeour C (1998) Carbon transfer between plants and its control in networks of arbuscular mycorrhizas. Funct Ecol 12:406–412

Givnish TJ (1994) Does diversity beget stability? Nature 371:113–114

Grime JP, Mackey JML, Hillier SH, Read DJ (1987) Floristic diversity in a model system using experimental microcosms. Nature 328:420–422

Hartnett DC, Wilson GWT (1999) Mycorrhizae influence plant community structure and diversity in tallgrass prairie. Ecology 80:1187–1195

Hector A (1998) The effect of diversity on productivity: detecting the role of species complementarity. Oikos 82:597–599

Hector A, Schmid B, Beierkuhnlein C, Caldeira MC, Diemer M, Dimitrakopoulos PG, Finn JA, Freitas H, Giller PS, Good J, Harris R, Hogberg P, Huss-Danell K, Joshi J, Jumpponen A, Korner C, Leadley PW, Loreau M, Minns A, Mulder CPH, O'Donovan G, Otway SJ, Pereira JS, Prinz A, Read DJ, Scherer-Lorenzen M, Schulze E-D, Siamantziouras A-SD, Spehn EM, Terry AC, Troumbis AY, Woodward FI, Yachi S, Lawton JH (1999) Plant diversity and productivity experiments in European grasslands. Science 286:1123–1127

Helgason T, Daniell TJ, Husband R, Fitter AH, Young JPW (1998) Ploughing up the wood-wide web. Nature 394:431

Hooper DU (1998) The role of complementarity and competition in ecosystem responses to variation in plant diversity. Ecology 79:704–719

Hooper DU, Vitousek PM (1998) Effects of plant composition and diversity on nutrient cycling. Ecol Monogr 68:121–149

Huston MA (1997) Hidden treatments in ecological experiments: re-evaluating the ecosystem function of biodiversity. Oecologia 110:449–460

Hutchinson GE (1959) Homage to Santa Rosalia; or why are there so many kinds of animals. Am Nat 93:145–159

Jakobsen I, Abbot LK, Robson AD (1992) External hyphae of vesicular-arbuscular mycorrhizal fungi associated with *Trifolum subterraneum* L. I. Spread of hyphae and phosphorus inflow into roots. New Phytol 120:371–380

Janos DP (1980a) Mycorrhizae influence tropical succession. Biotropica 12:56–64

Janos DP (1980b) Vesicular-arbuscular mycorrhizae affect lowland tropical rain forest plant growth. Ecology 6:151–162

Jastrow JD, Miller RM, Lussenhop J (1998) Contributions of interacting biological mechanisms to soil aggregate stabilization in restored prairie. Soil Biol Biochem 30:905–916

Johnson NC (1993) Can fertilization of soil select less mutualistic mycorrhizae? Ecol Appl 3:749–757

Johnson NC, Zak DR, Tilman D, Pfleger FL (1991) Dynamics of vesicular-arbuscular mycorrhizae during old-field succession. Oecologia 86:349–358

Johnson NC, Graham JH, Smith FA (1997) Functioning of mycorrhizal associations along the mutualism-parasitism continuum. New Phytol 135:575–585

Jones CG, Lawton JH (1995) Linking species and ecosystems. Chapman and Hall, New York

Klironomos JN (2000) Host-specificity and functional diversity among arbuscular mycorrhizal fungi. In: Bell CR, Brylinsky M, Johnson-Green P (eds) Microbial biosystems: new frontiers. Proceedings of the 8th International Symposium on Microbial ecology, Atlantic Canada Society for Microbial Ecology, Halifax, Canada, pp 845–851

Klironomos JN, Kendrick WB (1993) Research on mycorrhizas: trends in the past 40 years as expressed in the 'MYCOLIT' database. New Phytol 125:595–600

Klironomos JN, McCune J, Hart MM, Neville J (2000) The influence of arbuscular mycorrhizae on the relationship between plant diversity and productivity. Ecol Lett 3:137–141

Loreau M (1998a) Biodiversity and ecosystem functioning: a mechanistic model. Proc Natl Acad Sci USA 95:5632–5636

Loreau M (1998b) Separating sampling and other effects in biodiversity experiments. Oikos 82:600–602

Morton JB, Benny GL (1990) Revised classification of arbuscular mycorrhizal fungi (Zygomycetes): a new order, Glomales, two new suborders, Glomineae and Gigasporineae, and two new families, Acaulosporaceae and Gigasporaceae, with an emendation of Glomaceae. Mycotaxon 37:471–491

Naeem S, Thompson LJ, Lawler SP, Lawton JH, Woodfin RM (1994) Declining biodiversity can alter the performance of ecosystems. Nature 368:734–737

Naeem S, Thompson LJ, Lawler SP, Lawton JH, Woodfin RM (1995) Empirical evidence that declining species diversity may alter the performance of terrestrial ecosystems. Philos Trans R Soc Lond B 347:249–262

Newsham KK, Fitter AH, Watkinson AR (1995a) Multi-functionality and biodiversity in arbuscular mycorrhizas. Trends Ecol Evol 10:407–411

Newsham KK, Fitter AH, Watkinson AR (1995b) Arbuscular mycorrhiza protect an annual grass from root pathogenic fungi in the field. J Ecol 83:991–1000

Remy W, Taylor TN, Haas H, Kerp H (1994) Four hundred-million-year-old vesicular-arbuscular mycorrhizae. Proc Natl Acad Sci USA 91:11841–11843

Rillig MC, Wright SF, Allen MF, Field CB (1999) Rise in carbon dioxide changes soil structure. Nature 400:628

Sanders IR, Fitter AH (1992) The ecology and function of vesicular-arbuscular mycorrhizas in co-existing grassland species. II. Nutrient uptake and growth of vesicular-arbuscular mycorrhizal plants in a semi-natural grassland. New Phytol 120:525–533

Sanders IR, Koide RT (1994) Nutrient acquisition and community structure in co-occurring mycotrophic and non-mycotrophic old-field annuals. Funct Ecol 8:77–84

Simon L, Bousquet J, Levesque RC, Lalonde M (1993) Origin and diversification of endomycorrhizal fungi and coincidence with vascular land plants. Nature 363:67–69

Tilman D (1993) Species richness of experimental productivity gradients: how important is colonization limitation? Ecology 74:2179–2191

Tilman D, Downing JA (1994) Biodiversity and stability in grasslands. Nature 367:363–365

Tilman D, Wedin D, Knops J (1996) Productivity and sustainability influenced by biodiversity in grassland ecosystems. Nature 379:718–720

Tilman D, Lehman CL, Thomson KT (1997) Plant diversity and ecosystem productivity: theoretical considerations. Proc Natl Acad Sci USA 94:1857–1861

van der Heijden MGA, Klironomos JN, Ursic M, Moutoglis P, Streitwolf-Engel R, Boller T, Wiemken A, Sanders IR (1998a) Mycorrhizal fungal diversity determines plant biodiversity, ecosystem variability and productivity. Nature 396:69–72

van der Heijden MGA, Boller T, Wiemken A, Sanders IR (1998b) Differential arbuscular mycorrhizae species are potential determinants of plant community structure. Ecology 79:2082–2091

van der Heijden MGA, Klironomos JN, Ursic M, Moutoglis, Streitwolf-Engel R, Boller T, Wiemken A, Sanders IR (1999) "Sampling effect", a problem in biodiversity manipulation? A reply to David A. Wardle. Oikos 87:408–410

Walker B (1992) Biodiversity and ecological redundancy. Conserv Biol 6:18–23

Wardle DA (1999) Is "sampling effect" a problem for experiments investigating biodiversity-ecosystem functioning relationships? Oikos 87:403–407

Wilkinson DM (1998) The evolutionary ecology of mycorrhizal networks. Oikos 82:407–410

Wilson JM (1984a) Competition for infection between vesicular-arbuscular mycorrhizal fungi. New Phytol 97:427–435

Wilson JM (1984b) Comparative development of infection by three vesicular-arbuscular mycorrhizal fungi. New Phytol 97:413–426

Wilson JM, Trinick MJ (1983) Infection development and interactions between vesicular-arbuscular mycorrhizal fungi. New Phytol 93:543–553

Yachi S, Loreau M (1999) Biodiversity and ecosystem productivity in a fluctuating environment: the insurance hypothesis. Proc Natl Acad Sci USA 96:1463–1468

Zar JH (1984) Biostatistical analysis, 2nd edn. Prentice-Hall, Englewood Cliffs

10 Arbuscular Mycorrhizal Fungi as a Determinant of Plant Diversity: in Search of Underlying Mechanisms and General Principles

MARCEL G.A. VAN DER HEIJDEN

Contents

10.1	Summary	243
10.2	Introduction	244
10.3	Arbuscular Mycorrhizal Fungi as a Determinant of Plant Diversity	245
10.3.1	Importance of Plant Species Composition and Nutrient Availability	245
10.3.2	Mycorrhizal Dependency	249
10.3.3	Underlying Mechanisms and Explaining Models	252
10.4	Ecological Significance of Arbuscular Mycorrhizal Fungal Diversity	255
10.4.1	Influence on Plants and Plant Communities	255
10.4.2	Thoughts on Underlying Mechanisms	256
10.4.3	Mycorrhizal Species Sensitivity	258
10.5	Conclusions	260
References		261

10.1 Summary

Recently, several studies have reported that arbuscular mycorrhizal fungi (AMF) enhance plant diversity of grasslands by specifically stimulating the growth of subordinate, often rare plant species. The underlying mechanisms by which AMF promote growth of such mycorrhizal-dependent plant species are reviewed in this chapter. These mechanisms can be used to explain how AMF promote plant diversity. Here I show that a positive relationship exists between the mycorrhizal dependency of a plant and the amount of phosphorus obtained from AMF. In addition, a re-analysis of previously published material shows that interplant carbon transport through a mycorrhizal hyphal network, from one plant to another, is directed towards plant species with the highest mycorrhizal dependency. Plant species with high mycorrhizal dependency, therefore, receive much more resources from AMF than plant species with a lower dependency. The inclusion of these results in a conceptual model shows that the supply of additional resources by AMF can enable mycorrhizal-dependent plant species to establish and coexist with other plant species and this can explain how AMF enhance plant diversity. In some instances, AMF can also reduce diversity. This has been observed in tall grass prairies that are dominated by mycorrhizal-dependent plant species and in which the majority of plants had a low mycorrhizal dependency. That AMF influences on plant diversity depend on the plant species composition is discussed.

The species composition and diversity of AMF communities also affects the structure, composition and diversity of plant population and communities. The available evidence is discussed. Theoretical and empirical evidence presented in this chapter suggests that a positive relationship exists between the mycorrhizal dependency of a plant species and the degree to which such a plant species responds differentially to different AMF species. It appears that the differential responses of plants to different AMF species are much higher when plants have a high mycorrhizal dependency. Based on these results, it is proposed that the mycorrhizal dependency is an appropriate measure that can be used to estimate how plants respond to changes in the abundance and species composition of AMF and AMF communities. Finally, I will show that the ratio and abundance of mycorrhizal-dependent versus low/non mycorrhizal-dependent plant species comprising a community can be used as a measure to predict the extent to which AMF and AMF diversity influence the diversity, structure and functioning of natural communities.

10.2 Introduction

Arbuscular mycorrhizal fungi are among the most abundant fungi in many ecosystems (Read 1991). These fungi form mutualistic symbioses with the roots of approximately 60 % of all plant species (Trappe 1987), thereby acting as extensions of plant root systems. With their extensive hyphal networks which spread into the soil, AMF acquire nutrients, especially poorly mobile nutrients such as phosphorus. AMF supply these nutrients to their plant hosts (Smith and Read 1997; Jakobsen et al., Chap. 3, this Vol.) and the growth of many plant species is enhanced when AMF are present (e.g. McGonigle 1988). Many ecosystems are dominated by plant species which associate with AMF and in soils of such ecosystems up to 25 m of AMF hyphae occur per gram soil (Smith and Read 1997). This length is equivalent to 20,000 km hyphae m^{-3}

soil. Their abundance and ubiquity has been known for a long time but it is only recently that some of their ecological functions have been elucidated. AMF are a determinant of plant diversity, they alter plant species composition and contribute to a range of important ecosystem functions. This will be discussed in this chapter. Moreover, it has been shown that AMF improve plant nutrition (Koide 1991), enhance productivity (Hart and Klironomos, Chap. 9, this Vol.), contribute to soil aggregation and soil structure (Miller and Jastrow 2000), improve the water relations of plants (Smith and Read 1997), suppress root pathogens and reduce plant disease (Sharma et al. 1992; Newsham et al. 1994). Moreover, by influencing plants, and due to the production of extensive hyphal networks in the soil, AMF directly and indirectly interact with many other organisms including herbivores (Gehring and Whitham, Chap. 12, this Vol.), soil invertebrates (Gange and Brown, Chap. 13, this Vol.), saprotrophs (Leake et al., Chap. 14, this Vol.), soil bacteria (Schreiner et al. 1997; Timonen et al. 1998) and other mutualists (Daft and El-Giahmi 1974; Xie et al. 1995). These examples show that AMF should be considered to understand the ecology and complexity of natural communities.

Processes and ecological mechanisms by which AMF influence plant diversity and affect plant species coexistence are presented in this chapter. Special emphasis is given to the mycorrhizal dependency of the plant species comprising a community because it can be shown that the ratio and abundance of mycorrhizal-dependent versus non/low mycorrhizal-dependent plant species comprising a community can be used to predict how much AMF influence plant diversity and functioning of natural communities.

10.3 Arbuscular Mycorrhizal Fungi as a Determinant of Plant Diversity

10.3.1 Importance of Plant Species Composition and Nutrient Availability

An increasing number of studies show that AMF have a significant, often drastic influence on productivity and diversity of plant communities. However, results are often conflicting and repeated patterns do not always occur. It has, for instance, been shown in field and greenhouse experiments that AMF promote plant diversity in European grasslands (Grime et al. 1987; Gange et al. 1990, 1993; Exp. 1 in van der Heijden et al. 1998a). However, AMF can also reduce diversity, as has recently been observed in American tall grass prairies (Hartnett and Wilson 1999). Furthermore, some studies report that AMF enhance community productivity (Fig. 9.4 in: Hart and Klironomos, Chap. 9, this Vol.; Gange et al. 1993; Klironomos et al. 2000), while others do not show

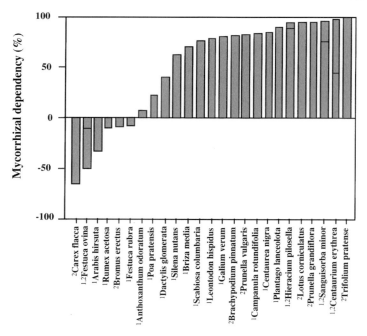

any difference in productivity between mycorrhizal and non-mycorrhizal communities (Grime et al. 1987; Sanders and Koide 1994; Exp. 1 in van der Heijden et al. 1998a; Hartnett and Wilson 1999). These differences seem, at first glance, contradictory, but they can easily be explained when the mycorrhizal dependency of the plant species comprising these communities are considered (see below). It must be mentioned here that different methods are employed to study AMF effects (e.g. use of fungicides in the field; addition of AMF to plants and plant communities growing in sterilized soil). The advantages and disadvantages of these methods are discussed by Read in Chap. 1 (this Vol.).

It is proposed here that the number and relative abundance of mycorrhizal-dependent plant species in the species pool of a community can be used to predict how AMF affect communities. The mycorrhizal dependency or mycorrhizal responsiveness of a plant shows the extent to which a plant benefits from the presence of AMF compared to when it is absent (e.g. see Gerdeman 1975; Plenchette et al. 1983; Johnson et al. 1997; van der Heijden et al. 1998b). A mycorrhizal dependency >0 means that a plant benefits from AMF and a mycorrhizal dependency <0 means that AMF reduces the growth of a plant. Moreover, plants with a mycorrhizal dependency of +100% are fully dependent on AMF while plants with a mycorrhizal dependency of –100% are unable to grow when AMF are present. Details about the calculation of mycorrhizal dependency are given in the legends of Fig. 10.1. Usually, plant com-

Fig. 10.1. Mycorrhizal dependency of 25 plant species occurring in European calcareous grassland (Data modified after Grime et al. 1987 and van der Heijden et al. 1998a). The mycorrhizal dependency of each plant species is calculated as follows: If biomass of $\sum_{1}^{n} a_n > bn$, then mycorrhizal dependency = $\left(1 - \dfrac{bn}{\left(\sum_{1}^{n} a_n\right)}\right) \times 100\%$, (Eq. 1). If biomass $\sum_{1}^{n} a_n < bn$, then mycorrhizal dependency = $\left(-1 + \dfrac{\left(\sum_{1}^{n} a_n\right)}{bn}\right) \times 100\%$ (Eq. 2), where a is the mean plant dry mass of a treatment inoculated with AMF, n is the number of treatments where plants were inoculated with AMF and b is the mean plant dry mass of the non-AMF treatment. Plant species that occurred in both studies had two mycorrhizal dependencies, one is shown by the *bar* and one by the *division* within the bar. The formula to calculate the mycorrhizal dependency has originally been defined by Plenchette et al. (1983) (Percentage mycorrhizal dependency = [(dry mass mycorrhizal plant – dry mass non-mycorrhizal plant)/dry mass mycorrhizal plant] × 100%). This formula was adapted by van der Heijden et al. (1998b) where the dry mass of the mycorrhizal plant was substituted by the mean dry mass across all treatments with AMF. This was done since it was observed that plants respond differently to different AMF species. The formula presented by van der Heijden et al. (1998a) needs, however further modification since the authors did not consider that the growth of some plants is reduced by AMF. The mycorrhizal dependency of plants as calculated by their formula (equation 1) is asymmetrical since values for mycorrhizal dependency range from $-\infty$ to +100%. Two equations are, therefore required, one for plants which grow better with AMF (Eq. 1) and one for plants which perform better without AMF (Eq. 2). The use of Eqs. (1) and (2) ensures that positive and negative values for mycorrhizal dependency are symmetric and comparable (e.g. values for mycorrhizal dependency range from –100 to +100%). The original formula for mycorrhizal dependency needs, thus, to be adapted if plants grow better without mycorrhizal fungi and, if plants respond differently to treatments with different AMF. The mycorrhizal dependency of 90 plant species characteristic for North American tall grass prairie are given in Wilson and Hartnett (1998)

---

munities comprise plant species that vary in their mycorrhizal dependency and, thus, differ in their response to AMF. For example, a calculation of mycorrhizal dependency of characteristic species from European calcareous grasslands shows that they comprise plant species with a wide range in mycorrhizal dependencies from –65 to 99% (Fig. 10.1). A comparison of published studies where the abundance or presence of AMF has been manipulated in plant communities reveals that AMF enhance plant diversity when the majority of the plant species present in a community have a high mycorrhizal dependency. For example, 9 of the 18 plant species growing in the simulated grasslands of Grime et al. (1987) and 8 of the 11 plant species occurring in the simulated grasslands of van der Heijden et al. (1998a) were almost completely dependent on AMF (Fig. 10.1). AMF enhanced plant diversity in these studies. However, AMF reduce diversity if the majority of plant species have a negative, or low mycorrhizal dependency. This reduction in diversity was observed in tall grass prairies studied by Hartnett and Wilson (1999). In that study, only three of the eight species that responded to AMF had a high mycorrhizal dependency. AMF consequently reduced plant diversity. Similar results were obtained in grasslands dominated by grasses where shoots were infected by a host-specific fungal endophyte. The presence of this fungal endophyte in the dominant grass decreased plant diversity by stimulating growth of the already dominant grass (Clay and Holah 1999). Moreover, the increase in productivity of an American old-field community by AMF and the absence of such an increase in European calcareous grassland in the two studies by van der Heijden et al. (1998a) can be explained by differences in the abundance of high mycorrhizal-dependent versus low mycorrhizal-dependent plant species. The American old-field communities were dominated by several plant species with a high mycorrhizal dependency (J. Klironomos, pers. observ.), while the European grasslands were dominated by *Bromus erectus*, a plant species with a low mycorrhizal dependency (Fig. 10.1). It is, thus, likely that AMF enhance productivity when communities are dominated by mycorrhizal-dependent plant species that benefit greatly from AMF, such as in the American old-field community. Furthermore, plant communities comprised of only non-mycorrhizal plant species, or plant communities without AMF, might not fully use all available resources, leading to reduced productivity and possibly to nutrient losses (Klironomos et al. 2000). The way in which AMF influence plant communities, therefore depends on the plant species composition or, more precisely, on the mycorrhizal dependencies of the plant species composing the community. The composition and diversity of AMF communities also affect plant communities as will be discussed in Section 10.4.

The impact of AMF on plant communities is not only dependent on plant species composition. Many pot experiments have shown that the importance of AMF ceases when nutrient availability, in particular phosphorus supply exceeds the level of demand (Koide 1991). The mycorrhizal dependency of many plant species also decreases when they are grown under nutrient-rich conditions (Habte and Manjunath 1991; Hetrick et al. 1992). It is, therefore, likely that the importance of mycorrhizal fungi as a determinant of plant diversity ceases when nutrient availability and ecosystem productivity (which is an indirect measure of nutrient availability) increases. This can been derived from Fig. 10.2. This figure shows the humped-back relationship between productivity and plant diversity that has been observed for European grasslands (Al-Mufti et al. 1977) and many other ecosystems (reviewed in Tilman and Pacala 1993). It is hypothesized that this humped-back curve, which is based on observations from natural communities with AMF, will be lowered when AMF are absent. This hypothesis is based on the observation

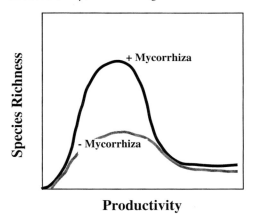

Fig. 10.2. Hypothesized relationship between productivity and plant species richness of European grasslands with and without mycorrhizal fungi. A typical humped-back curve is observed when mycorrhizal fungi are present (the natural situation). However, the humped-back curve will be lowered if mycorrhizal fungi are absent. Note: Mycorrhizal fungi are naturally present in grassland. Grasslands without AMF are rarely observed. The curve without mycorrhizal fungi is, therefore, only drawn to illustrate the ecological significance of these mutualistic soil fungi

that AMF enhance diversity in nutrient-poor European grassland while it is expected that AMF have a negligible role in productive, nutrient-rich grasslands. Two additional sources of evidence underline this hypothesis. Firstly, the number of plant species that respond positively to mycorrhizal fungi is high in unproductive calcareous grassland (Fig. 10.1). In contrast, productive grasslands are normally dominated by grasses, most of which hardly respond to AMF (see below). Secondly, the limited importance of mycorrhizal fungi in productive environments can be derived from the fact that the abundance and activity of AMF (measured as hyphal length or amount of root colonization) is much lower in productive grasslands (I. Kotorova and M. Henselmans; pers. comm.). In this respect, it is important to mention that the few experiments with plant communities in which the abundance/presence of AMF has been manipulated were executed in grassland communities on relatively nutrient-poor soils. Additional experiments should be conducted on a wider range of ecosystems with varying levels of nutrient availability. Such experiments are essential to obtain a general view about the functional importance of AMF on a more global scale.

10.3.2 Mycorrhizal Dependency

Differences in the mycorrhizal dependencies of plants comprising plant communities could explain why community responses to AMF are so variable. However, what makes plants vary in their mycorrhizal dependency? The data presented in Fig. 10.3 show that highly mycorrhizal-dependent plant species derive more benefits from AMF in comparison with plant species with a low mycorrhizal dependency. Plant species with a high mycorrhizal dependency, such as *Hieracium pilosella*, *Prunella vulgaris* and *Brachypodium pinnatum*

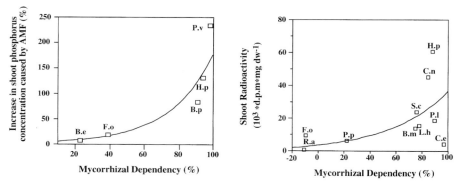

Fig. 10.3.a Relationship between the mycorrhizal dependency (*x-axis*) and increase in shoot phosphorus concentration caused by AMF (*y-axis*). The mycorrhizal dependency is calculated as in Fig. 10.1. The following plant species are shown: *B.e Bromus erectus*(data modified after van der Heijden et al. 1998b), *F.o Festuca ovina* (data modified after van der Heijden et al. 1998b), *H.p Hieracium pilosella* (data modified after van der Heijden et al. 1998b), *B.p Brachypodium pinnatum* (data modified after van der Heijden et al. 1999), *P.v Prunella vulgaris* (data modified after van der Heijden et al. 1999) Curve fit: $y=0.031 \times 10^{0.017x}$ ($r^2=0.85$; $P=0.02$). **b** Relationship between the mycorrhizal dependency (*x-axis*) and the shoot radioactivity (*y-axis*) of plants growing in the neighbourhood of a *Festuca ovina* plant which was sealed into a transparent box and supplied with 3,700 kBq $^{14}CO_2$ (data modified after Grime et al. 1987). The mycorrhizal dependency is calculated as in Fig. 10.1. This figure shows that transfer of assimilates from *Festuca* to other plants via a common mycorrhizal hyphal network was directed towards plant species with the highest mycorrhizal dependency. The shoot radioactivity is given in $10^3 \times$d.p.m. per mg dry weight, as presented in Grime et al. (1987). The following plant species are shown: B.m = *Briza media*, C.e = *Centarium erythrea*, C.n = *Centaurea nigra*, F.o = *Festuca ovina*, H.p = *Hieracium pilosella*, L.h = *Leontodon hispidus*, P.l = *Plantago lanceolota*, P.p = *Poa pratensis*, R.a = *Rumex acetosa* S.C = *Scabiosa columbaria*. Curve fit: $y=3.1576 \times 10^{0.936x}$ ($r^2=0.43$; $P=0.045$)

have a much higher phosphorus concentration when AMF are present compared to the non-mycorrhizal state (Fig. 10.3a). In contrast to these species, the plant phosphorus concentration of plants with a low dependency, such as *Bromus erectus* and *Festuca ovina*, is hardly enhanced by AMF. This suggests that one of the mechanisms by which AMF stimulate the growth of mycorrhizal-dependent plant species is the supply of additional phosphorus. Studies with agricultural plants species confirm this relationship (Plenchette et al. 1983; Haas et al. 1987). These studies found that the phosphorus content of mycorrhizal-dependent plant species was increased by AMF as well. The phosphorus concentration is shown in Fig. 10.3a because this measure is independent from plant biomass, unlike phosphorus content.

A re-analysis of the data from Grime et al. (1987) gives another explanation as to why plants vary in their mycorrhizal dependency. In that study, shoots of the dominant plant species, *Festuca ovina*, were labelled with radioactive carbon (^{14}C). They showed that below ground transport of carbon probably occurred through a common mycorrhizal hyphal network from the dominant species, *Festuca*, to the subordinates (for more information about interplant carbon transport see Simard et al., Chap. 2, this Vol.). Plant species with the highest mycorrhizal dependency obtained most carbon from *Festuca* and, thus, these plants benefited most from connection to the hyphal network (Fig. 10.3b). This carbon supply could be especially important for the establishment of small seedlings that are often light-limited when they germinate under an established canopy (e.g. Dale and Causton 1992; Hubbell et al. 1999; Kiers et al. 2000). Several studies have shown that AMF also enhance the supply of iron, zinc, copper, manganese and, in some cases, water and nitrogen (reviewed in Smith and Read 1997). Considering the effects shown in Fig. 10.3a, b, it is likely that mycorrhizal-dependent plant species obtain more of these benefits from AMF compared to species with a lower mycorrhizal dependency. Experimental studies are needed to confirm the existence of such relationships. The observations made in Fig. 10.3 suggest, thus that mycorrhizal-dependent plant species form a more intimate association with AMF enabling them to use the resources acquired by the fungal mycelium.

There may be a reason why some plants have such a high mycorrhizal dependency as illustrated by a study of Hetrick et al. (1992). They observed, for 23 tall grass prairie forbs that a positive relationship exists between the mycorrhizal dependency of a species and its root fibrousness (Hetrick et al. 1992). The roots of plant species with a high mycorrhizal dependency are often thick and unbranched, have a few root hairs, and are not well adapted to acquire nutrients (Baylis 1975; Hetrick et al. 1992; Newsham et al. 1995). This suggests that mycorrhizal-dependent plant species need AMF to obtain sufficient amounts of nutrients. This contrasts with plant species that have a low mycorrhizal dependency, like many grasses. These species have a well-developed, fibrous root system consisting of many fine roots adapted to forage for nutrients. They are less dependent on AMF. However, it has been suggested that such plants still benefit from the protective role of AMF against pathogenic soil fungi (Newsham et al. 1995). The traits that influence the mycorrhizal dependency of a plant are summarized in Table 3.2 (Jakobsen et al., Chap. 3, this Vol.). There are other strategies to cope with low nutrient availability. Several species, including some of the non-mycorrhizal sedges, form cluster or proteoid roots (Dinkelaker et al. 1995; Skene 1998; Miller et al. 1999), roots that are extremely efficient in nutrient uptake. Moreover, many plant species associate with ericoid mycorrhizal fungi and ecto-mycorrhizal fungi, fungi that are well adapted to cope with nutrient stressed, sometimes, harsh environments (e.g. see Aerts, Chap. 5, this Vol.; Fig. 17.2 in: van der Heijden and Sanders, Chap. 17, this Vol.).

10.3.3 Underlying Mechanisms and Explaining Models

In the previous section it has been shown that AMF supply additional amounts of resources (e.g. phosphorus) to the mycorrhizal-dependent plant species. The supply of these additional nutrients by AMF could enable the mycorrhizal-dependent plants to grow at resource availabilities where they would otherwise be unable to grow. This can be illustrated with the help of a model developed by Tilman (1988). According to this model, the population growth of a plant species can be represented by isoclines that show the amount of growth in relation to the availability, or supply, of limiting resources. The resource supply levels where plant populations stop growing, define the zero population growth isoclines. The supply of additional resources by AMF could reduce the growth isoclines of a mycorrhizal-dependent plant, thus, broadening its potential niche. This is shown for the hypothetical plant species (plant B) in Fig. 10.4. In this case, AMF enhance the supply of phosphorus, thereby lowering the isocline of plant B (Fig. 10.4). This model can be extended to predict the outcome of competition between two plant species (Tilman 1988). The model predicts that if two plant species compete for resources, the one with the lowest zero population growth isocline, i.e. the one with the lowest resource requirements, will be competitively superior. Furthermore, the model predicts that plants can coexist if their zero population growth isoclines cross. Coexistence could then be possible because the growth of each plant species is limited by a different resource. As a consequence, intraspecific competition is greater than interspecific competition (Tilman 1988). The reduction of the zero population growth isocline of mycorrhizal-dependent plant species by AMF, as explained above, can alter their

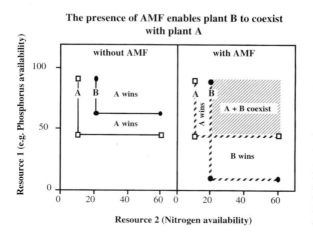

Fig. 10.4. Resource dependent zero growth isoclines of two hypothetical plant species, plant A (□) and B (●) without AMF (*solid line*) and with AMF (*hatched line*). The *y*-axis and *x*-axis show phosphorus and nitrogen availability respectively. The positions of the isoclines determine the outcome of competition between plant A and B. In the absence of AMF (*left*), plant B is outcompeted by plant A at each resource level. Plant B is able to coexist with plant A when AMF are present (*right*)

competitive ability with other plant species. For example plant B will be outcompeted and replaced by plant A in the absence of AMF because the zero growth isocline of plant B is always inside that of plant A (Fig. 10.4). However, if AMF are present, the zero growth isocline of plant A is unaltered but that of plant B is reduced and it crosses the zero growth isocline of plant A, so that coexistence occurs (Fig. 10.4). Several studies have indeed shown that AMF alter plant competition (Fitter 1977; Hetrick et al. 1989; Allen and Allen 1990; West 1996; Marler et al. 1999). For instance, Hetrick et al. (1989) showed that the outcome of competition between two tall grass prairie grasses was reversed depending on whether AMF were present or absent. Moreover, Marler et al. (1999) showed that AMF indirectly enhanced competitive effects of an invasive forb on a native bunchgrass. Finally, several plant species are only able to coexist with other plants if AMF are present indicating that AMF enhance their competitive ability (Grime et al. 1987; Hetrick et al. 1989; van der Heijden et al. 1998a).

The above-mentioned model can also be used to explain how AMF enhance or reduce diversity in plant communities. The simulated calcareous grassland communities studied by Grime et al. (1987) and experiment 1 in van der Heijden et al. (1998a) were dominated by *Festuca ovina* and *Bromus erectus*, two plant species that have a low mycorrhizal dependency (Fig. 10.1). The majority of the plant species in both studies had, however, a high mycorrhizal dependency (Fig. 10.1). AMF could, therefore have reduced their growth isoclines as is postulated in Fig. 10.5. Such a decrease could have enabled these species to coexist with *Festuca* and *Bromus* due to the crossing of their growth isoclines (Fig. 10.5). This, in turn leads to a higher number of plant species that can coexist when AMF are present. This provides an explanation as to how AMF enhanced plant diversity. The opposite is also possible. AMF can reduce the growth isoclines of plant species that are already dominant. This can increase their competitive ability and that can consequently lead to a reduction of plant diversity if the other plant species have a low mycorrhizal dependency. This could explain why AMF reduced plant diversity in American tall grass prairies (Hartnett and Wilson 1999).

The results presented in Fig. 10.5 are based on the concept of resource competition and growth isoclines of plant populations. Accordingly, plant species coexistence and plant diversity is explained by differential resource use (also modified by AMF).

Seedling recruitment into established vegetation is another important factor that determines plant diversity (Grubb 1977; Tilman 1997). Several authors have suggested that AMF promote seedling establishment by integrating emerging seedlings into extensive hyphal networks and by supplying nutrients to the seedlings (Francis and Read 1995; Zobel et al. 1997). Many investigators have tested AMF effects on plants by using seedlings and juvenile plants, thereby, indirectly stressing the importance of AMF for seedling establishment. My own experiments have even indicated that AMF are most important for the seedling

AMF enhance the number of plant species that can coexist

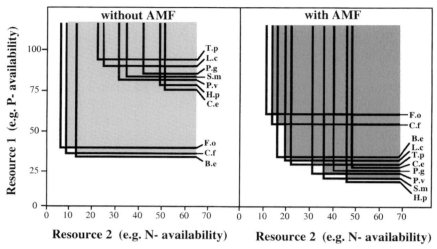

Fig. 10.5. Hypothetical resource dependent growth isoclines of the 11 plant species that were grown in simulated calcareous grasslands (Exp. 1 of van der Heijden et al. 1998a). The positions of the isoclines determine the outcome of competition between the plants at each resource level. Coexistence between plants is only possible when their growth isoclines cross (further details see text). The y-axis and x-axis show phosphorus and nitrogen availability respectively. Growth isoclines of each plant species are shown when AMF are absent (*left*) or present (*right*). Without AMF only three species can coexist (*light grey shading*). With AMF all 11 species can coexist (*dark grey shading*). Plant species shown are abbreviated as follows: *B.e Bromus erectus, B.p Brachypodium pinnatum, C.e Centaurium erythrea, C.f Carex flacca, F.o Festuca ovina, H.p Hieracium pilosella, L.c Lotus corniculatus, P.g Prunella grandiflora, P.v Prunella vulgaris, S.m Sanguisorba minor, T.p Trifolium pratense*

and juvenile stage of several plant species (van der Heijden, unpubl. data). The combination of two major theories which explain plant species coexistence, by resource partitioning (Ricklefs 1977; Aarssen 1983; Tilman 1988) and by opportunities for seedling establishment (Grubb 1977), therefore provides an appropriate explanation how AMF enhance plant diversity.

Plant species can also coexist because of spatial and temporal variation in resource heterogeneity (Grubb 1977; Grime 1979; Huston 1979; During and Willems 1984). With this, most ecologists intuitively assume that the abiotic resources vary across and within habitats. However, the availability of biotic resources such as AMF also varies across and within habitats. AMF are absent or low in abundance in disturbed habitats or patches and spatial variations in patterns of mycorrhizal infectiveness have been observed (Read 1991; Brundrett and Abott 1995; Boerner et al. 1996). The growth of non-mycorrhizal species, such as *Carex flacca* and *Rumex acetosa*, and ruderal plant species such as, *Arabis hirsitua*, is reduced by AMF (Grime et al. 1987; Francis and

Read 1995; van der Heijden et al. 1998a). For such plants, patches where AMF are absent or low in abundance, are preferable sites for colonization. This contrasts with mycorrhizal-dependent plant species that need AMF for growth and establishment. Spatial variation in the distribution and abundance of AMF within a habitat could, therefore, enhance diversity by allowing mycorrhizal and non-mycorrhizal plant species to coexist in patches with and without AMF, respectively (Janos 1980; Allen 1991).

10.4 Ecological Significance of Arbuscular Mycorrhizal Fungal Diversity

10.4.1 Influence on Plants and Plant Communities

Not only are AMF present in most plant communities, but communities of AMF occur which vary in species composition and number (Walker et al. 1982; Morton et al. 1995). Up to 25 different AMF species have been recorded within plant species-rich, grassland communities (Johnson et al. 1991; Bever et al. 1996). Other communities like intensively used agricultural fields contain only a few AMF species and have, generally a lower AMF diversity (Fig. 8.4 in: Clapp et al., Chap. 8, this Vol.; Cuenca et al. 1998; Helgason et al. 1998). Descriptive and experimental studies suggest that the abundance and distribution of AMF species within a community is affected by edaphic factors such as pH, phosphorus and nitrogen availability, moisture content (Stahl and Smith 1984; Ernst et al. 1984; Hayman and Tavares 1985; Louis and Lim 1988; Miller and Bever 1999; Egerton-Warburton and Allen 2000) as well as by plant species composition (Johnson et al. 1992; Sanders and Fitter 1992; Bever et al. 1996; Eom et al. 2000). Differences in AMF species composition could be important for the plants inhabiting such communities because it has been shown that AMF species differ from each other in the way they stimulate plant growth (e.g. Powell et al. 1982; Jensen 1984; Haas et al. 1987; Raju et al. 1990). Different AMF species have also been shown to provide plants with different amounts of phosphorus (Jakobsen et al. 1992a,b; van der Heijden et al. 1998b), zinc (Burkert and Robson 1994) and manganese (Arines et al. 1989). Moreover, different AMF species show spatial differences in acquisition of soil phosphate (Smith et al. 2000). This could lead to complementary resource and enhanced plant productivity if several AMF species are simultaneously present. Different AMF species also alter spatial growth patterns of clonal plants (Streitwolf-Engel et al. 1997). The variation in clonal growth caused by different AMF species is ecologically important since it is greater than growth differences caused by genotypic variation for several fitness related traits and because it can alter the plant population size by effects on clonal plant reproduction (Streitwolf-Engel et al. 2001). Co-occurring plant species respond differently to different AMF species (Roldan-Fajardo 1994; van der Heijden et al. 1998b; Klironomos 1999; Kiers et al. 2000) and we have recently shown that AMF species composition controls plant community structure and diversity (van der Heijden et al. 1998a). We observed that the biomass of several plant species co-occurring in simulated grasslands changed by a factor of five due to the presence of a different AMF species. We also showed a two-fold increase in plant diversity and productivity of artificial assembled plant communities when AMF diversity was enhanced from one to 14 AMF species (Fig. 9.4). Hart and Klironomos discuss this in detail by in Chap. 9. These observations emphasize the importance of AMF species composition and they suggest that the success of a plant within a community depends on the AMF species present.

10.4.2 Thoughts on Underlying Mechanisms

The basic model adapted from Tilman (1988) to explain how the presence of AMF affected plant species coexistence and diversity (Sect. 10.3.3) could also be used to explain how different AMF species benefit different plants species and affect plant community structure. For instance, the extent to which zero growth isoclines of plant species are lowered by AMF may also depend on the identity of the AMF species because different AMF species supply different amounts of resources. Growth isoclines of plant species, such as *Hieracium pilosella* or *Prunella vulgaris*, which acquire large amounts of nutrients from AMF (Fig. 10.3a) are, therefore, likely to vary when AMF species that supply different amounts of nutrients are present. Coexistence between such plants can, as explained in Fig. 10.4, vary depending on AMF species composition. In one study we showed that different AMF species indeed alter the way in which two plant species coexisted (van der Heijden et al. 1999). The proportional biomass that each plant species contributed to the total biomass varied by a factor of six due to the presence of a different AMF species. The differential growth response of plants to different AMF is greatest when different AMF genera are present (Fig. 9.5 in Hart and Klironomos, Chap. 9, this Vol.). It is, therefore, likely that spatial variation in the presence of AMF genera may have a profound impact on the coexistence of plants. Moreover, since plant species vary in their resource requirements (e.g. Aerts and Chapin 2000), it is likely that certain AMF species that supply a particular combination of nutrients, would be a better match to specific plant species. This can result in functional compatibility between pairs of plant and AMF species (Ravnskov and Jakobsen 1995; van der Heijden et al. 1998a,b; Klironomos 1999). An increase in the number of AMF species would then lead to an increase of plant species that benefit from AMF. The added beneficial effect of each additional AMF species could, in this way, explain how AMF diversity enhanced plant diversity (van der Heijden et al. 1998a).

The isocline approach to explain plant species coexistence and plant diversity (Fig. 10.5) can also be applied here. In that case, Fig. 10.5 should be refined with the assumption, based on the existence of functional compatibility between specific plant and AMF species that AMF diversity is required to reduce the growth isoclines of several plant species and to enhance plant diversity. The presence of one AMF species would, in this scenario, be insufficient for an increase in plant diversity since growth isoclines of only a few plant species would be reduced substantially.

Positive feedback between pairs of plant and AMF species could also explain why enhanced AMF diversity leads to increased plant diversity, phosphorus capture, hyphal length and productivity (for more information about positive feedback, see Bever et al., Chap. 11, this Vol.). The observations discussed above are based on the assumption that optimal matching between specific plant-fungal combinations in multi-species communities occurs (e.g. each plant becomes colonized by those AMF species that provide the greatest benefit). This assumption might be true, however to date, there are, as I far as I know, no data showing this. The reality is even more complex since multiple colonization of plant roots by several AMF species has been observed (Clapp et al., Chap. 8, this Vol.; Rosendahl et al. 1990; Clapp et al. 1995). It is not known if each of those AMF species that colonizes the plant root has a beneficial effect on plant growth. Optimal matching suggests also that host specificity or selectivity occurs between specific plant-fungal combinations. It has not yet been shown that plant or AMF species are able to select a suitable partner. Moreover, it is thought that AMF lack any degree of host specificity since each plant species can become colonized by each AMF species (Law 1988; Fitter 1990). Further details about specificity in the mycorrhizal symbiosis are given by Sanders in Chap. 16.

Another possible explanation for the positive relationship between AMF diversity, plant diversity and productivity is the so called "sampling effect" (Wardle 1999). The sampling effect occurs when productivity or other ecosystem functions increase with diversity because the high diversity treatments have a higher probability of containing the most productive species. In this case that would mean that plant diversity and productivity increase with AMF diversity since the chance of including an AMF species that is beneficial/productive to all or most plant species is highest in treatments with a high AMF diversity. The existence of an "all-productive" AMF species (i.e. one that stimulates the growth of all or most plant species) is unlikely since the productivity induced by an AMF species depends on the identity of the plant host it associates with (van der Heijden et al. 1999).

Spatial variations in the presence of different AMF species could also increase diversity of whole communities when the AMF species, prevalent at each patch, enhance the growth of those plants that are functionally most compatible. The influence of such spatial variations in AMF species composition are further discussed by Hart and Klironomos (Chap. 9, this Vol.; e.g. see Fig. 9.3).

10.4.3 Mycorrhizal Species Sensitivity

The variation in the growth response of a plant species to different AMF species can be defined as its mycorrhizal species sensitivity (MSS). Plant species, like *Hieracium*, thus, have a high MSS because they respond so differently to different AMF species and species like *Bromus* have a low MSS since they show hardly any differential response (van der Heijden et al. 1998b). Consequently, these species exhibit high sensitivity and low sensitivity towards the species composition of AMF communities. It is likely that a relationship between the mycorrhizal dependency and mycorrhizal species sensitivity exists (Fig. 10.6). If the growth stimulation by AMF is small, then there is also a small potential margin for differences in growth response to different AMF species (arrow 1 in Fig. 10.6). In contrast, plant species with a high mycorrhizal dependency have a large potential margin for differential effects by different AMF species (arrow 2 in Fig. 10.6). The MSS of a plant species can be estimated by calculating the coefficient of variation on the biomass data of a plant species in response to treatments with different AMF species. The coefficient of variation gives a measure for the amount of variation among different treatments (Sokal and Rolph 1981). The MSS is not a fixed value since it can vary when the response of a plant to other (more or less beneficial) AMF species is included. MSS as such gives only an indication as to the extent in which plants respond differently to different AMF species. The existence of MSS, of variation in the growth response of a plant to different AMF species, implies also that the mycorrhizal dependency of plants can be variable (van der Heijden et al. 1998b).

I have determined the MSS from the plant species in experiment 1 of van der Heijden et al. (1998a). This analysis (Fig. 10.7) gives some further evidence for the existence of a relationship between mycorrhizal dependency and mycorrhizal species sensitivity. It appears that plant species with the highest mycor-

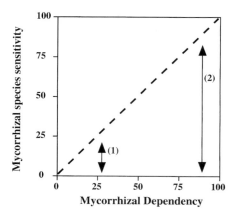

Fig. 10.6. Hypothesized relationship between mycorrhizal dependency and mycorrhizal species sensitivity. The *arrows* indicate the margin for variation in growth response of a plant to different AMF species for (*1*) plant species with low mycorrhizal dependency and (*2*) plant species with a high mycorrhizal dependency. The margin for variation in response is greater for species at position (*2*) than for species at position (*1*)

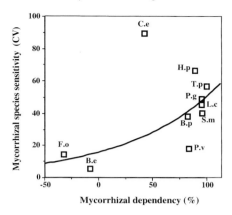

Fig. 10.7. Relationship between the mycorrhizal dependency and mycorrhizal species sensitivity of ten plant species that form an association with AMF. These plant species were growing in the simulated calcareous grasslands of van der Heijden et al. (1998a). Plant species are abbreviated as in Fig. 10.5. The mycorrhizal dependency of these plants species is based on the biomass and is determined as explained in Fig. 10.1. The coefficient of variation (CV) is used to estimate the mycorrhizal species sensitivity of each plant species (for details see text). Curve fit: $y=10^x$ ($r^2=0.45$; $P=0.03$)

rhizal dependency also have the highest MSS (Fig. 10.7). The observations in Fig. 10.7 form two separate clusters along the regression line, one group of plants with a high mycorrhizal dependency and a high MSS and one group of plants with a low mycorrhizal dependency and a low MSS. This points to the need for additional research to confirm the validity of the relationship between mycorrhizal dependency and MSS. Other plant species (especially those with intermediate mycorrhizal dependency) need to be included as well. Interestingly, the relationship between mycorrhizal dependency and mycorrhizal species sensitivity (Fig. 10.7), and the relationship between mycorrhizal dependency and increase in shoot phosphorus concentration and below ground carbon transport (Fig. 10.3) are both exponential, suggesting that these measures are affected by mycorrhizal dependency in a similar way. This similarity is not surprising since it is likely that plant species that acquire large amounts of resources from AMF are greatly affected when AMF species that supply different amounts of resources occupy the roots (see also above). The position of plant species with a very negative mycorrhizal dependency on this relationship, such as ruderals or non-mycorrhizal plant species, is unclear and it needs further investigation since these species are affected by AMF in a different way. They do not gain advantages from AMF but perceive AMF as antagonistic (Francis and Read 1995). The degree in which AMF inhibit the growth of these species could vary among AMF species.

The relationship between mycorrhizal dependency and mycorrhizal species sensitivity (Figs. 10.6, 10.7) suggests that plant communities in which the majority of plants are mycorrhizal-dependent are most likely to be strongest affected by AMF species composition or by changes in AMF species composition. A comparison between an American old-field community and a European calcareous grassland community support this observation. The American old-fields were dominated by several plant species with a high mycorrhizal dependency (J. Klironomos, pers. observ.) and variations in the species composition of AMF communities greatly influenced productivity and diversity of this community (van der Heijden et al. 1998a). The European grassland was dominated by a plant species with a low mycorrhizal dependency and AMF species composition hardly affected productivity and diversity of this community (van der Heijden et al. 1998a). However, it should be notable that AMF species composition can still be of utmost importance for such a community, since it can change growth and abundance of the subordinate or rare plant species. The two studies mentioned above are the first to have investigated how plant communities respond to AMF communities. Additional studies are required to obtain a more general view about the ecological significance of AMF communities.

The degree in which AMF species composition influences plant diversity and other ecosystem functions could thus, depend on the ratio and abundance of mycorrhizal-dependent versus low mycorrhizal-dependent plant species present in the species pool of a community. The relationship between mycorrhizal dependency and MSS suggests that knowledge of the mycorrhizal dependency of a plant is sufficient to predict ecosystem functioning in response to AMF. I suggest however, the determination of both the mycorrhizal dependency and the mycorrhizal species sensitivity of plants until the existence of a relationship between MSS and mycorrhizal dependency has repeatedly been shown.

10.5 Conclusions

The results presented in this review show that the mycorrhizal dependency is a unifying concept, summarizing mechanisms that explain why plants respond so differently to AMF. The evidence presented here suggests that the mycorrhizal dependencies of the plants composing a community can be used to estimate how much, and in which way, AMF and AMF species composition influences plant population structure, plant community diversity and ecosystem functioning. Moreover, mycorrhizal-dependent and-non-mycorrhizal-dependent plant species could be classified in different functional groups. Ecologists that manipulate plant species diversity in artificial assembled plant communities should, therefore, include both mycorrhizal-dependent plant species and non-mycorrhizal plant species or plant species with a low mycorrhizal dependency. Finally, this chapter shows that ecologists need to consider AMF as a determinant of plant community structure, plant diversity and ecosystem functioning.

Acknowledgements. I would like to thank Rien Aerts, Keith Clay and Ian Sanders for fruitful discussions and helpful comments on this manuscript.

References

Aarssen LW (1983) Ecological combining ability and competitive combining ability in plants: toward a general evolutionary theory of coexistence in systems of competition. Am Nat 122:707-731

Aerts R, Chapin FS III (2000) The mineral nutrition of wild plants revisited: a re-evaluation of processes and patterns. Adv Ecol Res 30:1-67

Al-Mufti MM, Sydes CL, Furness SB, Grime JP, Band SR (1977) A quantitative analysis of shoot phenology and dominance in herbaceous vegetation. J Ecol 65:759-791

Allen EB, Allen MF (1990) The mediation of competition by mycorrhizae in successional and patchy environments. In: Grace JB, Tilman D (eds) Perspectives on plant competition. Academic Press, San Diego, pp 367-385

Allen MF (1991) The ecology of mycorrhizae. Cambridge Univ Press, Cambridge

Arines J, Vilarino A, Sainz M (1989) Effect of different inocula of vesicular-arbuscular mycorrhizal fungi on manganese content and concentration in red clover (*Trifolium pratense*) plants. New Phytol 112:215-219

Baylis GTS (1975) The magnolioid mycorrhiza and mycotrophy in root systems derived from it. In: Sanders FE, Moss B, Tinker PB (eds) Endomycorrhiza. Academic Press, London, pp 378-389

Bever JD, Morton JB, Antonovics J, Schultz PA (1996) Host-dependent sporulation and species diversity of arbuscular mycorrhizal fungi in a mown grassland. J Ecol 84:71-82

Boerner RE.J, DeMars BG, Leicht PN (1996) Spatial patterns of mycorrhizal infectiveness of soils along a successional chronosequence. Mycorrhiza 6:79-90

Brundrett MC, Abott LK (1995) Mycorrhizal fungus propagules in the jarrah forest. II Spatial variability in inoculum levels. New Phytol 131:461-469

Burkert B, Robson AD (1994) ^{65}Zn uptake in subterranean clover (*Trifolium subterraneum*) by three vesicular-arbuscular mycorrhizal fungi in a root-free sandy soil. Soil Biol Biochem 26:1117-1124

Clapp JP, Young JPW, Merryweather JW, Fitter AH (1995) Diversity of fungal symbionts in arbuscular mycorrhizas from a natural community. New Phytol 130:259-265

Clay K, Holah J (1999) Fungal endophyte symbiosis and plant diversity in successional fields. Science 285:1742-1744

Cuenca G, De Andrade Z, Escalante G (1998) Diversity of Glomalean spores from natural, disturbed and revegetated communities growing on nutrient-poor tropical soils Soil Biol Biochem 30:711-719

Daft MJ, El-Giahmi AA (1974) Effect of *Endogone* mycorrhizal on plant growth. VII. Influence of infection on the growth and nodulation in French Bean (*Phaseolus vulgaris*). New Phytol 73:1139-1147

Dale MP, Causton DR (1992) The ecophysiology of *Veronica chamaedrys* V *Montana* and *V. officinalis*. III effects of shading on the phenology of biomass allocations - a field experiment. J Ecol 80:505-515

Dinkelaker B, Hengeler C, Marschner H (1995) Distribution and function of proteoid roots and other root clusters. Bot Acta 108:183-200

During HJ, Willems JH (1984) Diversity models applied to a chalk grassland. Vegetatio 57:103-114

Egerton-Warburton LM, Allen EB (2000) Shifts in arbuscular mycorrhizal communities along an anthropogenic nitrogen deposition gradient. Ecol Appl 10:484-496

Eom AH, Hartnett DC, Wilson GWT (2000) Host plant species effects on arbuscular mycorrhizal fungal communities in tallgrass prairie. Oecologia 122:435-444

Ernst WHO, Van Duin WE, Oolbekking GT (1984) Vesicular arbuscular mycorrhiza in dune vegetation. Acta Bot Neerland 33:151-160

Fitter AH (1977) Influence of mycorrhizal infection on competition for phosphorus and potassium by two grasses. New Phytol 79:119–125

Fitter AH (1990) The role and ecological significance of vesicular-arbuscular mycorrhizas in temperate ecosystems. Agric Ecosyst Environ 29:257–265

Francis R, Read DJ (1995) Mutualism and antagonism in the mycorrhizal symbiosis, with special reference to impact on plant community structure. Can J Bot 73 [Suppl 1]:1301–1309

Gange AC, Brown VK, Farmer LM (1990) A test of mycorrhizal benefit in an early successional plant community. New Phytol 115:85–91

Gange AC, Brown VK, Sinclair GS (1993) Vesicular-arbuscular mycorrhizal fungi: a determinant of plant community structure in early succession. Funct Ecol 7:616–622

Gerdemann JW (1975) Vesicular-arbuscular mycorrhizae. In: Torrey JG, Clarkson DT (eds.) The development and function of roots. Academic Press, New York, pp 575–591

Grime JP (1979) Plant strategies and vegetation processes. Wiley, Chichester

Grime JP, Mackey JML, Hillier SH, Read DJ (1987) Floristic diversity in a model system using experimental microcosms. Nature 328:420–422

Grubb P (1977) The maintenance of species richness in plant communities: the importance of the regeneration niche. Biol Rev 52:107–145

Haas JH, Bar-Yosef B, Krikun J, Barak R, Markovitz T, Kramer S (1987) Vesicular-arbuscular-mycorrhizal-fungus infestation and phosphorus fertigation to overcome pepper stunting after methyl bromide treatment. Agron J 79:905–910

Habte M, Manjunath A (1991) Categories of vesicular-arbuscular mycorrhizal dependency of host species. Mycorrhiza 1:3–12

Hartnett DC, Wilson WT (1999) Mycorrhizae influence plant community structure and diversity in tall grass prairie. Ecology 80:1187–1195

Hayman DS, Tavares M (1985) Plant growth responses to vesicular-arbuscular mycorrhiza. XV. Influence of soil pH on the symbiotic efficiency of different endophytes. New Phytol 100:367–377

Helgason T, Daniell TJ, Husband R, Fitter AH, Young JPY (1998) Ploughing up the woodwide web? Nature 394:431

Hetrick BAD, Wilson GWT, Hartnett DC (1989) Relationships between mycorrhizal dependency and competitive ability of two tall grass prairie grasses. Can J Bot 67:2608–2615

Hetrick BAD, Wilson GWT, Todd TC (1992) Relationship of mycorrhizal symbiosis, rooting strategy and phenology among tall grass prairie forbs. Can J Bot 70:1521–1528

Hubbell SP, Foster RB, O'Brien ST, Harms KE, Condit R, Wechsler B, Wright SJ, de Lao SL (1999) Light-cap disturbances, recruitment limitation, and tree diversity in a neotropical forest. Science 283:554–557

Huston MA (1979) General hypothesis of species diversity. Am Nat 113:81–101

Jakobsen I, Abbott LK, Robson AD (1992a) External hyphae of vesicular-arbuscular mycorrhizal fungi associated with *Trifolium subterraneum* L. I: Spread of hyphae and phosphorus inflow into roots. New Phytol 120:371–380

Jakobsen I, Abbott LK, Robson AD (1992b) External hyphae of vesicular-arbuscular mycorrhizal fungi associated with *Trifolium subterraneum* L. I. Hyphal transport over defined distances. New Phytol 120:509–516

Janos DP (1980) Mycorrhizae influence tropical succession. Biotropa [Suppl] 12:56–64

Jensen A (1984) Responses of barley, pea and maize to inoculation with different vesicular-arbuscular mycorrhizal fungi in irradiated soil. Plant Soil 78:315–323

Johnson NC, Zak DR, Tilman D, Pfleger FL (1991) Dynamics of vesicular arbuscular mycorrhizae during old field succession. Oecologia 86:349–358

Johnson NC, Tilman D, Wedin D (1992) Plant and soil controls on mycorrhizal fungal communities. Ecology 73:2034–2042

Johnson NC, Graham JH, Smith FA (1997) Functioning of mycorrhizal associations along the mutualism-parasitism continuum. New Phytol 135:575–586

Kiers ET, Lovelock CE, Krueger EL, Herre EA (2000) Differential effects of tropical arbuscular mycorrhizal fungal inocula on root colonization and tree seedling growth: implications for tropical forest diversity. Ecol Lett 3:106–113

Klironomos J (2000) Host-specificity and functional diversity among arbuscular mycorrhizal fungi. In: Bell CR, Brylinsky M, Johnson-Green P (eds) Microbial biosystems: new frontiers. Proceedings of the 8th international symposium on microbial ecology, Society for Microbial Ecology, Halifax, Canada pp. 845–851

Klironomos JN, McCune J, Hart M, Neville J (2000) The influence of arbuscular mycorrhizae on the relationship between plant diversity and productivity. Ecol Lett 3:137–141

Koide R (1991) Nutrient supply, nutrient demand and plant response to mycorrhizal infection. New Phytol 117:365–386

Law R (1988) Some ecological properties of intimate mutualisms involving plants. In: Davy AJ, Hutchings MJ, Watkinson AR (eds) Plant population ecology. Blackwell, Oxford, pp. 315–341

Louis L, Lim G (1988) Differential response in growth and mycorrhizal colonization of soybean to inoculation with two isolates of *Glomus clarum*, in soils of different P availability. Plant Soil 112:37–43

Marler MJ, Zabinski CA, Callaway RM (1999) Mycorrhizae indirectly enhance competitive effects of an invasive forb on a native bunchgrass. Ecology 80:1180–1186

McGonigle TP (1988) A numerical analysis of published field trials with vesicular-arbuscular mycorrhizal fungi. Funct Ecol 2:473–478

Miller RM, Jastrow JD (2000) Mycorrhizal fungi influence soil structure. In: Kapulnik Y, Douds DD (eds) Arbuscular Mycorrhizae: Physiology and Function. Kluwer, Dordrecht, pp 3–18

Miller RM, Smith CI, Jastrow JD, Bever JD (1999) Mycorrhizal status of the genus *Carex* (Cyperaceae). Am J Bot 86:547–553

Miller SP, Bever JD (1999) Distribution of arbuscular mycorrhizal fungi in stands of the wetland grass *Panicum hemitomon* along a wide hydrologic gradient. Oecologia 119:586–592

Morton JB, Bentivenga SP, Bever JD (1995) Discovery, measurement and interpretation of diversity in symbiotic endo-mycorrhizal fungi (Glomales, Zygomycetes). Can J Bot 73 [Suppl]:25–32

Newsham KK, Fitter AH, Watkinson AR (1994) Root pathogenic and arbuscular mycorrhizal fungi determine fecundity of asymptomatic plants in the field. J Ecol 82:805–814

Newsham KK, Fitter AH, Watkinson AR (1995) Multi-functionality and biodiversity in arbuscular mycorrhizas. Trends Ecol Evol 10:407–411

Perotto S, Bonfante P (1997) Bacterial associations with mycorrhizal fungi: close and distant friends in the rhizosphere. Trends Microbiol 5:496–501

Plenchette C, Fortin JA, Furlan V (1983) Growth response of several plant species to mycorrhizae in a soil of moderate P-fertility. I. Mycorrhizal dependency under field conditions. Plant Soil 70:199–209

Powell CL, CLark GE, Verberne NJ (1982) Growth response of four onion cultivars to isolates of VA mycorrhizal fungi. NZ J Agric Res 25:465–470

Raju PS, Clark RB, Ellis JR, Maranville JW (1990) Effects of species of VA-mycorrhizal fungi on growth and mineral uptake of sorghum at different temperatures. Plant Soil 121:165–170

Ravnskov S, Jakobsen I (1995) Functional compatibility in arbuscular mycorrhizas measured as hyphal P transport to the plant. New Phytol 129:611–618

Read D (1991) Mycorrhizas in ecosystems. Experientia 47:376–391

Ricklefs RW (1977) Environmental heterogeneity and plant species diversity: a hypothesis. Am Nat 111:376–381

Roldan-Fajardo BE (1994) Effect of indigenous arbuscular mycorrhizal endophytes on the development of six wild plants colonizing a semi arid area in south-east Spain. New Phytol 127:115–121

Rosendahl S, Rosendahl CN, Søchting U (1990) Distribution of VA mycorrhizal endophytes amongst plants of a Danish grassland community. Agric Ecosyst Environ 29:329–336

Sanders IR, Fitter AH (1992) Evidence for differential responses between host-fungus combinations of vesicular-arbuscular mycorrhizas from a grassland. Mycol Res 96:415–419

Sanders IR, Koide RT (1994) Nutrient acquisition and community structure in co-occurring mycotrophic and non-mycotrophic old-field annuals. Funct Ecol 8:77–84

Schreiner RP, Mihara KL, McDaniel H, Bethlenfalvay GJ (1997) Mycorrhizal fungi influence plant and soil functions and interactions. Plant Soil 188:199–209

Sharma AK, Johri BN, Gianinazzi S (1992) Vesicular-arbuscular mycorrhizae in relation to plant disease. World J Microbiol Biotechnol 8:559–563

Skene KR (1998) Cluster roots: some ecological considerations. J Ecol 86:1060–1064

Smith FA, Jakobsen I, Smith SE (2000) Spatial differences in acquisition of soil phosphate between two arbuscular mycorrhizal fungi in symbiosis with *Medicago truncatula*. New Phytol 147:357–366

Smith SE, Read DJ (1997) Mycorrhizal symbioses, 2nd edn. Academic Press, London

Sokal, RR, Rohlf FJ (1981) Biometry: the principles and practice of statistics. Freeman, New York

Stahl PD, Smith WK (1984) Effects of different geographic isolates of *Glomus* on the water relations of *Agropyron smithii*. Mycologia 76:261–267

Streitwolf-Engel R, Boller T, Wiemken A, Sanders IR (1997) Clonal growth traits of two *Prunella* species are determined by co-occurring arbuscular mycorrhizal fungi from a calcareous grassland. J Ecol 85:181–191

Streitwolf-Engel R, van der Heijden MGA, Wiemken A, Sanders IR (2001) The ecological significance of arbuscular mycorrhizal fungal effects on clonal plant growth. Ecology 82:2846–2859

Tilman D (1988) Plant strategies and the dynamics and structure of plant communities. Princeton Univ Press, Princeton

Tilman D (1997) Community invasibility, recruitment limitation, and grassland biodiversity. Ecology 78:81–92

Tilman D, Pacala S (1993) The maintenance of species richness in plant communities. In: Ricklefs RE, Schluter D (eds) Species diversity in ecological communities. Univ Chicago Press, Chicago, pp 13–25

Timonen S, Jorgensen KS, Haahtela K, Sen R (1998) Bacterial community structure at defined locations of *Pinus sylvestris Suillus bovinus* and *Pinus sylvestris Paxillus involutus* mycorrhizospheres in dry pine forest humus and nursery peat. Can J Microbiol 44:499–513

Trappe JM (1987) Phylogenetic and ecological aspects of mycotrophy in the angiosperms from an evolutionary standpoint. In: Safir GR (ed) Ecophysiology of VA mycorrhizal plants. CRC Press, Boca Raton, USA, pp 5–25

van der Heijden MGA, Klironomos JN, Ursic M, Moutoglis P, Streitwolf-Engel R, Boller T, Wiemken A, Sanders IR (1998a) Mycorrhizal fungal diversity determines plant biodiversity, ecosystem variability and productivity. Nature 396:69–72

van der Heijden MGA, Boller T, Wiemken A, Sanders IR (1998b) Different arbuscular mycorrhizal fungal species are potential determinants of plant community structure. Ecology 79:2082–2091 van der Heijden MGA, Klironomos JN, Ursic M, Moutoglis P, Streitwolf-Engel R, Boller T, Wiemken A, Sanders IR (1999) "Sampling effect" a problem in biodiversity manipulation? A reply to David A Wardle. Oikos 87:408–410

van der Heijden MGA (1999) Ecological significance of mycorrhizal diversity: on the role of arbuscular mycorrhizal fungi as a determinant of plant community structure and diversity, chapter 3. Ph-D thesis, Basel, Switzerland

Walker C, Mize CW, McNabb HS (1982) Populations of endogonaceous fungi at two locations in central Iowa. Can J Bot 60:2518–2529

Wardle DA (1999) Is sampling effect a problem for experiments investigating biodiversity-ecosystem function relationships. Oikos 87:403–407

West HM (1996) Influence of arbuscular mycorrhizal infection on competition between *Holcus lanatus* and *Dactylis glomerata*. J Ecol 84:429–438

Wilson GWT, Hartnett DC (1998) Interspecific variation in plant response to mycorrhizal colonization in tallgrass prairie. Am J Bot 85:1732–1738

Xie ZP, Staehelin C, Vierheilig H, Wiemken A, Jabbouri S, Broughton WJ, Vogellange R, Boller T (1995) Rhizobial nodulation factors stimulate mycorrhizal colonization of nodulating and nonnodulating soybeans. Plant Physiol 108:1519–1525

Zobel M, Moora M, Haukioja E (1997) Plant coexistence in the interactive environment: arbuscular mycorrhiza should not be out of mind. Oikos 78:202–208

11 Dynamics within the Plant – Arbuscular Mycorrhizal Fungal Mutualism: Testing the Nature of Community Feedback

James D. Bever, Anne Pringle, Peggy Ann Schultz

Contents

11.1	Summary	268
11.2	Introduction	268
11.3	Mutual Interdependence of Plant and Arbuscular Mycorrhizal Fungal Growth Rates Cause Feedback Dynamics	270
11.3.1	Positive Feedback	271
11.3.2	Negative Feedback	272
11.4	Identifying Positive Versus Negative Feedback: Complimentary Approaches	274
11.4.1	Interdependence of Plant and Fungal Population Growth Rates: A Mechanistic Approach	274
11.4.2	Plant Response to Manipulated Arbuscular Mycorrhizal Fungal Communities: A Phenomenological Approach	275
11.5	Evidence for Interdependence of Plant and Fungal Population Growth Rates	276
11.5.1	Specificity of Plant Response to Arbuscular Mycorrhizal Fungal Species	277
11.5.2	Specificity of Arbuscular Mycorrhizal Fungal Response to Plant Species	278
11.6	Testing Feedback Between Plant and Arbuscular Mycorrhizal Fungal Communities	279
11.6.1	Evaluation of Feedback Using the Mechanistic Approach	279
11.6.2	Evaluation of Feedback Using the Phenomenological Approach	283
11.6.3	A Comparison of Mechanistic and Phenomenological Approaches to Testing Feedback	285
11.7	Implications and an Extension of the Feedback Framework	286
11.7.1	Negative Feedbacks and the Evolutionary Maintenance of Mutualism	287

11.7.2	Spatial Structure and the Dynamics of Feedback	288
11.8	Conclusion	289
References		290

11.1 Summary

A growing body of work demonstrates that arbuscular mycorrhizal fungal communities can be diverse and that individual fungal species within fungal communities are ecologically distinct. Specifically, arbuscular mycorrhizal fungi (AMF) have been shown to differ in their effect on plant hosts and differ in their response to plant hosts. The mutual interdependence of plant and AMF relative growth rates can result in an active dynamic interaction between plant and AMF communities. In this chapter, we describe two possible types of dynamics: that of positive feedback and that of negative feedback. In the case of positive feedback, initial differences in the community composition are reinforced and the dynamic is predicted to lead to the loss of diversity from the community. In the case of negative feedback, the dynamic between the plant and fungal community directly contributes to the maintenance of diversity within both communities.

We develop two complimentary approaches to testing these dynamics: a mechanistic approach that builds on detailed knowledge of plant-fungal interactions and a more phenomenological approach where many of the details of AMF biology are considered as a black-box. We first identify the criteria necessary for positive and negative feedback, then we review the published evidence supporting the community feedback model. Using examples from a grassland in North Carolina, we demonstrate the potential for negative feedback within the plant-arbuscular mycorrhizal relationship. Finally, we discuss the broader implications of the community feedback for our understanding of the ecology and evolution of the plant-AMF mutualism.

11.2 Introduction

The interaction between plants and arbuscular mycorrhizal fungi is a classic example of an interspecific mutualism (Boucher 1986). For many plant species, the uptake of an essential soil nutrient, phosphorus, is enabled by plants' intimate associations with arbuscular mycorrhizal fungi (AMF). In turn, AMF are obligately dependent upon plants for their carbohydrates (information on the ecophysiology of the arbuscular mycorrhizal symbiosis is given by Jakobsen et al., Chap. 3, this Vol.). Historically, the study of mycorrhizae, as well as the study of other mutualisms (Janzen 1986; Bronstein 1994), has been dominated by a natural history perspective, which focused on the morphology, physiology and taxonomic range of the association. However, the field is increasingly driven by experimental assessments of the ecological importance of AMF (see Read, Chap. 1, this Vol.). Experiments have illustrated the tremendous impact of these fungi on broad themes of plant ecology, including the dynamics of plant populations (Koide et al. 1994), the outcome of interspecific competition (Hartnett et al. 1993), the trajectory of succession (Janos 1980; Allen and Allen 1990), and the stabilization of soil aggregates (Miller and Jastrow 2000).

Recent experimental studies have also revealed that species of AMF differ greatly in their individual ecologies. For example, plant species have been found to differ in their responses to individual isolates and species of AMF (Adjoud et al. 1996; Streitwolf-Engel et al. 1997; van der Heijden et al. 1998a,b). Because growth promotion can be host-specific, AMF diversity and identity can have strong effects on plant-plant interactions. This fact has been recently illustrated by two experimental tests that found strong effects of both the identity and diversity of the AMF community on plant composition and productivity (van der Heijden et al. 1998b). Chapters 9 and 10 (this Vol.) provide additional information on the impact of AMF diversity on plant diversity and ecosystem functioning (e.g., Fig. 9.4). AMF have also been shown to differ in their growth responses to plant species (Johnson et al. 1992a; Sanders and Fitter 1992; Bever et al. 1996; Eom et al. 2000) and in their response to disturbance (Johnson 1993; Helgason et al. 1998; Douds and Millner 1999; Egerton-Warburton and Allen 2000). Factors that influence AMF community composition are also discussed by Clapp et al. (in Chap. 8, this Vol.; e.g., Fig. 8.4). Because of accumulated evidence of the unique ecologies of AMF, there is growing appreciation of the impacts of AMF diversity and dynamics on plant ecology.

However, the ability to generalize and extend our knowledge of the importance of the ecological diversity among AMF species on plant ecology is limited by the nascent state of empirical work on the ecology of individual AMF species, and also by the lack of appropriate conceptual frameworks on the dynamics of mutualism. The development of theory on the ecology and evolution of mutualistic interactions has lagged behind that of antagonistic interactions (Boucher 1986; Bronstein 1994) and hindered recognition of the potential importance of mutualists in such critical ecological processes as the maintenance of diversity.

We have developed a general model of mutualism that identifies the potential for dynamics within mutualisms to contribute to the maintenance of diversity (Bever 1999). This community model builds on the observation that plant and fungal population growth rates are mutually interdependent and identifies two very different dynamical outcomes. A relatively symmetric delivery of benefit between plants and fungi results in a positive feedback dynamic, while negative feedback can result if the delivery of benefit between plants and fungi is highly asymmetric (Bever et al. 1997; Bever 1999). These two dynamics have very different effects: Positive feedback strengthens mutualism, but reduces diversity in the system. Negative feedback weakens the mutualism, but maintains diversity within the plant and fungal communities (Bever 1999).

The feedback model may allow us to better understand the dynamics between plants and AMF. In this chapter, we first review the basic model of community feedback and identify the very different predictions of positive and negative feedback. We then develop two alternative approaches to evaluate these feedbacks: a mechanistic approach that builds on detailed knowledge of plant-fungal interactions (Bever 1999), and a more phenomenological approach where many of the details of AMF biology are black-boxed (Bever et al. 1997). We first identify the criteria necessary for positive versus negative feedback within the mechanistic and phenomenological approaches. We then review the published evidence supporting the essential features of the community feedback model, and illustrate the utility of the two approaches using data from plants and fungi that coexist within a grassland in North Carolina. Using these examples, we demonstrate the potential for negative feedback within this mutualism. Finally, we discuss the broader implications of the community feedback for our understanding of the ecology and evolution of the plant – AMF mutualism.

11.3 Mutual Interdependence of Plant and AM Fungal Growth Rates Cause Feedback Dynamics

We begin with the essential assumption of the feedback model: that plant and AMF population growth rates are mutually interdependent. That is, the relative rates of increase of two plant populations depend upon the identity of the AMF species with which they associate and, in turn, the relative rates of increase of two fungal populations depend upon the identity of the plants with which they associate. We draw the distinction between the ability to colonize or be colonized and the responses of plant and fungi to the colonization. While the associations between plants and AMF are thought not to be host-specific, the response of the plants to fungi and of the fungi to plants may depend upon the specific plant-fungal combination. That is, the benefits from the association are host/fungus-specific (for a comprehensive discussion on specificity in the arbuscular mycorrhizal symbiosis, see Sanders, Chap. 16, this Vol.).

When plant and fungal population growth rates are mutually interdependent, the long-term dynamics depend upon both plant and fungal perspectives. That is, we cannot predict the outcome of long-term dynamics among two plant species without knowing what interactions are at play within the associated AMF community. Two very different types of dynamics are described below.

11.3.1 Positive Feedback

We have identified that under symmetric fitness relations, in which the fungus that delivers the greatest benefit to a particular plant also receives the greatest benefit from that plant, the dynamic will be characterized by positive feedback (Fig. 11.1a; Bever et al. 1997; Bever 1999). In this scenario, an initial abundance of one plant-fungal combination will be reinforced and will ultimately lead to the extinction of the other types. In considering the fitness set described in Fig. 11.1a, for example, if plant A is initially abundant, then fungus X will have a higher average growth rate than fungus Y and will, therefore, increase in frequency. As a result of this increase in fungus X, the growth rate of plant A will increase relative to plant B, resulting in a further reduction in the representation of plant B. This process will continue until plant B and fungus Y are excluded from the community (depicted in Fig. 11.2a). Therefore, in the case of symmetric fitness relations the dynamics within the AMF community will contribute to the loss of diversity within the plant community, and vice versa. Moreover, we have also demonstrated that coadaptation of mutualists, the maximization of benefit to both participants, is a possible, but not a necessary, result of positive feedback dynamics (Bever 1999).

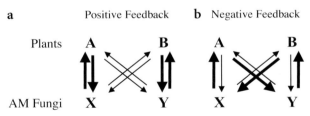

Fig. 11.1. Two potential consequences of the mutual interdependence of plant and AMF population growth rates. The direction of benefit between the two plant species, A and B, and two fungal species, X and Y, is depicted with the direction of the *arrows*. Thickness of arrows represents the degree of benefit received, with a thicker line meaning greater benefit than a thin one. **a** The symmetric fitness relations between the plants and fungi will result in a positive feedback dynamic in which an initial abundance of one plant species, plant A for example, will result in an increase in the best matched fungus, fungus X, which will thereby increase the growth rate of plant A relative to plant B. **b** The highly asymmetric fitness relations between the plants and fungi will result in a negative feedback dynamic in which an initial abundance of plant A would cause the composition of the AMF community to change in a manner which decreases the growth rate of plant A relative to plant B. (Redrawn from Bever et al. 1997)

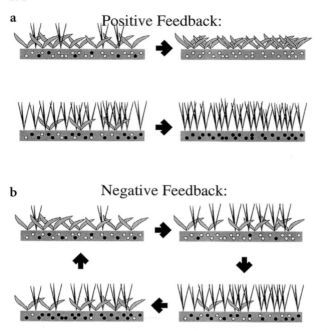

Fig. 11.2. Community dynamics under positive and negative feedback. **a** Under positive feedback, the initial abundance of one type, the broad-leaved grass, for example, will result in an increase in the light-spored fungal species with which it performs best. As a result, the growth rate of the broad-leaved grass will increase leading to further reductions in frequency and the eventual exclusion of the narrow-leaved grass from the community. If we happened to have begun with an initial abundance of the narrow-leaved grass, this same process could lead to the eventual exclusion of the broad-leaved grass. **b** Under negative feedback, the initial abundance of the broad-leaved grass again leads to an increase in the light-spored species of AM fungus. However, in this case, the narrow-leaved grass performs best with the light-spored fungus species and as a result, the narrow-leaved grass increases its representation in the community. As the narrow-leaved grass attains high abundance, the dark-spored species of AMF begins to increase. As a result of this change in the fungal community composition, the relative growth rate of the broad-leaved grass increases. This negative feedback process thereby maintains both plant species and both fungal species within the community

11.3.2 Negative Feedback

It is also possible that the fitness relationships between plants and AMF are highly asymmetric, as depicted in Fig. 11.1b (Bever et al. 1997; Bever 1999). In this case, the fungus that delivers the greatest benefit to a particular plant may have a greater growth rate on a second plant species (Fig. 11.1b). This asymmetry in the fitness relations results in an active, sustained dynamic between plants and AMF communities, in which no single plant species is able to replace the other (Bever 1999). As one plant species becomes increasingly abundant, the fungal community changes in a manner that increases the growth rate of the second species (i.e., an indirect facilitation). If we consider the fitness set depicted in Fig. 11.1b, for example, an initial abundance of plant A will cause fungus Y to have higher growth rates than fungus X and, therefore, increase in frequency. This increased frequency of fungus Y will then cause the growth rate of plant A to decrease relative to that of plant B, thereby allowing plant B to increase in the community (depicted in Fig. 11.2b). Therefore, negative feedback through changes in the composition of the AMF community can contribute to the maintenance of diversity within both plant and fungal communities. Moreover, because multiple types are maintained within both plant and fungal communities, the co-adaptation of mutualists is not possible within this negative feedback dynamic (Bever 1999).

These two dynamics illustrate that dynamics within the AMF community can have very different impacts on the plant community (Table 11.1). However, the model necessarily relies on simplifying assumptions. Specifically, the relative growth rates of the plants and AMF are assumed to result wholly from their mycorrhizal associations. Therefore, the plant species, for example, are assumed to be equivalent in other aspects of their biology, including competitive ability, niche, seasonality, resistance to herbivory, etc. Clearly, differences in these other factors could alter the predicted dynamics. Plants and fungi are also assumed to associate with each other in proportion to their relative abundance in their respective communities. That is, we assume that there is no

Table 11.1. Summary of the conditions for, and predictions from, positive and negative feedback dynamics between plants and AMF communities. The mutual interdependence of plant and fungal population growth rates will result in positive feedback between plant and fungal communities when the sign of the product of the plant and fungal interaction coefficients (I_P and I_F) are positive (where $I_P=a-b-c+d$ and $I_F=k-l-m+n$) and individual plant and fungal species are not universally superior (as derived in Bever 1999). Negative feedback will then result when the product of the interaction coefficients is negative. Under positive feedback dynamics, diversity would be lost from the community at a local scale and the efficiercny of growth promotion may increase over time (i.e., coadaptation is possible). Negative feedback dynamics will maintain diversity within the plant and fungal communities, however the efficiency of growth promotion will not be maximized (i.e., coadaptation is not possible)

	Positive Feedback ($I_P \times I_F > 0$)	Negative Feedback ($I_P \times I_F < 0$)
Diversity	Lost	Maintained
Coadaptation	Possible, not necessary	Not possible

specificity in association, though there are differential growth effects. We are also assuming that the plant and fungal communities are well-mixed and randomly arranged across the community. We will discuss consequences of these assumptions in the final sections of the chapter.

11.4 Identifying Positive Versus Negative Feedback: Complimentary Approaches

We have used two parallel conceptual approaches to identify the conditions for positive versus negative feedback between plants and AMF communities. The first approach builds on knowledge of the interdependence of plant and fungal population growth rates for a specific pair of plants and a specific pair of fungi. The second approach assumes that interdependence exists, but assesses the dynamic consequences of these interactions by relying on measures of the plant's responses to manipulated fungal communities. The first approach is more mechanistic, while the second approach is more phenomenological. There are also subtle differences in the underlying assumptions of the two approaches, which we will address in detail in the section comparing the two methods.

11.4.1 Interdependence of Plant and Fungal Population Growth Rates: A Mechanistic Approach

We begin by imagining that we know the growth rates of two plant species in association with two fungal species and vice versa, as presented in Fig. 11.3. That is, a population of plant A grows at a rate of a in association with fungus X and at a rate of b with fungus Y, while a population of plant B grows at a rate of c and d with fungi X and Y, respectively. Similarly, a population of fungus X grows at a rate of k and l in association with plants A and B, respectively and a population of fungus Y grows at a rate of m and n in association with plants A and B, respectively. Provided that no one plant species or one fungal species has superior growth rates in all associations (i.e., there is a reversal in their rank performance in association with different partners), the nature of the community dynamics will depend upon the signs of plant and fungal interaction coefficients (I_P and I_F, respectively), where $I_P=a-b-c+d$ and $I_F=k-l-m+n$ (as derived in Bever 1999). The plant interaction coefficient, I_P, measures the specificity of response of plants to AMF, and the fungal interaction coefficient, I_F, measures the specificity of response of AMF to plants. They correspond to the interaction terms in two-way analyses of variance (as discussed below). When no single plant or fungus is universally superior, the community dynamics will be characterized by positive feedback when the product of the

Plant Growth Response:

	Fungus X	Fungus Y
Plant A	a	b
Plant B	c	d

Fungal Growth Response:

	Plant A	Plant B
Fungus X	k	l
Fungus Y	m	n

Fig. 11.3. Symbolic representation of the interdependent growth response of plants and fungi

interaction coefficients is positive (i.e., $I_P \times I_F > 0$). If the product of the interaction coefficients is negative (i.e., $I_P \times I_F < 0$), the community dynamics will be characterized by negative feedback (Bever 1999). The dynamics, conditions and expectations can then be summarized as in Table 11.1.

11.4.2 Plant Response to Manipulated AM Fungal Communities: A Phenomenological Approach

One could focus solely on the plant point of view and put the fungal community dynamics within a black box. Using this approach, we have demonstrated that one can test for feedback by first training a common fungal community on two plant species and then evaluating the growth of these two plant species in association with the differentiated fungal communities (Bever et al. 1997). This approach is depicted in Fig. 11.4. When the plant species grow better with their own fungal communities than each other's fungal communities, then positive feedback dynamics result. In contrast, negative feedback dynamics result when plants grow best with each other's fungal communities. Specifically, the nature of the feedback is determined by the sign and magnitude of the home versus away contrast, which corresponds to the difference between the average growth of the two plant species in their own fungal communities and the average growth of the two species in each other's fungal communities (Fig. 11.4; Bever 1994; Bever et al. 1997). Within this experimental framework, a positive contrast indicates positive feedback, while a negative contrast indicates negative feedback.

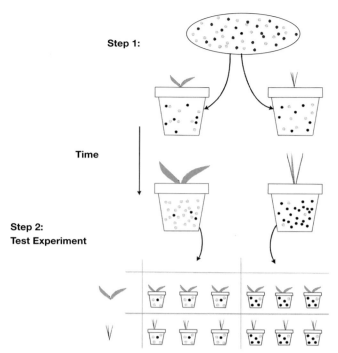

Fig. 11.4. Experimental protocol for phenomenological testing of the feedback between plants and AMF. In order to focus on feedbacks through the AMF community, the initial inoculum is made up of a diverse mixture of pure cultures of different AMF and is therefore free of root pathogens. This mixture is then distributed into replicate pots into which sterile replicates of two plant species are planted and grown for an initial training period. The composition of the AMF community may differentiate during this time due to host-specific differences in the growth rates or competitive abilities of the AMF species. The feedback on plant growth due to this differentiation in the AMF community can then be evaluated by using these soil communities as inoculum in a separate test experiment as depicted in the second step

11.5 Evidence for Interdependence of Plant and Fungal Population Growth Rates

The essential assumption of the community feedback model is that while any plant can associate with any AM fungus, the benefits of associating with one species versus another are variable. Indeed, abundant evidence demonstrates that associations between plants and AMF lack specificity (e.g., Sanders, Chap. 16, this Vol.; Smith and Read 1997; Hoeksema 1999), yet the benefits from the association to both plants and fungi depend upon the specific combination of plant and fungus.

11.5.1 Specificity of Plant Response to Arbuscular Mycorrhizal Fungal Species

Agronomists have long been interested in identifying the optimal fungal isolate to inoculate particular crops and, as a result, abundant literature has accumulated demonstrating that plant growth promotion depends upon the matching of plant and fungal species (e.g., Nemec 1978; Powell et al. 1982; Wilson 1988). Several recent papers have demonstrated similar specificity of plant responses between plants and fungi co-occurring within semi-natural ecosystems (Adjoud et al. 1996; Streitwolf-Engel et al. 1997; van der Heijden et al. 1998a,b). For example, *Eucalyptus dives* grew taller when inoculated with *Glomus caledonium* than with *G. intraradices*, while *Eucalyptus viminalis* grew taller in association with *G. intraradices* than with *G. caledonium* (Adjoud et al. 1996). The specificity of fungal benefit to hosts remains strong even in the presence of interspecific competition between plants. For example, van der Heijden and coworkers (1998b) found strong effects of fungal identity on the relative success of plant species grown in mixtures. Moreover, the differential response of plants to different fungi depends strongly on the identity of the plant and the fungus involved in the interaction (see Figs. 9.2, 10.7, 10.8 in Chaps. 9, 10, this Vol.).

The majority of experimental tests of plant response to fungal species measure growth rates of individual plants, while the model developed in Bever (1999) builds on differences in population growth rates. There are actually little direct data on the specificity of fungal impacts on long-term dynamics of plant populations. For annual plants, seed set at the end of one growing season provides a direct measure of fitness, though there can also be strong maternal effects on fitness (Lewis and Koide 1990; Koide and Lu 1995). However, for perennial plants, the choice of an appropriate measure from a single growing season is much more difficult. Biomass can be allocated to flower, fruit and seed production, or multiple forms of long-lived vegetative structures (e.g., shoots, roots, rhizomes, bulbs, tubers, corms, etc.). These structures are unlikely to contribute equally to the long-term dynamics of the plant population and, therefore, the empiricist is faced with a difficult task of assigning value to the measures. Population growth rates, however, are likely to be correlated with measures of plant growth rates, provided other aspects of plant growth form, competitive ability, and life-history are similar (as they are likely to be within a plant species or genera). Therefore, comparisons of plant growth rates are common surrogates of population growth rates within studies of plant ecology.

11.5.2 Specificity of Arbuscular Mycorrhizal Fungal Response to Plant Species

Relative to tests of specificity of plant response to AMF, tests of the specificity of fungal response to plant species are uncommon. In fact, an evaluation of the AMF response to plant hosts is considerably more difficult than the converse because the fungal structures are embedded within the roots of plants and soil. Fungi, like perennial plants, can reproduce vegetatively through hyphal extension and by the production of dormant propagules in the form of asexual spores. While the intra and extraradical hyphal structures can be measured, at present they cannot be easily identified to species (for more information on AMF identification see Clapp et al., Chap. 8, this Vol.). Moreover, the hyphal walls of some species may persist for some time after the hyphae are dead, making the distinction between living and dead hyphae difficult. Therefore, evaluations of fungal biomass of particular fungal species (the comparable measure to the biomass of perennial plants) are very difficult, particularly within a community context.

As a consequence of the difficulties in measuring biomass of different AMF, a majority of previous efforts to measure host-specific growth rates of AMF have focused on counts of the asexually produced spores. A growing body of evidence suggests that sporulation by AMF is related to the growth of a fungus within plant roots (see Abbott and Gazey 1994). Gazey et al. (1992) describe both a critical level of root colonization necessary to trigger sporulation, and a strong correlation between length of colonized root and sporulation for two species of *Acaulospora*. Douds and Schenk (1990) found a similar relationship between the extent of colonization and sporulation for two of three species examined. Although Pearson and Schweiger (1993) did not find a correlation between root length colonization and sporulation for *Scutellospora calospora*, sporulation did require a critical amount of root length colonization. In agricultural systems, sporulation is often associated with plant death (see references in Gemma et al. 1989). AMF spores are relatively large structures replete with lipids and nuclei; they represent considerable carbon investments. Moreover, as plant roots die back (due to seasonality or senescence), spore density will accurately represent the population size of individual fungal species. Therefore, spore production at the end of a plant's growing season should be strongly correlated with overall fungal population growth rates and fitness. There are however descriptive studies of AMF frequencies in roots and sporulation in the soil that do not support this and these are reviewed in Clapp et al. (Chap. 8, this Vol.). Given that spores of different fungal species vary over many orders of magnitude in size, comparisons across fungal species are best made in terms of total spore biovolume.

Using spore counts, workers have found strong evidence for the host-specificity of fungal population growth rates. Much of this work has evaluated the growth responses of fungi mixed within diverse soil communities (e.g., Johnson et al. 1992a; Sanders and Fitter 1992; Bever et al. 1996; Eom et al. 2000). For example, in a recent study within tallgrass prairie of the central United States, *Glomus diaphanum* sporulated more in association with *Solidago missouriensis* than with *Panicum virgatum*. Conversely, *Gl. mosseae* was found to sporulate more profusely with *Panicum* than with *Solidago* (Eom et al. 2000). In these experiments, the differences in fungal responses to hosts could be due to host-specificity in their basal growth rates or to host-specificity in their competitive abilities. Efforts to find differences in fungal response to host species when sporulation is evaluated in isolation (i.e.,. one fungus and one plant per pot) have been less successful in identifying host-specificity. For example, Hetrick and Bloom (1986) compared the sporulation of three AMF species in association with five crop species. While the crops differed in their overall quality as hosts, the response to these hosts was generally consistent across fungal species.

11.6 Testing Feedback Between Plant and AMF Communities

To evaluate feedback on plant growth through the mycorrhizal fungal community, ecologists must separate the effects of the AMF community from the effects of other soil organisms (see Read, Table 1.2, Chap. 1, this Vol.). We discuss two approaches to test the nature of feedback, mechanistic and phenomenological, both of which require culturing of the fungi. For each approach, we review the relevant published work and then discuss an empirical test.

11.6.1 Evaluation of Feedback Using the Mechanistic Approach

While there is evidence for the mutual interdependence of plant and fungal population growth rates, there have been few tests of community feedback between plants and AMF. In fact, there has only been one study that of Hetrick and Bloom (1986) that evaluated both plant and AMF response using the same plant and fungal species combinations. We had previously reanalyzed the data of Hetrick and Bloom in the context of our model and found a general tendency towards positive feedback (Bever 1999). The approach used in this analysis, however, was limited by access to the original data. Below, we illustrate a more rigorous statistical evaluation of feedback using a portion of a larger data set on the interactions of plants and AMF (Pringle, Bever, Antonovics and Kuo, unpubl.) within a well-studied grassland in North Carolina (description contained in Fowler and Antonovics 1981; Bever et al. 1996).

11.6.1.1 An Example of Negative Feedback

In this experiment, we evaluated the interactions between six plant species and five fungal species all of which co-occur within the grassland. These plant species were grown with the fungal species in all pair-wise combinations within pots. The plants were grown in sterilized soil and the fungal species were derived from spores collected in the field. Other soil organisms such as Phytium and Bacillus were not present in this experiment. Day lengths and temperatures mimicked field conditions. At the end of 5 months, both plant mass and fungal sporulation were measured, giving meaningful measures of both plant and fungal performance. Here, we report the results for the interactions between two plant species, *Allium vineale* and *Plantago lanceolata*, and two fungal species, an undescribed species of *Glomus* (*Gl.* "*d1*" in Bever et al. 1996) and *Scutellospora pellucida*. We begin with a simple description of the results and then present a statistical evaluation of the dynamics.

Plantago grew larger with *S. pellucida* than it did with *Glomus* "*d1*", while *Allium* grew larger with *Glomus* "*d1*" than it did with *S. pellucida* (Fig. 11.5a). Although *Plantago* grew larger with *S. pellucida*, *S. pellucida* sporulated preferentially with *Allium* (Fig. 11.5b). Similarly, although *Allium* grew larger with *Glomus* "*d1*", *Glomus* "*d1*" sporulated preferentially with *Plantago* (Fig. 11.5b). This asymmetry between the relative performance of the two plants and the relative performance of the two fungi is suggestive of the negative feedback fitness relations illustrated in Figs. 11.1 and 11.2.

Plantago grew much larger than *Allium* with both fungal species ($F_{1,36}=443$, $P<0.0001$, Fig. 11.5a). However, given the very different growth forms and life-histories of these two species, it is not obvious that this overall difference in mass translates into differences in population growth rates (in fact, these species have co-occurred at our study site for at least 25 years; Fowler and Antonovics 1981). Our inference into plant-fungal dynamics assumes that the overall population growth rates of the two plant species is roughly equivalent. Given this assumption, the rank performance of the two species would be dependent on the identity of the fungus with which they were associated. We are particularly interested in evaluating whether the relative growth responses of the two plant species to the fungi depends upon the specific combination of plants and fungi. This is tested as the interaction term within a standard two-way analysis of variance. In the case of an analysis with two plants and two fungi, the plant species by AMF species interaction term corresponds precisely to a test of the plant interaction coefficient (I_p) derived in Bever (1999). That is, in the hypothetical fitness set presented in Fig. 11.3, the interaction term would test the significance of the quantity, $I_p=a-b-c+d$. In the example of *Plantago* and *Allium* being presented here (Fig. 11.5a), this interaction term is significant ($F_{1,36}=7.65$, $P<0.009$). Precisely, $I_p=0.204\pm0.074$. From this, we infer that the relative rates of population growth of these two plant species depends upon the identity of the fungus with which they associ-

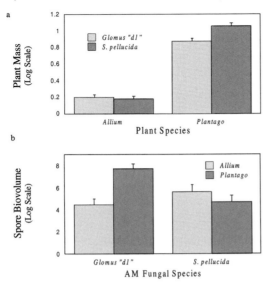

Fig. 11.5. An example of the mutual interdependence of plant and fungal growth rates among co-occurring plant and fungal species from a North Carolina grassland (from Pringle, Bever, Antonovics and Kuo, unpubl.). **a** The growth rate of *Allium vineale* relative to *Plantago lanceolata* depends upon whether they were inoculated with the undescribed AMF species, *Glomus* "d1", or *Scutellospora pellucida* (plant × AMF interaction $F_{1,36}=7.65$, $P<0.009$). **b** Conversely, the growth rate of *Glomus* "d1" relative to *Scutellospora pellucida* depends upon whether they were grown in association with *Allium* or *Plantago* (AMF × plant interaction $F_{1,22}=14.37$, $P<0.001$). This mutual interdependence would be predicted to result in negative feedback dynamics (see Fig. 11.6)

ate and the long-term stability of their interaction depends upon the dynamics within the fungal community.

The relative performance of the *Glomus* "d1" and *S. pellucida* also depend on the particular plant species with which they associate (interaction of plant and AMF species, $F_{1,22}=14.37$, $P<0.001$). Precisely, $I_F=-4.20\pm1.10$. From this, we infer that the relative rates of population growth of these two fungal species depends upon the identity of the plant with which they associate.

The nature of the community dynamics between the plants, *Allium* and *Plantago*, and the AMF, *Glomus* "d1" and *S. pellucida*, would then depend upon the product of the plant and fungal interaction coefficients. In this example, $I_P \times I_F=-8.57$, which is significantly less than zero (Fig. 11.6). Therefore, the community dynamics between these four species would be predicted to result in negative feedback. As *Plantago*, for example, becomes increasingly common, *Glomus* "d1" would be predicted to increase relative to *S. pellucida*, thereby decreasing the growth rate of *Plantago* relative to that of *Allium*. This dynamic would result in oscillations in the composition of the plant community over time (Figs. 11.2b, 11.7), and would contribute to the maintenance of all four species within the system. Therefore, in this example, the dynamics within the mutualism would be expected to maintain diversity within both plant and fungal communities. However, we would not expect *Plantago* to maximize the benefit it receives from its association with the AMF community.

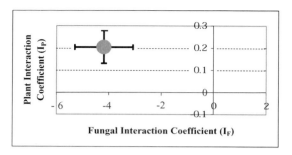

Fig. 11.6. Evidence of negative feedback between the plant species, *Allium vineale* and *Plantago lanceolata*, and the AMF species, *Glomus* "*d1*" and *Scutellospora pellucida*. Both plant and fungal interaction coefficients are significantly different from 0 ($P<0.009$ and $P<0.001$, respectively) and their product is negative ($I_P \times I_F < 0$). Therefore, the dynamics between the plant and fungal communities would be predicted to directly contribute to the coexistence of both plant species and both fungal species

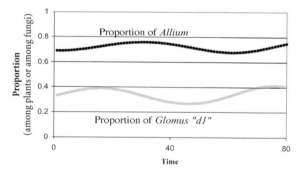

Fig. 11.7. Example of potential dynamics predicted within North Carolina grassland. The proportion of *Allium* among the plants and the proportion of *Glomus* "*d1*" among the fungi are plotted against time. Here, we present the discrete version of the model analyzed in Bever (1999) assuming that the maximum fitness of the two plant species is equal and their relative fitness is given in Fig. 11.5a. The fitness of the fungi was taken from Fig. 11.5b. Under negative feedback, the proportion of plants and of fungi oscillate out of phase from each other, with both plant species and both fungal species being maintained in the community (e.g., the proportion of *Allium* goes neither to zero or one). Note that although the two plant species are assumed to have equivalent maximal fitness, *Allium* is maintained in greater abundance because *S. pellucida*, with which *Allium* grows best, has higher average fitness than *Glomus* "*d1*"

11.6.2 Evaluation of Feedback Using the Phenomenological Approach

The phenomenological approach was first described in the context of testing feedbacks through changes in the entire soil community (Bever 1994; Bever et al. 1997). For this purpose, whole soil inocula, representing the entire soil community, is first trained and then tested (as depicted in Fig. 11.4). This approach has proven very useful in identifying examples in which the dynamics within the soil community, as a whole, can directly contribute to the maintenance of diversity within the plant community (see Bever 1994; Bever et al. 1997). We argue here that the approach can also provide a powerful tool to evaluate feedbacks on plant growth in particular, through changes in the composition of the AMF communities. The major difficulty is in separating the effects of differentiation of the AMF community from the effects of the differentiation of other components of the soil community.

11.6.2.1 Correlation of Arbuscular Mycorrhizal Fungal Patterns with Processes of Soil Community Feedback

The negative feedback observed through changes in the soil community composition (e.g., Bever 1994; Bever et al. 1997) could have been generated by host-specific differentiation within the AMF community. In fact, we have previously trained whole soil communities on four plant species (*Allium vineale, Anthoxanthum odoratum, Panicum sphaerocarpon*, and *Plantago lanceolata*), and observed strong host-specific differentiation of the AMF community (described in Bever et al. 1996). We then used these trained soils as inocula in a test experiment (as in Fig. 11.4) and observed strong negative feedback between three pairs of these same species (between *Anthoxanthum* and *Panicum, Anthoxanthum* and *Plantago*, and *Panicum* and *Plantago*, Bever et al. 1997). Therefore, we have observed both AMF community differentiation and the process of negative soil community feedback in the same set of experiments. These two observations are suggestive of causation, i.e., the differentiation of the AMF community may have generated the negative feedback observed through the soil community. Johnson and co-workers (1991, 1992b) have also found AMF community differentiation in a system in which yield-decline, due to repeated monocropping of corn or soybeans, has been well established. Finally, Kiers and co-workers (2000) have found that several species of tropical forest trees respond differentially to inoculation with field-collected roots of different plant species. They observed higher levels of AMF colonization when grown with their own root-inocula and yet the overall growth results suggested negative feedback. These results might suggest that the AMF community differentiation contributed to the soil community feedback.

Great caution must be exercised in inferring causation from these correlations. In other experiments within our system, we have demonstrated that both the accumulation of root pathogens in the genus *Pythium* (Mills and Bever 1998), and the differentiation of rhizosphere bacteria in the genus *Bacillus* (Westover and Bever 2001), can generate the negative feedbacks observed between our plant species. Similarly, accumulation of root feeding nematodes is a likely alternative cause of the yield-decline due to corn monocultures (Agrios 1997). Note that simply using washed-roots as inocula (as done in Exp. 2 of Bever 1994 and Kiers et al. 2000) does not eliminate potentially confounding effects of these other organisms, since these root pathogens are unlikely to be removed without simultaneously killing the AMF. Moreover, this problem cannot be easily resolved with biocide treatments, since a standard chemical treatment for fungal root pathogens, the fungicide Benomyl, also knocks back AMF. Because of the potentially confounding effects of root pathogens, tests of feedback between plants and AMF must use pure cultures of AMF.

11.6.2.2 Phenomenological Evaluation of Arbuscular Mycorrhizal Fungal Community Feedback

Simply testing feedbacks through trained whole soil (or root) inocula is not an adequate test of feedbacks through changes in the AMF community. Nevertheless, we still argue that the phenomenological approach to investigating soil community feedbacks can be of great value in deciphering feedbacks between plants and AMF. The problems of confounding effects of soil pathogens can be alleviated by starting the process using pure cultures of AMF. Multiple species of AMF must first be isolated and cultured from the same site as the plants to be tested. Pure cultures of these fungi can then be mixed to re-establish a diverse community of AMF that can then be trained, evaluated, and tested as depicted in Fig. 11.4. The feedback obtained from this experiment can then be attributed to differentiation observed within the AMF community.

We have used this approach to confirm that the AMF community differentiation does contribute to the whole soil negative feedback observed between *Plantago* and *Panicum* (Bever et al. 1997). In evaluating the AMF community composition prior to the feedback test, we found that the AM fungus, *Scutellospora calospora*, accumulated more with *Plantago* than *Panicum*, while the fungi, *Acaulospora morrowiae* and *A. trappei*, grew best with *Panicum*. Testing of plant response to single species isolates of these AMF subsequently showed that *Plantago* grew better with these two species of *Acaulospora* than with *S. calospora*, thereby generating the negative feedback on plant growth (Bever, unpubl.).

11.6.3 A Comparison of Mechanistic and Phenomenological Approaches to Testing Feedback

While both the mechanistic and phenomenological approaches can be used to test feedback on plant growth through changes in the AMF community, these two approaches may incorporate important differences in the biology underlying the processes being tested, depending upon how the relative fitness of AMF is evaluated. In the example of mechanistic observation of negative feedback between the plant species, *Allium* and *Plantago*, the relative population growth rates of the fungi, *Glomus* "*d1*" and *S. pellucida*, were evaluated through their host-specific differences in sporulation within one plant-one fungus pairings (Fig. 11.5b). Through these measures, host-specific differences were observed in the basal growth rates of these fungi. Measurement of relative growth rates of AMF within the phenomenological approach is necessarily inclusive of other potential mechanisms for host-specific differentiation of the AMF community. These include host-specific differences in the competitive ability of the AMF and the possibility of allocation preferences in the host. The net result of these processes, i.e., the net host-specific differentiation of the AMF community, determines the ultimate feedback dynamics, as demonstrated in the phenomenological model of feedback (Bever et al. 1997).

In this context, it is interesting to note that while we observed greater sporulation by *Glomus* "*d1*" when grown alone with *Plantago* than when grown alone with *Allium* and higher sporulation by *S. pellucida* when grown alone with *Allium* than when grown alone with *Plantago* (Fig. 11.5b), we observed the opposite patterns when these two fungi were grown within diverse AMF and soil communities. That is, within a diverse AMF community and within a full complement of other soil organisms, *Glomus* "*d1*" grew poorest with *Plantago* while *S. pellucida* grew best with *Plantago* (Bever et al. 1996). Clearly, the outcome of host-specific differentiation of the AMF community can be context-dependent, i.e., it depends upon the environment in which the plant and fungi interact, including the presence of other soil organisms. Therefore, our prediction of the ultimate feedback dynamics is also context-dependent. In this case, the ultimate predicted dynamics are reversed from negative feedback between these species using estimates of the relative growth rates of fungal species when grown alone with their host, to a prediction of positive feedback using estimates of the relative growth rates of fungal species when grown within diverse soil communities.

A second difference between the two approaches is that the phenomenological approach is less reliant on measurement of AMF population growth rates. Given an initial mixture of clean cultures of AMF species, feedback through the AMF community can be measured (as described in Fig. 11.4) without knowledge of the growth rates of AMF populations. Moreover, both infective hyphal fragments and viable spores can be included in the inoculum that contributes to the test of feedback (Fig. 11.4), and, therefore, the method is less reliant on the accuracy with which spore counts reflect AMF fitness. Of course, in practice, we would very much want to know how the AMF community differentiated in order to interpret the results and we are still largely dependent upon spore counts for this measure (see Clapp et al., Chap. 8, this Vol.).

Community dynamics may best be understood when the mechanistic and phenomenological approaches are combined. The mechanistic approach results in a detailed knowledge of individual interactions, building confidence in the resulting predictions. However, the phenomenological approach is inclusive of a broader array of biological processes. Therefore, the dynamical predictions may be more robust than those from mechanistic approaches. Perhaps an ideal understanding of the interactions within a community results from the combination of mechanistic and phenomenological data. If the same species are used in both phenomenological and mechanistic experiments, we can begin to distinguish when an environmental context makes a difference – e.g. some feedbacks remain constant in any environment, and others are affected by the presence or absence of a particular AMF species, soil organism, phosphorous levels, or other environmental factor. The context-dependence of AMF community feedback needs to be explicitly evaluated.

11.7 Implications and Extensions of Feedback Framework

An increasing amount of empirical work has focused on the potential role of AMF in maintaining diversity within plant communities (Grime et al. 1987; Wilson and Hartnett 1997; Zobel and Moora 1997; van der Heijden et al. 1998a,b; Hartnett and Wilson 1999; Smith et al. 1999; Kiers et al. 2000). There are many potential mechanisms through which AMF could contribute to plant species coexistence, including suppression of competitive dominants, alteration of plant niche breadth of individual species, and fungal-specificity of plant response. The mechanisms that can explain how AMF affect plant species coexistence and plant diversity are discussed in Chapter 10 (this Vol.). We have argued that the dynamics between the plant and AMF community can itself contribute to the maintenance of diversity within the plant community (Fig. 11.1; Bever et al. 1997; Bever 1999). In this chapter, we demonstrate that the dynamics between plant and fungal species can generate negative feedbacks and that negative feedbacks can directly contribute to the long-term maintenance of diversity within both the plant and fungal community (Figs. 11.5–11.7). While we have provided evidence for negative feedback between plant and AMF communities in one system, we emphasize that much work remains to be done before generalizations about plant – AMF feedbacks can be made. Positive feedbacks are also likely to occur. In this chapter, we have outlined an approach to evaluating these feedbacks.

There is, however, a consistency between the evidence of negative feedback through changes in the community of AMF (Figs. 11.5–11.7) and the observations that inoculation of microcosms with diverse communities of AMF can contribute to the coexistence of multiple plant species (Hart and Klironomous, Chap. 9, this Vol.; van der Heijden, Chap. 10, this Vol.; Grime et al. 1987). The feedback dynamic results from fungal-specific differences in plant species response to AMF. As more fungal species are added to the system, there is increased likelihood of including individual pairs of plants and fungi that exhibit strong specificity in responses. The likelihood of negative feedback dynamics that would contribute to the coexistence of plant species may also increase. Such negative feedback would generate a negative frequency dependence that prevents any one plant species from becoming too rare, thereby contributing to overall evenness within the plant community. One should also note, however, positive feedback may not significantly reduce diversity over the short time frame of these experiments.

While negative feedback through the AMF community is a potential mechanism generating the plant community responses to AMF diversity treatments, it only builds upon a limited set of the differences among the plant species and differences among the fungal species. Differences among the plant species in their competitive ability, niche space, growth habit, and life-histories and how they are regulated by AMF, could also undoubtedly play critical roles in generating the plant community response to AMF diversity treatments. These processes have not been integrated within the simple community feedback framework. Integrating and testing these processes remains an enormous task. One contribution of the work presented here to this effort, however, is that the dynamics within the AMF community need to be integrated into these mechanisms to understand their long-term stability.

11.7.1 Negative Feedbacks and the Evolutionary Maintenance of Mutualism

The potential for negative feedbacks to develop between plants and AMF appears to add fuel to a long identified evolutionary quandary. Under a negative feedback dynamic, neither the benefit derived by the plant or fungi can be maximized (Bever 1999). In fact, the process of negative feedback would seem to lead to the eventual dissolution of the mutualism and an abundance of data suggest that the plants do not always benefit from AMF (reviewed in Johnson et al. 1997). What, then, maintains the mutualism? Spatial structure generated by local dispersal has been suggested as an explanation for the maintenance of mutualism within mycorrhizal fungi in the face of potential "cheating" AMF (Wilkinson 1997). This mechanism requires a very high level of spatial association (basically a one plant-one fungus pairing), or else the "cheating" fungus will spread. While AMF communities are known to be spatially structured (Fig. 9.3 in Hart and Klironomos, Chap. 9, this Vol.; Friese and Koske 1991; Bever et al. 1996), the necessary spatial structure for this mechanism to operate, however, does not exist, since multiple fungal species (which likely differ in their level of mutualism) are known to occupy the root systems of individual plants (Clapp et al. 1995; Bever et al. 1996; Merryweather and Fitter 1996). Moreover, spatially explicit simulations of this process suggest that negative feedback can quickly overwhelm spatial-structure, even when generated by very local dispersal (Molofsky et al. 1999; Molofsky, Bever and Antonovics, unpubl.).

Alternatively, the evolution of specificity of association might also contribute to the evolutionary maintenance of mutualism (see Sanders, Chap. 16, this Vol.). In the face of uneven benefit from individual fungal isolates, plants might be expected to evolve resistance to inferior mutualists, thereby leading to higher specificities of associations (e.g., Law 1985; Law and Koptur 1986). Such specificities could stem the decline of benefit to individual plants due to negative feedback dynamics. We note, however, that a mycorrhizae-dependent plant is better off associating with an inferior mutualist than with no mutualist at all. For example, while *Plantago* would be better off associating with *S. pellucida* than with *Glomus "d1"* (Fig. 11.5a), it is still better off associating with *Glomus "d1"* than with no AM fungus (Pringle, Bever, Antonovics and Kuo, unpubl.). Therefore, given the high level of spatial variability in the distribution of individual AMF species known to be present at this site (Bever et al. 1996; Schultz 1996), it could be strongly disadvantageous for a plant species to be overly selective in choosing its AMF symbiont. The maintenance of the plant-AMF mutualism will likely depend upon a different mechanism.

11.7.2 Spatial Structure and the Dynamics of Feedback

The model of feedback developed within this chapter assumes that the plants and fungi interact within a well-mixed community (Bever et al. 1997; Bever 1999). Such mixing would result from high levels of dispersal by all participants. Both plants and fungi however, are likely to disperse locally and this local dispersal may alter predictions of the feedback model. Spatially explicit simulations suggest that the prediction that negative feedback will contribute to plant species coexistence is very robust to the level of dispersal (Molofsky et al. 1999; Molofsky, Bever, Antonovics, unpubl.). However, local dispersal can alter the prediction of loss of diversity due to positive feedback. Under local dispersal of plants and fungi and positive feedback, diversity will still be lost at the local scale with the formation of patches of single plant-fungal combinations. For example, in the positive feedback fitness sets illustrated in Fig. 11.1, some patches will form with all plant A and fungus X and other patches will form with all plant B and fungus Y. These patches can then be stable for long periods. Diversity can then be maintained in these discrete patches across a larger landscape (Bever et al. 1997; Molofsky et al. 1999, 2001).

11.8 Conclusion

In this chapter, we have reviewed a conceptual framework that builds on the unique ecologies of individual AMF species. Specifically, this framework identifies the two diametrically opposed dynamics of positive and negative feedback through changes in the AMF community composition (Fig. 11.1, Table 11.1). We have then developed two alternative approaches to evaluating these feedbacks: A mechanistic approach that builds on detailed knowledge of plant-fungal interactions (Fig. 11.3) and a more phenomenological approach in which much of the details of the AMF biology is black-boxed (Fig. 11.4). The fundamental assumptions of this feedback framework are well supported by available data. As a way of illustrating the mechanistic approach, we have presented evidence suggesting that negative feedback may operate between a pair of plants and a pair of fungi that coexist within a grassland in North Carolina (Figs. 11.5–11.7). We have then described a second example of negative feedback from this same study site obtained from the phenomenological approach. We feel that these examples provide dramatic illustration of the value of incorporating AMF community dynamics into our understanding of the impact of these fungi on plant populations and communities.

The current framework, however, is far from a complete description of the dynamics of the plant-AMF interaction. While the framework contributes to the discussion of the role of AMF diversity in plant community processes, it in no way provides a complete description of all of the relevant issues. The current framework, for example, cannot predict the context-dependence of feedback dynamics or explain the evolutionary maintenance of mutualism. Inclusion of additional details of the biology of plants and AMF can alter expectations and provide further insights into aspects of the plant-AMF mutualism. For example, inclusion of spatial processes identifies that positive feedback, while leading to the loss of diversity at a local scale, can contribute to the long-term maintenance of diversity at larger scales through the creation and reinforcement of monomorphic patches. Within the study of the plant-AMF interaction, there remains plenty of room for creative insights.

Acknowledgements. Our approach was refined by extensive conversations with many people, including our lab groups and the lab group of Dr. Ricardo Herrera. This manuscript was improved by the thoughtful comments of I. Sanders and M. van der Heijden. The work was supported by USDA grants 92-37101-7461 and 94-37101-0354 and NSF grants DEB-9996221 and DEB-9985272 to JDB, and NSF grant DEB-9801551 to AP

References

Abbott LK, Gazey C (1994) An ecological view of the formation of VA mycorrhizas. Plant Soil 159:69–78
Adjoud D, Plenchette C, Halli-Hargas R, Lapeyrie F (1996) Response of 11 eucalyptus species to inoculation with three arbuscular mycorrhizal fungi. Mycorrhiza 6:129–135
Agrios GN (1997) Plant pathology. Academic Press, San Diego
Allen MF, Allen EB (1990) The mediation of competition by mycorrhizae in successional and patchy environments. In: Grace JB, Tilman D (eds) Perspectives on plant competition. Academic Press, New York, pp 367–389
Bever JD (1994) Feedback between plants and their soil communities in an old field community. Ecology 75:1965–1977
Bever JD (1999) Dynamics within mutualism and the maintenance of diversity: inference from a model of interguild frequency dependence. Ecol Lett 2:52–62
Bever JD, Morton J, Antonovics J, Schultz PA (1996) Host-dependent sporulation and species diversity of mycorrhizal fungi in a mown grassland. J Ecol 75:1965–1977
Bever JD, Westover KM, Antonovics J (1997) Incorporating the soil community into plant population dynamics: the utility of the feedback approach. J Ecol 85:561–573
Boucher DH (1986) The biology of mutualisms. Croom Helm, London
Bronstein JL (1994) Our current understanding of mutualism. Q Rev Biol 69:31–51
Clapp JP, Young JPW, Merryweather JW, Fitter AH (1995) Diversity of fungal symbionts in arbuscular mycorrhizas from a natural community. New Phytol 130:259–265
Douds DD, Millner PD (1999) Biodiversity of arbuscular mycorrhizal fungi in agroecosystems. Agric Ecosyst Environ 74:77–93
Douds DD, Schenck NC (1990) Relationship of colonization and sporulation by VA mycorrhizal fungi to plant nutrient and carbohydrate contents. New Phytol 116:621–627
Egerton-Warburton LM, Allen EB (2000) Shifts in arbuscular mycorrhizal communities along an anthropogenic nitrogen deposition gradient. Ecol Appl 10:484–496
Eom AH, Hartnett DC, Wilson GWT (2000) Host plant species effects on arbuscular mycorrhizal fungal communities in tallgrass prairie. Oecologia 122:435–444
Fowler N, Antonovics J (1981) Competition and coexistence in a North Carolina Grassland. J Ecol 69:825–841
Friese CF, Koske RE (1991) The spatial dispersion of spores of vesicular-arbuscular mycorrhizal fungi in a sand dune: microscale patterns associated with the root architecture of American beachgrass. Mycol Res 95:952–957
Gazey C, Abbott LK, Robson AD (1992) The rate of development of mycorrhizas affects the onset of sporulation and production of external hyphae by two species of *Acaulospora*. Mycol Res 96:643–650
Gemma JN, Koske RE, Carreiro M (1989) Seasonal dynamics of selected species of V-A mycorrhizal fungi in a sand dune. Mycol Res 92:317–321
Grime JP, Macky JM, Hillier SH, Read DJ (1987) Mechanisms of floristic diversity: evidence from microcosms. Nature 328:420–422
Hartnett DC, Wilson GWT (1999) Mycorrhizae influence plant community structure in tallgrass prairie. Ecology 80:1187–1195
Hartnett DC, Hetrick BAD, Wilson GWT, Gibson DJ (1993) Mycorrhizal influence on intra- and inter-specific neighbour interactions among co-occurring prairie grasses. J Ecol 81:787–795
Helgason T, Daniell TJ, Husband R, Fitter AH, Young JPW (1998) Ploughing up the wood-wide web? Nature 394:431–431
Hetrick BAD, Bloom J (1986) The influence of host plant on production and colonization ability of vesicular-arbuscular mycorrhizal spores. Mycologia 78:32–36

Hoeksema JD (1999) Investigating the disparity in host specificity between AM and EM fungi: lessons from theory and better-studied systems. Oikos 84:327–332
Janos DP (1980) Mycorrhizae influence tropical succession. Biotropica 12S:56–64
Janzen DH (1986) The natural history of mutualisms. In: Boucher D (ed) The biology of mutualism. Croom Helm, London, pp 40–99
Johnson NC (1993) Can fertilization of soil select for less mutualistic mycorrhizae? Ecol Appl 3:749–757
Johnson NC, Pfleger FL, Crookston RK, Simmons SR, Copeland PJ (1991) Vesicular arbuscular mycorrhizas respond to corn and soybean cropping history. New Phytol 117:657–663
Johnson NC, Tilman D, Wedin D (1992a) Plant and soil controls on mycorrhizal fungal communities. Ecology 73:2034–2042
Johnson NC, Copeland PJ, Crookston RK, Pfleger FL (1992b) Mycorrhizae – possible explanation for the yield decline with continuous corn and soybean. Agron J 84:387–390
Johnson NC, Graham JH, Smith FA (1997) Functioning of mycorrhizal associations along the mutualism-parasitism continuum. New Phytol 135:575–585
Kiers ET, Lovelock CE, Krueger EL, Herre EA (2000) Differential effects of tropical mycorrhizal fungal inocula on root colonization and tree seedling growth: implications for tropical forest diversity. Ecol Lett 3:106–113
Koide RT, Lu XH (1995) On the cause of offspring superiority conferred by mycorrhizal infection of *Abutilon theophrasti*. New Phytol 131:435–441
Koide RT, Shumway DL, Mabon SA (1994) Mycorrhizal fungi and reproduction of field population of *Abutilon theophrasti* Medic. (Malvaceae). New Phytol 126:123–130
Law R (1985) Evolution in a mutualistic environment. In: Boucher D (ed) The biology of mutualism. Croom Helm, London, pp 145–170
Law R, Koptur S (1986) On the evolution of non-specific mutualism. Biol J Linn Soc 27:251–267
Lewis JD, Koide RT (1990) Phosphorus supply, mycorrhizal infection and plant offspring vigor. Funct Ecol 4:695–702
Merryweather J, Fitter A (1996) The arbuscular mycorrhizal fungi of *Hyacinhoides non-scripta*. II. Seasonal and spatial patterns of fungal populations. New Phytol 138:131–142
Miller MA, Jastrow JD (2000) Mycorrhizal fungi influence soil structure. In: Douds D (ed) Arbuscular Mycorrhizas: molecular biology and physiology. Kluwer, Dordrecht
Mills KE, Bever JD (1998) Maintenance of diversity within plant communities: soil pathogens as agents of negative feedback. Ecology 79:1595–1601
Molofsky J, Durrett R, Dushoff J, Griffeath D, Levin S (1999) Local frequency dependence and global coexistence. Theor Popul Biol 55:1–13
Molofsky J, Bever JD, Antonovics J (2001) Coexistence under positive frequency dependence. Proc R Soc Lond (in press)
Nemec S (1978) Response of six citrus rootstocks to three species of *Glomus*, a mycorrhizal fungus. Proc Florida State Hortic Soc 91:10–14
Pearson JN, Schweiger P (1993) *Scutellospora calospora* (Nicol. & Gerd.) Walker & Sanders associated with subterranean clover: dynamics of colonization, sporulation and soluble carbohydrates. New Phytol 124:215–219
Powell CL, Clark GE, Verberne NJ (1982) Growth response of four onion cultivars to isolates of VA mycorrhizal fungi. N Z J Agric Res 25:465–470
Sanders IR, Fitter AH (1992) Evidence for differential responses between host-fungus combinations of vesicular-arbuscular mycorrhizas from a grassland. Mycol Res 96:415–419
Schultz PA (1996) Arbuscular mycorrhizal species diversity and distribution in an old field community. Diss, Duke University, Durham, North Carolina, USA

Smith MD, Hartnett DC, Wilson GWT (1999) Interacting influence of mycorrhizal symbiosis and competition on plant diversity in tallgrass prairie. Oecologia 121:574–582

Smith SE, Read DJ (1997) Mycorrhizal symbiosis, 2nd edn. Academic Press, San Diego

Streitwolf-Engel R, Boller T, Wiemken A, Sanders IR (1997) Clonal growth traits of two *Prunella* species are determined by co-occurring arbuscular mycorrhizal fungi from a calcareous grassland. J Ecol 85:181–191

Van der Heijden MGA, Boller T, Wiemken A, Sanders IR (1998a) Different arbuscular mycorrhizal fungal species are potential determinants of plant community structure. Ecology 79:2082–2091

Van der Heijden MGA, Klironomos JN, Ursic M, Moutogolis P, Streitwolf-Engel R, Boller T, Wiemken A, Sanders IR (1998b) Mycorrhizal fungal diversity determines plant biodiversity, ecosystem variability and productivity. Nature 396:69–72

Westover KM, Bever JD (2001) Mechanisms of plant species coexistence: complementary roles of rhizosphere bacteria and root fungal pathogens. Ecology 82:3285–3294

Wilkinson DM (1997) The role of seed dispersal in the evolution of mycorrhizae. Oikos 78:394–396

Wilson DO (1988) Differential plant response to inoculation with two VA mycorrhizal fungi isolated from a low-pH soil. Plant Soil 110:69–75

Wilson GWT, Hartnett DC (1997) Effects of mycorrhizae on plant growth and dynamics in experimental tallgrass prairie microcosms. Am J Bot 84:478–482

Zobel M, Moora M (1997) Plant coexistence in the interactive environment: arbuscular mycorrhiza should not be out of mind. Oikos 78:202–208

Section D:

Multitrophic Interactions

Section D:

Multitrophic Interactions

12 Mycorrhizae-Herbivore Interactions: Population and Community Consequences

CATHERINE A. GEHRING, THOMAS G. WHITHAM

Contents

12.1	Summary	295
12.2	Introduction	296
12.3	Effects of Aboveground Herbivory on Mycorrhizal Fungi	297
12.3.1	Population Level Responses	297
12.3.2	Community Level Responses	299
12.3.3	Carbon Limitation as a Mechanism of Herbivore Impacts on Mycorrhizae	301
12.3.4	Conditionality in Mycorrhizal Responses to Herbivory	302
12.4	Effects of Mycorrhizal Fungi on the Performance of Herbivores	305
12.4.1	Patterns of Interaction	305
12.4.2	Mechanisms of Mycorrhizal Impacts on Herbivores	307
12.4.3	Not All Mycorrhizae Are Equal	308
12.5	Ecological and Evolutionary Implications	311
12.5.1	Herbivore Effects on Mycorrhizae	311
12.5.2	Mycorrhizae Effects on Herbivores	312
12.5.3	Combined Effects	314
12.6	Suggestions for Future Research	315
References		316

12.1 Summary

Herbivores and mycorrhizal fungi commonly occur together on host plants and thus are likely to interact in ways that could indirectly affect each other's performance. In this chapter, we review the published studies on these interactions and their implications for plant, herbivore and fungal populations and communities. Six major patterns emerge from our review. First, in the majority of cases, aboveground herbivores reduce mycorrhizal colonization and alter mycorrhizal fungal community composition. Second, mycorrhizae also affect herbivores, but these effects are highly variable and range from positive to negative. Third, both of these interactions are conditional. For example, mycorrhizae are more likely to be affected by herbivores as the herbivory and/or the cumulative effects of environmental stress intensify. Under relatively benign conditions, herbivory may have no effect on mycorrhizae, but as plants are challenged, the impacts of herbivores on mycorrhizae can be great. Fourth, all mycorrhizae are not equal; the impacts of mycorrhizae on herbivores differ greatly between arbuscular mycorrhizal (AM) and ecto-mycorrhizal (EM) fungi and among species. Because different mycorrhizal species vary in their mutualistic effects on plants, such variation in impacts on herbivores should also be expected. Fifth, different types of herbivores are affected differently by mycorrhizae. Generalist herbivores appear to be affected more than specialists, and leaf chewers are more likely to be negatively affected, while sap feeders and their allies are more likely to be positively affected. Sixth, few studies have examined mycorrhizal effects on herbivores and herbivore effects on mycorrhizae in the same system. However, if interactions between aboveground herbivores and mycorrhizal fungi are similar to those observed between above- and belowground herbivores, they are likely to be highly asymmetrical with aboveground herbivores having greater negative effects on mycorrhizae than vice versa.

12.2 Introduction

Both herbivores and mycorrhizal fungi can have significant effects on plant performance, plant community dynamics and ecosystem processes (*herbivores* – Naiman et al. 1986; Choudhury 1988; Pastor et al. 1988; Huntly 1991; Pastor and Naiman 1992; Findlay et al. 1996; Ritchie et al. 1998; *mycorrhizae* – this Vol.; Fogel 1979; Grime et al. 1987; Perry et al. 1989; Miller and Allen 1992; van der Heijden et al. 1998a,b). Although a great deal is known about the independent effects that herbivores and mycorrhizal fungi have on their host plants, comparatively few studies have examined their interactions. These interactions are likely to be widespread and may have different effects on plant communities and ecosystems than either herbivores or mycorrhizae acting alone. Here, we review the interactions of herbivores and mycorrhizae, and then examine their consequences for populations and communities at both ecological and evolutionary levels. Interactions of mycorrhizal fungi with other heterotrophic organisms such as soil invertebrates and saprotrophic soil fungi are reviewed respectively by Gange and Brown (Chap. 13, this Vol.) and Leake et al. (Chap. 14, this Vol.).

This chapter is divided into three major sections. The first section reviews the literature on how herbivory affects mycorrhizal populations and results in a shift in mycorrhizal community structure. We also explore how this interaction is conditional and dependent upon combined challenges to the plant. In the second section, we examine how mycorrhizae influence herbivore performance; how these effects vary with specialist and generalist insects, feeding mode, and how mycorrhizal types or species vary in their impacts on herbivores. Finally, in section three, we examine the implications of the three-way interaction among mycorrhizae, plants and their herbivores. We explore how herbivores may reduce the benefits of having mycorrhizal mutualists, how herbivores and mycorrhizae may vary in their relative importance under different environmental conditions, and how mycorrhizal impacts on keystone herbivores could influence biodiversity and community structure.

12.3 Effects of Aboveground Herbivory on Mycorrhizal Fungi

12.3.1 Population Level Responses

Gehring and Whitham (1994a) reviewed the relationship between aboveground herbivory and mycorrhizal colonization, finding that herbivory reduced mycorrhizal colonization in 23 of the 37 plant species that had been studied. Arbuscular mycorrhizal (AM) fungi colonized most of the plant species, with only two species colonized by ecto-mycorrhizal (EM) fungi. Mycorrhizal colonization was not affected by herbivory in ten of the plant species while positive effects were observed in two cases, one involving several unnamed plant species in the Serengeti grasslands of Africa (Wallace 1981) and another involving soybeans (*Glycine max*) (Bayne et al. 1984). Variable effects, either negative or neutral, were observed in two grass species *Agropyron desertorum* and *Bouteloua gracilis* (cited in Gehring and Whitham 1994a).

More recent studies have increased the sample sizes of EM plant species and yielded similar patterns. Five additional plant species have been examined (four EM species and one AM species) and mycorrhizal colonization was reduced following herbivory in all but one species (Markkola 1996; Gange and Bower 1997; Rossow et al. 1997; Kolb et al. 1999; Saikkonen et al. 1999). For example, Gange and Bower (1997) documented that herbivory by a leaf-chewing lepidopteran (*Arctia caja*) reduced levels of AM colonization in *Plantago lanceolata*. Similarly, defoliation by western spruce budworm (*Choristoneura occidentalis*) reduced the EM of Douglas fir (*Pseudotsuga menziesii*) (Kolb et al. 1999). In the first study to quantify the effects of vertebrate herbivores on EM fungi, Rossow et al. (1997) used exclusion fences to document that winter browsing by moose and snowshoe hare reduced the EM colonization of wil-

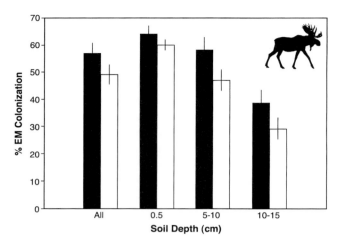

Fig. 12.1. The negative effects of winter browsing by moose and snowshoe hare on the ecto-mycorrhizal colonization of balsam poplar (*Populus balsamifera*) and willow (*Salix* spp.) at varying depths in the soil profile. *Solid bars* represent plots where herbivores were excluded and *open bars* represent control plots. *Bars* represent mean ± 1 SE. (Adapted from Rossow et al. 1997 with permission)

lows (*Salix* spp.) and poplars (*Populus balsamifera*) in a taiga ecosystem (Fig. 12.1). This reduction was observed at all three root depths sampled, but was larger in magnitude in the two deeper samples (Fig. 12.1).

In contrast, two separate studies demonstrated that manual defoliation had no significant effect on levels of EM colonization in Scots pine (*Pinus sylvestris*) (Markkola 1996; Saikkonen et al. 1999). Saikkonen et al. (1999) observed that EM colonization of Scots pine was nearly 100% despite severe manual defoliation that left plants with only a single year of needles. Root biomass declined with increasing levels of defoliation, but EM colonization and fungal biomass in roots remained relatively constant.

An overall summary combining the earlier studies reviewed by Gehring and Whitham (1994a) and the newer research, demonstrates that, in general, aboveground herbivores negatively affect mycorrhizal colonization. Mycorrhizal colonization was reduced in 64.3% of the 42 plant species examined, with no effect observed in 26.1% of the plant species. Species in which mycorrhizae responded positively to herbivory were rare, accounting for only 4.8% of those examined. Similarly, species that showed variable responses to herbivory were also rare, accounting for the remaining 4.8% of species. Grasses still predominated among the plant species studied, but while semi-arid ecosystems were best represented in the earlier review, additional ecosystems have been represented in recent work, including the Alaskan taiga (Rossow et al. 1997). A wide array of types of herbivory was studied, ranging from manual defoliation in the laboratory to natural levels of insect and vertebrate herbivory in the field. The breadth of these studies suggests that the patterns observed are not unique to particular types of herbivores or ecosystems. However, studies of aboveground herbivore effects on mycorrhizae have been limited in their measurements of mycorrhizal function, focusing entirely on measurements of internal mycorrhizal colonization. Because of the importance of external hyphae to the functioning and biomass of mycorrhizal associations, it would be valuable for future studies to extend beyond changes in mycorrhizal colonization to include changes in extramatrical hyphal production.

12.3.2 Community Level Responses

In addition to the effects of herbivores on mycorrhizal populations (i.e., levels of mycorrhizal colonization), herbivores may also affect the species composition of mycorrhizal fungi that colonize host plant roots. Fewer studies have investigated this possibility or its consequences, but one study on AM fungi and two studies on EM fungi demonstrate that such effects may occur. Bethlenfalvay and Dakessian (1984) used bait plants to estimate AM fungal spore diversity in a Nevada rangeland and found more species of AM fungi in plots that had been free of ungulate grazing for 19 years than in grazed plots. In this case, reductions in both AM colonization and AM species richness were observed in the grazed plots.

We have found similar results in pinyon pine (*Pinus edulis*), where chronic herbivory by a needle-feeding scale insect (*Matsucoccus acalyptus*) reduces EM colonization (Del Vecchio et al. 1993; Gehring et al. 1997) and may cause shifts in the fungal community (Gehring et al., unpubl. data). We compared the communities of EM fungi that colonized two groups of trees using restriction fragment length polymorphism (RFLP) analysis of individual root tips (methods as in Gehring et al. 1998). These RFLP analyses generally distinguish among species of EM fungi (Gardes and Bruns 1996; Gehring et al. 1998). The first group consisted of resistant trees upon which scale survival and density was low, while the second group consisted of susceptible trees where scale survival and density was high. We sampled five resistant and five susceptible trees over a 2-year period (at least 25 root tips per tree collected from all compass directions of the tree) and sampled 14 other trees per group less intensively (five to nine tips per tree) in one of the two years. In both cases, we found that resistant and susceptible trees supported a different community of EM fungi. Figure 12.2 shows the results of an ordination analysis using non-metric multidimensional scaling (NMDS – Kruskal 1964; Minchin 1987a,b; Clarke 1993) that illustrates the position of five resistant and five susceptible trees sampled over 2 years in two-dimensional space based on the composition of their fungal communities. An analysis of similarity (ANOSIM) of the communities of resistant and susceptible trees demonstrated that they differed significantly from one another ($R=0.25$, $P=0.002$; Fig. 12.2). For example, *Tricholoma terreum* and *Geopora cooperi* were more

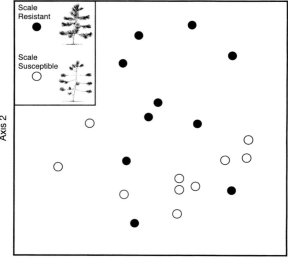

Fig. 12.2. An NMDS ordination plot showing that the ecto-mycorrhizal fungal community composition differs between pinyon pines (*Pinus edulis*) that are resistant versus susceptible to an insect herbivore, the pinyon needle scale (*Matsucoccus acalyptus*). Analyses are based upon 11 EM species distributed among five scale-resistant (*Matsucoccus acalyptus*; *closed circles*) and five scale-susceptible (*open circles*) trees (see text). In the plot, *circles* are placed in a configuration where the distances between *circles* are in rank order with their ecological dissimilarities using the Bray-Curtis index. Because any rotation of this configuration that preserves the distance/dissimilarity relationship is equally good, the axes themselves have no real meaning

abundant on the resistant trees, while two unidentified ascomycetes were more prevalent on susceptible trees. No significant differences were observed in the overall species richness of either individual trees (mean = 3.2 species per tree for resistant trees and 3.1 for susceptible trees), or of the tree types as a whole, with a total of 11 species observed in both resistant and susceptible groups. These data indicate that resistant and susceptible trees support different EM fungal communities, but the causal mechanism is unknown. We suspect that these differences are likely caused by scale herbivory. Alternatively, inherent differences among tree types may favor particular EM communities. Analysis of the EM fungal communities of susceptible trees from which the scale insects have been removed for several years are underway to help distinguish between these alternative hypotheses. Besides, abiotic soil factors and anthropogenic disturbances such as nitrogen deposition also have a big impact on EM fungal species composition as is nicely illustrated in Fig. 7.2 (Erland and Taylor, Chap. 7, this Vol.).

In contrast to the above studies where herbivores affected both colonization and community structure, Saikkonen et al. (1999) observed no shift in

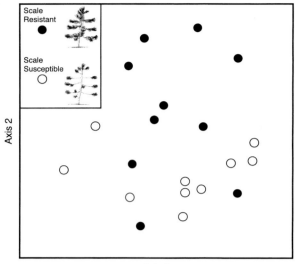

and Whitham 1995) suggest that environmental stress can be an important factor. Removal of 25% of pinyon shoots by clipping, a typical level of herbivore damage observed in the field, caused reductions in EM colonization at a water and nutrient-stressed site. However, the same clipping experiment at a less stressful site had no effect on mycorrhizal colonization. The less stressful site had significantly higher levels of soil ammonium, nitrate, phosphorous and moisture as well as increased levels of tree growth and cone production (Gehring and Whitham 1994b). Although seedlings from both site types benefited from EM colonization, trees at the less stressful site generally had lower levels of mycorrhizal colonization than trees at the more stressful site, perhaps due to the higher nutrient status of the soils (Gehring and Whitham 1994b). The results of the clipping experiment suggest that environmental conditions play a significant role in determining mycorrhizal responses to herbivory (Fig. 12.3).

Although we are not aware of any studies which directly link herbivore impacts on belowground carbon allocation with changes in mycorrhizal colonization, plant species can vary in their patterns of belowground energy allocation in response to herbivory. In most plant species, defoliation results in immediate reductions in root growth (Wilson 1988). However, exceptions to this situation may help explain the cases of neutral or positive effects of herbivores on mycorrhizae. For example, aboveground herbivory can increase root biomass and root exudation in some plants (e.g., Holland et al. 1996), potentially leading to subsequent increases in mycorrhizal colonization. In other plant species, belowground allocation does not shift in response to herbivory. In *Agropyron spicatum*, root growth continued unabated following severe defoliation (85% foliage removal), suggesting that belowground carbon allocation was not reduced (Richards 1984). However, this allocation strategy resulted in slower aboveground recovery in *A. spicatum*, compared to a congener, *A. desertorum* that did reduce carbon allocation belowground following herbivory (Richards 1984). Regardless of the efficacy of these strategies in facilitating recovery from herbivory, plants clearly vary in their carbon allocation strategies and this may represent another condition that influences mycorrhizal responses to herbivory.

In combination, the above studies of herbivore intensity, environmental stress and plant allocation strategies support the hypothesis that mycorrhizal responses to herbivory are conditional. A general model of how two of these factors (abiotic environmental stress and intensity of herbivory) interact to affect mycorrhizal colonization results in four predictions (Fig. 12.4). First, plants will have higher levels of mycorrhizal colonization at abiotically stressful sites, particularly if the source of that stress is low levels of soil nutrients and/or moisture. This prediction is supported by theoretical and empirical data from a variety of systems (e.g., Marx et al. 1977; Chapin 1980; MacFall et al. 1991). Second, mycorrhizal colonization is likely to decline progressively as the intensity of herbivory increases at both abiotically stressful and less

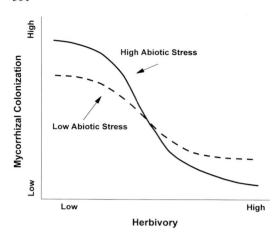

Fig. 12.4. A predicted relationship between intensity of herbivory, degree of environmental stress and levels of mycorrhizal colonization in host plants (see text for discussion)

stressful sites. This prediction follows from the above review of studies that examined mycorrhizal colonization in plants experiencing a range of herbivore damage. Third, with increasing herbivory, the decline in mycorrhizal colonization will be greatest in high stress sites. We feel this is likely because plants growing in stressful locations may reduce carbon allocation belowground more quickly in response to herbivory than plants growing in less stressful areas. This prediction is supported by our studies of pinyon pine summarized in Fig. 12.3. Fourth, we predict that in stressful environments, the addition of another stress (i.e., herbivory), will result in a reversal in the relationship between mycorrhizal colonization and environmental stress. At low levels of herbivory, mycorrhizal colonization will be higher in stressful environments than in more favorable environments. However, as herbivore damage increases, the sites with the highest levels of mycorrhizal colonization should reverse (see crossed lines in Fig. 12.4). In stressful environments, the combined effects of environmental stress and herbivory should lead to the lowest levels of mycorrhizal colonization.

Although we predict that mycorrhizal colonization will usually decline as the intensity of herbivory increases, shifts in mycorrhizal fungal communities could alter the dynamics of this decline or eliminate it entirely. For example, changes in mycorrhizal fungal community composition to species that are energetically less costly in response to herbivory (Saikkonen et al. 1999) could result in declines in mycorrhizal colonization only at the very highest levels of herbivory and no change in colonization at moderate or low levels of herbivory. Alternatively, if only energetically costly fungal species are present in a given habitat, mycorrhizal colonization could begin to decline at very low levels of herbivory.

12.4 Effects of Mycorrhizal Fungi on the Performance of Herbivores

12.4.1 Patterns of Interaction

Herbivores affect mycorrhizae, but is the reverse also true? Mycorrhizal fungi may affect the ability of plants to resist herbivory in three potential ways (Gehring et al. 1997). First, by increasing the vigor of host plants, mycorrhizae may improve the performance of those herbivores that perform better on more vigorous host plants (Price 1991). Secondly, because herbivore performance is negatively associated with plant vigor in many plant species (White 1984; Waring and Cobb 1992), mycorrhizal colonization could improve plant resistance by increasing plant performance. Third, mycorrhizae could alter plant carbon to nutrient ratios, allowing increased investment in carbon-based antiherbivore defenses, which then reduce herbivore performance (Jones and Last 1991). These latter two possibilities are not mutually exclusive. Furthermore, even if the higher levels of plant growth and tissue nutrient concentrations often associated with mycorrhizae do not directly affect herbivore performance, increased plant vigor might improve the ability of plants to tolerate herbivory by allowing them to sustain herbivore damage without loss of productivity (Gehring and Whitham 1994a; Borowicz 1997).

Only a small number of studies have examined the effects of mycorrhizal colonization on herbivore performance and fewer still have addressed the question of tolerance. Table 12.1 summarizes the studies of 17 herbivorous insect species on ten species of host plants. Of these plants, two support EM, seven support AM and one supports both AM and EM. The majority of these studies were conducted on potted plants, but a few were conducted in the field (Gange and West 1994; Gehring et al. 1997). Most focused on herbivores that fed on aboveground plant parts, but two focused on root herbivores (Gange et al. 1994; Gange and Bower 1997).

Mycorrhizal colonization affected the performance of 12 out of 17 insect species (70.6%), but the direction of these effects was highly variable (Table 12.1). Four (33%) responded positively, five responded negatively (42%) and three (25%) showed variable responses depending upon the type or species mix of mycorrhizal fungi, the insect performance parameter measured or other factors. Arbuscular mycorrhizal fungi affected herbivore performance in more cases (90.9%) than EM fungi (42.9%). However, only three plant species that form EM have been examined.

Specialist and generalist insect species appear to respond differently to mycorrhizal inoculation. The performance of specialist insect herbivores was influenced by mycorrhizal colonization in 55.6% of interactions, as opposed to 87.5% of interactions among generalists. Specialists have been hypothe-

Table 12.1. Summary of mycorrhizal effects on herbivore performance; AM, arbuscular mycorrhizae; EM, ecto-mycorrhizae

Insect species	Mode of feeding	Host range	Plant species	Type of mycorrhizae	Effect of mycorrhizae	Reference
Arctia caja	Leaf chewing	Generalist	*Plantago lanceolata*	AM	Negative	Gange and West (1994)
Heliothis zea	Leaf chewing	Generalist	*Glycine max*	AM	Negative	Rabin and Pacovsky (1985)
Myzus ascalonicus	Phloem sap	Generalist	*Plantago lanceolata*	AM	Positive	Gange et al. (1999)
Myzus persicae	Phloem sap	Generalist	*Plantago lanceolata*	AM	Positive	Gange et al. (1999)
Otiorhynchus sulcatus	Root feeding	Generalist	*Taraxacum officinale*	AM	Negative	Gange et al. (1994)
Otiorhynchus sulcatus	Root feeding	Generalist	*Fragaria X ananassa*	AM	Variable[a]	Gange and Bower (1997); Gange (2001)
Schizaphis graminum	Phloem sap	Generalist	*Sorghum* spp.	AM	No effect	Pacovsky et al. (1985)
Spodoptera frugiperda	Leaf chewing	Generalist	*Glycine max*	AM	Negative	Rabin and Pacovsky (1985)
Lygus rugulipennis	Meristem	Generalist	*Pinus sylvestris*	EM	Variable[b]	Manninen et al. (1998, 1999b)
Epilachna varivestis	Leaf scraping	Specialist	*Glycine max*	AM	Positive	Borowicz (1997)
Polyommatus icarus	Leaf chewing	Specialist	*Lotus corniculatus*	AM	Positive	Goverde et al. (2000)
Urophora carduii	Stem galling	Specialist	*Cirsium arvense*	AM	Negative	Gange and Nice (1997)
Chaetophorus populicola	Phloem sap	Specialist	*Populus* hybrids	AM/EM	Variable[c]	Gehring and Whitham (this Chap.)
Cinara pinea	Phloem sap	Specialist	*Pinus sylvestris*	EM	No effect	Manninen et al. (1999b)
Gilpinia pallida	Leaf-chewing	Specialist	*Pinus sylvestris*	EM	No effect	Manninen et al. (1999b)
Matsucoccus acalyptus	Mesophyll	Specialist	*Pinus edulis*	EM	No effect	Gehring et al. (1997)
Schizolachnus pineti	Phloem sap	Specialist	*Pinus sylvestris*	EM	Variable[d]	Manninen et al. (1999a,b)

[a] Variability depended upon the mix of mycorrhizal fungal species.
[b] Oviposition was positively affected but growth was negatively affected.
[c] Effects were different for AM and EM (see Fig. 12.5).
[d] Variability depended upon the study.

sized to be less susceptible to the defensive compounds produced by their host plants than generalists (van der Meijden 1996). Because mycorrhizae can alter the defensive compound production of their host plants (Jones and Last 1991; Gange and West 1994), mycorrhizae may have greater effects on the generalist insects that are more sensitive to these compounds (Manninen 1999).

The mode of insect feeding may also be important (Borowicz 1997; Gange et al. 1999). The performance of insects that fed on phloem sap, leaf mesophyll, or scraped leaf tissue was rarely negatively affected by inoculation with mycorrhizal fungi (10% of cases), but frequently positively affected (50% of cases). In contrast, leaf chewers were affected in four cases (66%), and showed a negative response in three of the four cases. These patterns held even within separate studies on the same plant species; in soybeans, two leaf-chewing insects performed better on nonmycorrhizal plants (Rabin and Pacovsky 1985), while a leaf-scraping beetle performed better on mycorrhizal plants (Borowicz 1997).

12.4.2 Mechanisms of Mycorrhizal Impacts on Herbivores

Earlier we demonstrated that the effects of herbivores on mycorrhiza could be conditional, and based upon the patterns described above, we suggest that the same may be true for mycorrhizal effects on herbivores. Whether mycorrhizae will have an effect on herbivore performance is likely to depend upon the combined effects of mycorrhizal colonization on plant nutrient and defensive chemical concentrations, and host plant growth. In cases where mycorrhizae have little effect on plant growth or plant chemistry, we expect few effects on herbivore performance. If mycorrhizae improve plant vigor and plant tissue nutrient concentrations without concomitant changes in plant defensive chemistry, we predict that herbivores will respond positively to mycorrhizal inoculation. The interactions are also likely to depend upon the type of herbivore and whether or not chemical defenses are effective against it. For example, specialist insects that have evolved mechanisms to deal with many of the chemical defenses mounted by their host plants (van der Meijden 1996) may not respond to mycorrhizally mediated increases in plant defensive chemicals. Abiotic environmental conditions are also likely to play a role as hypothesized by Jones and Last (1991) who predicted that EM colonization would improve anti-herbivore defense in well-watered trees only if light was not limiting and soil nutrient levels were low.

Many of the available data support the hypothesis that the effects of mycorrhizal fungi on herbivores are likely to vary conditionally with the effects of mycorrhizal fungi on plant growth and plant chemistry. For example, AM colonization of *Plantago lanceolata* increased the concentrations of iridoid glycosides, one class of known feeding deterrents, and negatively affected the larval performance of a generalist insect, *Arctia caja* (Gange and West 1994). In contrast, EM colonization of Scots pine resulted in only slight changes in growth or chemical composition of EM plants compared with non-inoculated controls, and only one of several insect species showed any change in herbivore performance (Manninen et al. 1998, 1999a,b). Borowicz (1997) observed that Mexican bean beetle (*Epilachnis verivestis*) survival and mass at pupation were greatest in plants with mycorrhizae and intermediate growth rates. She hypothesized that with intermediate stress, plant defenses were reduced, but the presence of mycorrhizae made them nutritionally superior for beetles. Gange et al. (1999) also observed that aphid performance on *Plantago lanceolata* was improved by mycorrhizae at low and intermediate levels of phosphorous fertilization, but not at high levels, supporting this intermediate stress hypothesis.

Although the above studies are consistent with the hypothesis that mycorrhizal effects on herbivores are conditional, the results of other studies are less clear. For example, Rabin and Pacovsky (1985) found that AM fungi reduced the performance of two generalist insects feeding on soybeans, but this effect was not related to changes in host plant vigor or concentrations of nutrients or defensive compounds in plant tissues. This latter example indicates that we do not understand the full complement of conditions that influence the interactions between mycorrhizae and herbivores. However, we maintain that further study of the effects of mycorrhizal fungi on plant defensive chemistry, combined with plant-herbivore theory will refine our knowledge of the conditions under which mycorrhizae will affect herbivore performance.

12.4.3 Not All Mycorrhizae Are Equal

Although many studies have examined herbivore responses to one plant species inoculated with one fungal species, many mycorrhizal species are involved in the wild. At least 150 species of fungi form AM associations and >5000 species are hypothesized to form EM associations (Smith and Read 1997). Recent data suggest that different types of mycorrhizal fungi (EM and AM) or species of AM fungi are not functionally equivalent in terms of their effects on herbivores.

To determine if AM and EM fungi had similar effects on herbivores, we examined the performance of a specialist phloem-feeding aphid (*Chaitophorous populicola*) on naturally occurring hybrid cottonwoods (*Populus angustifolia* × *P. fremontii*). Cottonwoods are members of the Salicaceae, one of the plant families that routinely forms both EM and AM associations in nature (Smith and Read 1997). We rooted cuttings of the same cottonwood genotype in moist, steam-sterilized sand. Once the cuttings had rooted, we inoculated them with either an EM fungus, *Pisolithus tinctorius*, ($n=20$) or with a mix of AM fungi (*Glomus* spp., *G. etunicatum*, *G. clarum*, and *Entrophosphora colombiana*) ($n=21$) obtained from Plant Health Care, Inc.

Another group of cuttings ($n=23$) were left non-inoculated as controls. Small variations in the sample sizes were due to the quantities of inoculum available. Cuttings were planted in 4 l pots in a mixture of steam-sterilized sand, peat and vermiculite (2:1:1) and allowed to overwinter in a greenhouse in Flagstaff, Arizona. We transported them to a field site in northern Utah when they were 8 months old. At this time, we verified the treatments by determining levels of mycorrhizal colonization on five plants per group (Brundrett et al. 1996). Five mature aphids were collected from cottonwoods in the field and caged onto each ramet where they reproduced parthenogenetically.

After 3 weeks, we found that the aphid population was 6.9 times higher on EM plants than AM plants, while controls had intermediate values (Fig. 12.5). AM plants averaged only slightly more aphids at the conclusion of the experiment (mean = 5.44) than at the beginning, with populations going to extinction on 63 % of the plants ($F_{2,46}=10.783$, $P<0.0001$). These large differences in aphid performance on AM and EM plants demonstrate that these groups of fungi provide different benefits to cottonwoods. Because the same clone was used throughout, differences in aphid performance were not due to differences in plant genotype.

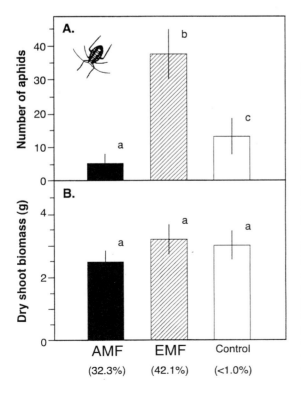

Fig. 12.5. Aphid populations of *Chaitophorus populicola* were greatly affected by the fungal types of their host plants. **A** Aphid abundance after 3 weeks of parthenogenetic reproduction. **B** Dry shoot biomass of cottonwood ramets inoculated with either arbuscular mycorrhizal fungi (*AMF*), ecto-mycorrhizal fungi (*EMF*), or non-inoculated as controls. Data are means + 1 SE and *different letters above bars* denote significant differences at $P<0.05$ (according to a one-way ANOVA). *Values in parentheses* are the mean levels of mycorrhizal colonization of five plants per group that were sacrificed before aphids were added to plants to verify the treatment groups

The differences in aphid performance were also not associated with variation in plant growth. Although AM plants were slightly smaller than EM or control plants, these differences were not statistically significant ($F_{2,40} = 2.153$, $P=0.132$; Fig. 12.5). This suggests that the effects of mycorrhizae on herbivore performance were largely independent of mycorrhizal effects on plant growth and that neither increased host stress or vigor was associated with aphid performance.

Even within AM or EM associations, different mycorrhizal species can vary in their effects on herbivore performance. Goverde et al. (2000) studied the effect of varying species of AM fungus on the performance of the common blue butterfly (*Polyommatus icarus*) feeding on *Lotus corniculatus*. Plants were inoculated with two species of AM fungi, alone and in combination, and the performance of butterfly larvae that were fed leaves from these plants was assessed relative to nonmycorrhizal plants. Larval survival increased (1.6- to 3.8-fold) and larval weights significantly increased when fed mycorrhizal plants relative to controls regardless of the species of fungus. However, larval food consumption, larval food use and adult lipid concentrations varied significantly with the species of mycorrhizal fungus used as inoculum. Adult lipid concentrations were 15% higher when larvae were fed plants inoculated only with AM species #1 than when they were inoculated with a mix of AM species #1 and #2. This difference is potentially important because high lipid concentrations are associated with improved fecundity and longevity in butterflies (Goverde et al. 2000).

Gange and Hamblin (cited in Gange and Bower 1997) and Gange (2001) also observed that species of mycorrhizal fungi interact with one another in complex ways that affect herbivore performance. They found that survival of the root-feeding black vine weevil (*Otiorhynchus sulcatus*) was lower in strawberry plants inoculated with either of two AM fungi, *Glomus mosseae* or *Glomus intraradices,* than in non-inoculated plants. The reductions in larval survival were similar for both species of fungi. However, when strawberry plants were inoculated with both species of fungi together, larval survival was similar to that in controls, suggesting that interactions among the fungi influenced herbivore performance.

These studies suggest that variation in mycorrhizal fungal type and species composition in nature influences plant performance indirectly through effects on herbivore performance. The cottonwood example illustrates that herbivore performance can be dramatically affected even when plant biomass is not altered by mycorrhizal colonization. Mycorrhizae not only alter plant growth, but also the chemical composition of plant tissues, and are thus likely to affect herbivore performance in diverse ways. Considering that tens of AM fungi and hundreds of EM fungi may be associated with a single plant species (Kendrick 1992), it is important to evaluate the effects of this fungal community and its composition on plant-herbivore interactions.

12.5 Ecological and Evolutionary Implications

12.5.1 Herbivore Effects on Mycorrhizae

Given that herbivores and mycorrhizal fungi are prevalent in most terrestrial environments, the fact that herbivores generally cause reductions in mycorrhizal colonization has important implications for both the fungi and their host plants. Gange and Bower (1997) proposed that the lack of an effect of AM on plant performance in the field was due to reductions in mycorrhizal colonization caused by herbivory. Gange and Brown (Chap. 13, this Vol.) show in a field experiment that positive effects of soil fungi (including AM fungi) on plant species richness disappear when soil insects are present (Fig. 13.2). This suggests that soil herbivores might disrupt the functioning of mycorrhizal associations under field conditions. Similarly, in a greenhouse study comparing clipped and non-clipped plants, Borowicz and Fitter (1990) observed that non-clipped plants benefited more from mycorrhizal inoculation than clipped plants. Thus, it is important for future studies to examine if herbivory reduces the effectiveness of mycorrhizae as plant mutualists.

Because all mycorrhizae are not equal mutualists and some may even be neutral or parasitic (Johnson et al. 1997), it is important to determine how herbivory may differentially affect this continuum of interactions. For example, when herbivory is high, photosynthetic tissue losses to herbivores could result in the loss of important mutualistic species and cause a shift towards parasitic species. If chronic grazing by ungulates reduces AM fungal species richness in rangelands around the world as it does in Nevada (Bethlenfalvay and Dakessian 1984), fungal taxa may be lost from vast areas. Commercial agriculture has caused dramatic reductions in AM fungal diversity (Fig. 8.4 in Chap. 8, this Vol.; Helgason et al. 1998), and it is possible that grazing may have similar impacts. In a study of a Swiss grassland community, van der Heijden et al. showed that the survival of certain rarer plant species was promoted by increased fungal species richness. They showed that plant species varied in the growth benefits they received from different taxa of AM fungi, with one plant species receiving a growth benefit only when it occurred in association with one of the four fungal taxa used in the experiment (van der Heijden et al. 1998b). Severe grazing could also lower plant species richness by promoting weedy plant species that do not require mycorrhizal fungi, and by eliminating plant species that require specific fungal taxa for growth and survival.

Although the above studies argue that the landscape level impacts of severe herbivory can negatively affect mycorrhizal and plant species richness, herbivory interacting with genetic variation in the resistance traits of individual plant species may increase fungal richness. For example, at cinder sites in northern Arizona, pinyon pines vary in their susceptibility to two keystone insect herbivores, creating a mosaic of resistant and susceptible trees

(Whitham and Mopper 1985; Cobb and Whitham 1993). The composition of the EM fungal community of trees resistant vs. susceptible to herbivory by these insects differs (Fig. 12.3). As a consequence, the species richness of the community is greater with both tree types than with only resistant or susceptible trees ($n=11$ EM species on scale-resistant trees, 11 EM species on scale-susceptible trees, and a total of 15 EM species when both groups are combined). Although it is unlikely that the species of fungi associated with resistant and susceptible trees are found exclusively on those tree types, the existence of resistant and susceptible trees may allow otherwise rare fungal species to be better represented in the community.

The above examples suggest that the effects of herbivores on mycorrhizal fungal communities can affect plant and fungal biodiversity. These interactions may also affect the abundance and community composition of other organisms. For example, with severe defoliation, the shifts in EM morphotypes from those with extensive hyphal development to those with less extensive hyphal development (Saikkonen et al. 1999) may affect the community of organisms that feed on fungal tissue (i.e., soil microarthropods such as mites and collembola). The community structure of soil microarthropods can be influenced by both the total length of AM fungal hyphae and the types of AM fungi in the soil (Klironomos and Kendrick 1995). Similarly, microarthropods show preferences for certain types of saprophytic fungi (Klironomos et al. 1992) and are likely to show preferences among types of EM fungal hyphae. A great variety of soil-dwelling organisms occur in association with plant roots (Gange and Brown, Chap. 13, this Vol.). For example, an estimated 600 bacterial, actinomycete and fungal taxa occur in the mycorrhizosphere of pinyon pine (Kuske et al., unpubl. data). Shifts in energy allocation belowground are thus likely to impact the soil food web linked to plant roots and their associated mycorrhizal fungi.

12.5.2 Mycorrhizae Effects on Herbivores

Differences among mycorrhizal types or species in their interactions with herbivores have the potential to affect community and ecosystem level processes. For example, Fig. 12.5 showed that in a common garden experiment, cottonwoods with EM supported nearly seven times more aphids than nonmycorrhizal controls. Thus, EM fungi had a positive effect on these aphids, while aphids on cottonwoods with AM fungi had a high probability of becoming extinct. Clearly, these different groups of mycorrhizae have very different effects on this herbivore.

Why might the different effects of mycorrhiza on an aphid be important at community and ecosystem levels? In this cottonwood system, the aphid, *C. populicola* is a keystone species, i.e., a species whose effect on the community is disproportionate to its biomass (Paine 1966; Power et al. 1996). Because it

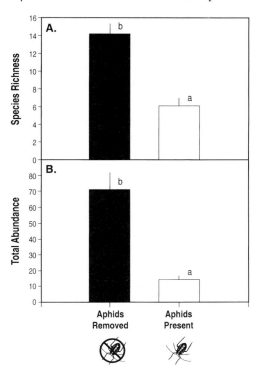

Fig. 12.6. Mycorrhizal type greatly affects the performance of the aphid, *Chaitophorus populicola* (Fig. 12.5), whose presence, in turn, greatly affects the biodiversity of the surrounding arthropod community. **A** Species richness per tree, and **B** arthropod abundance are shown for trees with aphids compared to trees where aphids were experimentally removed. Shown are means + 1 SE, *different letters* indicate significant differences between means. Overall, 90 arthropod species are affected. (Adapted from Wimp and Whitham 2001)

secretes honeydew, these aphids are tended by ants, which fiercely protect the aphid as a valuable energy resource (Wimp and Whitham 2001). In the presence of the ant-aphid mutualism, the associated community of 90 other arthropod species (65 were considered rare) was affected (Fig. 12.6). In the presence of ant-aphid mutualists, species richness decreased by 57% and abundance decreased by 80% relative to trees without ants and aphids. Thus, this ant-aphid interaction acts as a keystone mutualism that affects arthropod biodiversity and community structure.

At the ecosystem level, studies of other aphids have shown that the honeydew produced by aphids caused a reduction in available nitrogen, nitrogen mineralization rates, aboveground net primary production and nitrogen uptake by trees (Grier and Vogt 1990). Llewellyn (1972) calculated that the effects of the honeydew produced by aphids on the carbon cycle of trees could be as great as the effects of beef cattle on grasslands. In combination, these studies argue that mycorrhizal interactions (either positive or negative) with keystone species are likely to have important implications for other trophic levels, which ultimately affect community and ecosystem level processes.

If mycorrhizae-herbivore interactions affect the ecology and distribution of diverse species in other trophic levels, then we must also consider the possibility that over time, they may influence their evolution as well. Gange and

Bower (1997) suggested that generalist insect herbivores respond to plant defensive chemicals, whereas specialist herbivores that feed on a single plant family should be much less affected by changes in plant defense. By combining data on the distributions of generalist vs. specialist insects from 30 plant families in the United Kingdom with data on AM incidence in the same families (percent of species within a family), they found a positive association between specialist insects and the plant species that form mycorrhizal associations. They suggested that mycorrhizal colonization improved plant defenses against generalist herbivores, resulting in a higher proportion of specialists in highly mycorrhizal taxa. Thus, generalist insects should be more negatively affected by mycorrhizal colonization than specialist insects.

In support of this hypothesis, the summary of studies in Table 12.1 shows that generalists were affected by mycorrhizae more often than specialists (87.5 vs. 55.6 %). Also, the effects of mycorrhizae on generalists were negative in 57 % of cases as opposed to only 20 % of specialists. However, the hypothesis of Gange and Bower (1997) seems to assume that all effects of mycorrhizae on herbivore performance are negative, yet in the studies conducted to date, this is true only 42 % of the time.

Gange and Bower (1997) also argued that the nonmycorrhizal habit of some plants may have evolved in response to insect herbivory. They showed that herbivore load (total number of associated insects within a family/total number of plant species in that family) is positively correlated with the incidence of mycorrhizal species in a family. Thus, nonmycorrhizal plant species may have lost the mycorrhizal habit because mycorrhizae did not confer a selective advantage in anti-herbivore defense. Regardless of whether or not these hypotheses are correct, they are important because they examine mycorrhiza-herbivore interactions in a broader context.

12.5.3 Combined Effects

Very few studies have simultaneously examined the effects of herbivores and mycorrhizae on one another to determine if the net effect is symmetrical or asymmetrical. One set of studies showed that the interaction was asymmetrical; the herbivore negatively affected mycorrhizae, but not vice versa. Scale herbivory reduced EM colonization of pinyon roots by approximately 30 % (Del Vecchio et al. 1993; Gehring et al. 1997), while increased mycorrhizal colonization had no corresponding effect on scale performance (Gehring et al. 1997).

In another study, the interaction was more balanced. Arbuscular mycorrhizal colonization reduced the performance of *Arctia caja* larvae on *Plantago lanceloata* (Gange and West 1994), but *A. caja* herbivory resulted in 33 % reductions in mycorrhizal colonization over time in the same plant species (Gange and Bower 1997). In this latter system, timing of fungal colonization and/or herbivory as well as other environmental factors may substantially affect whichever of these interactions has the greatest impact on plant, fungus and herbivore populations. With so few simultaneous studies to evaluate the combined interactions, no generalizations can be made.

However, Gehring et al. (1997) suggested that the more numerous studies of interactions between aboveground and belowground herbivores might provide insights into the less studied field of herbivore-mycorrhiza interactions. Although belowground herbivores and mycorrhizae exhibit very different ecological roles, because both groups are dependent upon a common host and act as substantial belowground carbohydrate sinks, their interactions with aboveground herbivores may be comparable. Interestingly, all documented cases of competition between root-feeding and foliage-feeding insects were asymmetrical, in which the foliage feeders negatively affected the root feeders, while the root feeders either had no effect or a beneficial effect on the performance of foliage feeders (Denno et al. 1995; Masters and Brown 1995). If these interactions are analogous to those between mycorrhizal fungi and herbivores, we predict that aboveground herbivore-mycorrhiza interactions will frequently be asymmetrical, with herbivores having negative effects on mycorrhizae while mycorrhizae would be more likely to have neutral or positive effects on aboveground herbivores. We expect this to be particularly likely in cases where mycorrhizal fungi had little or no effect on plant defensive chemistry. Feedback models as presented by Bever et al. (Chap. 11, this Vol.) could help to unravel feedbacks in fungus-plant-herbivore interactions.

12.6 Suggestions for Future Research

Based on the above patterns of interaction between herbivores and mycorrhizal fungi, we suggest three avenues for future research. First, because species of mycorrhizal fungi are not functionally equivalent, it is important to understand the implications of herbivore-caused changes in the mycorrhizal community, and to quantify how these changes affect plant performance. For example, do the fungal species that remain, following severe herbivory, have low carbon demands as predicted (Saikkonen et al. 1999), and what is the relationship between the cost of a mutualist and its benefits to the host plant? How do the changes in fungal community with herbivory affect host plant performance, i.e., are the fungi that remain, inferior mutualists relative to those that were lost or reduced in abundance? Because herbivory and drought affect mycorrhizal community composition, we predict that the cumulative impacts of climate change and herbivory could substantially alter mycorrhizal communities and have undesirable ecological and economic consequences.

Second, we believe that it is important to simultaneously examine the impacts of mycorrhizae and herbivores on each other's fitness over a wide range of conditions. Although we predicted that aboveground herbivores would generally have greater effects on mycorrhizal fungi than vice versa, is this the case? In addition, how do the observed patterns of mycorrhizal effects on herbivores interact with herbivore effects on mycorrhizae? For example, generalist insects are more likely to be affected by mycorrhizal fungi than specialist insects. Are generalist insects then less likely to have negative effects on mycorrhizal fungi? Or, if mycorrhizal abundance is reduced through the feeding of a specialist insect, does the plant become more susceptible to a generalist herbivore?

Third, it may be especially important to consider how mycorrhizal fungi affect keystone herbivores, which in turn affect a much larger community involving diverse taxa and trophic levels. These community-level impacts are then likely to affect nutrient cycling, carbon flow, productivity and other ecosystem-level process. Similarly, the impact of herbivory on the mycorrhizal community may also have ecosystem level consequences. The magnitude of these interactions has led some authors to suggest that mycorrhizae and herbivores can influence one another's evolution, and preliminary data suggest this is an important avenue of future study.

Acknowledgements. We thank M. van der Heijden and I. Sanders for the invitation to contribute to this volume, K. Haskins for conducting the DECODA analysis and producing Fig. 12.2, and T. Prideaux of the CSIRO Tropical Forest Research Centre library for help locating references. The manuscript was improved by the comments of M. van der Heijden, I. Sanders, K. Haskins, J. Schweitzer, P. McIntyre, J. Bailey, P. Morrow, T. Theimer, J. Ruel, T. Trotter, G. Wimp and two anonymous reviewers. This research was supported by NSF grants DEB-9726648, 9615313, 950372, and USDA grant 95-37302-1801.

References

Allen MF, Richards JH, Busso CA (1989) Influence of clipping and soil water status on vesicular-arbuscular mycorrhiza of two semi-arid tussock grasses. Biol Fertil Soils 8:285–289

Andersen CP, Rygiewicz PT (1991) Stress interactions and mycorrhizal plant response: understanding carbon allocation priorities. Environ Pollut 73:217–244

Bayne HG, Brown MS, Bethlenfalvay GJ (1984) Defoliation effects on mycorrhizal colonization, nitrogen fixation and photosynthesis in the *Glycine-Glomus-Rhizobium* symbiosis. Physiol Plant 62:576–580

Bethlenfalvay GJ, Dakessian S (1984) Grazing effects on mycorrhizal colonization and floristic composition of the vegetation on a semiarid range in northern Nevada. J Range Manage 37:312–316

Bethlenfalvay GJ, Evans RA, Lesperance AL (1985) Mycorrhizal colonization of crested wheatgrass as influenced by grazing. Agron J 77:233–236

Borowicz V (1997) A fungal root symbiont modifies plant resistance to an insect herbivore. Oecologia 112:534–542

Borowicz V, Fitter AH (1990) Effects of endomycorrhizal infection, artificial herbivory, and parental cross on growth of *Lotus corniculatus* L. Oecologia 82:402–407

Bronstein J (1994) Conditional outcomes in mutualistic interactions. Trends Ecol Evol 9:214–217

Brundrett M, Bougher N, Dell B, Grove T, Malajczuk N (1996) Working with mycorrhizas in forestry and agriculture. ACIAR Monogr 32, 374 pp

Chapin FS III (1980) The mineral nutrition of wild plants. Annu Rev Ecol Syst 11:233–260

Choudhury D (1988) Herbivore induced changes in leaf-litter resource quality: a neglected aspect of herbivory in ecosystem nutrient dynamics. Oikos 51:389–393

Clarke KR (1993) Non-parametric multivariate analyses of changes in community structure. Aust J Ecol 18:117–143

Cobb NS, Whitham TG (1993) Herbivore deme formation on individual trees: a test case. Oecologia 94:496–502

Daft MJ, El-Giahmi AA (1978) Effect of arbuscular mycorrhiza on plant growth. VIII. Effect of defoliation and light on selected hosts. New Phytol 80:365–372

Del Vecchio TA, Gehring CA, Cobb NS, Whitham TG (1993) Negative effects of scale insect herbivory on the ectomycorrhiza of juvenile pinyon pine. Ecology 74:2297–2302

Denno RF, McClure MS, Ott JR (1995) Interspecific interactions in phytophagous insects: competition reexamined and resurrected. Annu Rev Entomol 40:297–332

Dungey HS, Potts BM, Whitham TG, Li H-F (2000) A genetic component to community structure. Evolution 54:1938–1946

Findlay S, Carreiro M, Krischik V, Jones CG (1996) Effects of damage to living plants on leaf litter quality. Ecol Appl 6:269–275

Fogel R (1979) Mycorrhiza and nutrient cycling in natural forest ecosystems. New Phytol 86:199–212

Gange AC (2001) Species specific responses of a root- and shoot feeding insect to arbuscular mycorrhizal colonization of its host plant. New Phytologist (in press)

Gange AC, Bower E (1997) Interactions between insects and mycorrhizal fungi. In: Gange AC, Brown VK (eds) Multitrophic interactions in terrestrial systems. Blackwell, Oxford, pp 115–132

Gange AC, Nice HE (1997) Performance of the thistle gall fly, *Urophora carduii*, in relation to host plant nitrogen and mycorrhizal colonization. New Phytol 137:335–343

Gange AC, West HM (1994) Interactions between arbuscular mycorrhizal fungi and foliar-feeding insects in *Plantago lanceolata* L. New Phytol 128:79–87

Gange AC, Brown VK, Sinclair GS (1994) Reduction of black vine weevil larval growth by vesicular-arbuscular mycorrhizal infection. Entomol Exp Appl 70:115–119

Gange AC, Bower E, Brown VK (1999) Positive effects of an arbuscular mycorrhizal fungus on aphid life history traits. Oecologia 120:123–131

Gardes M, Bruns TD (1996) ITS primers with enhanced specificity for basidiomycetes: application to the identification of mycorrhiza and rusts. Mol Ecol 2:113–118

Gehring CA, Whitham TG (1991) Herbivore-driven mycorrhizal mutualism in insect susceptible pinyon pine. Nature 353:556–557

Gehring CA, Whitham TG (1994a) Interactions between aboveground herbivores and the mycorrhizal mutualists of plants. Trends Ecol Evol 9:251–255

Gehring CA, Whitham TG (1994b) Comparisons of ectomycorrhiza on pinyon pines (*Pinus edulis*; Pinaceae) across extremes of soil type and herbivory. Am J Bot 81:1509–1516

Gehring CA, Whitham TG (1995) Duration of herbivore removal and environmental stress affect the ectomycorrhiza of pinyon pines. Ecology 76:2118–2123

Gehring CA, Cobb NS, Whitham TG (1997) Three-way interactions among ecto-mycorrhizal mutualists, scale insects, and resistant and susceptible pinyons. Am Nat 149:824–841

Gehring CA, Theimer TC, Whitham TG, Keim P (1998) Ecto-mycorrhizal fungal community structure of pinyon pines growing in two environmental extremes. Ecology 79:1562–1572

Goverde M, van der Heijden MGA, Wiemken A, Sanders IR, Erhardt A (2000) Arbuscular mycorrhizal fungi influence life history traits of a lepidopteran butterfly. Oecologia 125:362–369

Grier CC, Vogt DJ (1990) Effects of aphid honeydew on soil nitrogen availability and net primary production in an *Alnus rubra* plantation in western Washington. Oikos 57:114–118

Grime JP, Mackey JML, Hillier SH, Read DJ (1987) Floristic diversity in a model system using experimental microcosms. Nature 328:420–422

Helgason T, Daniell TJ, Husband R, Fitter AH, Young JPW (1998) Ploughing up the wood-wide web? Nature 394:431

Holland JN, Weixin C, Crossley DA Jr (1996) Herbivore-induced changes in plant carbon allocation: assessment of below-ground C fluxes using carbon-14. Oecologia 107:87–94

Huntly N (1991) Herbivores and the dynamics of communities and ecosystems. Annu Rev Ecol Syst 22:477–503

Johnson NC, Graham JH, Smith FA (1997) Functioning of mycorrhizal associations along the parasitism-mutualism continuum. New Phytol 135:575–585

Jones CG, Last FT (1991) Ectomycorrhiza and trees: implications for aboveground herbivory. In: Barbosa P, Krischik VA, Jones CG (eds) Microbial mediation of plant-herbivore interactions. Wiley, New York, pp 65–103

Kendrick B (1992) The fifth kingdom. Mycologue Publications, Waterloo

Klironomos JN, Kendrick B (1995) Relationships among microarthropods, fungi and their environment. Plant Soil 170:183–197

Klironomos JN, Widden P, Deslandes I (1992) Feeding preferences of the collembolan *Folsomia candida* in relation to microfungal successions on decaying litter. Soil Biol Biochem 24:685–692

Kolb TE, Dodds KA, Clancy KM (1999) Effect of western spruce budworm defoliation on the physiology and growth of potted Douglas-fir seedlings. For Sci 45:280–291

Kruskal JB (1964) Multidimensional scaling by optimizing goodness of fit to a nonmetric hypothesis. Psychometrika 29:1–27

Llewellyn M (1972) The effects of the lime aphid, *Eucallipterus tiliae* L. (Aphididae) on the growth of the lime *Tilia x vulgaris* Hayne. J Appl Ecol 9:261–282

MacFall JS, Slack SA, Iyer J (1991) Effects of *Hebeloma arenosa* on and phosphorous fertility on growth of red pine (*Pinus resinosa*) seedlings. Can J Bot 69:372–379

Manninen A-M (1999) Susceptibility of Scots pine seedlings to specialist and generalist insect herbivores – importance of plant defense and mycorrhizal status. Natural and Environmental Sciences 100. PhD, Univ Kuopio, Kuopio Univ Publ C

Manninen A-M, Holopainen T, Holopainen JK (1998) Susceptibility of ecto-mycorrhizal and nonmycorrhizal Scots pine (*Pinus sylvestris*) seedlings to a generalist insect herbivore, *Lygus rugulipennis*, at two nitrogen availability levels. New Phytol 140:55–63

Manninen A-M, Holopainen T, Holopainen JK (1999a) Performance of grey pine aphid, *Schizolachnus pineti*, on ecto-mycorrhizal and non-mycorrhizal Scots pine seedlings at different levels of nitrogen availability. Entomol Exp Appl 9:117–120

Manninen A-M, Holopainen T, Lyytikäinen-Saarenmaa P, Holopainen JK (1999b) The role of low level ozone exposure and mycorrhizas in chemical quality and insect herbivore performance on Scots pine seedlings. Global Change Biol 5:1–11

Markkola AM (1996) Effect of artificial defoliation on biomass allocation in ecto-mycorrhizal *Pinus sylvestris* seedlings. Can J For Res 26:899–904
Marx DH, Hatch AB, Mendicino JF (1977) High soil fertility decreases sucrose content and susceptibility of loblolly pine roots to ecto-mycorrhizal infection by *Pisolithus tinctorius*. Can J Bot 55:1569–1574
Masters GJ, Brown VK (1995) Host-plant mediated interactions between spatially separated herbivores: effects on community structure. In: Gange AC, Brown VK (eds) Multitrophic interactions in terrestrial system. Blackwell, Oxford, pp 217–237
Miller SL, Allen EB (1992) Mycorrhiza, nutrient translocation, and interactions between plants. In: Allen MF (ed) Mycorrhizal functioning: and integrative plant-fungal process. Routledge/Chapman and Hall, New York, pp 301–332
Minchin PR (1987a) An evaluation of the relative robustness of techniques for ecological ordination. Vegetatio 69:89–107
Minchin PR (1987b) An evaluation of the relative robustness of techniques for ecological ordination. Vegetatio 71:145–456
Naiman RJ, Melillo JM, Hobbie JE (1986) Ecosystem alteration of boreal forest streams by beaver (*Castor canadensis*). Ecology 67:1254–1269
Pacovsky RS, Rabin LB, Montllor CB, Waiss AC Jr (1985) Host-plant resistance to insect pests altered by *Glomus fasciculatum* colonization. In: Molina R (ed) Proceedings of the 6th North American Conference on Mycorrhiza. Oregon State University, Corvallis, p 288
Paine RT (1966) Food web complexity and species diversity. Am Nat 100:65–75
Pastor J, Naiman RJ (1992) Selective foraging and ecosystem processes in boreal forests. Am Nat 139:690–705
Pastor J, Naiman RJ, Dewey B, McIness P (1988) Moose, microbes and the boreal forest. Bioscience 38:770–777
Perry DA, Margolis H, Choquette C, Molina R, Trappe JM (1989) Ecto-mycorrhizal mediation of competition between coniferous tree species. New Phytol 112:501–511
Power ME, Tilman D, Estes JA, Menge BA, Bond WJ, Mills LS, Daily G, Castilla JC, Lubchenco J, Paine RT (1996) Challenges in the quest for keystones. Bioscience 46:609–620
Price PW (1991) The plant vigor hypothesis and herbivore attack. Oikos 62:244–251
Rabin LB, Pacovsky RS (1985) Reduced larva growth of two lepidoptera (Noctuidae) on excised leaves of soybean infected with a mycorrhizal fungus. J Econ Entomol 78:1358–1363
Richards JH (1984) Root growth response to defoliation in two Agropyron bunchgrasses: field observations with an improved root periscope. Oecologia 64:21–25
Ritchie ME, Tilman D, Knops JMH (1998) Herbivore effects on plant and nitrogen dynamics in oak savanna. Ecology 79:165–177
Rossow LJ, Bryant JP, Kielland K (1997) Effects of above-ground browsing by mammals on mycorrhizal infection in an early successional taiga ecosystem. Oecologia 110:94–98
Saikkonen K, Ahonen-Jonnarth U, Markkola AM, Helander M, Tuomi J, Roitto M, Ranta H (1999) Defoliation and mycorrhizal symbiosis: a functional balance between carbon sources and below-ground sinks. Ecol Lett 2:19–26
Smith SE, Read DJ (1997) Mycorrhizal symbiosis, 2nd edn. Academic Press, London
Trent JD, Wallace LL, Svejcar TJ, Christiansen S (1988) Effect of grazing on growth, carbohydrate pools, and mycorrhiza in winter wheat. Can J Plant Sci 68:115–120
van der Meijden E (1996) Plant defence, an evolutionary dilemma: contrasting effects of (specialist and generalist) herbivores and natural enemies. Entomol Exp Appl 80:307–310

van der Heijden MGA, Boller T, Wiemken A, Sanders IR (1998a) Different arbuscular mycorrhizal fungal species are potential determinants of plant community structure. Ecology 79:2082–2091

van der Heijden MGA, Klironomos JN, Ursic M, Moutoglis P, Streitwolf-Engel R, Boller T, Wiemken A, Sanders IR (1998b) Mycorrhizal fungal diversity determines plant biodiversity, ecosystem variability and productivity. Nature 396:69–72

Wall DH, Moore JC (1999) Interactions underground: Soil biodiversity, mutualism and ecosystem processes. Bioscience 49:109–117

Wallace LL (1981) Growth, morphology and gas exchange of mycorrhizal and nonmycorrhizal *Panicum coloratum* L., a C_4 grass species, under different clipping and fertilization regimes. Oecologia 49:272–278

Waring GW, Cobb NS (1992) The impact of plant stress on herbivore population dynamics. In: Bernays E (ed) Insect-plant interactions, vol 4. CRC, Boca Raton, pp 168–226

White TCR (1984) The abundance of invertebrate herbivores in relation to the availability of nitrogen in stressed food plants. Oecologia 63:90–105

Whitham TG, Mopper S (1985) Chronic herbivory: impacts on architecture and sex expression of pinyon pine. Science 228:1089–1091

Wilson JB (1988) A review of evidence on the control of shoot:root ratio, in relation to models. Ann Bot 61:433–449

Wimp GM, Whitham TG (2001) Biodiversity consequences of predation and host plant hybridization on an aphid-ant mutualism. Ecology 82:440–452

13 Actions and Interactions of Soil Invertebrates and Arbuscular Mycorrhizal Fungi in Affecting the Structure of Plant Communities

ALAN C. GANGE and VALERIE K. BROWN

Contents

13.1	Summary	321
13.2	Introduction	322
13.3	Soil Invertebrate Groups	323
13.3.1	Earthworms	324
13.3.2	Nematodes	325
13.3.3	Mites	327
13.3.4	Insects	328
13.3.5	Other Invertebrates	333
13.4	Field Studies	333
13.5	Conclusions	340
References		341

13.1 Summary

Soil invertebrates and arbuscular mycorrhizal (AM) fungi co-occur in all ecosystems of the world. However, the interactions between these organisms are rarely studied. We present some reasons for this lack of interest and suggest that this is an area of ecology in which much useful research could be done. There are four main groups of subterranean invertebrates that can interact with AM fungi; these are the earthworms, nematodes, mites and the insects. Earthworms are likely to be beneficial to AM fungi, they aid in the dispersal of spores and hyphal fragments in soil. However, their burrowing activities may disrupt the fragile fungal mycelium. The literature regarding plant parasitic nematodes is confusing, some studies show antagonistic effects between the organisms, while others show null or positive effects. Virtually nothing is known about the effects of mycophagous nematodes on AM functioning in natural communities. Contrary to popular opinion, we suggest that the Collembola are beneficial to AM fungi, through feeding on competing fungi in the rhizosphere. The interactions between mites and AM fungi in field situations have never been studied. The meagre laboratory evidence suggests they may be selective feeders and act in a similar way to Collembola. Interactions between insects and AM fungi are variable. In a recent study we show that the effect of AM fungi on rhizophagous insects appears to be one in which single fungi reduce insect performance while colonization by more than one species has no effect. The results of a large field experiment are presented which show that the insects and fungi act and interact in determining plant community structure. A mechanism for these interactions is given and a conceptual model involving seedling establishment proposed. Much further work is required to test this model and to extend it to include the other groups of subterranean invertebrates that interact with AM fungi.

13.2 Introduction

In any ecosystem, the soil fauna is composed of a diverse array of invertebrate groups. Many of these groups have the potential to interact with arbuscular mycorrhizal fungi (AM) fungi, though few of these interactions have actually been studied. There are several reasons for this lack of work. The first is the old cliché that many ecologists regard the soil ecosystem as a 'black box' in which it is best not to peer too closely. There is a lot of truth in this statement, for example in a review of subterranean insect herbivores, Brown and Gange (1990) found that only 2% of the insect-plant literature for the previous 20 years contained studies devoted to below-ground insects. The situation has not changed greatly in the last 10 years.

A second reason is the difficulty of working with systems in which the majority of the study organisms are invisible. It is difficult to monitor the progress of a laboratory experiment with rhizophagous insects or AM fungi without disrupting the experimental units too much. With both organisms in the system, the problems are compounded and the chances of experimental failure increase dramatically.

The life histories of the organisms can result in an avoidance of their use in experiments. Many subterranean insects have long life cycles, often remaining in the soil as a larva for several years (Brown and Gange 1990). Maintaining a laboratory culture of these organisms is difficult and new culture methods often have to be developed before an experiment is attempted (Carter 1975; Gange and Brown 1991a).

Finally, we believe that traditional methods of teaching, in which students graduate from universities specialising in entomology or mycology have led to a lack of collaborative studies. This situation is certainly changing as interdisciplinary research must be the way of the future. The critical fact is that these problems are not insurmountable. The study of the interactions between soil invertebrates and AM fungi is possible in the laboratory and field, and is exciting and rewarding. One major theme, which we hope will emerge from this chapter, is that there are many interesting opportunities for research in this area. We urge other ecologists to join those already in the 'black box'; it is not as dark in there as one might expect!

13.3 Soil Invertebrate Groups

The soil invertebrate groups that have received most attention in their interactions with AM fungi are the earthworms, nematodes and insects (Gehring and Whitham, Chap. 12, this Vol.; Fitter and Sanders 1992), but see also Brussaard et al. (1997) for other taxa, including mites and protozoa. Five types of interaction may be observed with these groups (Table 13.1). Invertebrates may ingest spores or hyphal fragments and by their movement transport these through the soil. Depending on whether the spores and hyphae are killed by passage through the gut, this interaction may have a negative effect on the fungi or it may be positive, leading to their dispersal. The nematodes and insects contain rhizophagous species and, by their feeding action, the functioning of the AM symbiosis in the root may be disrupted. Meanwhile, the presence of AM fungal colonization in roots may lead to chemical changes that may affect the feeding and growth of the rhizophage (Chap. 12, this Vol.). Mycophagous invertebrates, such as collembola or mites, may feed on non-mycorrhizal fungi in the rhizosphere and thus allow enhanced root colonization by the mycorrhiza. However, it is possible that these organisms may also feed on the mycorrhizal hyphae, thus disrupting the symbiosis. Many species of earthworm and subterranean insect are relatively large and are active burrowers in the soil. Their movements may physically disrupt the mycorrhizal mycelium, leading to breakages in hyphal links between this mycelium and the portion of the AM fungus present within the root.

Table 13.1. Possible interactions between subterranean invertebrates and arbuscular mycorrhizal fungi and the consequences for their host plants

Activity	Invertebrate group	Outcome for the plant	Examples
Spore ingestion and dispersal	Earthworms	Positive	Rabatin and Stinner (1989); Gange (1993); Harinikumar and Bagyraj (1994); Lee et al. (1996)
	Insects	Positive or negative	Gange (1994); Harinikumar and Bagyraj (1994)
	Isopods	Positive	Rabatin and Stinner (1988, 1991)
	Diplopods	Positive	Rabatin and Stinner (1988, 1991)
Effect of mycorrhiza on root herbivory	Nematodes	Positive	Roncadori (1997); Jaizme Vega et al. (1997); Siddiqui and Mahmood (1998); Habte et al. (1999)
	Insects	Positive or no effect	Gange et al. (1994); Gange (1996, 2001)
Feeding on competing rhizosphere micro organisms	Collembola	Positive	Klironomos et al. (1999); Gange (2000)
	Mites	Positive	Klironomos and Kendrick (1995, 1996)
Feeding on the mycorrhizal mycelium	Nematodes	Negative	Salawu and Esty (1979); Smith et al. (2000)
	Collembola	Negative?	Warnock et al. (1982); Klironomos and Kendrick (1995); Gange (2000)
	Mites	Negative?	Unknown
Physical breakage of the mycelium	Earthworms	Negative?	Pattinson et al. (1997)
	Insects	Negative?	Unknown

13.3.1 Earthworms

Earthworms are exceptionally important in maintaining soil structure and fertility (e.g. Wolters and Stickan 1991; Scheu and Parkinson 1994), and a number of studies have also shown them to be vectors of AM fungi. Earthworms will feed on live or newly senesced roots (Piearce 1978), but generally ingest decaying organic matter. By feeding on roots or other matter in the rhizosphere, they can ingest spores or hyphal fragments. Studies have shown that spores can pass unharmed through earthworm guts (Rabatin and Stinner 1989; Reddell and Spain 1991; Harinikumar and Bagyraj 1994). Indeed, earthworms appear to concentrate spore numbers, so that their faecal material (surface casts) contains higher spore numbers than does surrounding field soil (Reddell and Spain 1991; Lee et al. 1996). Gange (1993) recorded the total infective propagule numbers in worm cast material and surrounding field soil in a range of differently aged plant communities. It was found that casts and soil contained very low numbers of propagules in a recently disturbed community and there was no difference in propagule number between the two types of material. However, propagule number increased with successional age of the community and, in communities of 5 years and older, casts contained significantly higher numbers of propagules. Cast material also successfully initiated mycorrhizal colonization in *Plantago lanceolata*. This study therefore suggests that earthworms are potentially important vectors of AM fungi, but this depends on the successional status of the plant community studied.

Earthworms can be divided into three ecological groups: epigeic species which live on the surface and which do not burrow, anecic species which make burrows which open to the surface and endogeic species which rarely if ever come to the surface and which burrow within the soil profile. Individual burrows can be extensive, for example the common species *Lumbricus terrestris* can make burrows up to 3 m in length and the total burrow length beneath 1 ha of English pasture grassland, to a depth of 1 m, may be as high as 888 m (Edwards and Lofty 1977). Anecic and endogeic worms therefore have the potential to physically disrupt the mycorrhizal mycelium by their burrowing actions. The study of Pattinson et al. (1997) is interesting because it is the only one to address this problem. Here, it was found that earthworms appeared to have no significant disruption effect on an established mycorrhizal mycelium. However, this was measured by recording the AM colonization of 'bait plants' as a mycorrhizal measurement and unfortunately the amount of extra radical hyphal biomass was not measured. There is still a need for a well-designed experiment that addresses the effect of earthworm presence on the structure of the mycorrhizal mycelium. We suggest this is an important area for future research.

13.3.2 Nematodes

13.3.2.1 Plant Parasitic Nematodes

Nematodes that affect or are affected by AM fungi may be split into two groups – those which attack roots and those which feed on the fungi themselves. A comprehensive review of interactions between soil-dwelling plant parasitic nematodes and AM fungi is that of Roncadori (1997). A feature of this review is the inconsistency of the effects seen in various studies. Colonization of plants by AM fungi may reduce (the most common scenario), have no effect or even stimulate nematode abundance. Likewise, nematode effects on colonization may be negative, null or positive. One reason for these inconsistent responses is likely to be the preliminary nature of the experiments that have, of necessity, been performed in the past. More recent experiments have investigated the confounding role of soil P, inoculation time of both fungus and nematode, and mycorrhizal and plant species identity. These experiments suggest that mycorrhizal presence reduces the incidence of attack by root-feeding nematodes (e.g. Jaizme Vega et al. 1997; Siddiqui and Mahmood 1998). In general, mycorrhizal protection against nematode attack is linked to improved P status of the plant (Roncadori 1997). However, of equal importance is the sequence of colonization events of the plant by fungus and nematode. Thus, plants inoculated with AM fungi well before that of nematode introduction show increased tolerance to attack. However, if fungi and nematodes are inoculated together, there is little evidence of protection by the mycorrhiza (Smith et al. 1986). Such differences could be important in natural communities such as grasslands, where populations of soil nematodes may be very high (Curry 1994; Popovici and Ciobanu 2000). For an establishing grass seedling in temperate grasslands, the outcome of the interaction may depend on whether AM fungi or nematodes colonize the root system first. Such colonization sequences are unknown in natural communities. We regard this as an important avenue for future research, given the fact that plant-parasitic nematodes are likely to be extremely important in structuring natural plant communities (cf. Blomqvist et al. 2000).

Host-specific responses of plants to AM fungi are also important in affecting plant community structure (van der Heijden et al. 1998a,b) and it is becoming clear that the response of nematodes to mycorrhizal colonization also depends on the identity of the fungi in the root system. For example, Habte et al. (1999) have shown that three different AM fungi produced very different degrees of tolerance of white clover (*Trifolium repens*) to the root-knot nematode *Meloidogyne incognita*. In any plant community, researchers must first identify the co-occurring nematode and fungal species before meaningful experiments can be performed. An excellent example of this approach is provided by Little and Maun (1996) in which it was shown that the naturally occurring mycorrhizal associates of *Ammophila breviligulata* were able to protect the roots of this plant against attack by nematodes. The performance of the plant in field situations is thus determined, to an extent, by the occurrence of AM fungi and rhizophagous nematodes.

13.3.2.2 Mycophagous Nematodes

Many species of nematodes are mycophagous, but the consequences of mycorrhizal feeding by this group for the structure of plant communities are

uncertain. As with plant-parasitic nematodes, early work with this group and AM fungi produced conflicting results (Salawu and Esty 1979; Hussey and Roncadori 1981; Rabatin and Stinner 1991). These experiments generally used the nematode *Aphelenchus avenae* and, at very high densities of nematodes per plant, a reduction in mycorrhizal functioning has been observed in soybean (Salawu and Esty 1979) and cotton (Hussey and Roncadori 1981). However, at densities more realistic of those found in field situations, no effect on the mycorrhiza was found. Similarly, Giannakis and Sanders (1990) failed to record any effects of feeding by *A. composticola* on the mycorrhiza formed between *Glomus clarum* and *Trifolium pratense*. No recent experiments appear to have clarified the situation. However, a study by Smith et al. (2000) suggests that this too is an important gap in our understanding of plant community ecology. In their study, the fungicide benomyl was applied to tallgrass prairie over a seven-year period. This resulted in a reduction of 80 % in root colonization by AM fungi. However, there was a concomitant reduction in fungal-feeding nematode abundance of 12 %. Assuming that the fungicide had no direct effects on the nematodes (not tested in this study), this suggests that mycorrhizal feeding by nematodes was an important component of the below-ground food web in this system. Furthermore, the fact that these workers could find no effect of benomyl on total fungal biomass suggests that the reduction in AM abundance led directly to the reduction in nematode numbers. Additional evidence for mycophagous nematodes disrupting the functioning of the AM symbiosis comes from the experiments of Ingham et al. (1986) and Ingham (1988). In these studies, applications of nematicides to grassland resulted in large increases in mycorrhizal colonization of plant roots, suggesting that mycophagous nematodes were interfering with the symbiosis. There is an urgent need for well-designed field experiments that address the interactions of different nematode groups with AM fungi in a variety of plant communities. These experiments must be performed with naturally occurring AM associates of plants and nematodes. This is because it is highly likely that nematodes will show preferences for certain fungal species in their diets and using 'off the shelf' cultures of either could be meaningless. Ruess et al. (2000) show that one mycophagous nematode (*Aphelenchoides* spp.) has distinct preferences for certain species of ecto-mycorrhizal fungi, but that these preferences change over time. At this stage, it is impossible to speculate on how important mycophagous nematodes are in structuring plant communities, because we do not know the relative contributions of mycorrhizal and non-mycorrhizal fungi to their diets.

13.3.3 Mites

Mites are often extremely abundant in field soils (Petersen and Luxton 1982) and many of these are mycophagous. The interactions between mites and AM fungi have rarely been studied and the consequences of these interactions (if any) for plant communities are unknown. The possibility exists that by feeding on the mycorrhizal mycelium, mites may disrupt the functioning of the symbiosis, in a similar manner to that suggested for nematodes (above). However, only one study has examined the feeding preferences of mites when offered AM and non-AM fungi (Klironomos and Kendrick 1996). In this work, it was found that three species of mites consistently preferred to feed on conidial fungi rather than the mycorrhizal species, *Glomus macrocarpum*. Furthermore, the mites did not even show a preference for the mycorrhiza over sterile root fragments. In the same study, when offered no choice, some mite feeding was observed on mycorrhizal hyphae, but this was generally on the thinner structures, less than 5 µm in diameter. The limited evidence therefore suggests that mites will feed on mycorrhizal hyphae, but not through choice. There is one study that has investigated the effects of mites in combination with Collembola on AM functioning (Klironomos and Kendrick 1995) and this is discussed below, in the section on Collembola. Currently, all we can say is that mites have no known negative effects on the AM symbiosis, but they may well have positive interactions (see below).

13.3.4 Insects

13.3.4.1 Collembola

Collembola (springtails) are primitive insects and are abundant in virtually all soils and in the litter layer on soil surfaces. Most of the subterranean species feed (at least in part) on fungi (Hopkin 1997), and it has been suggested that they are important regulators of the symbiosis (Fitter and Sanders 1992). Gange (2000) provides a synthesis of the information on Collembola-mycorrhizal interactions and concluded that negative effects of the insects on the symbiosis are more likely to be the result of poor experimental design, rather than biological fact. There are some experiments that have shown that Collembola will feed on AM fungi in laboratory conditions (Moore et al. 1985; Thimm and Larink 1995). Early experiments appeared to show that feeding by Collembola could seriously impair the functioning of AM symbiosis. Warnock et al. (1982) grew seedling leek (*Allium porrum*) plants with and without the AM fungus *Glomus fasciculatum* and the Collembolan *Folsomia candida* in a factorial design. The fungus alone caused an increase in plant growth. However, Collembola presence appeared to reduce mycorrhizal functioning, as plants with the fungus and Collembolan grew no better than non-mycorrhizal plants. As fragments of fungal material were found in the guts of the insects, the conclusion drawn was that feeding by Collembola on external hyphae reduced the efficacy of the mycorrhiza. This experiment was extended by Finlay (1985) and, in a range of trials with *A. porrum* and *Trifolium pratense* (red clover), it was found that mycorrhizal colonization increased plant biomass when Collembola were absent but not so when they were present.

However, the laboratory feeding trials and the experiments of Warnock et al. (1982) and Finlay (1985) are open to criticism because in all cases the only fungus type presented to the Collembola was arbuscular mycorrhizal. When realistic (i.e. involving AM and non-AM fungi) fungal preference trials have been performed (Klironomos and Kendrick 1996; Klironomos and Ursic 1998; Klironomos et al. 1999), it has been shown that Collembola consistently graze on other soil fungi, in preference to AM fungal species. Therefore, as with mites discussed above, Collembola will graze on AM fungi, but not through choice. The pot trials appear to show an effect of Collembola on the mycorrhiza because in these cases the insects had little choice but to feed on the fungus. At the reasonably high densities used, a detrimental effect on the fungus was seen, but this is a situation that may never occur in the field, because there is no known plant community in which the only soil fungi present are arbuscular mycorrhizal. We believe it is worth emphasizing the point that future experiments involving the interactions of mites and AM fungi should not fall into the same trap as the early Collembolan-mycorrhizal ones did.

An elegant, realistic laboratory experiment (Klironomos and Kendrick 1995) examined the effects of feeding by a variety of microarthropods on AM fungi and growth of sugar maple (*Acer saccharum*). In this case, addition of three mite and three Collembolan species together had no effect on maple growth. However, addition of the microarthropods and some decaying maple leaf litter resulted in a 59% increase in arbuscular colonization. The explanation was that the Collembola and mites preferred to feed on the non-AM fungi associated with the litter in the experiment, and that providing them with an alternative food source allowed mycorrhizal growth. The mechanism is unknown, but it is possible that these microarthropods feed on fungi that might compete with the mycorrhiza for root space. In this way, Collembola and mites may have a positive effect on AM colonization, rather than a negative one. Another possibility is that when microarthropods feed on these other fungi, there is an enhanced uptake of mineral nutrients by the mycorrhiza as these are released from faeces (McGonigle 1995). With the increased root colonization by the mycorrhiza through lack of competition, there is then a consequent benefit to plant growth.

One field experiment (Lussenhop 1996) has also shown a positive effect of Collembola on AM fungi. Here, increasing Collembola numbers by 26% in field soil around soybean plants resulted in an increase in colonization of the roots. The most likely explanation was that the insects fed on fungi that may have been antagonistic to, or competitive with, the AM fungal species in the rhizosphere. No effects on plant growth were seen in this experiment, however. Therefore, the idea that Collembola have the potential to restrict mycorrhizal functioning in the field is attractive, but there is little good evidence for it. Instead, the intriguing possibility exists that the opposite scenario is true; Collembola may have the potential to benefit AM fungi (Gange 2000). This is because, given the choice, these insects prefer to feed on non-mycorrhizal species, thereby allowing nutrient release or release from competition for the mycorrhizal species. As Collembola (and indeed mites) can occur at extraordinarily high densities in some field soils (Petersen and Luxton 1982), the potential for their interactions with AM fungi in plant communities is great. We see this also as an important area of research that needs clarifying.

13.3.4.2 Rhizophagous Insects

A comprehensive review of the interactions between foliar- and root-feeding insects and mycorrhizal fungi is presented by Gehring and Whitham (Chap. 12, this Vol.). Insects can suck, mine, gall or chew roots and while the densities of subterranean species may be low compared to their foliar-feeding counterparts (Brown and Gange 1992, 1999), their effects on plants can be devastating. Root herbivory can seriously impair plant water and mineral nutrient uptake, affect stability and storage reserves and lead to dramatic reductions in above-ground biomass (Brown and Gange 1990). There are many examples of subterranean herbivory significantly reducing root biomass, but only a few studies have examined the effect of root herbivory on mycorrhizal colonization of a plant (Gange 2001). In this work, feeding by larvae of black vine weevil (*Otiorhynchus sulcatus*) significantly reduced the root biomass of strawberry (*Fragaria x ananassa*) but this had no effect on the extent of arbuscular colonization. This may be because the mycorrhiza colonized the roots before they were subjected to attack by the larvae. It is likely that, as with phytophagous nematodes (above), the outcome of this interaction will depend on the sequence of events of AM colonization and insect attack. Nevertheless, it is likely that most rhizophagous insects are likely to encounter roots already colonized by a mycorrhiza and it is surprising that very few experiments have been performed which ask what effect the presence of the mycorrhiza has on the performance of the insect.

Gange et al. (1994) showed that colonization of *Taraxacum officinale* plants by *Glomus mosseae* resulted in a halving of the growth rate and survival of black vine weevil (*Otiorhynchus sulcatus*) larvae. Concomitant effects were seen in above-ground plant biomass, so that a density of 150 larvae m^{-2} had no effect on biomass in mycorrhizal plants but significantly reduced it in non-mycorrhizal plants. The experiment was repeated by Gange (1996), using strawberry (*Fragaria x ananassa*) as a host plant, but in this case two AM fungi (*G. mosseae* and *G. intraradices*) were inoculated in a factorial design. Curiously, while each fungal species on its own reduced black vine weevil lar-

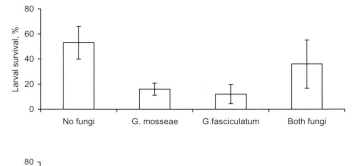

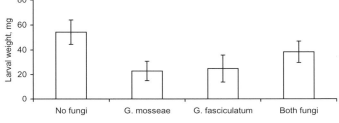

Fig. 13.1. Survival of black vine weevil larvae when reared on roots of strawberry inoculated with either *Glomus mosseae*, *G. fasciculatum*, both or neither fungi. Bars represent means ±1 SE; for statistical details see text. (Redrawn from Gange 2001)

val performance, the addition of both fungi had no effect. Plant biomass was reduced in the 'fungi + larvae' treatment to a similar degree to that in the 'no fungi + larvae' treatment. This unexpected result has been repeated in a recent experiment (Gange 2001), again using black vine weevil and strawberry, but in this case using *G. mosseae* and *G. fasciculatum*. These two fungi were found to be the dominant members of the mycorrhizal flora in a commercial strawberry field and to co-occur in 91% of spore samples from the field. Strawberry plants were grown from runners in sterilised soil and an inoculum of either, or both, fungi added to 20 pots in each treatment. Control pots received sterilised inoculum. Half of the 20 pots in each treatment were inoculated with ten eggs of black vine weevil and larvae left to grow for 10 weeks in a constant environment of 20 °C. At the end of this period, colonization of plants (measured by % root length colonized by arbuscules) was 9.9% in *G. mosseae* pots, 14.9% in *G. fasciculatum* pots and 11.3% in dual-inoculated pots. A summary of the data is presented in Fig. 13.1, in which the effect of fungal colonization on larval survival and size is presented. Colonization by either fungus reduced larval survival by about 70% (*G. mosseae*: $F_{1,36}=4.78$, $P<0.05$; *G. fasciculatum*: $F_{1,36}=6.01$, $P<0.05$) while there was only a 30% reduction when both fungi were inoculated, resulting in a significant interaction term between the fungi ($F_{1,36}=76.48$, $P<0.001$) (Fig. 13.1). Similarly, inoculation with either fungal species reduced larval weight by about 55% (*G. mosseae*: $F_{1,36}=11.29$, $P<0.01$; *G. fasciculatum*: $F_{1,36}=4.77$, $P<0.05$), but the dual inoculation treatment only resulted in a 20 % reduction, with another significant interaction term ($F_{1,36}$=61.65, P<0.001) (Fig. 13.1). Concomitant effects were seen in plant biomass, with this being increased by either fungal species, although plants inoculated with both fungi were no larger than those with either single inoculation. Furthermore, AM effects on larval performance were also manifest in plant biomass, with significant reductions in biomass being caused by weevil feeding in the control and dual fungal treatment, but not in either single inoculation.

The mechanism by which AM fungi affect root-feeding insects is unknown. It has been shown that AM colonization results in chemical changes in roots, and it is the alteration of root biochemistry that is the most likely explanation for the detrimental effects on the larvae. Many of the chemicals present in mycorrhizal roots may have activity against insect herbivores, for example phenolics (Morandi 1996), terpenoids (Peipp et al. 1997) and isoflavonoids (Vierheilig et al. 1998). All of these compounds have been shown to exhibit activity against phytophagous insects (Mullin et al. 1991; Dakora 1995). Why dual inoculation of fungi results in different effects on insects compared to single inoculations is even less clear. One possible reason is that the different mycorrhizal fungi compete for space in the root system and, in so doing, any fungal-induced chemical changes are disrupted. Whatever the mechanism is for induction or production of chemicals as a result of mycorrhizal colonization, such competition results in a lowering of chemical activity. These data do suggest that insect herbivores may be useful bioassay tools for studies involving chemical changes in roots as a result of mycorrhizal colonization and that future insect-mycorrhizal studies need to include a consideration of plant biochemistry.

The laboratory data on interactions between AM fungi and root-feeding insects are therefore limited. However, it does suggest that the responses seen in an insect and the consequent effects on plant growth are determined by the combination of fungal species present. As molecular methods become routine, exciting possibilities will open up for the study of these interactions (see Clapp et al., Chap. 8, this Vol.). For example, in the above studies, the extent of colonization by each fungal species in the root systems of dual-inoculated plants is unknown. As roots of naturally occurring plants may be colonized by a number of different fungal species (Clapp et al. 1995), it is important to understand the responses of insects to these different fungal combinations with carefully controlled laboratory experiments, which are realistic mimics of field situations. Experiments such as that of Gange (2001) in which naturally occurring fungal combinations are identified and then used in controlled studies are desperately required. At present, all we can do is speculate that if roots are generally colonized by more than one AM fungus, then the effects seen on the insects in the field may be limited.

13.3.5 Other Invertebrates

Soil-dwelling arthropods, like earthworms, have the potential to disperse mycorrhizal propagules through their feeding activities (see Sect. 13.3.1). The evidence is limited and reviewed by Gange and Bower (1997). It does appear that a variety of soil invertebrates, other than worms, can disperse mycorrhizal propagules and that these propagules can (but not always) withstand passage through the digestive tract. Thus, Rabatin and Stinner (1988) found that AM fungal spores could pass unharmed through the digestive tracts of wood lice (Isopoda) and millipedes (Diplopoda) and initiate colonization in plants. However, in this study, faecal material appears to have been collected and then immediately inoculated on to plant roots. In a similar study, Harinikumar and Bagyraj (1994) found that spores extracted from millipede faeces lost their viability after 4 days of storage. This contrasted with spores extracted from earthworm faecal material, which maintained viability for 12 months. There is no doubt that field-collected woodlice and millipedes contain significant numbers of spores (Rabatin and Stinner 1991) but the ecological significance of this fact is open to question. Certainly, other species of invertebrate appear to digest spores, for example, Harinikumar and Bagyraj (1994) found that no AM spores could withstand passage through the guts of termites. From the insect point of view, the most important dispersal agents are likely to be Diptera larvae (mainly Tipulidae and Bibionidae), Coleoptera larvae (Scarabaeidae), Lepidoptera larvae (Noctuidae) and ants (Formicidae) (Rabatin and Stinner 1991; Harinikumar and Bagyraj 1994). Tipulids consume much decaying organic matter in areas where AM fungi sporulate (Allen 1991) and Chafer beetle larvae have been shown to contain live mycorrhizal spores (Gange 1994). Ants can concentrate AM material in their nest mounds and this may even lead to selective plant establishment when the mounds are abandoned (Friese and Allen 1993). Multitrophic spore dispersal has been recorded, where intact spores were extracted from the guts of predatory beetles, which in turn had fed on noctuid larvae that had consumed the spores initially (Rabatin and Stinner 1991; Gange and Bower 1997). The importance of spore dispersal by these invertebrates, relative to that by earthworms or vertebrates (Allen 1987) or abiotic factors such as wind and rain is unknown.

13.4 Field Studies

Subterranean insect herbivores can have dramatic effects on the structure of plant communities. By acting as major agents of seedling mortality, they can significantly reduce vascular plant species richness during early succession

(Brown and Gange 1989a). In developing plant communities, soil dwelling insects have their greatest impact on the perennial forb life history group. Reduction of soil insect number by careful insecticide application leads to a dramatic increase in the abundance and diversity of perennial forbs and, in this way, soil insects accelerate the process of succession by hastening the disappearance of the forbs and allowing the ingress of perennial grasses (Brown and Gange 1989b). However, in mature grassland communities, the majority of root-feeding insects attack the perennial grasses and application of insecticide results in a decrease in species richness, as grasses are allowed to dominate the community (Brown and Gange 1993).

It is interesting that AM fungi have also been shown to affect plant diversity in early successional communities. In a highly controlled microcosm experiment (Grime et al. 1987) and in extensive field studies, AM fungi can promote plant species richness and this effect is also most noticeable in the perennial forbs (Gange et al. 1993). In the latter study, reducing AM fungal presence with a soil fungicide resulted in a halving of the perennial forb species number in one site and a reduction by a factor of 1.33 in another. A reanalysis of published data from European grasslands confirms this observation and shows that it is mainly the forbs that benefit greatly from AM fungi (Fig. 10.1; van der Heijden, Chap. 10, this Vol.). However, as with root-feeding insects, effects of AM fungi on plant species diversity appear to depend on the successional age and mycorrhizal status of plants in the community studied. For example, Hartnett and Wilson (1999) found that AM fungi depress plant species richness in tallgrass prairie, because their presence allowed strongly mycorrhizal grasses *(Andropogon gerardii, A. scoparius* and *Sorghastrum nutans)* to dominate, at the expense of weakly mycotrophic forbs. Furthermore, recent experiments by van der Heijden et al. (1998b) have shown that the structure of the plant community is related to the diversity of the AM fungal community (Hart and Klironomos, Chap. 9, this Vol.; van der Heijden, Chap. 10, this Vol.). Plant diversity increased with an increasing number of mycorrhizal taxa in experimental macrocosms (Fig. 9.4). It is therefore clear that root-feeding insects can depress plant species richness in early succession, but promote it in later successional communities. Meanwhile, AM fungi can promote it in early successional seres, but depress it in later seres, if the canopy dominants are mycorrhizal. Clearly, in any community, the intriguing fact appears to be that the actions of these organisms on plant species composition are contradictory. As both insects and fungi co-occur in all ecosystems of the world, we need to ask how these organisms act together in determining the structure of plant communities. Indeed, the same question needs to be asked in relation to the other subterranean invertebrates discussed above, but as we have seen, the basic knowledge is still rudimentary, to say the least.

Perhaps because of the difficulty of simultaneous manipulation of subterranean insect numbers and mycorrhizal fungi, there is very little published information on the interactions between these organisms in field conditions.

Any biocide used to reduce the numbers of either target group will have non-target effects (e.g. Smith et al. 2000) and it is important to interpret any results in the light of these problems. One example is provided by Gange and Brown (1993). In this study insect numbers were reduced by the application of a soil insecticide (chlorpyrifos) and mycorrhizal colonization reduced by the application of soil fungicide (iprodione). Both compounds have negligible side effects on seed germination (Gange et al. 1992). Iprodione does have small effects on total soil fungal biomass, but these effects are considerably less than the effect seen in mycorrhizal colonization. The plant community in question was an early successional one in southern England, developing from a buried seed bank after ploughing. Plant colonization was recorded over a two-year period and significant effects of both treatments were recorded. A recently disturbed community, such as this, is where populations of both the insects and fungi should be low (Brown and Gange 1990; Smith and Read 1997), yet insect feeding significantly decreased plant abundance and mycorrhizal presence increased it. Of most interest was the fact that significant interactions were found between the treatments. The benefit from mycorrhizal colonization was most clearly seen when the insects were absent, suggesting that the insects were disrupting the symbiosis in some way.

The study was repeated in a more extensive experiment, begun in 1990, at Silwood Park, Berkshire, UK. An area of recently ploughed land (450 m^2) was divided into plots, each 3 m×3 m, and assigned to one of four treatments, arranged in a randomised block design. Contact insecticide (chlorpyrifos) was applied to the soil to reduce insect numbers and contact fungicide (iprodione) applied in the same way to reduce mycorrhizal colonization. Insecticide and fungicide were applied in a factorial combination, giving four treatments. Five replicates of each treatment were sampled at two-weekly intervals for the first year and monthly thereafter. The data from the first 3 years of the experiment are presented here. On each sampling occasion, aboveground vegetation was sampled with the point quadrat method, which provides data on species number and relative abundance. As both root-feeding insects and AM fungi have been shown to affect plant species richness, this is used to assess effects on community structure. In this part of southern England, the year can be conveniently divided into a 'summer' (May-October) and 'winter' (November-April) growing season for purposes of the analysis. The data for three summer seasons are presented here.

Efficacy of the treatments is shown by the fact that at the end of year 3, insecticide application reduced numbers of soil-dwelling insects by 85%, but had no effect on mycorrhizal colonization in any plant species. At this stage, the dominant members of the rhizophagous insect community were Diptera (Tipulidae) and Coleoptera (Scarabaeidae and Elateridae). Similarly, application of fungicide reduced mycorrhizal colonization of *Plantago lanceolata* 'bait plants' from 35% root length colonized in control plots, to 3% in treated plots. Fungicide had no measurable effect on the soil insect numbers.

Trends in plant species richness in the four treatments during year 1 are shown in Fig. 13.2. In the first summer season, the community was dominated by annual forbs and both densities of root-feeding insects and mycorrhizal fungi were low, following the soil disturbance. A severe drought contributed to the decline in species richness in August of this season (Fig. 13.2a). However, both treatments began to have a significant effect on species richness during the second year (Fig. 13.2b). Application of insecticide resulted in an increase in plant species richness (Table 13.2), while fungicide application resulted in a decrease (Table 13.2). These data therefore confirm those of the previous studies in that subterranean insect herbivores act to reduce plant species richness while AM fungi enhance it. These trends were consistent and continued throughout the first three years of succession (Fig. 13.2; Table 13.2). Of greatest interest was the fact that in analyses from the second summer onwards, there was a significant interaction between the treatments. In each case, the nature of the interaction was the same and is perhaps most clearly seen in the data for the second summer (Fig. 13.2b). The effect of mycorrhizal reduction, when insects were absent, is seen by comparing the two lines of open symbols. Here, there was a large effect, in which addition of fungicide reduced species richness. If one compares the two lines of solid symbols, i.e. the effect of mycorrhizal reduction when insects were present, then the effect is much smaller. Therefore, the effect of AM fungi was most clearly seen when insects were

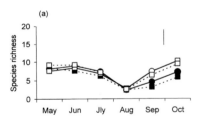
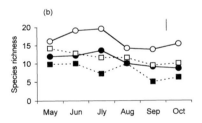
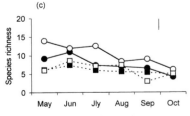

Fig. 13.2a–c. Trends in plant species richness during the first 3 years of secondary plant succession in southern England. ● Control (natural levels of soil insects and AM fungi); ■ application of soil fungicide; ○ application of soil insecticide; □ application of both compounds. *Vertical line* to *right* of graph indicates standard error for comparison from ANOVA. Statistical details see text and Table 13.2

Table 13.2. Summary of F values and significance from a repeated measures analysis of variance, testing for the effects of soil insecticide (I), soil fungicide (F) and the interaction between them (I×F) on plant species richness, during the summer season in each of the first 3 years of succession. All degrees of freedom: 1,16.
* $P<0.05$; ** $P<0.01$; *** $P<0.001$

	Year 1	Year 2	Year 3
I	8.0*	83.77***	11.11**
F	1.1	60.38***	39.82***
I×F	0.22	4.51*	7.58*

absent suggesting that the presence of these invertebrates somehow disrupt the functioning of the association.

One can look at the interaction another way and examine the effect of insects in the presence and absence of the fungi. If one compares the two solid lines (when fungi were present) then insects are seen to have a large effect. However, in the case of the dotted lines (when fungi were reduced), there was a much smaller effect of herbivory. Thus, the effect of insects was most clearly seen when fungi were present. This latter result may be tentative evidence that the null effect of mycorrhizal combinations, seen in laboratory trials, is also observable in the field. If mycorrhizal presence significantly reduced insect growth and survival in the field, then we would not expect to see an effect of herbivory only when the fungi were present.

The mechanism by which these changes appear to have occurred is through a control of perennial forb abundance, since few significant effects were found with annual forbs or perennial grasses. If we consider the mean total number of perennial forbs which occurred in the different treatment plots in summers 2 and 3 (Fig. 13.3), then it can be seen that insecticide application increased the number of forbs (year 2: $F_{1,16}=36.75$, $P<0.001$; year 3: $F_{1,16}=37.9$, $P<0.001$), while fungicide application reduced the number ($F_{1,16}=133.33$, $P<0.001$; $F_{1,16}=95.61$, $P<0.001$). An interaction again occurred between the treatments (year 2: $F_{1,16}=21.33$, $P<0.001$; year 3: $F_{1,16}=13.11$, $P<0.01$). We propose that the way in which forb species number is affected is by differential seedling establishment. Herbivory by root-feeding insects can decrease seedling number (Gange and Brown 1991b), while AM fungi can promote seedling establishment of mycorrhizal species (Grime et al. 1987). This leads to the production of a model for this system, depicted in Fig. 13.4. We suggest that in situations of high mycorrhizal abundance, many seedlings establish. If root-feeding insects are also present in this system, they consume

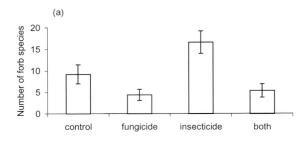
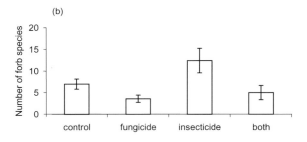

Fig. 13.3. Mean total number of perennial forbs found in each treatment in **a** summer of year 2 and **b** summer of year 3. Statistical details see text

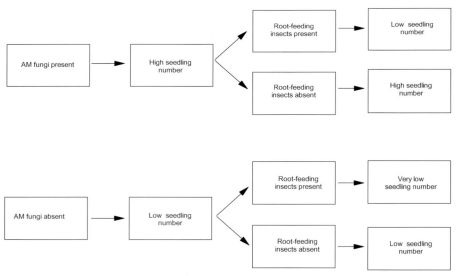

Fig. 13.4. A model for the interaction between soil-dwelling insects and AM fungi, mediated through effects on seedling establishment

many seedlings, resulting in a low number (this is the equivalent of the control treatment in the above experiment). If, instead, insects are absent then the seedling number remains high (the insecticide treatment above). Meanwhile, if AM fungi are rare and insects present, very few seedlings establish (the fungicide treatment above). Finally, if AM fungi are rare and insects absent then a low number of seedlings result (the dual pesticide treatment above).

An implicit assumption of this model is that all seedlings are equally vulnerable to insects, which may not be true and that all perennial forbs are mycorrhizal in this system, which is certainly not true. Furthermore, it currently implies that there is a linear sequence in time of AM fungi colonizing the seedling first, followed by insect attack. Clearly, this may not always be true either. Important objectives must be to understand the sequence and probabilities of colonization and attack events for seedlings in field conditions. However, until these are achieved, we believe that the model presents a reasonable explanation for the results found and should also be testable in a number of plant communities. Furthermore, the effects we have noted may also be modified by the spatial dynamics of the organisms. In the case of root-feeding insects, the spatial effects will be exacerbated by behavioural responses, host plant specificity and the interactions with other organisms, such as nematodes (Mortimer et al. 1999). Clearly, in order to fully understand plant community structure, we need to build similar models that involve the other soil invertebrate groups discussed in this chapter. Currently, this cannot be done, as the only manipulative field experiments involving AM fungi and subterranean invertebrates are those with insects, described here. Eventually, we need to produce a composite model that includes the interactions between the various groups themselves, as well as the fungi. Any change in plant community structure attributable to one group of organisms is likely to have significant 'knock-on' effects for others, through processes such as plant productivity, nutrient cycling or decomposition. An elegant study by Scheu et al. (1999) details the interactions between earthworms, Collembola and foliar-feeding aphids and shows how complex the linkages may be between aboveground and subterranean organisms, mediated through shared host plants. Meanwhile, a most interesting study by Bardgett et al. (1999) provides a tantalising glimpse of the complexity of the subterranean interactions. Here, it was found that low levels of root herbivory by clover cyst nematodes (*Heterodera trifolii*) resulted in increased root growth not only of the host plant (*Trifolium repens*), but also of the uninfected neighbouring grass, *Lolium perenne*. Clearly, such an effect could not only have consequences for AM colonization of these plants, but also for subterranean insects, which are believed to be resource-limited by the quantity of root available, rather than its quality (Brown and Gange 1990).

The limited evidence from field trials suggests that soil insects and AM fungi show significant interactions in developing plant communities. However, at this stage, it is impossible to separate the effects of macroarthropods, which may be detrimental to the mycorrhiza by severing hyphal links or by consumption of hyphae during the act of root feeding and microarthropods which may have positive effects on the symbiosis (Gange 2000). Insects appear to have a large effect on plant species when the fungi are also present, indicating that AM fungi probably do not have detrimental effects on insect growth as seen in laboratory trials. When AM fungi are reduced, insect herbivores have little effect on plant species, possibly because there are few seedlings for them to consume. All these interactions have come from early successional communities, where we would expect populations of both insects and fungi to be low. There is much to be done in communities of greater successional age, containing a higher diversity of both insects and fungi. In such communities, one would expect the influence of the subterranean insects to be large (and possibly larger than that of the AM fungi), due to their greater population sizes (Brown and Gange 1990).

We believe that laboratory experiments with subterranean invertebrates and AM fungi need to become better mimics of field situations. Future research strategies must include these mesocosm studies, which should provide controlled analogues of naturally occurring plant communities. It is clear that various soil-dwelling invertebrates and AM fungi all have the potential to radically alter the structure of plant communities. However, we are still a long way from understanding the relative strengths of these various forces in determining plant community composition and much more peering into the black box is required.

13.5 Conclusions

Subterranean invertebrates and AM fungi both act and interact to determine the structure of plant communities. Earthworms may affect nutrient cycling rates and appear to have positive effects on AM fungi (through dispersal). Plant parasitic nematodes are extremely abundant in field soils, and can alter community structure by weakening the performance of their host plants. Evidence suggests that these detrimental effects of nematodes on plants may be ameliorated by the presence of AM fungi. Mycophagous nematodes, mites and Collembola may reduce the functioning of the symbiosis, but there is little evidence for this. Indeed, Collembola appear to benefit the symbiosis, by feeding on non-AM fungi in the rhizosphere. The majority of field studies have involved subterranean insects and mycorrhizae. Clear interactions between these groups have been observed in early successional plant communities. Seedling establishment, particularly of perennial forbs seems to be an important mechanism by which these organisms structure plant communities.

We have attempted to highlight a number of areas that would benefit from further research. However, in all of these, we must caution against the use of unrealistic experiments, such as has happened in the past. Realistic experiments must involve naturally occurring combinations of invertebrates and fungi and controlled experiments must observe these at comparable densities to those seen in the field.

Acknowledgements. We are grateful to the Natural Environment Research Council for funding the field studies described in this chapter.

References

Allen MF (1987) Re-establishment of mycorrhizas on Mount St Helens: migration vectors. Trans Br Mycol Soc 88:413–417
Allen MF (1991) The Ecology of Mycorrhizae. Cambridge Univ Press, Cambridge
Bardgett RD, Denton CS, Cook R (1999) Below-ground herbivory promotes soil nutrient transfer and root growth in grassland. Ecol Lett 2:357–360
Blomqvist MM, Olff H, Blaauw MB, Bongers T, van der Putten WH (2000) Interactions between above- and belowground biota: importance for small-scale vegetation mosaics in a grassland ecosystem. Oikos 90:582–598
Brown VK, Gange AC (1989a) Herbivory by soil-dwelling insects depresses plant species richness. Funct Ecol 3:667–671
Brown VK, Gange AC (1989b) Differential effects of above- and below-ground insect herbivory during early plant succession. Oikos 54:67–76
Brown VK, Gange AC (1990) Insect herbivory below ground. Adv Ecol Res 20:1–58
Brown VK, Gange AC (1992) Secondary plant succession: how is it modified by insect herbivory? Vegetatio 101:3–13
Brown VK, Gange AC (1993) Subterranean insect herbivores in grassland ecosystems. In: Prestidge RA (ed) Proceedings of the 6th Australasian Grassland Invertebrate Ecology Conference. AgResearch, Hamilton, pp 26–31
Brown VK, Gange AC (1999) Plant diversity in successional grasslands: how is it modified by foliar insect herbivory? In: Kratochwil A (ed) Biodiversity in ecosystems. Kluwer, Dordrecht, pp 133–146
Brussaard L, Behan-Pelletier VM, Bignell DE, Brown VK, Didden W, Folgarait P, Fragoso C, Freckman DW, Gupta VVSR, Hattori T, Hawksworth DL, Klopatek C, Lavelle P, Malloch DW, Rusek J, Soderstrom B, Tiedje JM, Virginia RA (1997) Biodiversity and ecosystem functioning in soil. Ambio 26:563–570
Carter JB (1975) Rearing *Tipula oleracea* (Diptera, Tipulidae) in laboratory cultures. Plant Pathol 24:101–104
Clapp JP, Young JPW, Merryweather J, Fitter AH (1995) Diversity of fungal symbionts in arbuscular mycorrhizas from a natural community. New Phytol 130:259–265
Curry JP (1994) Grassland invertebrates: ecology, influence on soil fertility and effects on plant growth. Chapman and Hall, London
Dakora FD (1995) Plant flavonoids – biological molecules for useful exploitation. Aust J Plant Physiol 22:87–99
Edwards CA, Lofty JR (1977) Biology of earthworms. Chapman and Hall, London
Finlay RD (1985) Interactions between soil micro-arthropods and endomycorrhizal associations of higher plants. In: Fitter AH, Atkinson D, Read DJ, Usher MB (eds) Ecological interactions in soil. Blackwell, Oxford, pp 319–331
Fitter AH, Sanders IR (1992) Interactions with the soil fauna. In: Allen MF (ed) Mycorrhizal functioning. Chapman and Hall, New York, pp 333–354

Friese CF, Allen MF (1993) The interaction of harvester ants and VA mycorrhizal fungi in a patchy environment: the effects of mound structure on fungal dispersion and establishment. Funct Ecol 7:13–20

Gange AC (1993) Translocation of mycorrhizal fungi by earthworms during early succession. Soil Biol Biochem 25:1021–1026

Gange AC (1994) Subterranean insects and fungi: hidden costs and benefits to the greenkeeper. In: Cochran AJ, Farrally MR (eds) Science and Golf II: Proceedings of the World Scientific Congress of Golf. Spon, London, pp 461–466

Gange AC (1996) Reduction in vine weevil larval growth by arbuscular mycorrhizal fungi. Mitt Biol Bund Land Forst 316:56–60

Gange AC (2000) Arbuscular mycorrhizal fungi, Collembola and plant growth. TREE 15:369–372

Gange AC (2001) Species-specific responses of a root- and shoot-feeding insect to arbuscular mycorrhizal colonization of its host plant. New Phytol 150:611–618

Gange AC, Bower E (1997) Interactions between insects and mycorrhizal fungi. In: Gange AC, Brown VK (eds) Multitrophic interactions in terrestrial systems. Blackwell, Oxford, pp 115–131

Gange AC, Brown VK (1991a) Culturing root aphids using hydroponics. Entomol Gaz 42:165–169

Gange AC, Brown VK (1991b) Mechanisms of seedling mortality by subterranean insect herbivores. Oecologia 88:228–232

Gange AC, Brown VK (1993) Interactions between soil dwelling insects and mycorrhizas during early plant succession. In: Alexander IJ, Fitter AH, Lewis DH, Read DJ (eds) Mycorrhizas in ecosystems. CAB International, Wallingford, pp 177–182

Gange AC, Brown VK, Farmer LM (1992) Effects of pesticides on the germination of weed seeds: implications for manipulative experiments. J Appl Ecol 29:303–310

Gange AC, Brown VK, Sinclair GS (1993) VA mycorrhizal fungi: a determinant of plant community structure in early succession. Funct Ecol 7:616–622

Gange AC, Brown VK, Sinclair GS (1994) Reduction of black vine weevil larval growth by vesicular-arbuscular mycorrhizal infection. Entomol Exp Appl 70:115–119

Giannakis N, Sanders FE (1990) Interactions between mycophagous nematodes, mycorrhizal and other soil fungi. Agric Ecosyst Environ 29:163–167

Grime JP, Mackey JML, Hillier SH, Read DJ (1987) Floristic diversity in a model system using experimental microcosms. Nature 328:420–422

Habte M, Zhang YC, Schmitt DP (1999) Effectiveness of *Glomus* species in protecting white clover against nematode damage. Can J Bot 77:135–139

Harinikumar KM, Bagyraj DJ (1994) Potential of earthworms, ants, millipedes and termites for dissemination of vesicular-arbuscular mycorrhizal fungi in soil. Biol Fertil Soils 18:115–118

Hartnett DC, Wilson GWT (1999) Mycorrhizae influence plant community structure in tallgrass prairie. Ecology 80:1187–1195

Hopkin SP (1997) Biology of the springtails (Insecta: Collembola). Oxford Univ Press, Oxford

Hussey RS, Roncadori RW (1981) Influence of *Aphelenchus avenae* on vesicular-arbuscular endomycorrhizal growth response in cotton. J Nematol 2:48–52

Ingham ER (1988) Interactions between nematodes and vesicular-arbuscular mycorrhizae. Agric Ecosyst Environ 24:169–182

Ingham ER, Trofymow JA, Ames RN, Hunt HW, Morley CR, Moore JC, Coleman DC (1986) Trophic interactions and nitrogen cycling in a semi-arid grassland soil. II. System responses to removal of different groups of soil microbes or fauna. J Appl Ecol 23:615–630

Jaizme Vega MC, Tenoury P, Pinochet J, Jaumot M (1997) Interactions between the root-knot nematode *Meloidogyne incognita* and *Glomus mosseae* in banana. Plant Soil 196:27–35

Klironomos JN, Kendrick WB (1995) Stimulative effects of arthropods on endomycorrhizas of sugar maple in the presence of decaying litter. Funct Ecol 9:528–536

Klironomos JN, Kendrick WB (1996) Palatability of microfungi to soil arthropods in relation to the functioning of arbuscular mycorrhizae. Biol Fertil Soils 21: 43–52

Klironomos JN, Ursic M (1998) Density-dependent grazing on the extraradical hyphal network of the arbuscular mycorrhizal fungus, *Glomus intraradices*, by the collembolan, *Folsomia candida*. Biol Fertil Soils 26:250–253

Klironomos JN, Bednarczuk EM, Neville J (1999) Reproductive significance of feeding on saprobic and arbuscular mycorrhizal fungi by the collembolan, *Folsomia candida*. Funct Ecol 13:756–761

Lee KK, Reddy MV, Wani SP, Trimurtulu N (1996) Vesicular-arbuscular mycorrhizal fungi in earthworm casts and surrounding soil in relation to soil management of a semi-arid tropical Alfisol. Appl Soil Ecol 3:177–181

Little LR, Maun MA (1996) The 'Ammophila problem' revisited: a role for mycorrhizal fungi. J Ecol 84:1–7

Lussenhop J (1996) Collembola as mediators of microbial symbiont effects upon soybean. Soil Biol Biochem 28:363–369

McGonigle TP (1995) The significance of grazing on fungi in nutrient cycling. Can J Bot 73 [Suppl]:S1370–S1376

Moore JC, St John TV, Coleman DC (1985) Ingestion of vesicular-arbuscular mycorrhizal hyphae and spores by soil microarthropods. Ecology 66:1979–1981

Morandi D (1996) Occurrence of phytoalexins and phenolic compounds in endomycorrhizal interactions, and their potential role in biological control. Plant Soil 5:241–251

Mortimer SR, van der Putten WH, Brown VK (1999) Insect and nematode herbivory below ground: interactions and role in vegetation succession. In: Olff H, Brown VK, Drent RH (eds) Herbivores: between plants and predators. Blackwell, Oxford, pp 205–238

Mullin CA, Al-Fatafta AA, Harman JL, Serino AA, Everett SL (1991) Corn rootworm feeding on sunflower and other Compositae – influence of floral terpenoid and phenolic factors. Am Chem Soc Symp Ser 9:278–292

Pattinson GS, Smith SE, Doube BM (1997) Earthworm *Aporrectodea trapezoides* had no effect on the dispersal of a vesicular-arbuscular mycorrhizal fungi, *Glomus intraradices*. Soil Biol Biochem 29:1079–1088

Peipp H, Maier W, Schmidt J, Wray V, Strack D (1997) Arbuscular mycorrhizal fungus-induced changes in the accumulation of secondary compounds in barley roots. Phytochemistry 44:581–587

Petersen H, Luxton M (1982) A comparative analysis of soil faunal populations and their role in decomposition processes. Oikos 39:287–388

Piearce TG (1978) Gut contents of some lumbricid earthworms. Pedobiologia 18:153–157

Popovici I, Ciobanu M (2000) Diversity and distribution of nematode communities in grasslands from Romania in relation to vegetation and soil characteristics. Appl Soil Ecol 14:27–36

Rabatin SC, Stinner BR (1988) Indirect effects of interactions between VAM fungi and soil-inhabiting invertebrates on plant processes. Agric Ecosyst Environ 24:135–146

Rabatin SC, Stinner BR (1989) The significance of vesicular-arbuscular mycorrhizal fungal-soil macroinvertebrate interactions in agroecosystems. Agric Ecosyst Environ 27:195–204

Rabatin SC, Stinner BR (1991) Vesicular-arbuscular mycorrhizae, plant, and invertebrate interactions in soil. In: Barbosa P, Krischik VA (eds) Microbial mediation of plant-herbivore interactions. Wiley, Chichester, pp 141–168

Reddell P, Spain AV (1991) Earthworms as vectors of viable propagules of mycorrhizal fungi. Soil Biol Biochem 23:767–774

Roncadori RW (1997) Interactions between arbuscular mycorrhizas and plant parasitic nematodes in agro-ecosystems. In: Gange AC, Brown VK (eds) Multitrophic interactions in terrestrial systems. Blackwell, Oxford, pp 101–113

Ruess L, Zapata EJG, Dighton J (2000) Food preferences of a fungal-feeding *Aphelenchoides* species. Nematology 2:223–230

Salawu EO, Esty RH (1979) Observations on the relationships between a vesicular-arbuscular fungus, a fungivorous nematode, and the growth of soybeans. Phytoprotection 60:99–102

Scheu S, Parkinson D (1994) Effects of invasion of an aspen forest (Canada) by *Dendrobaena octaedra* (Lumbricidae) on plant growth. Ecology 75:2348–2361

Scheu S, Theenhaus A, Jones TH (1999) Links between the detritivore and the herbivore system: effects of earthworms and Collembola on plant growth and aphid development. Oecologia 119:541–551

Siddiqui ZA, Mahmood I (1998) Effect of a plant growth promoting bacterium, an AM fungus and soil types on the morphometrics and reproduction of *Meloidogyne javanica* on tomato. Appl Soil Ecol 8:77–84

Smith GS, Hussey RS, Roncadori RW (1986) Penetration and postinfection development of *Meloidogyne incognita* on cotton as affected by *Glomus intraradices* and phosphorus. J Nematol 18:429–435

Smith MD, Hartnett DC, Rice CW (2000) Effects of long-term fungicide applications on microbial properties in tallgrass prairie soil. Soil Biol Biochem 32:935–946

Smith SE, Read DJ (1997) Mycorrhizal symbiosis. Academic Press, San Diego

Thimm T, Larink O (1995) Grazing preferences of some collembola for endomycorrhizal fungi. Biol Fertil Soils 19:266–268

van der Heijden MGA, Boller T, Wiemken A, Sanders IR (1998a) Different arbuscular mycorrhizal fungal species are potential determinants of plant community structure. Ecology 79:2082–2091

van der Heijden MGA, Klironomos JN, Ursic M, Moutoglis P, Streitwolf-Engel R, Boller T, Wiemken A, Sanders IR (1998b) Mycorrhizal fungal diversity determines plant biodiversity, ecosystem variability and productivity. Nature 396:69–72

Vierheilig H, Bago B, Albrecht C, Poulin MJ, Piche Y (1998) Flavonoids and arbuscular-mycorrhizal fungi. Adv Exp Med Biol 9:9–33

Warnock AJ, Fitter AH, Usher MB (1982) The influence of a springtail *Folsomia candida* (Insecta, Collembola) on the mycorrhizal association of leek *Allium porrum* and the vesicular-arbuscular mycorrhizal endophyte *Glomus fasciculatus*. New Phytol 90:285–292

Wolters V, Stickan W (1991) Resource allocation of beech seedlings (*Fagus sylvatica* L.): relationship to earthworm activity and soil conditions. Oecologia 88:125–131

14 Interactions Between Ecto-mycorrhizal and Saprotrophic Fungi

J.R. Leake, D.P. Donnelly, L. Boddy

Contents

14.1	Summary	346
14.2	Introduction	346
14.2.1	The Importance of Ecto-mycorrhizal and Saprotrophic Fungi in Forest Soils	347
14.2.2	Structural and Functional Similarities Between Mycelia of Saprotrophic and Ecto-mycorrhizal Fungi	348
14.3	Competition for Nutrients Between Ecto-mycorrhizal and Saprotrophic Fungi	353
14.3.1	Evidence of Saprotrophic Nutrient Mobilising Activities of Ecto-mycorrhizal Fungi	354
14.3.2	The Effect of Short-Circuiting of the N and P Cycles by Ecto-mycorrhizal Fungi on the Activities of Saprotrophic Fungi	356
14.4	Interactions Between Ecto-mycorrhizal and Saprotrophic Fungi Observed in Microcosms Containing Natural Soil	359
14.4.1	Transfers of P Between Interacting Mycelia of Ecto-mycorrhizal and Saprotrophic Fungi	359
14.4.2	The Effect of Interaction with Saprotrophic Fungi on Growth and Carbon Allocation in Ecto-mycorrhizal Mycelium	360
14.4.3	The Effect of Interaction with Ecto-mycorrhizal Mycelium on the Growth and Morphology of Mycelial Cords of the Saprotrophic Fungus *Phanerochaete velutina*	364
14.5	Conclusions	366
	References	367

14.1 Summary

The microbiota of most forest soils is dominated by ecto-mycorrhizal and saprotrophic decomposer fungi which are the main organisms involved respectively in the supply of nutrients to trees and the decomposition of woody plant litter. In recognition of the pivotal role that these fungi play in plant nutrition and nutrient cycles in plantations and natural forests, the activities of their mycelial systems have been independently investigated in recent years. Such studies have revealed remarkable structural and functional similarities between these organisms, particularly in their active foraging for nitrogen and phosphorus in soils. This has highlighted the need to determine the nature and extent of interactions between the two groups of organisms. The use of radioactive tracers, time-sequence photography and digital image analysis of microcosms containing non-sterile soil in which the two groups of fungi are grown with their natural carbon resources have provided novel insights into their interactions. When mycelia of the fungi meet in soil they often interact strongly, resulting in cessation of growth or even die-back of mycelium. Such inhibition can be accompanied by loss of nutrients to the opposing fungus. This chapter reviews current evidence and presents some new data showing that interactions between the mycelial systems of mycorrhizal and saprotrophic fungi have major effects on the functioning of these organisms and upon nutrient and carbon cycling processes in temperate and boreal forest ecosystems.

14.2 Introduction

An understanding of the interactions between mycorrhizal and saprotrophic organisms, whether mutualistic, neutral or antagonistic is of paramount importance given their central roles in biogeochemical cycling in ecosystems. Both arbuscular (Dehne 1982; Newsham et al. 1995; Olsson et al. 1998; Filion et al. 1999) and ecto-mycorrhizal fungi (Chakravarty and Unestam 1987; Duchesne et al. 1988) have been shown to have inhibitory effects on root pathogenic fungi but their interactions with saprotrophic fungi have received surprisingly little attention. In part, this reflects the relative ease with which the effects of plant pathogens can be quantified in roots in contrast to the technical difficulties in the determination of the effects of interactions between mycelial systems of mycorrhizal and saprotrophic fungi that occur in soil. Competitive and combative interactions between wood-decay fungi have recently been reviewed (Boddy 2000) and interactions between these fungi and mycoparasites and bacteria have received some attention (Tsuneda and Thorn 1995), until recently, little was known of the effects of mycorrhizae on these saprotrophs.

In this chapter, we review our current knowledge of interactions between ecto-mycorrhizal and wood-decay saprotrophic fungi and their significance in forest soils and present some new observations on these interactions. We compare the structural and functional similarities between these two trophic groups of fungi and the nature and effect of their interactions. Attention is focused on interactions between saprotrophic and ecto-mycorrhizal fungi because we hypothesise that these interactions are of greater intensity than those involving arbuscular mycorrhizal fungi owing to the much higher biomass and greater saprotrophic capabilities of ecto-mycorrhizal basidiomycetes. Furthermore, the saprotrophic fungi, particularly those that effect wood-decomposition, are more active in forest soils under ecto-mycorrhizal trees than under herbaceous arbuscular mycorrhizal plant communities. There is insufficient information, at present, on interactions between mycorrhizal and saprotrophic fungi in forest ecosystems dominated by arbuscular mycorrhizal trees to include in this chapter.

14.2.1 The Importance of Ecto-mycorrhizal and Saprotrophic Fungi in Forest Soils

Two major groups of basidiomycetes often dominate the mycoflora of Boreal and Northern-Temperate forest soils, the saprotrophic wood and litter decomposers and the biotrophic ecto-mycorrhizal fungi. Their access to large pools of reduced carbon enables these organisms to play central roles in the mineralisation and cycling of carbon and nutrients in forest soils (Griffiths and Caldwell 1992; Boddy and Watkinson 1995). Wood-decomposer saprotrophic fungi are almost exclusively responsible for decomposition of lignocellulose, the most abundant organic compound in the terrestrial environment (Tanesaka et al. 1993) and this provides their principle source of carbon. Mycorrhizal fungi, through symbiotic associations with tree roots receive carbon, in the form of sugars, amounting to between 15 and 30% of the net photosynthate of their host plants (Simard et al., Chap. 2, this Vol.). By gaining access to such large pools of carbon, fungi dominate the microbial biomass in forest soils, their mycelium typically comprising between 15 and 20 mg g^{-1} soil dry weight (Markkola et al. 1995). This increases to up to 500 mg g^{-1} (i.e. 50% of soil dry weight) in areas with particularly dense proliferation of ecto-mycorrhizal mycelia (Ingham et al. 1991).

Both saprotrophic and mycorrhizal fungi contain much higher concentrations of the main plant-growth limiting macronutrients nitrogen (N) and phosphorus (P) than are typically found in tree litter (Abuzinadah and Read 1988; Boddy and Watkinson 1995). Consequently, in forest soils the quantities of these nutrients which are sequestered by mycelia are considerable. Bååth and Söderström (1979), for example, found that approximately 20% of the N and up to 18% of the organic P in a forest soil was contained in microbial (mainly fungal) biomass.

14.2.2 Structural and Functional Similarities Between Mycelia of Saprotrophic and Ecto-mycorrhizal fungi

The most aggressive fungi of both saprotrophic and ecto-mycorrhizal fungi are characterised by similar mycelial structure and function, including the production of linear aggregates of hyphae termed rhizomorphs (Cairney et al. 1991) or mycelial cords (see Table 14.1).

These hyphal aggregates develop behind the extending mycelial front enabling both mycorrhizal (Finlay and Read 1986a,b) and saprotrophic fungi (Boddy 1993, 1999) to forage for nutrients, particularly N and P, through large volumes of soil. At the same time, cords and rhizomorphs ensure the maintenance of access to their carbon resource bases (roots of their symbiotic partners and dead wood respectively). In both groups of fungi mycelial cords ensure efficient allocation and conservation of nutrients within the mycelia and minimal "leakage" into soil (Boddy and Watkinson 1995). The ability to bi-directionally translocate nutrients is fundamental to the success of the wood-decomposer fungi (Cairney 1992) which can transfer N and P into mycelia in wood in different parts of the systems thereby facilitating its decomposition, despite the high C:N and C:P ratio of the substrate. At the same time, the fungi allocate some of these nutrients to the hyphal tips foraging for, and colonising, new resources (Wells and Boddy 1990; Wells et al. 1998a). Similarly, ecto-mycorrhizal mycelial cords can transfer P absorbed by one part of the mycelium to the host plant and simultaneously to hyphal tips and areas of intensive mycelial proliferation (Finlay and Read 1986b). Further information about mycelial structures and foraging strategies of ecto-mycorrhizal fungi is given by Olsson et al. (Chap. 4, this Vol.).

The apparent structural and functional similarities between the mycelial systems of mycorrhizal and saprotrophic fungi (see Table 14.1), their common requirement for soil-derived nutrients and the foraging of their mycelia in the same locations in soil and litter are likely to bring them into competition and conflict. The outcomes of these interactions are of major importance as they will influence the ability, on the one hand, of the mycorrhizal plant to secure nutrients and, on the other, of the wood decomposers to facilitate decomposition and recycling of resources contained in their lignocellulose food bases. Interactions between the two groups of fungi are particularly important in both managed forests where seedlings are planted alongside the decaying stumps of harvested trees and in natural forests where fallen trees (nurselogs) are often the main sites of germination of ecto-mycorrhizal seedlings

Interactions Between Ecto-mycorrhizal and Saprotrophic Fungi 349

Table 14.1. Structural and functional similarities between mycelia of ecto-mycorrhizal and wood-decomposer saprotrophic fungi

Structure of mycelium	Ecto-mycorrhizal fungi	References	Wood-decomposer saprotrophic fungi	References
Mycelium extends into soil long distances from its carbon resource base (infected root tips in the case of mycorrhizae, decaying wood in the case of wood-decomposer saprotrophic fungi).	Some species produce mycelial networks which extend more than 20 cm from roots in undisturbed microcosms, but others produce little mycelium in soil. Genets of individual strains of mycorrhizal fungi can occasionally be found over distances 10–30 m apart.	Finlay and Read (1986a,b); Dahlberg and Stenlid (1994); Anderson et al. (1998)	Some saprotrophs have mycelial cords which extend more than 10 m and individual genotypes have been found over distances more than 100 m apart.	Boddy (1993); Wells and Boddy (1995a,c)
The ability to, on contact, sense nutrient- or carbon-rich patches in soil and allocate increased mycelial biomass to these regions.	The mycelial systems of some, but not all, ectomycorrhizal fungi show marked stimulation of mycelial growth in patches of fermenting litter which are relatively enriched in nutrients.	Read (1991); Bending and Read (1995a); Pérez-Moreno and Read (2000); Leake et al. (2001)	Many saprotrophic fungi allocate growth of mycelium in dynamic response to encounter of new carbon sources (pieces of decaying wood) or nutrient-rich patches.	Donnelly and Boddy (1998); Boddy (1999)
The ability to switch morphology from intensively foraging disaggregated mycelium to linear aggregates of mycelium as mycelial cords (where hyphae are undifferentiated) or rhizomorphs (in which differentiation of hyphae occurs).	Behind the actively advancing mycelial front (in regions which the mycorrhiza has depleted of available nutrients) mycelium frequently becomes aggregated into cords or rhizomorphs whose primary functions appear to be nutrient and carbon transport rather than absorption.	Finlay and Read (1986b); Bending and Read (1995a); Leake et al. (2001)	Some saprotrophic fungi grow through soil in mycelial cords or rhizomorphs but change morphology to intensively foraging disaggregated mycelium in response to local nutrient enrichment or to increased nutrient demand resulting from colonisation of new carbon-rich substrates.	Wells et al. (1997, 1998a,b); Boddy (1999)

Table 14.1 (*Continued*)

Structure of mycelium (*Continued*)	Ecto-mycorrhizal fungi	References	Wood-decomposer	References
Mycelial cords and rhizomorphs vary from simple undifferentiated linear aggregates of hyphae to the most complex rhizomorphs in which there is a clear distinction between cortex and medulla and in which very large diameter vessel hyphae are produced. These frequently lack cellular contents and often show evidence of septal wall breakdown so that the hyphae form long vessels.	Some ectomycorrhizal fungi do not form mycelial cords or rhizomorphs, but the most prominent species do form these structures. Mycorrhizas with very extensive external mycelial systems typically produce highly differentiated rhizomorphs, but these normally form behind a growing front of non-aggregated mycelium.	Agerer (1991); Cairney (1991); Franz and Acker (1995)	Most saprotrophic wood-decay fungi produce rhizomorphs, including the largest and most complex of these structures. In some cases (e.g. *Armillaria mellea* and *Marasmius* sp.), mycelial growth is highly aggregated and growth of the fungi through soil occurs primarily as rhizomorphs.	Cairney (1991, 1992); Cairney et al. (1989); Thompson and Rayner (1982)
Function of mycelium	Ecto-mycorrhizal fungi	References	Wood-decomposer saprotrophic fungi	References
Uptake and translocation of recources by mycelial cords and rhizomorphs				
Water	Ecto-mycorrhizal fungi have been shown potentially to play a major role in uptake and transport of water to plants (in laboratory studies).	Duddridge et al. (1980); Boyd et al. (1986)	Mass flow of water has been shown in cords and rhizomorphs.	Jennings (1987)

Carbon	Rhizomorphs provide major routes for rapid supply of carbon from mycorrhizal root-tips to the distal regions of the actively growing mycorrhizal front.	Finlay and Read (1986a); Read (1992); Finlay and Söderström (1992); Leake et al. (2001)	Carbon is rapidly moved through mycelial cords and rhizomorphs of saprotrophic fungi	Wells et al. (1995); Granlund et al. (1985)
Nitrogen	Ecto-mycorrhizal mycelia take up organic (amino acid) and inorganic nitrogen and transport them to roots and other parts of their mycelial networks.	Finlay et al. (1988); Finlay et al. (1989); Bending and Read (1995a)	Nitrogen translocation in cords and rhizomorphs of saprotrophic wood-decaying fungi has been demonstrated using ^{14}C-labelled amino acids and 3-O-methyl glucose and a-aminoisobutyric acid (both of which are not metabolised) and ^{15}N tracers.	Jennings (1987); Watkinson (1984)
Phosphorus	Phosphorus is taken up by foraging mycelium and is transported to both host plant roots and active mycorrhizal mycelium.	Finlay and Read (1986b); Timonen et al (1996); Lindahl et al. (1999)	Phosphorus uptake by diffuse mycelium and its transport through mycelial cords and rhizomorphs have been demonstrated in both the laboratory and in the field.	Watkinson (1971); Cairney (1992); Wells et al. (1990); Wells and Boddy (1995a,b,c); Olsson and Gray (1998)

Table 14.1 (*Continued*)

Function of mycelium (*Continued*)	Ecto-mycorrhizal fungi	References	Wood-decomposer saprotrophic fungi	References
Bi-directional transport. The ability simultaneously to transport resources to advancing new mycelium in soil and back to carbon resource bases (root-tips or wood).	Bi-directional transport has been recorded in individual hyphae and in mycelial systems comprising rhizomorphs and simple mycelium.	Shepherd et al. (1993); Finlay and Read (1986b)	Acropetal transport is typically greater than basipetal transport, but bi-directional transport is common in saprotrophic mycelial systems.	Granlund et al. (1985); Wells and Boddy (1995a,b)
Mobilisation of nitrogen and phosphorus from organic matter. Production of extracellular and cell-wall-bound enzymes:	The production of extracellular proteinases, phosphatases, phytases and chitinases by ecto-mycorrhizal fungi has been demonstrated in pure cultures of some species. Phospho-monoesterase appears to be ubiquitous in mycorrhizal fungi and activities are often comparable to those of saprotrophic fungi. Extracellular proteinase and phytase activities are found only in some species.	Colpaert and Van Laere (1996); Leake and Read (1997); Timonen and Sen (1998)	Studies of these enzymes in saprotrophic wood-decomposer fungi are limited, but they are well documented in saprotrophic fungi in general. Extracellular proteinase has been reported in *Serpula lacrymans* and *Coriolus versicolor*. There have been few studies of phosphatase in saprotrophic wood-decaying fungi, but where tested it has been found.	Wadekar et al. (1995); Jennings (1995)
Enzymic solubilisation and degradation of structural components of litter and dead wood: cell walls and cuticles.	The ability to degrade pectin has been reported for many but not all mycorrhizal fungi. Breakdown of cellulose and lignin and polycyclic aromatic compounds by ecto-mycorrhizal fungi may occur to varying extents.	Leake and Read (1997); Cairney and Burke (1994); Trojanoski et al. (1984); Durall et al. (1994); Gramss et al. (1999)	These are the activities which characterise wood-decomposer fungi. These fungi typically have exceptionally well-developed abilities to degrade hemicellulose, cellulose, lignin and polyphenolic compounds.	Jennings (1995); Boddy (1999)

(Maser 1988). In old-growth forests which contain naturally fallen trees, up to 98% of regeneration of trees occur in these situations (Hiura et al. 1996). Despite these similarities, there are important differences between the two trophic groups of fungi: the saprotrophs obtain their carbon from decaying organic matter whilst the mycorrhizal fungi obtain most of their carbon directly from their host plants.

14.3 Competition for Nutrients Between Ecto-mycorrhizal and Saprotrophic Fungi

There have been remarkably few studies of competitive interactions between mycorrhizal and saprotrophic fungi in soils. There have been a few laboratory studies of interactions between pure cultures of representatives of the two groups of fungi on agar plates (Shaw et al. 1995) and between mycorrhizal fungi (in symbiotic association with host plants) and saprotrophs in axenic microcosm systems (Dighton et al. 1987) but the relevance of such experiments to the field situation is likely to be very limited.

Gadgil and Gadgil (1971, 1975) were amongst the first workers to realise that, since ecto-mycorrhizal and decomposer fungi coexist in litter on the forest floor, the depletion of nutrients by mycorrhizal fungi may strongly affect the activity of the decomposers. To investigate effects of these interactions they trenched areas in a pine forest in order to exclude mycorrhizal roots, and observed that decomposition was faster in the absence of mycorrhizae. They attributed this effect to competition for nutrients or antagonistic effects of the mycorrhizal fungi on the saprotrophs.

A number of attempts have been made to repeat these observations using the same simple trenching approach (Berg and Lindberg 1980; Harmer and Alexander 1985; Staaf 1988) but these have not consistently lent support to what has been called the 'Gadgil effect'. However, the use of trenching to eliminate active mycorrhizal mycelium is highly invasive and will release large amounts of carbon and nutrient sources to the decomposer population. The quality and quantities of these nutrients are likely to differ appreciably from those that are released by more natural senescence processes, and it is inevitable that in some cases mineralisation may be retarded by immobilisation of nutrients required to decompose the fresh carbon sources released by the trenching process. Notwithstanding these limitations, there is some evidence (reviewed below) which lends support to the Gadgil hypothesis that mycorrhizal and saprotrophic micro-organisms compete for nutrients in forest soils and that competitive and antagonistic interactions often occur between these organisms.

14.3.1 Evidence of Saprotrophic Nutrient Mobilising Activities of Ecto-mycorrhizal Fungi

The intensity of competition for N and P between mycorrhizal and saprotrophic fungi will be controlled by the extent to which they depend upon the same sources of these nutrients. The traditional view that mycorrhizal fungi scavenge nutrients mineralised by saprotrophs implies limited competition between these two ecological groups of fungi. Saprotrophic micro-organisms, including many fungi, oxidise carbon in litter, thereby decreasing the C:N and C:P ratios of the substrates and releasing mineralised nutrients. These mineral nutrients have traditionally been regarded as the main sources utilised by mycorrhizal fungi. However, some saprotrophs, such as cord-forming wood decomposer fungi are very conservative in their nutrient use and employ their extensive mycelial systems to reallocate and redistribute nutrients, minimising their losses to the soil mineral pool (Boddy and Watkinson 1995). Translocation and redistribution of nutrients is highly developed in many saprotrophic wood-decay fungi which remove mineral nutrients from soil in order to facilitate the break down of lignocellulose which is very low in available N and P. This process contributes to decreasing C:N and C:P ratios in wood undergoing decay (Boddy and Watkinson 1995), whilst simultaneously increasing C:N and C:P ratios of the soil regions from which they have taken these nutrients. The ability to redistribute nutrients between compartments of the forest floor is a fundamental activity of many saprotrophic and mycorrhizal fungi (Entry et al. 1991).

The uptake of nutrients from the labile pools in soil by saprotrophs such as wood-decay fungi will inevitably lead them into competition with mycorrhizal fungi that access the same nutrient pools. Furthermore, it is now clear that competition for nutrients between the two groups of fungi is not limited to inorganic N and P but also includes organic compounds containing these elements (see Table 14.1; Leake and Read 1997). Although ecto-mycorrhizal fungi have a very limited capacity to use lignocellulose as a carbon source (Haselwandter et al. 1990; Tanesaka et al. 1993), they produce a range of extracellular, cell wall-bound and intracellular enzymes by which they can hydrolyse, assimilate and transform all of the major N- and P containing organic molecules in plant, microbial and animal detritus. The organic sources of N which are utilised by ecto-mycorrhizal fungi, include amino acids (Abuzinadah and Read 1988; Näsholm et al. 1998), peptides, proteins (Abuzinadah and Read 1986) and chitin (Leake and Read 1990; Hodge et al. 1996). They also use phosphomonoesters (Dighton 1983) and phosphodiesters (Griffiths and Caldwell 1992; Gianfrancesco 1999) as sole P sources, and a range of sparingly soluble inorganic P sources, such as aluminium phosphate (Cumming and Weinstein 1990). In a review of extracellular nutrient-mobilising enzymes produced by ecto-mycorrhizal fungi, Leake and Read (1997) found that out of 68 species (mainly studied in laboratory pure-culture experiments), 85% produced proteinases, all 47 species tested produced phosphatases, and of 20 species tested for phytase, 95% demonstrated this activity. In many cases, the activities of phosphatases (Dighton 1983; Colpaert and van Laere 1996) and proteinases (Ryan and Alexander 1992) in mycorrhizal fungi in pure culture are comparable to those found in saprotrophic fungi. Whilst these results reveal the potential of many ecto-mycorrhizal fungi to play a direct role in nutrient mobilisation in the field, the enzyme activities reported in pure culture systems should be viewed with great caution: most artificial culture media are very rich in nutrients, some of which are known to repress the production of many of these enzymes, and ecto-mycorrhizal mycelia grown in pure culture are structurally (and almost certainly functionally) different to those produced in symbiosis with host plants. There is increasing evidence that enzyme expression in ecto-mycorrhizal mycelial networks growing through soil is heterogeneous and localised to specific regions (Cairney and Burke 1996; Timonen and Sen 1998). The density and structure of mycelium which develops in litter patches colonised by ecto-mycorrhizal mycelium (see e.g. Leake et al. 2001), together with their intimate contact with, and penetration of, decaying litter particles (Ponge 1990) is likely to result in much higher local expression of enzyme activities than can be measured in pure cultures of the fungi in liquid media or on agar in the laboratory.

In addition to the enzymatic machinery necessary for the utilisation of relatively labile nutrient sources, a number of ecto-mycorrhizal fungi have recently been shown to produce a further 'artillery' of enzymes of the kinds required to degrade physical barriers such as cell walls and cuticles and mobilise nutrients rendered recalcitrant due to reaction with phenolic acids, tannins and lignins (Northup et al. 1995). These enzymes, many of which operate in concert, include: fatty acid esterase, polygalacturonase, xylanase, cellulase, tyrosinase, peroxidase, polyphenol oxidase, monophenol oxygenase, laccase and ligninase (Caldwell et al. 1991; Leake and Read 1997; Cairney and Burke 1999; Gramss et al. 1999).

The importance of the vegetative mycelium of mycorrhizal fungi in scavenging for, and mobilising nutrients from decomposing litter and transferring them to their associated plants has been clearly demonstrated (Read 1991; Bending and Read 1995a,b; Pérez-Moreno and Read 2000) in microcosms containing non-sterile soil. Localised patches of intense proliferation of ecto-mycorrhizal mycelium are initiated by the presence of partially decayed litter (Read 1991; Unestam 1991; Bending and Read 1995a,b; Leake et al. 2001), but cannot be elicited by enrichment with orthophosphate or ammonium (Read 1991). In a *Pinus sylvestris-Suillus bovinus* system, exploitation of patches of pine litter by the ecto-mycorrhizal fungus resulted in the selective removal of 23% of total N, 22% of P and 30% of K and greatly increased the C:N and C:P ratios of the residual material (Bending and Read 1995a). Similar results were obtained in the field by Entry et al. (1991, 1992). They found rates of nutrient removal from litter bags were increased by 20-30% in patches where ecto-mycorrhizal mycelium comprised 45–55% of the total soil biomass. The colonisation of litter by ecto-mycorrhizal mycelium was accompanied by an increase in activity of nutrient-mobilising enzymes such as phosphatase and proteinase, the activities of which were higher in litter patches intensively colonised by the mycorrhizal mycelium than in litter colonised only by saprotrophic fungi (Bending and Read 1995b) or containing much lower mycorrhizal mycelial biomass (Griffiths and Caldwell 1992). Pérez-Moreno and Read (2000) have now established the direct coupling between nutrient mobilisation from tree-litter by ecto-mycorrhizal fungi and nutrient gains by mycorrhizal tree seedlings. Between 35–40% of the total P in pine, beech and birch litter was removed following colonisation by mycorrhizal mycelium, and approximately 70% of this P was attributed to the utilisation of labile organic P by the fungi. The mycorrhizal mycelium appeared to retain about two-thirds of the P removed from the litter patches, this P being allocated throughout the mycelium outside the litter patches, the remaining 33% was transferred to the plants where it was divided almost equally between roots (52%) and shoots (48%).

14.3.2 The Effect of Short-Circuiting of the N and P Cycles by Ecto-mycorrhizal Fungi on Activities of Saprotrophic Fungi

The 'saprotrophic' capabilities of ecto-mycorrhizal fungi in the mobilisation of nutrients necessitates a new model of the N cycle (Fig. 14.1) for forest ecosystems in which these associations predominate.

The utilisation of organic N by ecto-mycorrhizal fungi (see pathway 1a in Fig. 14.1) effectively short-circuits the conventional N cycle. In turn, this process will have a number of important knock-on effects. In depleting the labile organic N pool, mycorrhizae compete directly with saprotrophs (cf. pathways 1 and 1a), thereby reducing the pool of labile N and C available to the latter. As a consequence, the supply of N to the mineralisation pathway (3) is reduced and the competition between mycorrhizal and saprotrophic fungi (6) for ammonium-N will be greater than originally envisaged. Rapid depletion of available ammonium limits the N supply to the nitrification pathway (4) and this, together with uptake of nitrate particularly by saprotrophic micro-organisms (6), minimises losses of N by denitrification (5) and leaching.

By depleting the litter of the most readily available N sources (Bending and Read 1995a), mycorrhizal fungi leave residues with a high lignocellulose content, the decomposition of which by micro-organisms may require import of N, including that from pathway 6, resulting in immobilisation of mineral nitrogen. The activities of the microbial oxidase enzymes involved in lignocellulose oxidation and phenolic-ring cleavage result in the accumulation of recalcitrant humic residues (Sivapalan 1982) containing N (pathway 7).

Interactions Between Ecto-mycorrhizal and Saprotrophic Fungi   357

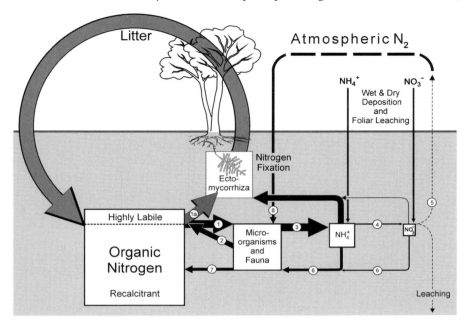

Fig. 14.1. The nitrogen cycle for forest ecosystems dominated by ecto-mycorrhizal trees. Ecto-mycorrhizal trees gain direct access to organic nitrogen in plant litter through activities of their fungal partners (pathway *1a*; *largest arrow*). Microbial and faunal litter and products provide another source of nitrogen for the trees (pathways *2* and *3*). Mycorrhizal and saprotrophic organisms compete for labile organic N (pathways *1* and *1a*). The main microbially driven pathways are: *1* microbial depolymerisation and assimilation of organic N, *2* release of microbial litter, *3* mineralisation (ammonification), *4* nitrification, *5* denitrification, *6* microbial immobilisation, *7* humification and *8* nitrogen fixation

The overall effect of ecto-mycorrhizal fungi on the N cycle is to increase the supply of N to plants whilst reducing the supply of both N and C to saprotrophs. Part of this process may be controlled by the production of polyphenol-rich litter by many trees (Northup et al. 1995). This inhibits N mineralisation, thereby restricting N availability to potential competitor plants lacking ecto-mycorrhizal associations (which can use dissolved organic N and N in protein-phenolic complexes (Northup et al. 1995). Atmospheric N deposition may change the pathways presented in Fig. 14.1. The importance of organic N for the N cycle will be reduced while that of inorganic N will increase. This can also lead to alterations in community structure of ecto-mycorrhizal fungi (Fig. 7.2; in Erland and Taylor, Chap. 7, this Vol.). In addition, it has been predicted that mycorrhiza-mediated shifts in plant species coexistence occur due to atmospheric N inputs (Figs. 5.1, 5.3 in: Aerts, Chap. 5, this Vol.).

In the P cycle, the abilities of ecto-mycorrhizal fungi to efficiently acquire P from all the main organic and inorganic pools in the soil (Fig. 14.2) are also

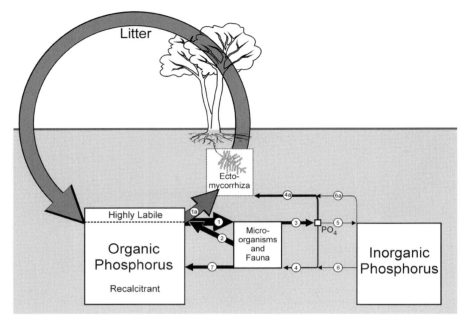

Fig. 14.2. The phosphorus cycle of forest ecosystems dominated by ecto-mycorrhizal trees. Ecto-mycorrhizal trees gain direct access to organic phosphorus in plant (*largest arrow*) and microbial litter (pathway 2) in soil through the phosphatase activities of their fungal partners (pathway *1a*). Mycorrhizal fungi (pathway *1a*) thus compete with saprotrophic micro-organisms and decomposer fauna (pathway *1*) for organic P. This reduces the importance of saprotrophic mineralisation of organic P to orthophosphate (pathway *3*). Both mycorrhizal (pathway *4a*) and saprotrophic micro-organisms (pathway *4*) compete for the very small pool of orthophosphate which remains in soil solution. The amounts of orthophosphate which are precipitated as insoluble inorganic P (pathway *5*) are minimised by the low concentration of orthophosphate in soil solution. Some mycorrhizal and saprotrophic micro-organisms have a limited ability to solubilise inorganic P (pathways *6a* and *6*). The process of humification (pathway *7*) of organic matter by microbial oxidase enzymes contributes to the accumulation of recalcitrant organic P in the soil

likely to bring them into direct competition with saprotrophic organisms such as the wood-decomposer fungi which also have a major demand for the nutrient from soil. Both groups of fungi forage extensively in soil for orthophosphate (pathway 4 and 4a in Fig. 14.2) as shown by Wells et al. (1990), Wells and Boddy (1995a,b) and Finlay and Read (1986b). They also form similar localised mycelial 'patches' in soil under conditions of nutrient limitation (Wells et al. 1997). For example, mycelial patches are induced on nutrient-poor soil by the demand for nutrients arising from the capture of new wood resources by the wood-decomposer *Phanerochaete velutina* (Wells et al. 1998a).

There are important parallels between the pathways of mobilisation of N and P in forest soils (Figs. 14.1, 14.2). The abilities of ecto-mycorrhizae to compete with saprotrophs for both organic N and P diminishes the importance of the pathways involving saprotrophic mineralisation of these nutrients prior to their uptake by mycorrhizae. Saprotrophs, as primary colonisers play a central role in the depolymerisation and assimilation of nutrients from lignocellulose-rich plant litter and at later stages of decomposition are important in mobilising nutrients from the more recalcitrant residues of decomposition (Persson et al. 1980). One of the most significant products of both of these activities is microbial litter (pathway 2 in Figs. 14.1, 14.2). This litter is relatively enriched in labile organic N and P in comparison to plant litter and thus provides one of the major potential sources of organic N and P utilised by ecto-mycorrhizae (Leake and Read 1997). Mycorrhizal mycelium develops most extensively in the fermenting litter of the forest floor (see e.g. Bending and Read 1995a), where the saprotrophic decay processes have enriched litter with nutrients and with microbial biomass and necromass that may then be utilised by the mycorrhizae.

14.4 Interactions Between Ecto-mycorrhizal and Saprotrophic Fungi Observed in Microcosms Containing Natural Soil

Much of the progress in our understanding of the structure and functioning of mycelial systems of mycorrhizal (Olsson et al., Chap. 4, this Vol.) and saprotrophic fungi (Boddy 1993, 1999) has come from studies in which the fungi have been grown in microcosms containing thin layers of non-sterile soil on which the fungi grow from their natural carbon sources (symbiotic association with tree roots or dead wood respectively). By growing representatives of the two groups of fungi alone and in confrontation in these kinds of systems, the responses of their mycelia to interaction can be assessed.

14.4.1 Transfers of P Between Interacting Mycelia of Ecto-mycorrhizal and Saprotrophic Fungi

Lindahl et al. (1999) grew *Pinus sylvestris* mycorrhizal with *Suillus variegatus* and *Paxillus involutus* in shallow trays of forest soil, into which single blocks of wood colonised by the saprotrophic wood-decay fungus *Hypholoma fasciculare* were introduced. Once the mycelia of the two fungi met, ^{32}P orthophosphate was added to the wood-blocks and its movement through the mycelial systems of *H. fasciculare* to the mycorrhizal plants was quantified by digital autoradiography followed by scintillation counting.

When the fungi met in the soil, the growth of *S. variegatus* was stimulated in contact with *H. fasciculare*, whereas the growth of *P. involutus* showed no apparent response (Lindahl et al. 1999). Of the radioactive P which was added, 7–8% of it was translocated by *H. fasciculare* out of the wood block, and of this 12–14% (0.84–1.12% of the total added P) was found in the plants having been taken up from the saprotrophic fungus and transferred to the plant by the mycorrhizal mycelium of *P. involutus* and *S. variegates*, respectively. To determine the extent of reciprocal transfer, mycorrhizal mycelium of each species, outside the zone of interaction, was supplied with ^{32}P and its movement to the saprotrophic fungus monitored. Only between 0.1 and 0.15% of the P supplied to the mycorrhiza was transferred to *H. fasciculare* mycelium in the wood-block.

Although the amounts of P transferred from the saprotroph to the mycorrhizal mycelium were higher than the amounts which were transferred in the reciprocal experiment from the mycorrhiza to the saprotroph, the quantities of P acquired by the plants from this route were very small. Nonetheless, it is clear that the transfer of nutrients can occur through the interacting mycelial systems. Whether this is due in part to leakage induced by the interaction, or simply the result of normal processes of mycelial senescence requires further investigation. That P can be transferred in the field from cord-forming saprotrophic fungi to ecto-mycorrhizal roots and even to AM herbs such as *Fragaria vesca* was shown by Wells and Boddy (1995c) after they added ^{32}P to wood-blocks that had been pre-inoculated with wood-decomposer fungi and which were left to establish extensive cord systems in a deciduous woodland for 13 months. These observations lend further support to the hypothesis of Leake and Read (1997) that some ecto-mycorrhizal fungi obtain significant amounts of nutrients from microbial biomass-necromass in soil.

14.4.2 The Effect of Interaction with Saprotrophic Fungi on Growth and Carbon Allocation in Ecto-mycorrhizal Mycelium

To study the patterns of growth and activity of the interacting mycelial systems we have developed microcosm systems of the type pioneered by Finlay and Read (1986a), of dimensions 30×25 cm, in which mycorrhizal plants are grown on a thin layer of compacted non-sterile peat (Fig. 14.3).

Mycorrhizal and saprotrophic fungi which were found in earlier experiments to grow exclusively on the peat surface, were selected so that their interactions could be observed and effects on mycelial functioning quantified non-destructively using radioactive tracers.

Plants of *Betula pendula*, mycorrhizal with *Paxillus involutus* were transplanted into peat based microcosms containing two small (3.2×3.2 cm) plastic trays (Fig. 14.3) of non-sterile partially decayed litter from the floor of a mixed coniferous-deciduous forest, following the method of Bending and

Read (1995a). Wood-blocks pre-colonised by *Phanerochaete velutina* were introduced into half of the microcosms. The other half of the microcosms (the controls) received wood-blocks without *Phanerochaete velutina*. Growth of the mycelial systems of the saprotrophic and mycorrhizal fungi were recorded by photography of the microcosms twice weekly for 6 weeks.

To determine the effect of interaction with *Ph. velutina* on the pattern of carbon allocation within the ecto-mycorrhizal mycelium, the shoots of the *Betula* seedlings were individually sealed in transparent Perspex boxes containing $^{14}CO_2$ and allowed to fix this carbon by photosynthesis during a 6 h incubation period in the light (see Leake et al. 2001 for details). At intervals of 8–48 h after commencing the ^{14}C labelling, the microcosms were covered with a thin Mylar film and the radioactivity in the central 20×24 cm area counted non-destructively for 50 min. in a Packard Instant Imager which provided false colour digital autoradiographs.

In the control microcosms, the mycorrhizal mycelium formed a dense and fairly uniform cover of the peat surface (see e.g. Fig. 14.3a). The allocation of radiolabelled photoassimilate from the plant to the mycorrhizal mycelium occurred rapidly, reaching a peak approximately 24 h after commencing labelling of the shoots (Fig. 14.3b,c). The activity in the mycorrhizal mycelium was concentrated at the advancing mycelial front and in a band approximately 2–3 cm wide behind it. 'Hot spots' of activity closer to the top of the microcosms were associated with clusters of mycorrhizal root-tips (Fig. 14.3b,c). Only a small proportion of the activity in the mycelium was used to maintain the older mycelium behind the mycelial front. By contrast, the allocation of radiolabelled photoassimilates to the mycorrhizal mycelium was disrupted in the interaction microcosms. Little or no tracer was detected in mycelium of *P. involutus* where it met that of *Ph. velutina* and the advance of the *P. involutus* mycelium down the microcosm was arrested in this region (Fig. 14.3d,e,f). This effect was seen in all three replicate microcosms in which the two fungi interacted.

The overall effect of interaction on carbon allocation to the advancing front of the mycorrhizal mycelium was quantified by running transects through each of the replicate digital autoradiographs and calculating the average radioactive counts detected at each position along these transects from the mycelial front, back towards the host tree. Three regions were compared within the same microcosm: mycelium to the far right of the interaction zone, in the interaction zone immediately above the wood-block and to the far left of the interaction (Fig. 14.4). The allocation of carbon to the mycelial front was greatest in the region to the left, farthest from the interaction zone. It was decreased to the left of the interaction zone and almost completely suppressed in the interaction zone (Fig. 14.4). The effect of proximity to the interacting fungus on total carbon allocation to mycorrhizal mycelium was tested by analysis of variance (ANOVA) of log transformed mean total number of ^{14}C counts detected in the 120 mm long transects from the advancing mycelial front, back towards the plant roots (Fig. 14.4). There was a highly significant

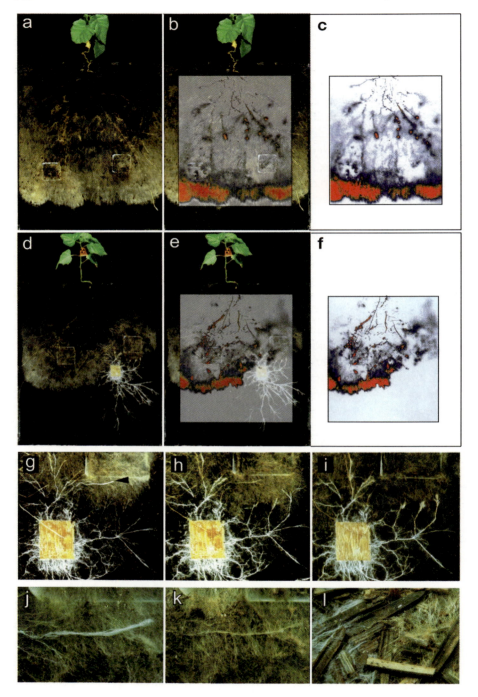

Fig. 14.3. Microcosms (30×25 cm) comprising a thin layer (0.2 cm deep) of non-sterile peat in which *Betula pendula* seedlings were grown in mycorrhizal association with *Paxillus involutus* alone or in interaction with the saprotrophic wood-decay fungus *Phanerochaete velutina* extending from an inoculated wood-block. **a** Control microcosm showing dense and fairly uniform *P. involutus* mycelial cover. **b** False colour digital autoradiograph superimposed on **a** showing the pattern of allocation of ^{14}C in the roots and mycorrhizal mycelium 24 h after ^{14}C pulse-labelling of the shoots of the *B. pendula*. Highest concentrations indicated by *red*, followed by *yellow*, *blue* and *grey*. **c** Digital autoradiograph showing the strong allocation of ^{14}C to the advancing mycelial front of *P. involutus*. The red threshold is set to 50 counts per pixel, the autoradiograph exposed to the microcosm for 50 min. **d** Interaction microcosm showing retarded advance of the mycorrhizal mycelium where confronted by *Ph. velutina*. **e** Digital autoradiograph superimposed on the photograph to show relationship between patterns of ^{14}C carbon allocation in the mycorrhizal mycelium and the presence of the saprotroph. **f** Digital autoradiograph showing the retarded advance and disruption of carbon allocation to the mycorrhizal mycelium in the zone of interaction with the saprotroph. **g** Detail of the microcosm shown in **d** showing vigorous mycelium of *P. involutus* in the litter-patch (plastic tray in the *right corner* of the photograph) with die-back of mycelial cords of the saprotroph to the left side of the litter patch and the production of a new cord (*arrowed*) immediately below it. **h** Further truncation and die-back of cords of the saprotroph in response to the mycorrhizal mycelium. **i** Local regrowth from thickened and truncated cords of the saprotroph coincident with a marked localised decrease in apparent vigour of the mycorrhizal mycelium. **j** Detail of the mycelial cord of the saprotroph below the litter patch shown in **g**. **k** Detail of the cord shown in **j** 14 days later. **l** Replacement of mycelium of *P. involutus* (brown mycelium on the *right*) in a litter patch by dense but diffuse white mycelium of *Ph. velutina* (on the *left*)

◀

overall effect of interaction with *P. velutina* on carbon allocation to the mycorrhizal mycelium (ANOVA, $P<0.001$; $F_{2,6}=71$). A Tukey multiple-comparison test revealed a highly significant reduction in carbon allocation to mycorrhizal mycelium in direct interaction with *P. velutina*, and in the mycorrhizal mycelium in close proximity on the right of the interaction, in comparison to the mycorrhizal mycelium furthest from the interaction on the left.

In a similar microcosm experiment in which the mycelium of *Suillus bovinus* mycorrhizal with *Pinus sylvestris* was grown alone and in interaction with *Ph. velutina* (Leake et al. 2001), the effect of the mycorrhiza on the growth of the saprotroph was limited. However, in this case the rate of growth, density of mycelium and amount of carbon (^{14}C-labelled photoassimilate) supplied to the mycorrhizal fungus and allocated to its external mycelium was reduced by two-thirds in the first 30 h after pulse-labelling of the plant shoots. Contrary to the situation seen in the *P. involutus* vs. *Ph. velutina* interaction (Figs. 14.3, 14.4), the inhibitory effect of the wood-decomposer in this study was not clearly evident where it contacted the mycorrhizal mycelium, but the vigour of the whole of the mycorrhizal mycelium decreased (see Leake et al. 2001). The widespread effect on the mycorrhizal mycelium of *Suillus bovinus* suggests effects of volatile inhibitory compounds produced by the saprotroph but this

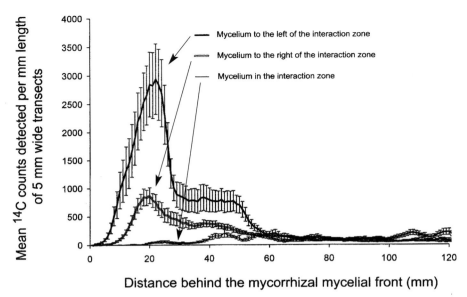

Fig. 14.4. The mean allocation of ^{14}C (cpm per 5 mm^2) to the mycelial front of *Paxillus involutus*. ^{14}C allocation has been measured by starting at the top of the mycelial front (d=0 mm) and moving 120 mm upwards to the plant in the direction of the soil surface. The values are for transects run parallel to the sides of the microcosms on either side of the zones of interaction with the saprotroph, and through the interaction zones. *Vertical bars* give standard error of means (n=3).

requires experimental verification. Small amounts of the ^{14}C-labelled photoassimilate were taken up and concentrated in mycelium of the wood-decomposer. The most likely source of this carbon was from the mycorrhizal mycelium, since the wood-decay fungi had contact with this mycelium and the ^{14}C-labelling experiment took place before the saprotroph grew into the region occupied by plant roots.

14.4.3 The Effect of Interaction with Ecto-mycorrhizal Mycelium on the Growth and Morphology of Mycelial Cords of the Saprotrophic Fungus *Phanerochaete velutina*

The growth of the *Ph. velutina* was also markedly affected by interaction with *P. involutus* (see e.g. Fig. 14.3d). Analysis of the growth of the saprotroph in all the replicate soil microcosms confirmed that the effects seen in Fig. 14.3 were typical. There was a significant ($P<0.05–P<0.001$) overall retardation of growth of the saprotroph towards the mycorrhizal mycelium in the interaction microcosms compared to the growth of the saprotroph in control microcosms containing non-mycorrhizal plants (Fig. 14.5). Growth of the sapro-

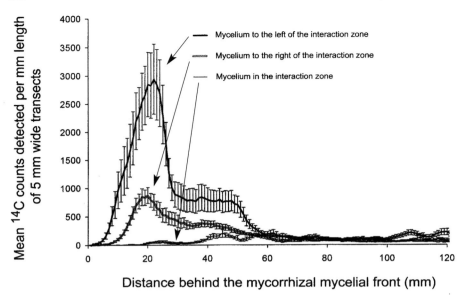

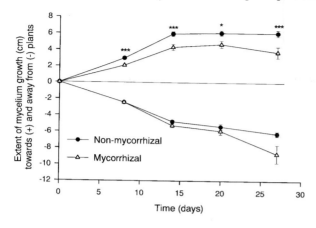

Fig. 14.5. The mean radial extent of mycelial cords of *Ph. velutina* growing towards (+) or away from (−) plants of *B. pendula* mycorrrhizal with *P. involutus* (*open triangle*) or non-mycorrhizal (*solid circle*). The extent of mycelial growth of the saprotroph is inhibited when growing towards mycorrhizal plants. Vertical bars indicate standard error of means ($n=3$). Significant differences between mycorrhizal and non-mycorrhizal systems are indicated (* $P<0.05$, *** $P<0.001$, Scheffé test)

troph away from the interaction zone was slightly (but not significantly) increased in the mycorrhizal microcosms (see Fig. 14.5). There were marked morphological changes to the mycelial systems in the interaction zone. The cords of *Ph. velutina* were truncated and branched repeatedly on meeting the mycorrhizal mycelium, these branches being deflected down the microcosm in the opposite direction of the advancing mycelial front (Fig. 14.3d). At a later stage, some of the mycelial cords of *Ph. velutina* in the interaction zone developed a curious truncated and highly branched 'fox tail' morphology and the mycorrhizal mycelium in the vicinity of these cords showed marked deterioration apparently due to senescence. These effects were only observed in the microcosms in which *Ph. velutina* interacted with mycorrhizal mycelium.

Of the three kinds of interactions which Dowson et al. (1988) described as occurring between different species and genotypes of saprotrophic fungi, namely 'replacement', 'deadlock' and 'intermingling' the interaction between the *P. involutus* and *Ph. velutina* initially appears to be of the 'deadlock' type since growth of both fungi was inhibited, but neither were replaced. As the interaction proceeded, a number of the cords of *Ph. velutina* which had grown over the surface of *P. involutus* (Fig. 14.3g,j) senesced and were replaced by the mycorrhizal mycelium (Fig. 14.3h,k). In the longer term, however, the saprotroph re-advanced and established extensive mycelial cords in the region behind the foraging front of the ecto-mycorrhizal fungus. In the initial part of the experiment, prior to mycelial interaction, the mycorrhizal mycelium grew into and colonised the litter patches. However, several weeks after the two mycelia met, the saprotroph entered the litter patches and replaced the (then degenerating) mycorrhizal mycelium.

In the litter the saprotroph changed from having a corded morphology to non-aggregated mycelium, a response which parallels the switch in mycorrhizal mycelium from loosely aggregated to densely foraging in response to litter-patch encounter (Bending and Read 1995a; Leake et al. 2001), which suggests that the resources remaining in the litter-patch after exploitation by the mycorrhizal mycelium (carbon, or nutrients or both) was being exploited by the saprotroph. This confirms the earlier indications (Leake and Read 1997) that saprotrophic and mycorrhizal fungi forage intensively through the same kind of resources in soil.

14.5 Conclusions

The intensity of interactions between mycelial systems of saprotrophic and ecto-mycorrhizal fungi revealed in microcosm studies highlights the potential importance of these interactions on the functioning of these organisms in forest ecosystems, where both groups of fungi will have access to even larger amounts of carbon than can be supplied by the small seedlings and woodblocks employed in microcosm studies. The results presented here support the view of Gadgil and Gadgil (1975) that interactions between these two ecological groups of fungi are important in nutrient cycling in boreal and temperate forest ecosystems. The abilities of many ecto-mycorrhizal fungi to use labile organic and mineral nutrients inevitably brings them into competition and conflict with some of the saprotrophs and this has major implications for the cycling of C, N and P in ecto-mycorrhizal forest ecosystems. However, interactions between the fungi are not limited to direct competition for nutrients but can also involve active combat, effected by a variety of mechanisms (Boddy 2000), which result in deadlock, where neither mycelium gains territory, or die-back of the opposing mycelium. During such interactions exchange of C and nutrients can occur. The outcome of interactions between the two groups of fungi can be finely poised so that while the mycorrhiza may replace the saprotroph in one microsite in the soil, at the same time, the positions can be reversed in other locations. From this, it is clear that mycelial interactions between ecto-mycorrhizal and saprotrophic fungi are highly dynamic, both temporarily and spatially.

It is apparent from the studies described in this chapter that residual litter components left after exploitation by mycorrhizal mycelium (typically with relatively high C:P and C:N ratios; see Bending and Read 1995a) may provide substrates suitable for some specialised saprotrophs like the wood-decay fungi. Such observations imply a dynamic inter-play between the activities of mycorrhizal and groups of saprotrophic fungi in the processes of nutrient cycling and litter decomposition. Ecto-mycorrhizal fungi appear to be rather ineffective as primary decomposers of litter (see e.g. Colpaert and van Tichelen 1996) but make a major contribution to nutrient cycling from partially fermented litter (Pérez-Moreno and Read 2000). The abilities of wood-decay saprotrophic fungi to translocate nutrients through their mycelial networks may enable them to extract both carbon and nutrients from the recalcitrant lignocellulose-rich residues left behind after fermented litter has been exploited by ecto-mycorrhizae.

Despite the recent advances, our knowledge of the spatial and temporal interactions between mycorrhizal and saprotrophic fungi remain incomplete. We are currently quantifying mycorrhizal and saprotrophic mycelial morphology by image analysis and fractal geometry, and transfer of C, N and P between them during mycelial interactions. The significance of these interactions for biogeochemical cycling, microbial biodiversity and ecosystem functioning need to be more fully elucidated. The effects revealed to date show that we can no longer ignore the interface between mycorrhizal and saprotrophic fungal systems in soil if we are to understand the functioning and role of the ecto-mycorrhizal wood-wide web in forest ecosystems.

Acknowledgements. We gratefully acknowledge funding by the Natural Environment Research Council GR3/11059 'Interactions between mycelia of ecto-mycorrhizal and saprotrophic wood-decomposer fungi in forest soil: effects on litter decomposition, nutrient mobilisation and tree nutrition'. Ms I Johnson provided skilled technical assistance throughout the microcosm studies.

References

Abuzinadah RA, Read DJ (1986) The role of proteins in the nitrogen nutrition of ecto-mycorrhizal plants. I. Utilization of peptides and proteins by ectomycorrhizal fungi. New Phytol 103:481–493

Abuzinadah RA, Read DJ (1988) Amino acids as nitrogen sources for ectomycorrhizal fungi: utilisation of individual amino acids. Trans Br Mycol Soc 91:473–479

Agerer R (1991) Characterisation of ectomycorrhiza. Methods Microbiol 23:25–73

Anderson IC, Chambers SM, Cairney JWG (1998) Use of molecular methods to estimate the size and distribution of mycelial individuals of the ectomycorrhizal basidiomycete *Pisolithus tinctorius*. Mycol Res 102:295–300

Bååth E, Söderström B (1979) Fungal biomass and fungal immobilisation of plant nutrients in Swedish coniferous forest soils. Rev Ecol Biol Soils 16:477–489

Bending GD, Read DJ (1995a) The structure and function of the vegetative mycelium of ectomycorrhizal plants. V. The foraging behaviour of ectomycorrhizal mycelium and the translocation of nutrients from exploited organic matter. New Phytol 130:401–409

Bending GD, Read DJ (1995b) The structure and function of the vegetative mycelium of ectomycorrhizal plants. VI. Activities of nutrient mobilising enzymes in birch litter colonised by *Paxillus involutus* (Fr.) Fr. New Phytol 130:411–417

Berg B, Lindberg T (1980) Is litter decomposition retarded in the presence of mycorrhiza in forest soil? Swedish Coniferous Forest Project Internal Report 95, p 10

Boddy L (1993) Cord forming fungi: warfare strategies and other ecological aspects. Mycol Res 97:641–655

Boddy L (1999) Saprotrophic cord-forming fungi: meeting the challenge of heterogeneous environments. Mycologia 91:13–32

Boddy L (2000) Interspecific combative interactions between wood-decaying basidiomycetes – a review. FEMS Microbiol Ecol 31:185–194

Boddy L, Watkinson SC (1995) Wood decomposition, higher fungi, and their role in nutrient redistribution. Can J Bot 73 [Suppl]:S1377–S1383

Boyd R, Furbank RT, Read DJ (1986) Ectomycorrhiza and water relations of trees. In: Gianinazzi-Pearson V, Gianinazzi S (eds) Physiological and genetical aspects of Mycorrhizae. INRA, Paris, pp 689–693

Cairney JWG (1991) Structural and ontogenetic study of ectomycorrhizal rhizomorphs. In: Varma AK, Read DJ, Norris JR (eds) Experiments with mycorrhizae. Academic Press, London

Cairney JWG (1992) Translocation of solutes in ectomycorrhizal and saprotrophic rhizomorphs. Mycol Res 96:135–141

Cairney JWG, Burke RM (1994) Fungal enzymes degrading plant cell walls: their possible significance in the ectomycorrhizal symbiosis. Mycol Res 98:1345–1356

Cairney JWG, Burke RM (1996) Physiological heterogeneity within fungal mycelia: an important concept for a functional understanding of the ectomycorrhizal symbiosis. New Phytol 134:685–695

Cairney JWG, Burke RM (1999) Do ecto- and ericoid mycorrhizal fungi produce peroxidase activity? Mycorrhiza 8:61–65

Cairney JWG, Jennings DH, Veltkamp CJ (1989) A scanning electron microscope study of the internal structure of mature linear mycelial organs of four basidiomycete species. Can J Bot 67:2266–2271

Cairney JWG, Jennings DH, Agerer R (1991) The nomenclature of fungal multi-hyphal linear aggregates. Cryptogam Bot 2:246–251

Caldwell BA, Castellano MA, Griffiths RP (1991) Fatty acid esterase production by ectomycorrhizal fungi. Mycologia 83:233–236

Chakravarty P, Unestam T (1987) Mycorrhizal fungi prevent disease in stressed pine seedlings. J Phytopathol 118:335–340

Colpaert JV, van Laere A (1996) A comparison of the extracellular enzyme activities of two ectomycorrhizal and leaf-saprotrophic basidiomycetes colonising beech leaf litter. New Phytol 134:133–141

Colpaert JV, van Tichelen KK (1996) Decomposition, nitrogen and phosphorus mineralization from beech leaf litter colonized by ectomycorrhizal or litter-decomposing basidiomycetes. New Phytol 134:123–132

Cumming JR, Weinstein LH (1990) Utilization of $AlPO_4$ as a phosphorus source by ectomycorrhizal *Pinus rigida*. New Phytol 116:99–106

Dahlberg A, Stenlid J (1994) Size, distribution and biomass of genets in populations of *Suillus bovinus* (L. Fr) Roussel revealed by somatic incompatibility. New Phytol 128:225–234

Dehne HW (1982) Interaction between vesicular-arbuscular mycorrhizal fungi and plant pathogens. Phytopathology 72:1115–1119

Dighton J (1983) Phosphatase production by mycorrhizal fungi. Plant Soil 71:455–462

Dighton J, Thomas ED, Latter PM (1987) Interactions between tree roots, mycorrhizas, a saprotrophic fungus and the decomposition of organic substrates in a microcosm. Biol Fertil Soils 4:145–150

Donnelly DP, Boddy L (1998) Developmental and morphological responses of mycelial systems of *Stropharia caerulea* and *Phanerochaete velutina* to soil nutrient enrichment. New Phytol 138:519–531

Dowson CG, Rayner ADM, Boddy L (1988) The form and outcome of mycelial interactions involving cord-forming decomposer basidiomycetes in homogeneous and heterogeneous environments. New Phytol 109:423–432

Duchesne LC, Peterson RL, Ellis BE (1988) Interaction between the ectomycorrhizal fungus *Paxillus involutus* and *Pinus resinosa* induces resistance to *Fusarium oxysporum*. Can J Bot 66:558-562

Duddridge JA, Malibari A, Read DJ (1980) Structure and function of mycorrhizal rhizomorphs with special reference to their role in water transport. Nature 287:834-836

Durall DM, Todd AW, Trappe JM (1994) Decomposition of ^{14}C-labelled substrates by ectomycorrhizal fungi in association with Douglas fir. New Phytol 127:725-729

Entry JA, Rose CL, Cromack K (1991) Litter decomposition and nutrient release in ectomycorrhizal mat soils of a Douglas-fir ecosystem. Soil Biol Biochem 23:285-290

Entry JA, Rose CL, Cromack K (1992). Microbial biomass and nutrient concentrations in hyphal mats of the ectomycorrhizal fungus *Hysterangium setchellii* in a coniferous forest soil. Soil Biol Biochem 24:447-453

Filion M, St-Arnaud M, Fortin JA (1999) Direct interaction between the arbuscular mycorrhizal fungus *Glomus intraradices* and different rhizosphere microorganisms. New Phytol 141:525-533

Finlay RD, Read DJ (1986a) The structure and function of the vegetative mycelium of ectomycorrhizal plants. I. Translocation of ^{14}C-labelled carbon between plants interconnected by a common mycelium. New Phytol 103:143-156

Finlay RD, Read DJ (1986b). The structure and function of the vegetative mycelium of ectomycorrhizal plants. II. The uptake and distribution of phosphorus by mycelium interconnecting host plants. New Phytol 103:157-165

Finlay RD, Ek H, Oldham G, Söderström B (1988) Mycelial uptake, translocation and assimilation of nitrogen from ^{15}N-labelled ammonium by *Pinus sylvestris* plants infected with four different ectomycorrhizal fungi. New Phytol 110:59-66

Finlay RD, Ek H, Oldham G, Söderström B (1989) Uptake, translocation and assimilation of nitrogen from ^{15}N-labelled ammonium and nitrate sources by intact ectomycorrhizal systems of *Fagus sylvatica* infected with *Paxillus involutus*. New Phytol 113:47-55

Franz F, Acker G (1995) Rhizomorphs of *Picea-abies* ectomycorrhizae – ultratructural aspects and elemental analysis (EELS and ESI) on hyphal inclusions. Nova Hedwigia 60:253-267

Gadgil RL, Gadgil PD (1971) Mycorrhiza and litter decomposition. Nature 233:133

Gadgil RL, Gadgil PD (1975) Suppression of litter decomposition by mycorrhizal roots of *Pinus radiata*. N Z J For Sci 5:35-41

Gianfrancesco RU (1999) Phosphorus nutrition of mycorrhizal and non-mycorrhizal plants of upland soils with special reference to the utilization of the phosphodiester DNA under sterile conditions. PhD Thesis, University of Sheffield, UK, 309 pp

Gramss G, Kirsche B, Voigt K-D, Günther T, Fritsche W (1999) Conversion rates of five polycyclic aromatic hydrocarbons in liquid cultures of fifty eight fungi and the concomitant production of oxidative enzymes. Mycol Res 103:1009-1018

Granlund HI, Jennings DH, Thompson W (1985) Translocation of solutes along rhizomorphs of *Armillaria mellea*. Trans Br Mycol Soc 84:111-119

Griffiths RP, Caldwell BA (1992) Mycorrhizal mat communities in forest soils. In: Read DJ, Lewis DH, Fitter AH, Alexander IJ (eds) Mycorrhizas in ecosystems. CAB International, Wallingford, pp 98-105

Harmer R, Alexander IJ (1985) Effect of root exclusion on nitrogen transformations and decomposition processes in spruce humus. In: Fitter AH, Atkinson D, Read DJ, Usher MB (eds) Ecological interactions in soils: plants, microbes and animals. Blackwell, Oxford, pp 277-279

Haselwandter K, Bobleter O, Read DJ (1990) Degradation of ^{14}C-labelled lignin and dehydropolymer of coniferyl alcohol by ericoid and ectomycorrhizal fungi. Arch Microbiol 153:352-354

Hiura T, Sano J, Konno Y (1996) Age structure and response to fine scale disturbances of *Abies sachalinensis, Picea jezoensis, Picea glehnii*, and *Betula ermanii* growing under the influence of a dwarf bamboo understory in northern Japan. Can J For Res 26:289-297

Hodge A, Alexander IJ, Gooday W, Killham K (1996) Carbon allocation patterns in fungi in the presence of chitin in the external medium. Mycol Res 100:1428-1430

Ingham ER, Griffiths RP, Cromack K, Entry JA (1991) Comparison of direct vs. fumigation incubation microbial biomass estimates from ectomycorrhizal mat and non-mat soils. Soil Biol Biochem 23:465-472

Jennings DH (1987) Translocation of solutes in fungi. Biol Rev 62:215-243

Jennings DH (1995) The physiology of fungal nutrition. Cambridge Univ Press, Cambridge, pp 158-250

Leake JR, Read DJ (1990) Chitin as a nitrogen source for mycorrhizal fungi. Mycol Res 94:993-995

Leake JR, Read DJ (1997) Mycorrhizal fungi in terrestrial habitats. In: Wicklow DT, Söderström B (eds) The Mycota. IV. Environmental and microbial relationships. Springer, Berlin Heidelberg New York

Leake JR, Donnelly DP, Saunders EM, Boddy L, Read DJ (2001) Rates and quantities of carbon flux to ectomycorrhizal mycelium following ^{14}C pulse labelling of *Pinus sylvestris* L seedlings: effects of litter patches and interaction with a wood-decomposer fungus. Tree Physiol 21:71-82

Lindahl B, Stenlid J, Olsson S, Finlay RD (1999) Translocation of ^{32}P between interacting mycelia of a wood-decomposing fungus and ectomycorrhizal fungi in microcosm systems. New Phytol 144:183-193

Markkola AM, Ohtonen R, Tarvainen O, Ahonen-Jonnarth U (1995) Estimates of fungal biomass in Scots pine stands on an urban pollution gradient. New Phytol 131:139-147

Maser C (1988) From the forest to the sea: a story of fallen trees. USDA Rep PNW-GTR-229, Portland, Oregon

Näsholm T, Ekblad A, Nordin A, Giesler R, Högberg M, Högberg P (1998) Boreal forest plants take up organic nitrogen. Nature 392:914-916

Newsham KK, Fitter AH, Watkinson AR (1995) Arbuscular mycorrhiza protect an annual grass from root pathogenic fungi in the field. J Ecol 83:991-1000

Northup RR, Zengshou Y, Dahlgren RA, Vogt K (1995). Polyphenol control of nitrogen release from pine litter. Nature 377:227-229

Olsson PA, Francis R, Read DJ, Soderstrom B (1998) Growth of arbuscular mycorrhizal mycelium in calcareous dune sand and its interaction with other soil microorganisms as estimated by measurement of specific fatty acids. Plant Soil 201:9-16

Olsson S, Gray SN (1998) Pattern and dynamics of ^{32}P-phosphate and labelled 2-aminoisobutyric acid (^{14}C-AIB) translocation in intact basidiomycete mycelia. FEMS Microbiol Ecol 26:109-120

Pérez-Moreno J, Read DJ (2000) Mobilization and transfer of nutrients from litter to tree seedlings via vegetative mycelium of ectomycorrhizal plants. New Phytol 145:301-309

Persson T, Bååth E, Clarholm M, Lundkvist H, Söderström BE, Solenius B (1980) Trophic structure, biomass dynamics and carbon metabolism of soil organisms in a Scots pine forest. In: Persson T (ed) Structure and function of northern coniferous forests. Ecol Bull (Stockh) 32:419-459

Ponge JF (1990) Ecological study of a forest humus by observing a small volume. I. Penetration of pine litter by mycorrhizal fungi. Eur J For Pathol 20:290-303

Read DJ (1991) Mycorrhizas in ecosystems. Experientia 47:376-391

Read DJ (1992) The mycorrhizal mycelium. In: Allen MF (ed) Mycorrhizal functioning. Chapman and Hall, London, pp 102-133

Ryan EA, Alexander IJ (1992) Mycorrhizal aspects of improved growth of spruce when grown in mixed stands on heathlands. In: Read DJ, Lewis DH, Fitter AH, Alexander IJ (eds) Mycorrhizas in ecosystems. CAB International, Wallingford, pp 237-245

Shaw TM, Dighton J, Sanders FE (1995) Interactions between ectomycorrhizal and saprotrophic fungi on agar and in association with seedlings of lodgepole pine (*Pinus contorta*). Mycol Res 99:159-165

Shepherd VA, Orlovich DA, Ashford AE (1993) Cell to cell transport via motile tubules in growing hyphae of a fungus. J Cell Sci 105:1173-1178

Sivapalan K (1982) Humification of polyphenol-rich plant residues. Soil Biol Biochem 14:309-310

Smith ML, Bruhn, JN, Anderson JB (1992) The fungus *Armillaria-bulbosa* is among the largest and oldest living organisms. Nature 356:428-431

Staaf H (1988) Litter decomposition in beech forests – effect of excluding the roots. Biol Fertil Soils 6:302-305

Tanesaka AE, Masuda H, Kinugawa K (1993) Wood degrading ability of basidiomycetes that are wood decomposers, litter decomposers or mycorrhizal symbionts. Mycologia 85:347-354

Thompson W, Rayner ADM (1982) Structure and development of mycelial cord systems of *Phanerochaete laevis* in soil. Trans Br Mycol Soc 78:193-200

Timonen S, Sen R (1998) Heterogeneity of fungal and plant enzyme expression in intact Scots pine *Suillus bovinus* and *Paxillus involutus* mycorrhizosphere developed in natural forest humus. New Phytol 138:355-366

Timonen S, Finlay RD, Olsson S, Söderström B (1996) Dynamics of phosphorus translocation in intact ectomycorrhizal systems: non destructive monitoring using a b-scanner. FEMS Microbiol Ecol 19:171-180

Trojanowski J, Haider K, Hüttermann A (1984) Decomposition of ^{14}C-labelled lignin, holocellulose and lignocellulose by mycorrhizal fungi. Arch Microbiol 139:202-206

Tsuneda A, Thorn RG (1995) Interactions of wood decay fungi with other microorganisms, with emphasis on the degradation of cell walls. Can J Bot 73 [Suppl]:S1325-S1333

Unestam T (1991) Water repellency, mat formation and leaf-stimulated growth of some ectomycorrhizal fungi. Mycorrhiza 1:13-20

Wadekar RV, North MJ, Watkinson SC (1995) Proteolytic enzymes in two wood-decaying basidiomycte fungi, *Serpula lacrymans* and *Coriolus versicolor*. Microbiology 141:1575-1583

Watkinson SC (1971) Phosphorus translocation in the stranded and unstranded mycelium of *Serpula lacrimans*. Trans Br Mycol Soc 57:535-539

Watkinson SC (1984) Inhibition of growth and development of *Serpula lacrymans* by the non-metabolised amino acid analogue 2-aminoisobutyric acid. FEMS Microbiol Lett 24:247-250

Wells JM, Boddy L (1990) Wood decay, and phosphorus and fungal biomass allocation in mycelial cord systems. New Phytol 116:285-295

Wells JM, Boddy L (1995a) Translocation of soil-derived phosphorus in mycelial cord systems in relation to inoculum resource size. FEMS Microbial Ecol 17:67-75

Wells JM, Boddy L (1995b) Effect of temperature on wood decay and translocation of soil derived phosphorus in mycelial cord systems. New Phytol 129:289-297

Wells JM, Boddy L (1995c) Phosphorus translocation by saprotrophic basidiomycete mycelial cord systems on the floor of a mixed deciduous woodland. Mycol Res 99:977-980

Wells JM, Hughes C, Boddy L (1990) The fate of soil-derived phosphorus in mycelial cord systems of *Phanerochaete velutina* and *Phallus impudicus*. New Phytol 114:595-606

Wells JM, Boddy L, Evans R (1995) Carbon translocation in mycelial cord systems of *Phanerochaete velutina* (DC.: Pers.) Parmasto. New Phytol 129:467-476

Wells JM, Donnelly DP, Boddy L (1997) Patch formation and developmental polarity in mycelial cord systems of *Phanerochaete velutina* on a nutrient-depleted soil. New Phytol 136:653-665

Wells JM, Boddy L, Donnelly DP (1998a) Wood decay and phosphorus translocation by the cord-forming basidiomycete *Phanerochaete velutina*: the significance of local nutrient supply. New Phytol 138:607–617

Wells JM Harris MJ, Boddy L (1998b) Temporary phosphorus partitioning in mycelial systems of the cord-forming basidiomycete *Phanerochaete velutina*. New Phytol 140:283–293

Section E:

Host Specificity and Co-evolution

15 Mycorrhizal Specificity and Function in Myco-heterotrophic Plants

D.L. Taylor, T.D. Bruns, J.R. Leake, D.J. Read

Contents

15.1	Summary	375
15.2	Introduction	376
15.3	Evidence for Specificity in Myco-heterotrophs	377
15.3.1	Overview of Specificity in the Orchidaceae	378
15.3.2	Overview of Specificity in the Monotropoideae	393
15.4	Influences on Specificity	395
15.4.1	Local Distribution of Fungi	395
15.4.2	Habitat and Genetic Influences on Specificity	396
15.4.3	Ontogenetic Influences on Specificity	397
15.5	Evolution of Specificity	399
15.6	Fungal Trophic Niches and Mycorrhizal Carbon Dynamics	399
15.7	Conclusions and Future Goals	405
References		407

15.1 Summary

We present an analysis of fungal specificity in myco-heterotrophic orchids and monotropes. We argue that specificity represents a continuum and can only be properly assessed using phylogenetic data. Several green orchids associate with wide phylogenetic arrays of *Rhizoctonia* species, and hence show little specificity, while other green orchids, and all studied achlorophyllous orchids and monotropes, associate with narrow phylogenetic groups of fungi, and hence show significant specificity. In several species, this tight specificity has been shown to apply from seed germination through adulthood under natural conditions, though not necessarily under in vitro conditions. Patterns of specificity have been correlated with patterns of fungal distribution and habitat variation in several myco-heterotrophs. However, studies of other myco-heterotrophs have shown that tight specificity is expressed even when diverse fungi co-exist with the plant. Moreover, in one case, genetic influences of the host plant have been shown to outweigh environmental influences over the patterns of specificity. Major host jumps and intraspecific host-race formation have contributed to the evolution of specialisation in several myco-heterotrophs. Some achlorophyllous orchids associate with wood-decay or parasitic fungi, but many recent studies have revealed associations with ecto-mycorrhizal fungi in orchids, monotropes, and a liverwort. Tracer studies show that autotrophic ecto-mycorrhizal host plants can provide the fixed carbon to nourish myco-heterotrophs linked by a shared fungal partner. Important outstanding questions concern recognition phenomena, the origins and evolution of specificity, the physiology and ecology of carbon exchange, and whether myco-heterotrophs interact with fungi in fundamentally different ways than do autotrophs.

15.2 Introduction

Leake (1994) defined plants that depend upon fungi for the supply of essential carbon sources, and in which the "normal" polarity of sugar movement from plant to fungus is reversed, as "myco-heterotrophs". Two classes of myco-heterotrophic plant (MHP) were recognised, one in which the ability to fix carbon has been completely lost (the "fully" myco-heterotrophic plants), and one in which, at least in later stages of the life cycle, some autotrophic capability is retained (the so-called "partial" myco-heterotrophs).

Dependence on fungal-derived energy sources has arisen independently on multiple occasions through the evolution of land plants. Full myco-heterotrophy occurs in roughly 400 species distributed through the Ericaceae (Monotropoideae), Polygalaceae, and Gentianaceae of the Dicotyledonae and the Burmanniaceae, Corsiaceae, Lacandoniaceae, Orchidaceae, Petrosaviaceae and Triuridaceae of the Monocotyledonae (Leake 1994). The achlorophyllous, gametophytic stages of some leptosporangiate ferns and lycopods as well as the sporophyte and gametophyte of the hepatic *Cryptothallus mirabilis*, are also myco-heterotrophic (Read et al. 2000). This assemblage of unrelated plants includes many taxa that have been shown to associate with ecto-mycorrhizal (EM) Basidiomycete fungi, and a number of taxa whose mycorrhizal organs contain vesicles and/or arbuscules, indicating associations with arbuscular mycorrhizal (AM) glomalean fungi. The repeated evolution of this habit across the two major mycorrhizal categories, and in most major lineages of vascular plants, argues against the view that myco-heterotrophy is a rare and anomalous strategy. Partial myco-heterotrophy, which encompasses all green orchids, appears to be even more widespread.

Seen from this perspective, full MHPs represent one end of an evolutionary continuum across which dependence upon fungi for supply of carbon moves from absolute to partial to none. This perception is contrary to that of Robinson and Fitter (1999) who consider full MHPs to be entirely distinct from all other mycorrhizal plants. Specificity toward particular fungi, a striking feature of at least some full MHPs, can also be viewed as a continuum, with possible ties to the autotroph–myco-heterotroph continuum. Here, we analyze environmental, genetic and evolutionary influences on specificity in the best-studied MHPs of the Orchidaceae and Monotropoideae from a continuum perspective. Specificity phenomena between plants and arbuscular mycorrhizal fungi or ecto-mycorrhizal fungi are reviewed respectively by Sanders (Chap. 16, this Vol.) and Molina et al. (1992).

Knowledge of the fine structure and cellular biology of the mycorrhizal interactions in several partial MHPs has progressed significantly in recent years (Peterson et al. 1998; Schmid and Oberwinkler 1993, 1994, 1995, 1996; Schmid et al. 1995; Uetake et al. 1992, 1997; Uetake and Ishizaka 1996; Uetake and Peterson 1997, 1998), but will not be covered in this review. Instead, we will focus on major gaps in the understanding of myco-heterotrophy that were pointed out in the conclusions of Leake (1994), namely, those concerning the identities and trophic niches of the fungal associates, the ecology of seed germination under natural conditions, and the dynamics of carbon transfer from fungus to plant.

15.3 Evidence for Specificity in Myco-heterotrophs

The fact that most plants display little evidence of specificity in their relationships with mycorrhizal fungi (Molina et al. 1992; also see Sanders, Chap. 16, this Vol.) make it all the more important to determine, in those cases where specificity is seen, the nature and impact of the specific associations. Both the evolution and ecology of specificity in symbioses are important areas of basic inquiry (Bernays 1988; Jaenike 1990; Berenbaum 1996), but to date, few model systems involving specificity *toward*, rather than *by,* fungi have been identified. From an applied standpoint, recognition of a requirement for a specific symbiont could prove critical to the conservation of MHPs.

There is a long and lively history of debate concerning the specificity of orchids toward their fungal symbionts (see e.g. Hadley 1970 versus Clements 1988). Some of the controversy over specificity arises simply from differences in the often unstated definitions used by different workers. A second major source of confusion has been the problematic identification and taxonomy of some of the fungal symbionts, especially the 'Rhizoctonia' species that frequently colonise orchids (Ramsay et al. 1986; Clements 1988). A third source of confusion has likely been the wide variation in fungal isolation procedures, and difficulties in distinguishing isolates that are mycorrhizal from those that colonise orchids as non-mycorrhizal endophytes or parasites (Andersen and Rasmussen 1996). New phylogenetic data and methods offer hope for clarifying these points of confusion.

Molecular-phylogenetic and ultrastructural methods are helping to resolve both *Rhizoctonia* systematics, and the problems of discriminating mycorrhizal from non-mycorrhizal isolates. DNA sequence data from ribosomal genes and spacer regions are revolutionising our understandings of fungal systematics and mycorrhizal ecology (Bruns et al. 1992; Swann and Taylor 1993; Berbee 1996; Gardes and Bruns 1996; Hibbett et al. 1997; Karen and Nylund 1997; O'Donnell et al. 1997). These methods have allowed the rapid identification of fungi that are difficult or impossible to isolate in pure culture, thus, circumventing the biases particular to fungal isolation procedures (Taylor and Bruns 1997; Taylor and Bruns 1999b). However, the polymerase chain reaction (PCR) method can introduce its own set of biases, so a combination of techniques is preferable (Taylor and Bruns 1999b), and fungal isolation is obviously a necessary step prior to conducting most experiments.

Problems associated with varying definitions of specificity can be overcome by adapting the modern, consensus definition of specificity from the general evolutionary and ecological literature (Thompson 1994). Specificity, as we define it, is not a binary categorical descriptor, but instead, represents a continuous axis, where the position of an organism on the axis is defined by "*the phylogenetic breadth of the mycorrhizal associations of that particular plant or fungus.*" We note that specificity in the fungal partner need not bear any particular relation to specificity in the plant; degree of specificity is a unique attribute of each partner. The description of an organism as 'more' or 'less' specific is best made by comparison with other organisms that are involved in similar interactions. Studies that have investigated specificity in myco-heterotrophic plants are discussed below and summarised in Table 15.1.

15.3.1 Overview of Specificity in Orchidaceae

Most orchids display an unusual life-history strategy in which minute "dust seeds" that lack substantial energy reserves are produced in great number and are typically highly adapted for wind dispersal (Ramsbottom 1922). This strategy is also typical of many MHPs outside the Orchidaceae (Leake 1994). Immediately following seed germination, in the pre-photosynthetic phase during which most plants utilise their seed reserves, orchids form mycorrhizal associations and extract energy-containing compounds needed for growth from their fungi. While most orchids eventually develop green, photosynthetic organs, fully myco-heterotrophic orchids have given up photosynthesis entirely, and rely upon fungal derived energy sources throughout their life cycle. Of the roughly 400 fully myco-heterotrophic angiosperms, approximately 35 % occur in the Orchidaceae (Leake 1994). These species are distributed across several tribes and many genera that are not sister taxa, showing that the transition to complete myco-heterotrophy has occurred independently on numerous occasions within this family (Dressler 1993).

Orchidaceous mycorrhizae are distinct from other major mycorrhizal categories both anatomically and in the taxonomy of the fungal symbionts. Hyphae proliferate abundantly within certain cortical cells, forming coils, known as pelotons that are reminiscent of Paris-type arbuscular (AM) or ericoid mycorrhizae (Burgeff 1959; Smith and Read 1997), but do not form any "mantle" outside the plant. However, AM vesicles and arbuscules are absent and all reliably described orchid fungi belong to the Basidiomycetes.

15.3.1.1 Rhizoctonia Systematics

Fungi isolated from mycorrhizal organs of adult orchids frequently belong to the anamorphic form-genus *Rhizoctonia* (Burgeff 1959; Currah et al. 1997; Hadley 1982). Higher fungi are often given two names which are based upon two different sets of characters: anamorphic names are based upon vegetative characteristics such as hyphal morphology and asexual spores, while teleomorphic names are based upon features of the sexual reproductive structures, i.e., macro and microscopic features of sporocarps. Sexual structures generally provide more informative characters for taxonomy and systematics than do vegetative structures. However, fungi placed in the form-genus *Rhizoctonia* seldom reveal their basidiocarps, and, hence, are often referred to and identified by their anamorphs. Production of chains of swollen, monilioid cells (asexual resistant propagules) is a uniting anamorphic feature among *Rhizoctonia* fungi. The *Rhizoctonia* species from orchids have often been treated as though they belong to a coherent taxonomic entity, perhaps due to their anamorphic similarities and ubiquity in soil, despite the fact that the teleomorphs differ sufficiently to suggest distant phylogenetic relationships.

Members of the following *Rhizoctonia* anamorph/teleomorph pairs, all of which belong to the Hymenomycetes, have been isolated from orchids: *Opadorhiza/Sebacina, Epulorhiza/Tulasnella, Ceratorhiza/Ceratobasidium* and *Rhizoctonia* DC/*Thanatephorus* (Moore 1987, 1996; Andersen and Rasmussen 1996). While they have not been isolated from orchids, there are a number of additional teleomorphs that have, at times, been linked to *Rhizoctonia*, including *Athelia, Botryobasidium, Cejpomyces, Helicobasidium (Ure-*

Table 15.1. Selected studies dealing with specificity in myco-heterotrophic plants

Taxon	Samples	Place	Fungal identification methods[a]
Green orchids			
Amerorchis rotundifolia	19 Adults	Field	*In planta* + isolation: vegetative morphology
Acianthus reniformis	26 Adults	Field	Isolation: morphology of vegetative and sexual stages
Acianthus caudatus	12 Adults	Field	Isolation: morphology of vegetative and sexual stages
Acianthus exsertus	12 Adults	Field	Isolation: morphology of vegetative and sexual stages
20 *Caladenia* spp.	98 Adults	Field	Isolation: morphology of vegetative and sexual stages
Calypso bulbosa	? Adults	Field	*In planta* + isolation: morphology of vegetative and sexual stages
Cypripedium candidum	? Adults	Field	Isolation: vegetative morphology
Cypripedium parviflorum	? Adults	Field	Isolation: vegetative morphology
Dactylorhiza purpurella	21 Adults	Field	Isolation: vegetative morphology
Dactylorhiza purpurella	Seedlings	Lab	In vitro germination tests
5 *Diuris* spp.	28 Adults	Field	Isolation: morphology of vegetative and sexual stages
Goodyera repens	Seedlings	Lab	In vitro germination tests
Goodyera oblongifolia	8 Adults	Field	*In planta* + isolation: vegetative morphology
Microtis parviflora	18 Adults + 72 seedlings	Field	Isolation: vegetative morphology
Microtis parviflora	Seedlings	Lab	In vitro germination tests
Platanthera hyperborea	13 Adults	Field	*In planta* + isolation: vegetative morphology
Platanthera hyperborea	15 Seedling packets	Field	*In planta* + isolation: vegetative morphology
Platanthera leucophaea	? Adults	Field	Isolation: vegetative morphology
Platanthera obtusata	14 Adults	Field	Isolation: vegetative morphology
Pogonia ophioglossoides	? Adults	Field	Isolation: vegetative morphology
Pterostylis barbata	6 Adults	Field	Isolation: vegetative morphology, anastomosis grouping
Pterostylis nana	8 Adults	Field	Isolation: vegetative morphology, anastomosis grouping
Pterostylis aff. *rufa*	7 Adults	Field	Isolation: vegetative morphology, anastomosis grouping
Spiranthes sinensis	37 Adults	Field	Isolation: vegetative morphology
Spiranthes sinensis	Seedlings	Lab	In vitro germination tests
Spiranthes sinensis	18 Adults + 27 seedlings	Field	Isolation: vegetative morphology, anastomosis grouping

Reference	Identified fungi[b]	Trophic group[c]
Zelmer et al. (1996)	*Epulorhiza* (7), *Moniliopsis* (3)	Unknown
Warcup (1981)	25 of 26 isolates were *Sebacina vermifera*	Possibly ecto-mycorrhizal, unknown
Warcup (1981)	*Sebacina vermifera* (2), *Tulasnella cruciata* (10)	Possibly ecto-mycorrhizal, unknown
Warcup (1981)	*Tulasnella calospora*	Unknown
Warcup (1971)	108 of 110 isolates were *Sebacina vermifera*	Possibly ecto-mycorrhizal
Currah et al. (1988)	*Rhizoctonia* spp., *Rhizoctonia anaticula*, *Thanatephorus pennatus*, unidentified clamped fungi	Unknown
Curtis (1939)	*Rhizoctonia subtilis*[d]	Saprotroph
Curtis (1939)	*Rhizoctonia subtilis*[d]	Saprotroph
Harvais and Hadley (1967)	A variety of unidentified *Rhizoctonia* spp., *R. repens*, *R. solani*, and other fungi	Unknown
Harvais and Hadley (1967)	Unidentified *Rhizoctonia* spp., *R. repens*, and *R. solani* (including pathogenic strains)	Unknown
Warcup (1971)	*Tulasnella calospora*	Unknown
Hadley (1970)	*Ceratobasidium cornigerum* (1/3), *Ceratobasidium* sp. (1/2), *Thanatephorus cucumeris* (3/9), *Rhizoctonia* sp. (2/4)	N/A
Zelmer et al. (1996)	*Epulorhiza* (1), *Ceratorhiza* (34), *Moniliopsis* (3)	Unknown
Perkins et al. (1995)	2 *Epulorhiza* spp. were isolated from seedlings and adults	Unknown
Perkins et al. (1995)	*Epulorhiza repens*, *Epulorhiza* sp., 3 *Ceratorhiza* spp.	N/A
Zelmer et al., (1996)	*Epulorhiza* (7), *Ceratorhiza* (7), *Moniliopsis* (5)	Unknown
Zelmer et al. (1996)	*Epulorhiza* (3), *Ceratorhiza* (1), unknown clamped fungus (4), unidentified (1)	Unknown
Curtis (1939)	*Rhizoctonia robusta*, *R. sclerotica*, *R. Stahlii*, *R. subtilis*[d]	Saprotrophs
Currah et al. (1990)	*Epulorhiza anaticula*, *Ceratorhiza goodyerae-repentis*, *Sistotrema* sp.	Unknown, saprotrophs
Curtis (1939)	*Rhizoctonia monilioides*, *R. repens*[d]	Saprotrophs
Ramsay et al. (1987)	Binucleate *Rhizoctonia* spp., mostly anastomosis group P3	Unknown
Ramsay et al. (1987)	Binucleate *Rhizoctonia* spp., anastomosis groups P1 and P4	Unknown
Ramsay et al. (1987)	Binucleate *Rhizoctonia* spp., anastomosis group P2	Unknown
Terashita (1982)	*Rhizoctonia repens* (32), *Rhizoctonia solani* (16)	Unknown
Masuhara et al. (1993)	22 out of 23 *Rhizoctonia* tester strains, including binucleate and multinucleate isolates	N/A
Masuhara and Katsuya (1994)	All plants contained *Rhizoctonia repens*, 2 also had *R. solani*	Unknown

Table 15.1 (*Continued*)

Taxon	Samples	Place	Fungal identification methods[a]
Achlorophyllous orchids			
Cephalanthera austinae	26 Adults	Field	*In planta* + isolation: molecular (ITS RFLPs, ITS sequences, ML5-6 sequences)
Corallorhiza maculata	9 Adults	Field	Isolation: vegetative morphology
Corallorhiza maculata	? Adults	Field	*In planta* + *ex planta*: vegetative morphology and hyphal tracing
Corallorhiza maculata	104 Adults	Field	*In planta*: molecular (ITS RFLPs, ML5-6 sequences)
Corallorhiza mertensiana	27 Adults	Field	*In planta*: molecular (ITS RFLPs, ML5-6 sequences)
Corallorhiza striata	? Adults	Field	*In planta* + *ex planta*: vegetative morphology and hyphal tracing
Corallorhiza striata	8 Adults	Field	*In planta* + isolation: molecular (ITS RFLPs, ML5-6 sequencing)
Corallorhiza trifida	? Adults	Field	*In planta* + *ex planta*: vegetative morphology and hyphal tracing
Corallorhiza trifida	18+ Adults	Field	*In planta* + isolation: vegetative morphology
Corallorhiza trifida	4 Adults + 24 seedlings	Field	*In planta* + isolation: molecular (ITS RFLPs and ITS sequences)
Galeola altissima	Seedlings	Lab	In vitro germination tests
Galeola septentrionalis	? Adults	Field	*In planta* + isolation: vegetative morphology
Gastrodia cunninghamii	? Adults	Field	*In planta* + *ex planta*: vegetative morphology and hyphal tracing
Gastodia minor	? Adults	Field	*In planta* + *ex planta*: vegetative morphology and hyphal tracing
Gastrodia sesamoides	? Adults	Field	*In planta* + *ex planta*: vegetative morphology and hyphal tracing
Neottia nidus-avis	8 Adults + 7 seedlings	Field	*In planta*: molecular (ITS RFLPs, ITS sequences, 28S sequences)
Rhizanthella gardneri	1 Adult	Field	Isolation: morphology of vegetative and sexual stages
Yoania	? Adults	Field	*In planta* + *ex planta* + isolation: vegetative morphology and hyphal tracing
Achlorophyllous monotropes			
Allotropa virgata	? Adults	Field	*Ex planta*: mycorrhiza morphology
Allotropa virgata	37 Soil cores near adults	Field	*In planta* + *ex planta*: mycorrhiza morphology + molecular (ITS RFLPs)
Hemitomes congestum	? Adults	Field	*Ex planta*: mycorrhiza morphology
Monotropa uniflora	30 Adults	Field	*In planta* + *ex planta*: vegetative morphology and hyphal tracing

Reference	Identified fungi[b]	Trophic group[c]
Taylor and Bruns (1997)	14 Species spanning the *Thelephora-Tomentella* group	Ecto-mycorrhizal
Zelmer et al. (1996)	7 *Moniliopsis* isolates	Parasitic
Campbell (1970b)	Mainly *Armillaria melea*, also two unknown fungi	Parasitic
Taylor and Bruns (1997); Taylor and Bruns (1999b)	20 Species spanning much of the Russulaceae	Ecto-mycorrhizal
Taylor and Bruns (1999b)	3 Closely related species in the Russulaceae	Ecto-mycorrhizal
Campbell (1970b)	Unknown brown fungus, unknown white fungus	Ecto-mycorrhizal
Taylor (1997)	A narrow clade within the *Thelephora-Tomentella* group	Ecto-mycorrhizal
Campbell (1970b)	*Mycena thuja*, unknown fungus	Unknown
Zelmer and Currah (1995)	Unknown yellow, clamped Basidiomycete	Ecto-mycorrhizal
McKendrick et al. (2000b)	7 ITS RFLP types, all in the *Thelephora-Tomentella* group	Ecto-mycorrhizal
Umata (1995)	*Erythromyces crocicreas, Ganoderma australe, Loweporus tephroporus, Microporus affinis, Phellinus* sp.	Wood decay saprotrophs
Hamada (1939)	*Armillaria mellea*	Parasitic + saprotrophic
Campbell (1962)	*Armillaria mellea*	Parasitic + saprotrophic
Campbell (1963)	Unknown, brown, clamped Basidiomycete	Ecto-mycorrhizal
Campbell (1964)	Possibly *Fomes mastoporus*	Wood decay saprotroph (white rot)
S.L. McKendrick, J.R. Leake, D.L. Taylor and D.J. Read (in prep.)	*Sebacina vermifera*-like fungi	Unknown
Warcup (1985, 1991)	The single isolate obtained was named *Thanatephorus gardneri*	Ecto-mycorrhizal
Campbell (1970a)	*Lycoperdon perlatum*	Saprotroph + ecto-mycorrhizal + parasitic
Castellano and Trappe (1985)	*Rhizopogon vinicolor*	Ecto-mycorrhizal
Lefevre et al. (1998)	*Tricholoma magnivelare*	Ecto-mycorrhizal
Castellano and Trappe (1985)	*Rhizopogon vinicolor, Cenococcoum geophilum*	Ecto-mycorrhizal
Campbell (1971)	*Armillaria mellea*	Parasitic

Table 15.1 (*Continued*)

Taxon	Samples	Place	Fungal identification methods[a]
Monotropa uniflora	23 Adults	Field[e]	*In planta*: morphology
Monotropa uniflora	6 Adults	Field	*In planta*: molecular (ML5–6 sequences)
Monotropa hypopitys	10 Adults	Field	*In planta* + *ex planta*: vegetative morphology and hyphal tracing
Monotropa hypopitys	11 Adults	Field	*In planta* + culture
Monotropa hypopitys	? Adults	Field	*Ex planta*: mycorrhiza-fruitbody connections, mycorrhiza morphology
Monotropa hypopitys	9 Adults	Field	*In planta*: molecular (TSOP, ML5–6 sequences)
Monotropastrum globosum	? Adults	Field	*In planta* + *ex planta*: vegetative morphology and hyphal connections
Pleuricospora fimbriolata	? Adults	Field	*Ex planta*: mycorrhiza-fruitbody connections, mycorrhiza morphology
Pterospora andromedea	? Adults	Field	*Ex planta* + isolation: mycorrhiza morphology
Pterospora andromedea	31 Adults	Field	*In planta*: molecular (ITS RFLPs, TSOP, ML5–6 sequences)
Pterospora andromedea	Seedlings	Lab	In vitro germination tests
Sarcodes sanguinea	Seedlings	Lab	In vitro germination tests
Sarcodes sanguinea	12 Adults	Field	*In planta*: molecular (TSOP, ML5–6 sequences)
Sarcodes sanguinea	57 Adults	Field	*In planta* + isolation: molecular (ITS RFLPs, ITS sequences)

[a] Isolation refers to the culturing of fungi from pelotons or from whole tissue sections. *In planta* refers to the identification of fungi by direct methods that do not require fungal isolation. *Ex planta* refers to direct observations of fungal morphology or hyphal connections from plants to identifiable fungal structures (fruit-bodies or ecto-mycorrhizal roots). For the molecular methods, ML5–6 refers to fungal mitochondrial large subunit ribosomal gene sequences in the region described by Bruns et al. (1998). TSOP refers to "taxon-specific oligonucleotide probe" and ITS-RFLP refers to restriction digests of the fungal nuclear internal transcribed spacer region.

[b] Fungal taxa are listed exactly as given in the cited publication; note that different authors have used different nomenclatures. For *Rhizoctonia* fungi, various names may refer to the same taxon (e.g. *Rhizoctonia repens*, *Epulorhiza repens*, and *Tulasnella calospora*). In the case of laboratory seed germination tests, only fungi that produced compatible interactions are listed. Fractions in parentheses show the number of compatible strains over the total number of strains tested for a given fungal species.

[c] The trophic niches are listed whether or not the fungi are currently accepted as the legitimate mycorrhizal symbionts. In some cases we infer the trophic category of a fungus from other sources. However, if the authors have made a definitive statement concerning the trophic niche, their conclusion is given preference.

[d] It is not clear how the *Rhizoctonia* epithets used or coined by Curtis fit into modern *Rhizoctonia* taxonomy.

[e] Based on study of herbarium specimens.

Reference	Identified fungi[b]	Trophic group[c]
Martin (1986)	*Russula* species, and unidentified Russulaceae	Ecto-mycorrhizal
Cullings et al. (1996)	Russulaceae	Ecto-mycorrhizal
Campbell (1971)	Possibly *Clitocybe squamulosa*	Parasitic + ecto-mycorrhizal
Martin (1985)	Several species of *Tricholoma*	Ecto-mycorrhizal
Castellano and Trappe (1985)	*Elaphomyces muricatus*	Ecto-mycorrhizal
Cullings et al. (1996)	Suilloid group (including *Rhizopogon*)	Ecto-mycorrhizal
Kasuya et al. (1995)	Unknown yellow, clamped Basidiomycete	Ecto-mycorrhizal
Castellano and Trappe (1985)	*Truncocolumella citrina, Rhizopogon vinicolor, Cenococcum geophilum*	Ecto-mycorrhizal
Castellano and Trappe (1985)	*Cenococcum geophilum*, unknown isolate	Ecto-mycorrhizal
Cullings et al. (1996)	*Rhizopogon subcaerulescens* group	Ecto-mycorrhizal
Bruns and Read (2000)	2 Closely related *Rhizopogon* spp.	Ecto-mycorrhizal
Bruns and Read (2000)	2 Closely related *Rhizopogon* spp.	Ecto-mycorrhizal
Cullings et al. (1996)	Cantharellaceae, *Rhizopogon, Suillus*, unknown fungus	Ecto-mycorrhizal
Kretzer et al. (2000)	*Rhizopogon ellenae* complex	Ecto-mycorrhizal

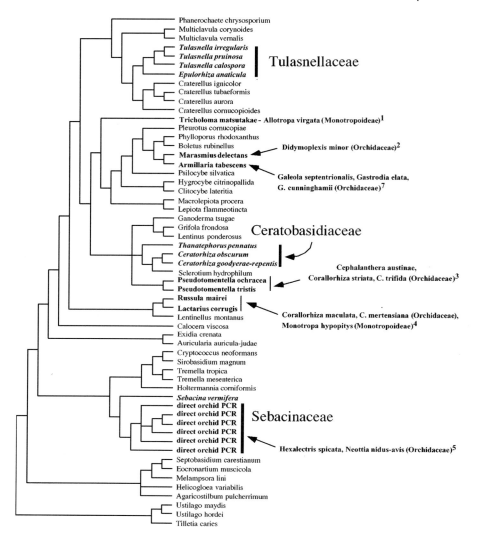

Fig. 15.1A,B. Maximum likelihood trees resulting from heuristic searches using the Hasegawa-Kishino-Yano (HKY) model of nucleotide substitution (unequal rates with gamma distribution shape parameter = 0.5 and Ti/Tv = 2.0) as implemented in PAUP*4.0 are shown. Swapping began using the most parsimonious trees; the cutoff for swapping on a tree was set at 1% worse than the maximum likelihood. Fungi that are members of *Rhizoctonia* sensu lato are indicated in *bold-italic* and are followed by the family of the associated teleomorph. The specific myco-heterotroph-fungal pairs are documented in the following references: *1* Lefevre et al. (1998); *2* Burgeff (1932); *3* Taylor (1997); Taylor and Bruns (1997); McKendrick et al. (2000b); *4* Cullings et al. (1996); Taylor and Bruns (1997, 1999b); *5* S.L. McKendrick, J.R. Leake, D.L. Taylor and D.J. Read, in prep.; D.L. Taylor, T.D. Bruns and S.A. Hodges, in prep.; *6* Cullings et al. (1996); Bidartondo et al. (2000); Kretzer et al. (2000); *7* Kusano (1911); Hamada (1939); Campbell (1962); Terashita (1996). **A** Strict consensus of the two maximum likelihood trees (-ln L scores = 9117.43441). The data set included 96 taxa and 451 bases from the 5′ end of the nuclear 28S ribosomal gene. The GenBank sequences are from the following studies:

Mycorrhizal Specificity and Function in Myco-heterotrophic Plants 387

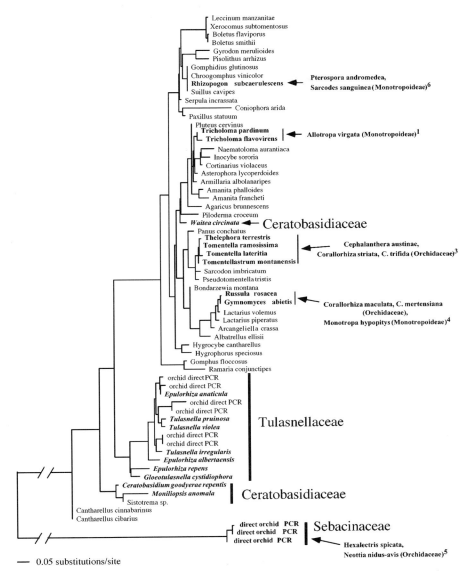

Berres et al. (1995); Feibelman et al. (1997); Hibbett et al. (1998); Johnson and Vilgalys (1998); Drehmel et al. (1999); Hopple and Vilgalys (1999); Mitchell and Bresinsky (1999); Taylor and Bruns (1999a) as well as D. L. Taylor (unpubl.). For ease of presentation, numerous taxa (mostly of the Agaricales) were pruned from the trees after the search was complete. The Ustilaginiomycetes were designated as a monophyletic outgroup. **B** Phylogram of one of the six maximum likelihood trees resulting from analysis of the 355 bp mitochondrial large subunit database of Bruns et al. (1998), with additional sequences from Kristiansen et al. (2001); D.L. Taylor, T.D. Bruns, S.A. Hodges (in prep.); and D.L. Taylor (unpubl.), for a total of 133 taxa. The six trees with scores of 4776.56233 were obtained from a heuristic search. Again, numerous taxa of less relevance were pruned from the resulting trees. Midpoint rooting was used *diniomycetes), Oliveonia, Scotomyces, Tofispora, Waitea* and *Ypsilonidium* (Andersen and Rasmussen 1996). The sexual structures of (1) *Sebacina,* (2) *Tulasnella,* and (3) *Ceratobasidium/Thanatephorus* differ markedly in the size, shape and septation of hypobasidia, basidia, sterigmata and basidiospores (Wells 1994; Andersen and Rasmussen 1996). In addition, the septal pore caps are unique in each of these three groups (Tu et al. 1977; Currah and Sherburne 1992; Andersen 1996; Muller et al. 1998). It is thus an unfortunate coincidence that a variety of soil-inhabiting Basidiomycetes were assigned to the same form-genus *and* were found associated with many orchids. This coincidence has caused confusion, and has likely slowed progress in understanding the specificity of orchids toward their mycorrhizal fungi.

Analysis of both nuclear (Fig. 15.1A) and mitochondrial large-subunit sequences (Fig. 15.1B) from an array of Basidiomycetes, including various orchid isolates, dramatically illustrate the large phylogenetic distances between the three major orchid *Rhizoctonia* clades, with *Waitea* comprising a fourth Hymenomycete clade that has not been found in orchids. Neither of these analyses resolves the placement of the major *Rhizoctonia* clades with confidence, which is not surprising, considering that there are very few representatives of the Corticiaceae, Aphyllophorales, or tremellaceous jelly fungi in these data sets, which were assembled primarily from available GenBank accessions (for fuller descriptions of the alignments and data sources, see Bruns et al. 1998; Taylor and Bruns 1999a). However, in addition to the wide divergences among the major *Rhizoctonia* clades, these analyses demonstrate that significant sequence variation is present among different orchid isolates *within* each major clade. Given that there are similar or lower levels of sequence divergence within other important Basidiomycete families (e.g. Russulaceae, Amanitaceae, Boletaceae; see discussion in Bruns et al. 1998), it appears that the major fungal clades associated with orchids are at least as old and diverse as these families, despite the fact that most orchid *Tulasnella* isolates have been referred to a single species, *T. calospora* (= *Rhizoctonia repens*). Hence, treating orchid *Rhizoctonia* strains as a homogeneous group is unsupportable from an evolutionary point of view.

Until recently, fungal specificity in orchids has been studied using only two approaches, the first being the analysis of seed germination and growth when paired with various fungal strains under controlled, monoxenic conditions on media in the laboratory, and the second being the morphological description and/or isolation of fungi from adult plants growing in the wild. We review the evidence about specificity obtained using these approaches in the next two sections.

15.3.1.2 Laboratory Seed Germination

Several of the most influential early workers in the field of orchid mycorrhiza, including Noel Bernard (1909) and Hans Burgeff (1932, 1936) found that only one or a few fungal strains were highly effective in promoting seed germination and growth in vitro. They interpreted these results as indicating a high degree of fungal specificity in these partially and fully MHPs. However, Curtis (1937, 1939), and a number of later workers, found that strains isolated from adults of one species were often ineffective in laboratory germination tests when paired with seeds of the same species, but stimulated germination of seeds from another orchid species (Hadley 1970; Harvais 1974; Harvais and Hadley 1967; Nishikawa and Ui 1976; Tokunaga and Nakagawa 1974). These inconsistent results led them to suggest a general lack of fungal specificity in green orchids. The few studies that have successfully obtained in vitro seed germination in achlorophyllous, fully myco-heterotrophic orchids have also suggested low specificity (Umata 1995, 1997a,b, 1998). *Ganoderma australe, Loweporus tephroporus, Microporus affinus, Phellinus* sp. and *Erythromyces crocicreas* all stimulated germination of the fully MHP *Galeola altissima* (Umata 1995), even though *E. crocicreas* is the only fungus known to associate with this orchid in the wild.

15.3.1.3 Isolation and Characterisation of Fungi from Wild Adults

Analyses of specificity in adult orchids based on the isolation of fungi and their identification by morphology have frequently suggested low specificity, at least in green species. Fungal isolates from single species of European, North American and Japanese terrestrial orchids often belong to several of the major *Rhizoctonia* clades (Currah et al. 1987, 1988, 1990; Curtis 1937, 1939; Harvais 1974; Harvais and Hadley 1967; Nishikawa and Ui 1976; Tokunaga and Nakagawa 1974). Furthermore, Curtis (1937) found that different orchid species growing in the same location often harboured the same *Rhizoctonia* strains, while a single orchid often harboured different *Rhizoctonia* strains in each distinct habitat in which it was found. Zelmer et al. (1996) examined squash mounts of mycorrhizal roots of 17 North American terrestrial species, and, like Curtis, observed different fungi in the same plant species from different sites. There are at least two potential limitations to these studies. First, because it has proven extremely difficult to induce fruit body formation in *Rhizoctonia* strains isolated from orchids, the identification of these isolates was based upon less reliable vegetative morphological features (see discussion in Andersen 1990; Andersen and Rasmussen 1996). Secondly, in most of these studies, isolates were obtained from slices of mycorrhizal tissue, which does not reliably discriminate between infections by mycorrhizal and other fungi.

In contrast to the above-mentioned studies, numerous green Australian orchids have yielded isolates that suggest significant specificity. Warcup and Talbot (1966, 1967, 1971, 1980) revolutionised orchid mycorrhizal research by obtaining fruit-body formation in numerous *Rhizoctonia* fungi isolated from orchids, enabling their teleomorphs to be determined. To avoid isolation of non-mycorrhizal root inhabiting fungi, individual pelotons were transferred onto isolation plates. Generally, isolates from several individuals of each orchid species were obtained, often across a wide geographic range. Several important patterns emerge from their analyses of hundreds of isolates and dozens of orchid species (Warcup 1971, 1981). First, some orchid species were associated almost exclusively with a single fungal species, while other orchids were restricted to associations within a single fungal genus, and only a few orchids consistently associated with *Rhizoctonia* species belonging to two or more of the major clades shown in Fig. 15.1A,B. Second, orchid species belonging to the same genus, or even subtribe, were often specific toward the same fungal species or genus. These patterns have been confirmed by several other groups working in Australia (Clements 1988; Perkins et al. 1995; Perkins and McGee 1995; Ramsay et al. 1986, 1987).

A further exception to the pattern of low specificity in adult green orchids was recorded in *Epipactis helleborine*. This orchid, which frequently occurs as albino forms intermixed with green plants, is consistently colonised by a morphologically distinct non-*Rhizoctonia* fungus with dark-coloured, unclamped, thick-walled verrucose hyphae over a wide geographic range in Eastern Europe (Salmia 1988, 1989). Although the dark fungus formed extensive pelotons, it could only be isolated when the most stringent sterilisation methods were employed (Salmia 1988). With less stringent sterilisation, numerous other soil inhabiting fungi appeared in culture and presumably out-competed the dark-coloured fungus.

15.3.1.4 Fully Myco-heterotrophic Orchids

While morphological studies of specificity in adult green orchids are contradictory, studies of adult achlorophyllous orchids have been more consistent, and suggest two interesting patterns. The first pattern is one of specific associations in myco-heterotrophic orchids (Table 15.1). The second pattern is that, with a few exceptions, the fungi involved are not *Rhizoctonia* species. Examples of these specific associations with non-*Rhizoctonia* fungi that have emerged from morphological identification methods include *Gastrodia elata* and *Galeola septentrionalis* with *Armillaria* spp. (Kusano 1911; Terashita 1996), *Galeola altissima* with *Erthryomyces crocicreas* (Hamada and Nakamura 1963), *Galeola hydra* with *Fomes* sp. (Burgeff 1959), *Didymoplexis minor* with *Marasmius coniatus*, and another *Gastrodia* sp. with a *Xerotus* sp. (Burgeff 1932, 1959). Campbell (1962, 1963, 1964, 1970a,b) also suggested specific fungal associations in a number of MHPs, although the reported identities of several of these fungi appear to be incorrect based upon later studies.

Studies in the last few years have applied two additional approaches to measuring and understanding specificity, namely the analysis of seed germination in the field, and the direct molecular identification, without isolation, of the fungi associated with both seedlings and adults from the field.

15.3.1.5 Seedling Germination in the Field

Gaku Masuhara and co-workers performed some of the earliest studies of seed germination in the field, and obtained evidence that the expression of specificity under laboratory conditions can differ strikingly from its expression in the field. As an aside, *Rhizoctonia* strains whose hyphae will fuse (anastomose) on Petri plates can potentially mate, while strains whose hyphae do not fuse appear to be reproductively isolated. Hence, the sorting of strains into "anastomosis groups" provides a useful technique for identifying related strains, especially within the large *Rhizoctonia solani* complex.

Masuhara et al. (1993) showed that in vitro germination and growth of the partial MHP *Spiranthes sinensis* was stimulated by *Rhizoctonia* tester strains from all but one of 23 multinucleate and binucleate anastomosis groups (mostly members of the *Ceratobasidium/Thanatephorus* clade), as well as *Rhizoctonia repens* (*Tulasnella*). In contrast, seeds that were planted into a field site in a rectangular array were colonised in 26 out of 27 cases by *Rhizoctonia repens* (Masuhara and Katsuya 1994). A similar predominance of *R. repens* was found in adults from the same site. Parsley stem baits were placed in the soil across the same grid, and isolations from these baits showed that seven different *Rhizoctonia* anastomosis groups were widely distributed in the field. Isolates of anastomosis group G were especially common, even though this fungus was never found in *Spiranthes* plants, while *R. repens* was not a dominant coloniser of the baits. The isolates obtained from parsley baits also induced *Spiranthes* germination and growth under sterile laboratory conditions, despite the fact that they were not associated with plants in the field. These authors proposed that the patterns in the field represent 'ecological specificity' while the in vitro results represent a much broader 'potential specificity'. Other studies have shown that the outcome of particular plant-fungus interactions in vitro are highly dependent on the exact nutrient conditions of the media (Beyrle and Smith 1993), and that fungal isolates tend to loose their symbiotic potential over time when maintained in culture (Ramsbottom 1922; Hadley 1982). These factors may have contributed to inconsistencies in laboratory germination studies.

15.3.1.6 Molecular Studies of Wild Plants

Recent studies of achlorophyllous orchids have improved our understandings of specificity by employing geographically widespread sampling and molecular tools for fungal identification. Zelmer and Currah (1995) showed that the leafless terrestrial orchid *Corallorhiza trifida* associates with a non-*Rhizoctonia* fungus displaying yellow, clamped hyphae in both Europe and Canada. Nuclear ribosomal internal transcribed spacer (ITS) sequence analysis has since shown that the specific symbionts of *C. trifida* are members of the Thelephoraceae (McKendrick et al. 2000b). Similar molecular methods, coupled with widespread sampling, demonstrated that *Cephalanthera austinae*, like *Corallorhiza trifida*, associates only with fungi in the Thelephoraceae, while *Corallorhiza maculata* and *C. mertensiana* associate only with fungi in the Russulaceae (Taylor and Bruns 1997, 1999b). Although these associations are not one-to-one, they indicate specificity toward fungi falling into single genera or families. The restriction of associations of achlorophyllous (and perhaps also green) orchids to discrete Basidiomycete families or genera represents narrow mycorrhizal specificity, in contrast to typical photosynthetic ecto-mycorrhizal plant species, which associate with numerous fungi representing wide phylogenetic arrays of Basidiomycetes and Ascomycetes (Molina et al. 1992).

While the trend of non-*Rhizoctonia* associations in achlorophyllous orchids is striking, there are some exceptions. The specific associates of the well-known achlorophyllous orchid *Neottia nidus-avis* are members of the *Sebacina* clade of *Rhizoctonia* fungi (S.L. McKendrick, J.R. Leake, D.L. Taylor and D.J. Read, in prep.). Similarly, a study of the achlorophyllous orchid *Hexalectris spicata* showed that the orchid was consistently associated with a *Sebacina*-like fungus on both the east and west coast of the United States, but that some east coast individuals also harboured *Thanatephorus pennatus* in both roots and non-mycorrhizal rhizomes (D.L. Taylor, T.D. Bruns and S.A. Hodges, in prep.).

The mixture of fungi found in *Hexalectris* serves to illustrate the potential dangers of evaluating specificity using isolation alone. The *Sebacina*-like fungi were slow growing and difficult to isolate, even from densely colonised roots, while the fast-growing *Thanatephorus* was easily isolated from roots and scantily colonised rhizomes. The authors suggest that the *Sebacina*-like fungi, which gave the dominant ITS restriction fragment length polymorphism (RFLP) patterns from peloton-filled roots of plants containing both fungi, are the specific associates of this orchid, while *Thanatephorus* occurs sporadically as a parasite or endophyte. Similarly, fast growing *Ceratorhiza* strains were isolated from *Corallorhiza maculata* (Zelmer et al. 1996), but considerable molecular evidence points to unculturable *Russula* species as the specific associates of this orchid (Taylor and Bruns 1997, 1999b). Orchids, like most plants, are probably hosts to numerous epiphytic and endophytic fungi (Carroll 1995; Clay 1993; Jumpponen and Trappe 1998; Petrini et al.

1995; Stone et al. 1996). Indeed, when isolations were carried out on roots and leaves of several epiphytic and lithophytic species of *Lepanthes*, very similar arrays of fungi, predominantly *Xylaria* and *Rhizoctonia* species, were isolated from *both* organs (Bayman et al. 1997). Since these fungi did not form pelotons in leaves, it seems that they were present as endophytes. Based on the patterns of isolation from *Lepanthes, Hexalectris* and *Corallorhiza*, we suggest that aggressive *Rhizoctonia* species from the *Ceratobasidium/Thanatephorus* clade may sometimes colonise orchids in an endophytic or parasitic manner while other fungi simultaneously function as the mycorrhizal symbionts of these orchids.

While narrow fungal specificity appears to be the rule in adult achlorophyllous orchids, generalisations cannot yet be made concerning specificity in green orchids due to the conflicting results of various studies. Molecular techniques may help to overcome the difficulties in identifying *Rhizoctonia*-like fungi and in dealing with mixed infections by both mycorrhizal and nonmycorrhizal fungi. Molecular techniques have so far been applied only to achlorophyllous species. However, a recent report demonstrated PCR-amplification of diagnostic fungal genes directly from single pelotons dissected from the green orchid *Dactylorhiza majalis* (Kristiansen et al. 2001). Such precise techniques offer hope for improved understandings of mycorrhizal interactions even in situations where multiple fungal species colonise a single plant.

15.3.2 Overview of Specificity in the Monotropoideae

The literature on specificity in the Monotropoideae is much less extensive than that on orchids. Monotropoid mycorrhizae are unique due to the penetration of plant epidermal cells by individual hyphal "pegs" which eventually lyse and eject their cytoplasm into the plant cell (Dudderidge and Read 1982). Monotropoid fine-roots are surrounded by a fungal mantle and contain Hartig-nets as in typical ecto-mycorrhizae (Smith and Read 1997).

Molecular studies have shown that individual species within the Monotropoideae associate with phylogenetically narrow groups of fungi, and thus far, all of these fungi have been ecto-mycorrhizal taxa. In the first such report, Cullings et al. (1996) used ITS-RFLPs, mitochondrial taxon-specific probes, and partial ITS sequence to show that *Pterospora andromedea*, throughout its range in western North America, was associated exclusively with fungi in the *Rhizopogon subcaerulescens* species group. They also examined three other species of monotropes, and, with the exception of *Sarcodes sanguinea*, found that all appeared to be specialists. However, their reports for species other than *P. andromedea* are in need of confirmation, as they were based on small sample sizes and no ITS data.

Two more recent reports confirm specificity for an additional species and remove the only apparent exception to specificity within the monotropes.

Lefevre et al. (1998) used molecular characterisation and morphology to show that *Allotropa virgata* associates exclusively with *Tricholoma magnivelare*. Kretzer et al. (2000) re-examined the associates of *Sarcodes sanguinea* from a much larger sample (76 plants versus 12) with a combination of ITS-RFLP and ITS sequence data obtained directly from roots; cultures were also isolated from the *Sarcodes* roots and used to synthesise mycorrhizae with pine. All sequences, RFLP patterns and cultures belonged to the *Rhizopogon ellenae* species complex. Additional sampling throughout most of the geographic range of *Sarcodes sanguinea* confirms this result (M. Bidartondo, pers. comm.). Interestingly, the *R. ellenae* complex is closely related to, but distinct from, the *Rhizopogon* species that associate with *P. andromedea*.

Kretzer et al. (2000) noted that the roots of *S. sanguinea* seemed to turn over fairly quickly, and those that were discoloured and had fragmented mantles were often difficult or impossible to amplify by PCR. They also noted, however, that fragmented mantles and associated rhizomorphs seen on older roots looked like those of *R. ellenae*. We suspect that the conflict with earlier molecular identification may have been based, at least in part, on amplifications from older roots by Culling et al. (1996). Nevertheless, six of the 12 mitochondrial large subunit (mt-LSU) ML5-6 region ribosomal gene sequences reported by Cullings et al. (1996) were suilloid, and are therefore consistent with a *Rhizopogon* identification. The other six are not good matches to any taxa in a fairly extensive database of ecto-mycorrhizal Basidiomycetes, although five of the six (reported as Cantharellaceae) are now known to be close to *Clavulina* (Bruns et al. 1998). Whatever the cause of the conflict, these earlier reports should now be viewed with skepticism given the greater sample size and higher resolution approach (ITS versus mt-LSU) used in the Kretzer et al. (2000) study.

All of the recent molecular studies have demonstrated specificity, and their results are therefore at odds with Castellano and Trappe (1985), who listed a different and more diverse set of fungal associates for several members of the Monotropoideae on the basis of attempts to trace mycelial connections between sporocarps and roots. It is noteworthy that the descriptions of the two fungal isolates from *Pterospora andromedea*, although not identified, resemble a *Rhizopogon* species. Using the same approach, Campbell (1971) claimed to have traced *Armillaria* rhizomorphs to *Monotropa uniflora* root balls, but the current molecular evidence suggests that the associates fall within the Russulaceae (Cullings et al. 1996). In contrast, morphological characterisation of the associates of *Monotropa uniflora* and *M. hypopithys* by Martin (1985, 1986) are compatible with recent molecular identifications. He reported that *M. uniflora* associated exclusively with several *Russula* species and that European collections of *M. hypopithys* associated with *Tricholoma* species. This agrees with the Cullings et al. (1996) report for *M. uniflora* and with recent unpublished work on *M. hypopithys* from M. Bidartondo (pers. comm.).

15.4 Influences on Specificity

In the preceding sections, we have summarised data showing that, unlike most autotrophic plants, a number of fully myco-heterotrophic species in the Orchidaceae and Monotropoideae are fungal specialists. We next consider the influences of genotype and environment on patterns of fungal specificity, and show how specificity has evolved in orchids and monotropes. We discuss evidence concerning the relative contributions of fungal distribution, habitat and genotype to the expression of specificity, and then take up the issue of whether changes in specificity occur through different developmental stages of the plants.

15.4.1 Local Distribution of Fungi

At present, the limited data available suggest that some MHPs that associate with specific fungi do so, even when surrounded by numerous fungal species. For example, a variety of fungal species, identified both as ecto-mycorrhizae and as fruit bodies, were observed in close proximity to, or intermingled with, the root balls of *Monotropa hypopitys, Allotropa virgata,* and *Pterospora andromedea* (Castellano and Trappe 1985). Combined with later evidence that these plants associate exclusively with single genera or species of fungi (Cullings et al. 1996; Lefevre et al. 1998), these observations are contrary to the hypothesis that specificity is simply due to an absence of alternate fungi. The studies described above of the *Rhizoctonia* fungi occurring in soil alongside the partial MHP *Spiranthes sinensis* (Masuhara and Katsuya 1994) repeat this pattern. Similarly, tree roots growing within a few centimetres of *Cephalanthera austinae* flower spikes were colonised by a variety of ecto-mycorrhizal fungi, often including species in the Russulaceae, while the orchid was associated exclusively with fungi in the Thelephoraceae (Taylor and Bruns 1997). Perhaps even more striking was the observation that every sampled individual of *Corallorhiza maculata*, growing in an area of several hundred square metres, was associated with a single *Russula* species which was never found fruiting on the plot, while mushrooms of six other *Russula* species were collected throughout the plot (Taylor and Bruns 1999b). These studies did not objectively quantify the diversity and abundance of fungal species at the sites, but do clearly show that numerous other fungi were present that were not associated with the specialised MHPs.

The only quantitative, belowground study of the fungal community surrounding a MHP revealed an immense diversity of ecto-mycorrhizal fungi and an intriguing fine-scale spatial patterning of these fungi (Bidartondo et al. 2000). ITS RFLP analysis revealed the presence of 80 different species of ecto-mycorrhizal fungi colonising red fir roots in only 36 soil cores harvested near *Sarcodes sanguinea* plants (Monotropoideae) at a single site. Within *Sarcodes* root balls, and in soil cores 10 cm away from the root balls, the exclusive *Sarcodes* symbiont, *Rhizopogon ellenae*, was the dominant fungus colonising fir roots (by mycorrhizal biomass). In contrast, *R. ellenae* was not a dominant fungus on fir roots in cores 100 cm from *Sarcodes* and was never found in cores 500 cm away. The authors argue that the *Sarcodes* plants must be promoting colonisation of fir roots by *Rhizopogon*, rather than occupying microsites already containing dense *Rhizopogon* mycorrhizae, due to the fact that the fir ecto-mycorrhizae occur surrounding the expanding *Sarcodes* root ball, rather than in its centre. This surprising observation is consistent with Bjorkman's finding that extracts from *Monotropa* plants stimulate growth of the *Monotropa* fungus, in vitro (Bjorkman 1960). Even more striking is the observation that fir root densities increase dramatically very close to *Sarcodes* plants (Bidartondo et al. 2000), suggesting that these plants are able to locally stimulate the growth of both the autotrophic host tree and the fungus. It would appear that these plants are able to significantly alter ecto-mycorrhizal community structure, at least on a fine spatial scale.

15.4.2 Habitat and Genetic Influences on Specificity

Studies of green Australian orchids have contributed to our understandings of geographic and habitat influences over specificity. Ramsay et al. (1987) isolated binucleate *Rhizoctonia* fungi (presumably *Ceratobasidium* spp.) from several *Pterostylis* species over a wide geographic area and from a variety of habitats, then determined the anastomosis group to which each isolate belonged. They found that one anastomosis group occurred over a wide area and range of habitats, while two others were common in dry habitats, and a fourth occurred only in a wetter coastal region. A fifth anastomosis group was also a dominant colonist, but only of *Pterostylis nana* and other species when they were growing in plantations of exotic pines, suggesting that this fungus is a habitat specialist, while isolates belonging to the first group were recovered from all *Pterostylis nana* plants growing outside the plantations in native plant communities. This appears to be a case where the pattern of fungal distribution across plant communities has impacts upon the observed patterns of specificity. Similar habitat and geographic patterns were uncovered by molecular analysis of the ecto-mycorrhizal *Russula* species associated with the fully MHP *Corallorhiza maculata* (Taylor and Bruns 1997, 1999b). Certain *Russula* species, defined by RFLP analysis of the ITS region, were the dominant symbionts of orchids growing in *Quercus* forests, but were never found in samples from nearby coniferous forests, thus, demonstrating a strong correlation between specificity and plant community. In addition, a single *Gymnomyces* species (Russulaceae) was found in all samples collected above 2000 m in elevation, but was never found at lower elevations.

An important outstanding issue concerns the determination of causation, as opposed to correlation, in producing the habitat and geographic patterns described above. Additional studies of *Corallorhiza maculata* have shown that much of the correlation between *Russula* species occurrences, in planta, and environmental factors could be due to co-variation with another critical parameter: plant genotype. Several sequence-characterised amplified region (SCAR) markers were developed for *C. maculata* and multilocus genotypes were determined for 122 plants (D.L. Taylor, T.D. Bruns and S.A. Hodges, in prep.). The *Russula* species associated with each of these plants was also determined, and placed in a phylogenetic context by sequencing of the fungal ITS region. Four of the six plant genotypes turned out to associate with their own separate clades of fungal species within the Russulaceae. The authors conclude that the genetic constituency of *C. maculata* individuals has a stronger influence on the occurrence of particular fungi, in planta, than does habitat, because different plant genotypes never shared a *Russula* species, even when growing together. Instead, each plant genotype was associated with a distinct fungal lineage, regardless of the presence of other plant genotypes and other *Russula* species at the same site. The correlations between fungal occurrences in *C. maculata* and habitat variables appear to be due to preferences of the various plant genotypes for different habitats. These sub-specialised plant genotypes appear to be analogous to the "host-races" commonly seen in herbivorous insects and phytopathogenic fungi (Price 1980; Farrell et al. 1992; Thompson 1994).

15.4.3 Ontogenetic Influences on Specificity

Symbiotic interactions often change in an organised fashion through the life of an organism. This concept is best illustrated by the many parasites that switch hosts between discrete life-cycle stages, such as host-alternating rust fungi. Due to the minute size of the seeds in most MHPs – a striking evolutionary convergence pointed out by Leake (1994) – the ontogenetic stages prior to the emergence of aboveground organs have been exceedingly difficult to observe in the field (Rasmussen and Whigham 1998). These difficulties have been largely overcome by encasing seeds in mesh bags or packets with pore sizes that retain the seeds but permit the entry of fungal hyphae (Rasmussen and Whigham 1993). This technique has been used to compare fungal associations in young protocorms, under nearly natural conditions, with fungal associations in wild adults, and to ask questions about the timing and stages of seedling development and the distribution of compatible fungi in the field.

The results from the majority of these seed-packet studies suggest that mycorrhizal specificity in MHPs is expressed from the very earliest stages of ontogeny. In the initial seed packet studies of Rasmussen and Whigham

(1993), they found that the achlorophyllous orchid *Corallorhiza odontorhiza* germinated only at sites where adults occurred naturally. Seeds that had germinated contained pelotons formed by a slow-growing, dark colored fungus bearing clamp connections. Similar fungi have been consistently observed in adults of this species, and found to belong to the Thelephoraceae based upon ITS sequence analysis (Taylor 1997). Thus, it seems likely that *C. odontorhiza* seedlings and adults are both specific toward fungi in the Thelephoraceae. Direct, PCR-based analyses of fungal ITS sequences in planta revealed that both minute seedlings and adults of *Corallorhiza trifida* from two continents associate exclusively with fungi in the Thelephoraceae (McKendrick et al. 2000b). Germination of *Corallorhiza maculata* seeds in packets occurred 2 years after planting, and involved the same *Russula* species that was found in most adults at that site (D.L. Taylor, unpubl. data). Molecular analyses also showed that seedlings from packets and wild adults of the fully MHP *Neottia nidus-avis* were associated with identical *Sebacina*-like species (S.L. McKendrick, J.R. Leake, D.L. Taylor and D.J. Read, in prep.). While these studies have found fungi belonging to the same genus or family in both seedlings and adults, seedlings were sometimes associated with fungal species that were not found in adults (McKendrick et al. 2000b). It is conceivable that a greater diversity of related species colonise wild seedlings than colonise adults, but the sample sizes of the present studies are too small to demonstrate this conclusively. An in vitro study has recently shown that *Pterospora* and *Sarcodes* (Monotropoideae) seeds germinate when challenged with closely related, but 'incorrect' species of *Rhizopogon*, while the seeds were unresponsive when challenged with an array of more distantly related ecto-mycorrhizal Basidiomycetes (Bruns and Read 2000). If similar phenomena occur in nature, it implies that specificity may narrow as plants mature subsequent to germination, due to compatibility phenomena. This narrowing could be accomplished by the replacement of incorrect with correct fungi during the life of the plant, or simply by the death of plants associated with the wrong fungus. The latter possibility is especially interesting because it would imply a large selection coefficient favouring 'targeting' by the plant of the most compatible fungi at the earliest possible stage. It will also be important to determine what role the fungi play in shaping observed patterns of specificity, e.g. whether fungal rejection of the plant could be responsible for incompatibility such as that seen in *Pterospora* seedlings associated with the wrong *Rhizopogon* species.

Since the seedlings of photosynthetic terrestrial orchids are initially heterotrophic, but become autotrophic at a later stage in ontogeny, it is worthwhile to ask whether changes in specificity coincide with the heterotrophic to autotrophic transition. Unfortunately, there is very little data on this subject. The field germination studies of *Spiranthes sinensis* clearly showed that this photosynthetic orchid targets the same fungus at seed germination and in adulthood (Masuhara and Katsuya 1994). In contrast, seedlings of the green orchids *Cypripedium calceolus*, *C. passerinum*, *Platanthera hyperborea* and

Spiranthes romanzoffiana from out-planted packets sometimes contained pelotons with clamp connections, while adults of these species never contained clamped hyphae, suggesting the occurrence of different fungal species in seedlings and adults (Zelmer et al. 1996). Furthermore, fungal hyphal morphologies in seedlings of *Platanthera hyperborea, Cypripedium passerinum* and *Spiranthes romanzoffiana* differed across sites (Zelmer et al. 1996). As with the studies of achlorophyllous orchids, larger sample sizes are needed to statistically test the apparent differences between seedlings and adults. Molecular identification of the fungi should also be pursued, as many of the fungi associated with seedlings in that study were not identified.

15.5 Evolution of Specificity

Molecular-phylogenetic studies are beginning to reveal evolutionary patterns of specialisation in MHPs, in addition to the ecological patterns described above. Representatives of only a few of the known specific associates of MHPs were available for inclusion in the phylogenetic trees presented in Fig. 15.1A,B. Yet, even these few examples demonstrate that a wide phylogenetic diversity of Basidiomycete fungi have become the targets of various orchids and monotropes. Furthermore, even closely related plants, such as different *Corallorhiza* species, target distantly related fungi (in this case the Russulaceae versus the Thelephoraceae), indicating that major "host-shifts" have occurred rapidly as measured on an evolutionary time-scale. Possibly, processes analogous to host-race formation have contributed to the evolution of mycorrhizal specialisation in other MHPs, as well as *Corallorhiza*. The recent finding of genetic variation in specificity within *Corallorhiza maculata* suggests that evolutionary changes in specificity are ongoing in this species. Such intraspecific genetic variation opens the way for studies of the selective pressures that are acting to shape specificity and other aspects of the mycorrhizal interaction.

15.6 Fungal Trophic Niches and Mycorrhizal Carbon Dynamics

The trophic niches (i.e. the major sources of energy) of the fungal symbionts of MHPs are of interest for several reasons. First, to the degree to which these plants are fungal-specific, they offer an indirect means of identifying, and perhaps quantifying, carbon flows through hyphal networks in soil. Second, the carbon sources of the fungi are likely to have strong impacts upon the ecology of these plants. Put more simply, if the fungus depends on a particular resource, the plant must also depend on it. Third, determination of fungal trophic niches may offer clues about the selective pressures that have shaped the evolution of specificity in MHPs. Of particular interest are differences in the trophic niches of fungi associated with autotrophic versus partially myco-heterotrophic versus fully myco-heterotrophic plants. It may be that one function of specificity in fully MHPs is to restrict associations to taxa belonging to a particularly suitable trophic niche. The trophic niches of fungal associations from several green orchids, achlorophyllous orchids and achlorophyllous monotropes are given in Table 15.1. The niches of ecto-mycorrhizal fungi and the impact of environmental factors on ecto-mycorrhizal fungi are discussed by Erland and Taylor (Chap. 7, this Vol.).

The *Rhizoctonia* species that have been found in wild orchid seedlings are of interest, since protocorms are non-photosynthetic, regardless of the later photosynthetic status of the plant. Fungi from each of the three *Rhizoctonia* clades found in adult orchids have also been reported in orchid protocorms from the field.

The *Ceratobasidium/Thanatephorus* clade has been most extensively studied due to the economic importance of the ubiquitous pathogen *Rhizoctonia solani* (*Thanatephorus cucumeris*). *Ceratobasidium* species have been isolated from numerous adult green orchids, and have also been found in a few wild seedlings (Zelmer et al. 1996; Hayakawa et al. 1999), most notably of orchids in the genus *Goodyera*. An elegant study by Downie demonstrated that the *Ceratobasidium* fungus of *Goodyera* occurred preferentially on dead pine needles in both the soil litter layer, and aboveground in the forest canopy (Downie 1943). Hence, it seems likely that seedlings of *Goodyera* acquire carbon indirectly from pine needles via the saprotrophic activities of the associated fungi. However, closely related or conspecific *Ceratobasidium* strains form endophytic associations with conifer roots (Sen et al. 1999).

The rare, achlorophyllous underground Australian orchids *Rhizanthella gardneri* and *R. slateri* form mycorrhizal associations with fungi which are also capable of forming ecto-mycorrhizal structures, including a mantle and Hartig net, on various photosynthetic hosts (Warcup 1985, 1991). Two isolates fruited in culture, but their taxonomic placement was problematic. Warcup eventually placed them in *Thanatephorus*, but stated that they have several atypical characters for that genus. Ecto-mycorrhiza formation has not been reported for any other *Thanatephorus* species. Further phylogenetic and ecological studies of these fungi are needed before concluding that *Thanatephorus* is a genus characterised by the ability to form ecto-mycorrhizae.

Rhizoctonia repens (*Tulasnella* clade) has been recorded in numerous wild orchid seedlings (Rasmussen and Whigham 1993; Masuhara and Katsuya 1994; Rasmussen 1995; Zelmer et al. 1996; Hayakawa et al. 1999). Strains of *R. repens* that were tested did have some cellulose degrading capabilities (Smith 1966), but several members of the genus *Tulasnella* were found to lack lignolytic enzymes (Worrall et al. 1997). There is no evidence that they form ecto-mycorrhizae. In summary, the trophic activities of fungi in this clade are almost completely unknown, which is unfortunate given their prominence as orchid mycorrhizal symbionts.

The last *Rhizoctonia* clade, which includes *Sebacina vermifera*, forms mycorrhizal associations with several genera of green orchids in Australia, and has recently been found to include specific associates of the achlorophyllous orchids *Neottia nidus-avis* (both seedlings and adults) and *Hexalectris spicata* (S.L. McKendrick, J.R. Leake, D.L. Taylor and D.J. Read, in prep.; D.L. Taylor, T.D. Bruns and S.A. Hodges, in prep). As with *Tulasnella*, the trophic activities of these fungi are unknown. The only clue is provided by the fact that *Sebacina vermifera* and related fungi have been isolated as secondary colonists of both ecto-mycorrhizae and arbuscular mycorrhizae of various autotrophs (Williams 1985; Milligan and Williams 1988; Warcup 1988; Williams and Thilo 1989). Whether they are plant root endophytes, plant parasites, saprotrophic rhizosphere fungi, or myco-parasites remains unclear.

As mentioned above, most achlorophyllous orchids do not associate with fungi that fall into the *Rhizoctonia* clades (Furman and Trappe 1971). These non-*Rhizoctonia* fungi are phylogenetically diverse (see Fig. 15.1A,B), but appear to share the attribute of exclusive access to large and reliable sources of fixed carbon. These fungi are aggressive pathogens, wood-decay fungi and ecto-mycorrhizal symbionts. It has been hypothesised that these fungi are preferred over *Rhizoctonia* species as targets for specificity by fully myco-heterotrophic plants due to their linkages to larger carbon sources (Taylor and Bruns 1997; McKendrick et al. 2000a). This hypothesis is currently difficult to evaluate given the uncertainty concerning the trophic activities of many *Rhizoctonia* species. Kusano (1911) made the radical claim that *Gastrodia elata* associates specifically with a species of *Armillaria*, a well-known genus of tree-killers and saprotrophs, and supported his claim with extremely detailed anatomical observations. Recent work in Japan has revealed the exact identities of several *Armillaria* species associated with another achlorophyllous orchid, *Galeola septentrionalis* (Cha and Igarashi 1996; Terashita 1996). *Gastrodia cunninghamii* in New Zealand has also been reported to associate with an *Armillaria* species (Campbell 1962). The only confirmed case of association with a wood-decay fungus occurs in the achlorophyllous liana, *Galeola altissima*, whose putative specific associate, *Erythromyces crocicreas*, decomposes the dead tree trunks on which the orchid is found (Hamada and Nakamura 1963; Umata 1995). There are two additional, unconfirmed cases. Burgeff reports the isolation of a *Xerotus* species, which could fall into the wood-decay category, from another *Galeola* species (Burgeff 1959), and Campbell believed the specific associate of *Gastrodia sesamoides* to be the wood-decay fungus *Fomes mastoporus* (Campbell 1964).

Several reports document specific associations between achlorophyllous orchids and ecto-mycorrhizal Basidiomycetes. Zelmer and Currah showed that isolates from *Corallorhiza trifida* produced ecto-mycorrhizae on *Pinus contorta* (Zelmer and Currah 1995). Molecular analyses were conducted on field-collected roots and rhizomes of the achlorophyllous orchids *Cephalanthera austinae* and *Corallorhiza maculata* and surrounding ecto-mycorrhizal tree roots (Taylor and Bruns 1997). PCR-amplified fungal ITS RFLP patterns and/or single strand conformational polymorphism (SSCP) "fingerprints" were identical in paired orchid and tree root samples, despite the strikingly different endomycorrhizal versus ecto-mycorrhizal anatomies observed.

A recent microcosm study (McKendrick et al. 2000a) using *C. trifida* provided the first confirmation both of the ability of this myco-heterotrophic orchid to act as a source of inoculum which enables ecto-mycorrhizae to form on the roots of autotrophs, and to facilitate the subsequent transfer of carbon between the plants through linking mycelium (Fig. 15.2a–i). *Betula pendula* plants were transferred to soil supporting naturally germinated seedlings of the orchid. Ecto-mycorrhizae formed rapidly on roots of *Betula* (Fig. 15.2a–c),

Fig. 15.2. A myco-heterotrophic orchid is linked to an ecto-mycorrhizal tree seedling through a shared fungal partner. **a–i** Microcosms in which the mycorrhizal fungus of the myco-heterotrophic orchid, *Corallorhiza trifida*, formed ecto-mycorrhizae on roots of *Betula pendula*. **a** The upper half of a representative microcosm containing *Corallorhiza* plants which were added as seedlings (*large arrow*), control plants in which hyphal connections to *Betula* were broken (*double arrow*), and a recruit of the orchid which developed from seed in situ (*small arrow*). Note the clusters of ecto-mycorrhizal tips on the *Betula* roots. **b** Detail of the "recruit" shown in **a**: fungal hyphal bridge (*large arrows*) between the developing orchid plant and adjacent mycorrhizal root tips; hyphae can be seen passing from the end of a rhizoid on the orchid seedling (*double arrow*) to adjacent ecto-mycorrhizal root tips. **c** Detail of the branched ecto-mycorrhizal root tips of *Betula* which developed in association with the mycorrhizal fungus of *C. trifida*. **d** The upper portion of a second replicate microcosm in which mycorrhizal links were established between *Corallorhiza* plants and *Betula* seedlings. For a key see **a**. Note the extensive cluster of mycorrhizal root tips in the lower right-hand quarter of the image. **e** Digital autoradiograph of the area shown in (d): counts detected in each pixel (0.25 mm²) are depicted on a linear 12-shade colour scale (0–23 counts per pixel; pixels with >23 counts displayed in the brightest red). Significant counts are seen in the recruit (*small arrow*), original orchid (*large arrow*) and in the ecto-mycorrhizal root tips of *Betula* (right hand quarter). **f** Details of area in **d** indicated by the *large arrow*, showing hyphal bridges between the original *Corallorhiza* plant and the *Betula* root which has grown horizontally across it. The *Betula* root has developed ecto-mycorrhizal short-roots linked to the orchid both by individual hyphal bridges (*white arrows*) and by multi-stranded rhizomorphs (*large black arrow*). One of the rhizomorphs (*double black arrow*) appears to connect to the group of rhizoids on the extreme right of the orchid rhizome. **g** The upper portion of a third replicate microcosm: the original orchid (*large arrow*) is partly buried in the soil but emerges in three other places below the *arrow*. Two recruits can be seen (*small arrows*). **h** Digital autoradiograph of the area shown in **g**: radioactivity can be visualised in the two recruits (*small arrows*), and in the original orchid (*large arrow* and *three circled areas* below this *arrow*) but none can be seen in the control orchid (*double arrow*). **i** Detail of the recruit shown in **g** and **h**, which is near the original plant, showing the developing orchid surrounded by young ecto-mycorrhizal roots tips of *Betula*

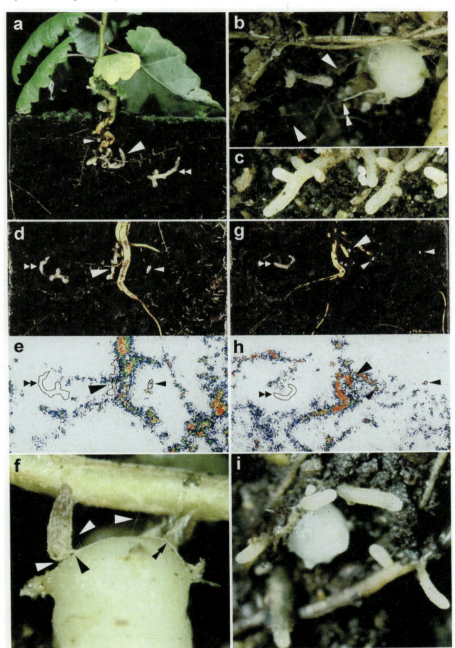

Fig. 15.3. A myco-heterotrophic, achlorophyllous hepatic is linked to an ecto-mycorrhizal tree seedling through a shared fungal partner. **A–D** Microcosms in which the mycorrhizal fungus of the achlorophyllous hepatic, *Cryptothallus mirabilis*, formed ecto-mycorrhizae with *Betula pendula*. **A** Microcosm supporting *Betula* seedlings growing on *Sphagnum* peat with plants of *C. mirabilis* (double arrows). Mycorrhizal fungal hyphae (*single arrows*) growing from *C. mirabilis* to colonise the peat. **B** A parallel microcosm with *Betula* grown on the same medium as in **A**, but without *Cryptothallus*. Note absence of fungal mycelia. **C** Close-up of *C. mirabilis* thallus grown with *B. pendula* as in **A** showing the conversion of roots of *Betula* to ecto-mycorrhizae (*single arrows*) in the vicinity of the hepatic. **D** Ecto-mycorrhizal laterals formed on monoxenically grown *Betula* seedlings inoculated with a pure culture of the mycorrhizal fungus of *C. mirabilis*. Sections of such roots (not shown) reveal a Hartig net and mantle which are typical ecto-mycorrhizal structures

and hyphae linking the myco-heterotroph to the autotroph could be readily observed (Fig. 15.2b,f). When shoots of the *Betula* plants linked in this way were fed $^{14}CO_2$ it was revealed, using digital autoradiography (Fig. 15.2e–h) followed by liquid scintillation counting, that significant quantities of carbon were transferred from the autotroph to the linked heterotroph. No transfer was detected in control systems lacking hyphal connections. When these experiments were repeated using *Salix repens* rather than *Betula* as the autotrophic associate, similar results were obtained demonstrating that the fungus lacked narrow specificity with regard to its autotrophic partners. The ecology of interplant carbon transport is further discussed in Chapter 2.

A similar relationship has also been demonstrated between the only known myco-heterotrophic bryophyte, the hepatic *Cryptothallus mirabilis*, and the autotroph *Betula*, with which *C. mirabilis* is consistently associated in nature (Read et al. 2000). Rhizotrons of the kind employed by McKendrick et al. (2000a) were again used. It was shown that hyphae emerging from the hepatic formed ecto-mycorrhizal associations with the *Betula* seedlings (Fig. 15.3A,C) while no such symbioses were produced in rhizotrons without the liverwort (Fig. 15.3B). Using fungal isolates obtained from pelotons found within the tissues of *C. mirabilis* thalli it was confirmed, under aseptic conditions, that the symbiont of the hepatic was able to produce mycorrhizae on *Betula* seedlings (Fig. 15.3D). Again, after exposing the autotroph to $^{14}CO_2$, significant quantities of carbon transfer to *C. mirabilis* were demonstrated (Read et al. 2000).

15.7 Conclusions and Future Goals

In this review, we have highlighted three novel patterns that have emerged from recent studies, primarily concerning members of the Orchidaceae and Monotropoideae. The first pattern is a consistently narrow specificity of these plants towards selected fungal families, genera and even species. In several of these taxa, this specificity has been shown to hold from the very earliest stages of ontogeny. The second pattern is that many of these specific associations are formed with fungi which simultaneously form ecto-mycorrhizae with neighboring autotrophic plants. This pattern is especially surprising in orchids and *Cryptothallus*, in view of the fact that their mycorrhizal structures involve internal penetration of cells in contrast to ecto-mycorrhizae in which the fungi are extracellular. The specificities shown between myco-heterotrophs of the Orchidaceae and Monotropoideae and their fungal partners are particularly striking, given the accumulating evidence that these plants often grow in the midst of diverse communities of ecto-mycorrhizal fungi. The third pattern is that carbon is often supplied to myco-heterotrophs from autotrophic plants through a shared mycorrhizal mycelial network. These patterns prompt two related questions, namely, *how* and *why* has specificity evolved in these plants?

Studies of the physiological mechanisms of recognition and rejection, on the parts of both the plant and the fungus, are needed in order to begin to understand *how* specificity is controlled at the cellular level. Recognition and rejection may be mediated by specific signal molecules and receptor genes, as with many plant-pathogen interactions. Comparative and phylogenetic studies are needed in order to determine *how* specificity has arisen from an evolutionary point of view. For example, it will be important to determine whether changes in specificity over evolutionary time are consistent and directional, (i.e., progressive narrowing) or chaotic, and to determine the frequency of host-jumps of different magnitudes (i.e., jumps between sister species, between genera, between families, etc.).

An understanding of *why* specificity exists will require rigorous analyses of the selective forces that act upon its expression. Tight specificity appears to be a more consistent feature of full MHPs than partial MHPs, suggesting that a plant's position on the autotrophy–myco-heterotrophy continuum may help to predict its place on the specificity continuum. While far more data are needed to test this pattern, if the pattern holds, it could help direct inquiries into the selective advantages of specificity. Experimental manipulations of MHPs in the field and microcosm should help to reveal the conditions under which different targets (i.e. different fungi) or levels of specificity (i.e. narrower or wider phylogenetic groups) are favoured. To understand selective pressures, it will be imperative to determine whether MHPs interact with their fungi, and their indirect autotrophic hosts, as parasites or as mutualists. The fact that MHPs acquire carbon from their fungi suggests that the plants may be acting as parasites. If so, the fungi would be expected to evolve defenses, thus setting the stage for an arms race which could easily favour specificity. On the other hand, it has long been postulated that MHPs supply some desirable metabolite to their fungi and thus act as mutualists. The stimulation of what we now believe to be *Tricholoma* by *Monotropa* extracts, and the proliferation of *Rhizopogon* mycorrhizae and fir roots around *Sarcodes*, support this view. However, stimulation does not necessarily imply benefit, as many parasites induce damaging proliferation (hypertrophy) of host tissues via hormonal manipulation.

Finally, we suggest that the question of broadest relevance concerning MHPs is whether they interact with fungal associates in a markedly different way than do photosynthetic plants, or whether they have simply modified existing interactions in ways that enable them to extract more carbon, and perhaps cheat the system? If the latter is the case, then MHPs represent only extremes in a continuum, and photosynthetic plants may also, at times, act as cheaters. Such a phenomenon would cast the mycorrhizal symbiosis in a less universally beneficent light, and demand more detailed analyses of costs and benefits, detection and regulation of cheaters, and the maintenance of fair exchanges.

Acknowledgements. DLT is grateful for fellowship support from the American Orchid Society and the University of California Office of the President. Financial support was provided to TDB by USDA grant 96-35101-3118. Financial support to JRL and DJR was provided by the Natural Environment Research Council (U.K.) grant GR3/10062. We thank Martin Bidartondo for comments on the manuscript.

References

Andersen TF (1990) A study of hyphal morphology in the form genus *Rhizoctonia*. Mycotaxon 37:25-46
Andersen TF (1996) A comparative taxonomic study of *Rhizoctonia* sensu lato employing morphological, ultrastructural and molecular methods. Mycol Res 100:1117-1128
Andersen TF, Rasmussen HN (1996) The mycorrhizal species of *Rhizoctonia*. In: Sneh B, Jabaji-Hare S, Neate S, Dijst G (eds) Rhizoctonia species: taxonomy, molecular biology, ecology, pathology and disease control, Second International Symposium on *Rhizoctonia*, Noordwijkerhout, Netherlands, 27-30 June 1995. Kluwer, Norwell, Massachusetts, pp 379-390
Bayman P, Lebron LL, Tremblay RL, Lodge DJ (1997) Variation in endophytic fungi from roots and leaves of *Lepanthes* (Orchidaceae). New Phytol 135:143-149
Berbee ML (1996) Loculoascomycete origins and evolution of filamentous ascomycete morphology based on 18S rRNA gene sequence data. Mol Biol Evol 13:462-470
Berenbaum MR (1996) Introduction to the symposium: on the evolution of specialization. Am Nat 148:S78-S83
Bernard N (1909) L'évolution dans la symbiose. Les orchidées et leur champignons commensaux. Ann Sci Nat Bot 9:1-196
Bernays EA (1988) Host specificity in phytophagous insects: Selection pressure from generalist predators. Entomol Exp Appl 49:131-140
Berres ME, Szabo LJ, McLaughlin DJ (1995) Phylogenetic relationships in auriculariaceous basidiomycetes based on 25S ribosomal DNA sequences. Mycologia 87:821-840
Beyrle HF, Smith SE (1993) Excessive carbon prevents greening of leaves in mycorrhizal seedlings of the terrestrial orchid *Orchis morio*. Lindleyana 8:97-99
Bidartondo MI, Kretzer AM, Pine EM, Bruns TD (2000) High root concentration and uneven ecto-mycorrhizal diversity near *Sarcodes sanguinea* (Ericaceae): a cheater that stimulates its victims? Am J Bot 87:1783-1788
Bjorkman E (1960) *Monotropa hypopitys* L. - an epiparasite on tree roots. Physiol Plant 13:308-327
Bruns TD, Read DJ (2000) Germination of *Sarcodes sanguinea* and *Pterospora andromedea* seeds is stimulated by the *Rhizopogon* species associated with the roots of adult plants. New Phytol 148:335-342
Bruns TD, Vilgalys R, Barns SM, Gonzalez D, Hibbett DS et al (1992) Evolutionary relationships within the fungi: analyses of nuclear small subunit rRNA sequences. Mol Phylogenet Evol 1:231-241
Bruns TD, Szaro TM, Gardes M, Cullings KW, Pan JJ et al (1998) A sequence database for the identification of ecto-mycorrhizal basidiomycetes by phylogenetic analysis. Mol Ecol 7:257-272
Burgeff H (1932) Saprophytismus und Symbiose. Studien an tropischen Orchideen. Fischer, Jena, 249 pp
Burgeff H (1936) Die Samenkeimung der Orchideen. Fischer, Jena, 312 pp
Burgeff H (1959) Mycorrhiza of orchids. In: Withner CL (ed) The orchids: a scientific survey. Ronald Press, New York, pp 361-395

Campbell EO (1962) The mycorrhiza of *Gastrodia cunninghamii* Hook Trans R Soc N Z 1:289–296

Campbell EO (1963) *Gastodia minor* Petrie, an epiparasite of Manuka. Trans R Soc N Z 2:73–81

Campbell EO (1964) The fungal association in a colony of *Gastrodia sesamoides* R. Br Trans R Soc N Z 2:237–246

Campbell EO (1970a) The fungal association of *Yoania australis*. Trans R Soc N Z Biol Sci 12:5–12

Campbell EO (1970b) Morphology of the fungal association in three species of *Corallorhiza* in Michigan. Mich Bot 9:108–113

Campbell EO (1971) Notes on the fungal association of two *Monotropa* species in Michigan. Mich Bot 10:63–67

Carroll G (1995) Forest endophytes: pattern and process. Can J Bot 73:S1316–S1324

Castellano M, Trappe J (1985) Mycorrhizal associations of five species of Monotropoidae in Oregon. Mycologia 77:499–502

Cha JY, Igarashi T (1996) *Armillaria jezoensis*, a new symbiont of *Galeola septentrionalis* (Orchidaceae) in Hokkaido. Mycoscience 37:21–24

Clay K (1993) The ecology and evolution of endophytes. Agric Ecosyst Environ 44:39–64

Clements MA (1988) Orchid mycorrhizal associations. Lindleyana 3:73–86

Cullings KW, Szaro TM, Bruns TD (1996) Evolution of extreme specialization within a lineage of ecto-mycorrhizal epiparasites. Nature 379:63–66

Currah RS, Sherburne R (1992) Septal ultrastructure of some fungal endophytes from boreal orchid mycorrhizas. Mycol Res 96:583–587

Currah RS, Sigler L, Hambleton S (1987) New records and new taxa of fungi from the mycorrhizae of terrestrial orchids of Alberta. Can J Bot 65:2473–2482

Currah RS, Hambleton S, Smreciu A (1988) The mycorrhizae and mycorrhizal fungi of *Calypso bulbosa* (Orchidaceae). Am J Bot 75:737–750

Currah RS, Smreciu EA, Hambleton S (1990) Mycorrhizae and mycorrhizal fungi of boreal species of *Platanthera* and *Coeloglossum* (Orchidaceae). Can J Bot 68:1171–1181

Currah RS, Zelmer CD, Hambleton S, Richardson KA (1997) Fungi from orchid mycorrhizas. In: Arditti J, Pridgeon AM (eds) Orchid biology: reviews and perspectives, vol VII. Kluwer, Boston, pp 117–170

Curtis JT (1937) Non-specificity of orchid mycorrhizal fungi. Proc Soc Exp Biol Med 36:43–44

Curtis JT (1939) The relation of specificity of orchid mycorrhizal fungi to the problem of symbiosis. Am J Bot 26:390–398

Downie DG (1943) Source of the symbiont of *Goodyera repens*. Trans Bot Soc Edinb 33:383–390

Drehmel D, Moncalvo J-M, Vilgalys R (1999) Molecular phylogeny of *Amanita* based on large-subunit ribosomal DNA sequences: implications for taxonomy and character evolution. Mycologia 91:610–618

Dressler RL (1993) Phylogeny and Classification of the Orchid Family. Dioscorides Press, Portland, 314 pp

Dudderidge JA, Read DJ (1982) An ultrastructural analysis of the development of mycorrhizas in *Monotropa hypopitys* L. New Phytol 92:203–214

Farrell BD, Mitter C, Futuyma DJ (1992) Diversification at the insect-plant interface. Bioscience 42:34–42

Feibelman TP, Doudrick RL, Cibula WG, Bennett JW (1997) Phylogenetic relationships within the Cantharellaceae inferred from sequence analysis of the nuclear large subunit rDNA. Mycol Res 101:1423–1430

Furman TE, Trappe JM (1971) Phylogeny and ecology of mycotrophic achlorophyllous angiosperms. Q Rev Biol 46:219–225

Gardes M, Bruns TD (1996) Community structure of ecto-mycorrhizal fungi in a *Pinus muricata* forest: above- and below-ground views. Can J Bot 74:1572-1583

Hadley G (1970) Non-specificity of symbiotic infection in orchid mycorrhiza. New Phytol 69:1015-1023

Hadley G (1982) Orchid Mycorrhiza. In: Arditti J (ed) Orchid biology: reviews and perspectives, vol II. Cornell Univ Press, Ithaca, pp 83-118

Hamada M (1939) Studien über die Mykorrhiza von *Galeola septentrionalis* Reichb. f. neuer Fall der Mykorrhiza-Bildung durch intraradicale Rhizomorpha. Jpn J Bot 10:151-211

Hamada M, Nakamura SI (1963) Wurzelsymbiose von *Galeola altissima* Reichb. F., einer chlorophyllfreien Orchidee, mit dem holzzerstörenden Pilz *Hymenochate crocicreas*. Berk Et Br Sci Rep Tohoku Univ Ser IV (Biol) 29:227-238

Harvais G (1974) Notes on the biology of some native orchids of Thunder Bay, their endophytes and symbionts. Can J Bot 52:451-460

Harvais G, Hadley G (1967) The relation between host and endophyte in orchid mycorrhiza. New Phytol 66:205-215

Hayakawa S, Uetake Y, Ogoshi A (1999) Identification of symbiotic rhizoctonias from naturally occurring protocorms and roots of *Dactylorhiza aristata* (Orchidaceae). J Fac Agric Hokkaido Univ 69:129-141

Hibbett DS, Pine EM, Langer E, Langer G, Donoghue MJ (1997) Evolution of gilled mushrooms and puffballs inferred from ribosomal DNA sequences. Proc Natl Acad Sci USA 94:12002-12006

Hibbett DS, Hansen K, Donoghue MJ (1998) Phylogeny and biogeography of Lentinula inferred from an expanded rDNA dataset. Mycol Res 102:1041-1049

Hopple JS, Vilgalys R (1999) Phylogenetic relationships in the mushroom genus Coprinus and dark-spored allies based on sequence data from the nuclear gene coding for the large ribosomal subunit RNA: divergent domains, outgroups, and monophylly. Mol Phylogenet Evol 13:1-19

Jaenike J (1990) Host specialization in phytophagous insects. Annu Rev Ecol Syst 21:243-273

Johnson J, Vilgalys R (1998) Phylogenetic systematics of *Lepiota* sensu lato based on nuclear large subunit rDNA evidence. Mycologia 90:971-979

Jumpponen A, Trappe JM (1998) Dark septate endophytes: a review of facultative biotrophic root-colonizing fungi. New Phytol 140:295-310

Karen O, Nylund J-E (1997) Effects of ammonium sulphate on the community structure and biomass of ecto-mycorrhizal fungi in a Norway spruce stand in southwestern Sweden. Can J Bot 75:1628-1642

Kasuya MCM, Masaka K, Igarashi T (1995) Mycorrhizae of *Monotropastrum globosum* growing in a *Fagus crenata* forest. Mycoscience 36:461-464

Kretzer AM, Bidartondo MI, Grubisha L, Spatafora JW, Szaro TM, Bruns TD (2000) Regional specialization of *Sarcodes sanguinea* (Ericaceae) on a single fungal symbiont from the *Rhizopogon ellenae* (Rhizopogonaceae) species complex. Am J Bot 87:1778-1782

Kristiansen KA, Taylor DL, Kjoller R, Rasmussen HN, Rosendahl S (2001) Identification of mycorrhizal fungi from *Dactylorhiza majalis* (Orchidaceae) based on PCR, SSCP and sequencing of Mitochondrial ribosomal LsDNA from single pelotons. Mol Ecol 10:2089-2093

Kusano S (1911) *Gastrodia elata* and its symbiotic association with *Armillaria mellea*. J Coll Agric Jpn 9:1-73

Leake JR (1994) The biology of myco-heterotrophic ('saprophytic') plants. Tansley review no 69. New Phytol 127:171-216

Lefevre CK, Carter CM, Molina R (1998) Morphological and molecular evidence of specificity between *Allotropa virgata* and *Tricholoma magnivelare*. In: Second International Conference on Mycorrhiza, Poster Presentation, Program and Abstracts. Uppsala, Sweden, p 107

Martin JF (1985) Sur la mycorhization de *Monotropa hypopithys* par quelques espèces du genre *Trichloma*. Bull Soc Mycol France 101:249–256

Martin JF (1986) Mycorhization de *Monotropa uniflora* L. par des Russulaceae. Bull Soc Mycol Fr102:155–159

Masuhara G, Katsuya K (1994) *In situ* and *in vitro* specificity between *Rhizoctonia* spp. and *Spiranthes sinensis* (Persoon) Ames. var. *amoena* (M. Bieberstein) Hara (Orchidaceae). New Phytol 127:711–718

Masuhara G, Katsuya K, Yamaguchi K (1993) Potential for symbiosis of *Rhizoctonia solani* and binucleate *Rhizoctonia* with seeds of *Spiranthes sinensis* var. *amoena* in vitro. Mycol Res 97:746–752

McKendrick SL, Leake JR, Read DJ (2000a) Symbiotic germination and development of myco-heterotrophic plants in nature: transfer of carbon from ecto-mycorrhizal *Salix repens* and *Betula pendula* to the orchid *Corallorhiza trifida* through shared hyphal connections. New Phytol 145:539–548

McKendrick SL, Leake JR, Taylor DL, Read DJ (2000b) Symbiotic germination and development of myco-heterotrophic plants in nature: ontogeny of *Corallorhiza trifida* Chatel and characterisation of its mycorrhizal fungi. New Phytol 145:523–537

Milligan MJ, Williams PG (1988) The mycorrhizal relationship of multinucleate rhizoctonias from non-orchids with *Microtis* (Orchidaceae). New Phytol 108:205–209

Mitchell AD, Bresinsky A (1999) Phylogenetic relationships of *Agaricus* species based on ITS-2 and 28S ribosomal DNA sequences. Mycologia 91:811–819

Molina R, Massicotte H, Trappe JM (1992) Specificity phenomena in mycorrhizal symbioses: community-ecological consequences and practical implications. In: Allen MF (ed) Mycorrhizal functioning: an integrative plant-fungal process. Chapman and Hall, New York, pp 357–423

Moore RT (1987) The genera of *Rhizoctonia*-like fungi: *Ascorhizoctonia, Ceratorhiza* gen. nov., *Epulorhiza* gen. nov., *Moniliopsis,* and *Rhizoctonia*. Mycotaxon 29:91–99

Moore RT (1996) The dolipore/parenthesome septum in modern taxonomy. In: Sneh B, Jabaji-Hare S, Neate S, Dijst G (eds) Rhizoctonia species: taxonomy, molecular biology, ecology, pathology and disease control. Second international symposium on *Rhizoctonia*, June 1995, The Netherlands. Kluwer, Boston, pp 13–34

Muller WH, Stalpers JA, Van Aelst AC, Van Der Krift TP, Boekhout T (1998) Field emission gun-scanning electron microscopy of septal pore caps of selected species in the *Rhizoctonia* s.l. complex. Mycologia 90:170–179

Nishikawa T, Ui T (1976) Rhizoctonias isolated from wild orchids in Hokkaido. Trans Mycol Soc Jpn 17:77–84

O'Donnell K, Cigelnik E, Weber NS, Trappe JM (1997) Phylogenetic relationships among ascomycetous truffles and the true and false morels inferred from 18S and 28S ribosomal DNA sequence analysis. Mycologia 89:48–65

Ogoshi A (1987) Ecology and pathogenicity of anastomosis and intraspecific groups of *Rhizoctonia solani* Kuhn. Annu Rev Phytopathol 25:125–143

Perkins AJ, McGee PA (1995) Distribution of the orchid mycorrhizal fungus, *Rhizoctonia solani*, in relation to its host, *Pterostylis acuminata*, in the field. Aust J Bot 43:565–575

Perkins AJ, Masuhara G, McGee PA (1995) Specificity of the Associations Between *Microtis parviflora* (Orchidaceae) and its mycorrhizal fungi. Aust J Bot 43:85–91

Peterson RL, Uetake Y, Zelmer C (1998) Fungal symbiosis with orchid protocorms. Symbiosis 25:29–55

Petrini O, Petrini LE, Rodrigues KF (1995) Xylariaceous endophytes: an exercise in biodiversity. Fitopatol Brasil 20:531–539

Price PW (1980) Evolutionary biology of parasites, vol 15. Princeton Univ Press, Princeton

Ramsay RR, Dixon KW, Sivasithamparam K (1986) Patterns of infection and endophytes associated with western Australian orchids. Lindleyana 1:203–214

Ramsay RR, Sivasithamparam K, Dixon KW (1987) Anastomosis groups among rhizoctonia-like endophytic fungi in southwestern Australian *Pterostylis* species (Orchidaceae). Lindleyana 2:161–166

Ramsbottom J (1922) Orchid Mycorrhiza. Trans Br Mycol Soc 12:28–61

Rasmussen HN (1995) Terrestrial orchids: from seed to mycotrophic plant. Cambridge Univ Press, New York

Rasmussen HN, Whigham DF (1993) Seed ecology of dust seeds *in situ*: a new study technique and its application to terrestrial orchids. Am J Bot 80:1374–1378

Rasmussen HN, Whigham DF (1998) The underground phase: A special challenge in studies of terrestrial orchid populations. Bot J Linnean Soc 126:49–64

Read DJ, Duckett JG, Francis R, Ligrone R, Russell A (2000) Symbiotic fungal associations in 'lower' land plants. Philos Trans R Soc Lond B 355:815–831

Robinson D, Fitter A (1999) The magnitude and control of carbon transfer between plants linked by a common mycorrhizal network. J Exp Bot 50:9–13

Salmia A (1988) Endomycorrhizal fungus in chlorophyll-free and green forms of the terrestrial orchid *Epipactis helleborine*. Karstenia 28:3–18

Salmia A (1989) Features of endomycorrhizal infection of chlorophyll-free and green forms of *Epipactis helleborine* (Orchidaceae). Ann Bot Fenn 26:15–26

Schmid E, Oberwinkler F (1993) Mycorrhiza-like interaction between the achlorophyllous gametophyte of *Lycopodium clavatum* L. and its fungal endophyte studied by light and electron microscopy. New Phytol 124:69–81

Schmid E, Oberwinkler F (1994) Light and electron microscopy of the host-fungus interaction in the achlorophyllous gametophyte of *Botrychium lunaria*. Can J Bot 72:182–188

Schmid E, Oberwinkler F (1995) A light- and electron-microscopic study on a vesicular-arbuscular host-fungus interaction in gametophytes and young sporophytes of the Gleicheniaceae (Filicales). New Phytol 129:317–324

Schmid E, Oberwinkler F (1996) Light and electron microscopy of a distinctive VA mycorrhiza in mature sporophytes of *Ophioglossum reticulatum*. Mycol Res 100:843–849

Schmid E, Oberwinkler F, Gomez LD (1995) Light and electron microscopy of a host-fungus interaction in the roots of some epiphytic ferns from Costa Rica. Can J Bot 73:991–996

Sen R, Hietala AM, Zelmer CD (1999) Common anastomosis and internal transcribed spacer RFLP groupings in binucleate *Rhizoctonia* isolates representing root endophytes of *Pinus sylvestris*, *Ceratorhiza* spp. from orchid mycorrhizas and a phytopathogenic anastomosis group. New Phytol 144:331–341

Smith SE (1966) Physiology and ecology of orchid mycorrhizal fungi with reference to seedling nutrition. New Phytol 65:488–499

Smith SE, Read DJ (1997) Mycorrhizal symbiosis, 2nd edn. Academic Press, San Diego

Stone JK, Sherwood MA, Carroll GC (1996) Canopy microfungi: function and diversity. Northwest Sci 70:37–45

Swann EC, Taylor JW (1993) Higher taxa of Basidiomycetes: an 18S rRNA gene perspective. Mycologia 85:923–936

Taylor DL (1997) The evolution of myco-heterotrophy and specificity in some North American orchids. PhD Thesis, Univ California, Berkeley

Taylor DL, Bruns TD (1997) Independent, specialized invasions of ecto-mycorrhizal mutualism by two non-photosynthetic orchids. Proc Natl Acad Sci USA 94:4510–4515

Taylor DL, Bruns TD (1999a) Community structure of ecto-mycorrhizal fungi in a *Pinus muricata* forest: minimal overlap between the mature forest and resistant propagule communities. Mol Ecol 8:1837–1850

Taylor DL, Bruns TD (1999b) Population, habitat and genetic correlates of mycorrhizal specialization in the 'cheating' orchids *Corallorhiza maculata* and *C. mertensiana*. Mol Ecol 8:1719–1732

Terashita T (1982) Fungi inhabiting wild orchids in Japan (II). Isolation of symbionts from *Spiranthes sinensis* var. *amoena*. Trans Mycol Soc Jpn 23:319–328

Terashita T (1996) Biological species of Armillaria symbiotic with *Galeola septentrionalis*. Nippon Kingakukai Kaiho 37:45–49

Thompson JN (1994) The coevolutionary process. Univ Chicago Press, Chicago

Tokunaga Y, Nakagawa T (1974) Mycorrhiza of orchids in Japan. Trans Mycol Soc Jpn 15:121–133

Tu CC, Kimbrough JW, Aldrich HC (1977) Cytology and ultrastructure of *Thanatephorus cucumeris* and related taxa of the *Rhizoctonia* complex. Can J Bot 55:2419–2436

Uetake Y, Ishizaka N (1996) Cytochemical localization of adenylate cyclase activity in the symbiotic protocorms of *Spiranthes sinensis*. Mycol Res 100:105–112

Uetake Y, Peterson RL (1997) Changes in actin filament arrays in protocorm cells of the orchid species, *Spiranthes sinensis*, induced by the symbiotic fungus *Ceratobasidium cornigerum*. Can J Bot 75:1661–1669

Uetake Y, Peterson RL (1998) Association between microtubules and symbiotic fungal hyphae in protocorm cells of the orchid species, *Spiranthes sinensis*. New Phytol 140:715–722

Uetake Y, Kobayashi K, Ogoshi A (1992) Ultrastructural changes during the symbiotic development of *Spiranthes sinensis* (Orchidaceae) protocorms associated with binucleate *Rhizoctonia* anastomosis group C. Mycol Res 96:199–209

Uetake Y, Farquhar ML, Peterson RL (1997) Changes in microtubule arrays in symbiotic orchid protocorms during fungal colonization and senescence. New Phytol 35:701–709

Umata H (1995) Seed germination of *Galeola altissima*, an achlorophyllous orchid, with aphyllophorales fungi. Mycoscience 36:369–372

Umata H (1997a) Formation of endomycorrhizas by an achlorophyllous orchid, *Erythrorchis ochobiensis*, and *Auricularia polytricha*. Mycoscience 38:335–339

Umata H (1997b) In vitro germination of *Erythrorchis ochobiensis* (Orchidaceae) in the presence of *Lyophyllum shimeji*, an ecto-mycorrhizal fungus. Mycoscience 38:355–357

Umata H (1998) A new biological function of shiitake mushroom, *Lentinula edodes*, in a myco-heterotrophic orchid, *Erythrorchis ochobiensis*. Mycoscience 39:85–88

Warcup JH (1971) Specificity of mycorrhizal association in some Australian terrestrial orchids. New Phytol 70:41–46

Warcup JH (1981) The mycorrhizal relationships of Australian orchids. New Phytol 87:371–381

Warcup JH (1985) *Rhizanthella gardneri* (Orchidaceae), its rhizoctonia endophyte and close association with *Melaleuca uncinata* (Myrtaceae) in Western Australia. New Phytol 99:273–280

Warcup JH (1988) Mycorrhizal associations of isolates of *Sebacina vermifera*. New Phytol 110:227–231

Warcup JH (1991) The *Rhizoctonia* endophytes of *Rhizanthella* (Orchidaceae). Mycol Res 95:656–659

Warcup JH, Talbot PHB (1966) Perfect states of some rhizoctonias. Trans Br Mycol Soc 49:427–435

Warcup JH, Talbot PHB (1967) Perfect states of rhizoctonias associated with orchids. New Phytol 66:631–641

Warcup JH, Talbot PHB (1971) Perfect states of rhizoctonias associated with orchids. II. New Phytol 70:35–40

Warcup JH, Talbot PHB (1980) Perfect states of rhizoctonias associated with orchids. III. New Phytol 86:267-272

Wells K (1994) Jelly fungi, then and now. Mycologia 86:18-48

Williams PG (1985) Orchidaceous rhizoctonias in pot cultures of vesicular-arbuscular mycorrhizal fungi. Can J Bot 63:1329-1333

Williams PG, Thilo E (1989) Ultrastructural evidence for the identity of some multinucleate rhizoctonias. New Phytol 112:513-518

Worrall JJ, Anagnost SE, Zabel RA (1997) Comparison of wood decay among diverse lignicolous fungi. Mycologia 89:199-219

Zelmer CD, Currah RS (1995) Evidence for a fungal liaison between *Corallorhiza trifida* (Orchidaceae) and *Pinus contorta* (Pinaceae). Can J Bot 73:862-866

Zelmer CD, Cuthbertson L, Currah RS (1996) Fungi associated with terrestrial orchid mycorrhizas, seeds and protocorms. Mycoscience 37:439-448

16 Specificity in the Arbuscular Mycorrhizal Symbiosis

IAN R. SANDERS

Contents

16.1	Summary	416
16.2	Introduction	416
16.3	Definitions of Specificity	417
16.4	Why Is Specificity in the Mycorrhizal Symbiosis Ecologically Interesting?	417
16.5	Theoretical Considerations on the Evolution of Specificity in Mutualistic Symbioses	419
16.6	Why Do We Not Already Know Whether Specificity Exists?	421
16.7	Evidence Supporting a Lack of Specificity	423
16.7.1	Arbuscular Mycorrhizal Fungi Have a Broad Host Range	423
16.7.2	A Systematic Perspective	424
16.7.3	Repeatable Patterns of Arbuscular Mycorrhizal Fungal Community Structure	425
16.7.4	Physiological Evidence	425
16.8	Evidence Supporting Specificity	425
16.8.1	Arbuscular Mycorrhizal Fungal Effects on Plant Fitness	425
16.8.2	Plant Species Effects on Fungal Fitness	426
16.9	What Information Is Missing and Which Experiments Are Needed?	428
16.9.1	Arbuscular Mycorrhizal Fungal Benefit from Specific Hosts	428
16.9.2	Reciprocal Benefit Between Plant Species and Fungal Species	428
16.9.3	A Genetic Basis for Specificity in Plants	429
16.9.4	A Genetic Basis for Specificity in Arbuscular Mycorrhizal Fungi	430
16.9.5	The Importance of the Hyphal Network for Understanding Specificity	433
16.10	Conclusions	434
	References	434

16.1 Summary

Different arbuscular mycorrhizal (AMF) fungal taxa have a differential effect on the growth of co-existing plant species. This means that in order to fully understand the role of these fungi in plant communities, information is needed on whether the symbiosis is specific. In this chapter, I briefly review the ecological consequences of specificity versus non-specificity in the arbuscular mycorrhizal symbiosis on plant ecology. Both from a theoretical approach, and based on observations, there has been an underlying assumption that no specificity exists in the arbuscular mycorrhizal symbiosis. I consider why these assumptions have been made. Direct evidence for or against specificity in the symbiosis is scant and the reason is mainly due to the difficulty in describing AMF community structure in natural communities (see Clapp et al., Chap. 8, this Vol.). Here, I take an evolutionary, as well as an ecological, approach to look at the evidence that predicts that evolution of specificity in the arbuscular mycorrhizal symbiosis could occur. I then consider alternative hypotheses and evidence that could explain why the evolution of specificity might not occur. These hypotheses are based on the growth habit, reproductive strategies and foraging behaviour of AMF and on new findings concerning AMF genetics.

16.2 Introduction

Given that the arbuscular mycorrhizal symbiosis is so widespread and that approximately 60 % of plant species are capable of forming this association (Trappe 1987) and, due to the capacity of the fungi to alter plant nutrition and provide other benefits to the plant, the symbiosis has the potential to influence ecosystem function (see Read, Chap. 1; Simard et al., Chap. 2; Hart and Klironomos, Chap. 9, this Vol.). In the last few years, a number of studies have indicated that the ecological effects of arbuscular mycorrhizal fungal (AMF) species on the growth and fitness of plant species are quite different. The relevance of these differential effects at the plant community level are discussed in this Volume (see Hart and Klironomos, Chap. 9; van der Heijden, Chap. 10; Bever et al., Chap. 11). Clearly, if different AMF species have the capability to make plants grow differently, then to know the true impact of AMF diversity on plant community structure and diversity and on plant population biology requires knowledge of whether there is any specificity in the arbuscular mycorrhizal symbiosis. Here, I consider specificity to be where a given AMF species or genotype forms a symbiosis with a particular plant species or genotype in natural ecosystems.

16.3 Definitions of Specificity

The existence of specificity in mycorrhizal symbioses has been reviewed in detail by Molina et al. (1992), covering both ecto-mycorrhizae and those formed by arbuscular mycorrhizal fungi. Specificity phenomena between myco-heterotrophic plants, on the one hand, and ecto mycorrhizal, orchid mycorrhizal fungi or other soil fungi, on the other, are discussed in Chapter 15 by Taylor et al. (this Vol.). Specificity has been considered from a plant-centric or myco-centric viewpoint. For example, plants could be specific with respect to their fungal partners by forming symbioses only with one or a limited number of particular fungal species. In this case, the fungi involved in this symbiosis may or may not be specific with respect to which plant species they form symbioses. AMF species may also be specific, in that they only form associations with given plant species, while the plants may or may not exhibit such specificity. In this chapter, I define specificity as where a given AMF species or genotype forms a symbiosis with a particular plant species or genotype in natural ecosystems and both plant-centric and myco-centric variations of this hypothesis are considered.

16.4 Why Is Specificity in the Mycorrhizal Symbiosis Ecologically Interesting?

It has been clearly demonstrated that different AMF that co-exist in natural communities can have differential effects on the growth of different plant species that make up the plant community (Streitwolf-Engel et al. 1997; van der Heijden et al. 1998a,b; van der Heijden, Chap. 10, this Vol.). The ecological significance of these findings depends very largely on whether any specificity in the arbuscular mycorrhizal symbiosis occurs in natural communities. Let us consider the two extreme hypothetical cases where the arbuscular mycorrhizal symbiosis is highly specific and where it exhibits low specificity (Fig. 16.1). If the symbiosis is highly specific and the fitness of a given plant species is dependent on one or a small number of AMF species, then distribution of that AMF may contribute to determining the distribution of that plant species (Fig. 16.1a). This could occur either by the ability or inability of the plant to establish from seed in an existing community, depending on the presence or absence of the appropriate AMF, or the inability to successfully proliferate by clonal growth into an area where the fungus is not present. In this highly specific case, because AMF are dependent on plants, it would likely mean that the distribution of the given plant species could also determine the distribution of the fungus. An outcome of such specificity would be a correlated pattern of patchiness of plant and fungal species in the community. A

(a)

Completely specific symbiosis

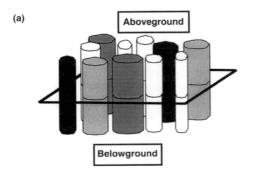

(b)

Completely unspecific symbiosis

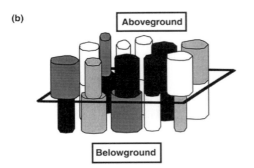

Fig. 16.1. a Distribution of different plant and AMF species in a scenario where the symbiosis is completely specific. Specificity results in correlated distributions of plant and fungal species, where the distribution of each partner in the symbiosis determines the size and distribution of the other partner. *Different colours* reflect different plant species (*aboveground*) or AMF species (*belowground*). b Scenario where the symbiosis is completely unspecific and distribution of one partner does not determine the distribution of the other partner. In this scenario different AMF have differential effects on plant growth causing size inequality in individuals of each species according to which AMF occupies its roots. Different *shading* above- and belowground represents different plant and fungal species respectively. Sizes of *cylinders* represent the size of the plant or fungal individuals

further prediction of the highly specific scenario is that increasing the number of AMF species should also result in an increase in the number of plant species that can become established in a community (Fig. 9.4 in Hart and Klironomos, Chap. 9, this Vol.). This should lead, therefore, to a relationship between the number of fungal species and the number of plant species that co-exist in a community.

In the scenario where the symbiosis is completely unspecific, the distribution of the different AMF in the soil is unlikely to play such a role in plant species distribution. However, because AMF species are known to differentially affect the size and fitness of different plant species, the distribution of the fungi could still affect plant community structure (Fig. 16.1b). A random distribution of the different AMF species may introduce a certain degree of apparently random distribution of plant size within a plant community. This could be a possible cause of size inequality in plant communities; a phenomenon which is thought to be a result of competitive interactions among individuals (Weiner and Thomas 1986; Shumway and Koide 1995). In the unspecific mycorrhizal symbiosis scenario, AMF diversity could also have strong effects on plant community structure, because a given AMF species is more beneficial for some plant species than for others. Therefore, the relative abundance of different AMF species in the community could still contribute to the relative abundance of different plants species comprising the community without any specificity existing. Different plant species have also been shown to determine life history traits of different AMF species (Sanders and Fitter 1992; Bever et al. 1996). Therefore, in both scenarios presented here, alterations in the plant community could determine the alteration of the AMF fungal community structure. Bever et al. (Chap. 11, this Vol.) discuss these interactions and how they could be coupled to form a feedback mechanism. Moreover, Hart and Klironomos (Chap. 9, this Vol.) show that spatial patterns in the distribution of different AMF occur on a local scale within plant communities (see Fig. 9.3).

A further scenario that has received little attention is that the phenomenon of specificity or non-specificity may not be a general one throughout all of the potential plant-fungal interactions comprising the arbuscular mycorrhizal symbiosis. There is also a possibility that some AMF species or genotypes may be generalists with respect to their host range and, therefore, be relatively unspecific and that others may be specialists that only form associations with one or a limited number of plant species or genotypes. I have not attempted to predict the ecological consequences of such a scenario but it is clear that this would make the impact of AMF diversity on plant ecology even more complex than the scenarios outlined above and in Fig. 16.1.

16.5 Theoretical Considerations on the Evolution of Specificity in Mutualistic Symbioses

Theory on the evolution of mutualistic symbioses has considered two separate stages of the evolutionary process: Firstly, how the transition from a non-mutualistic to a mutualistic symbiosis evolved. Secondly, how the partners have continued to evolve or co-evolve in a mutualistic environment, following the transition to a mutualistic association. The former of these has received considerably more attention than the latter (Clay 1988; Margulis and Fester 1991; Maynard-Smith and Szathmary 1995; Herre et al. 1999). In this chapter, I consider evolutionary processes that have occurred in the arbuscular mycorrhizal symbiosis once the partners have already evolved a mutualistic way of life. Law (1985) made a conceptual model for evolution in a mutualistic environment that considered the evolutionary consequences of a mutualistic lifestyle to be a reversal of those predicted by the red queen hypothesis in a parasitic association. The red queen hypothesis states that selection is greatest against the most frequent genotype of a host and a parasite in a population. The outcome of this selection is that the most frequent host genotype is parasitized by a specific parasite genotype and that, consequently, the density of that parasite genotype increases in the population. Selection then acts against the most frequent host genotype and favours less frequent host genotypes in the population that are resistant to the parasite. In turn, this favours other parasite genotypes in the population that are able to parasitize the commonest host genotypes. Thus, new and infrequent genotypes of parasite and host are favoured leading to rapid genetic change and potential evolution in the populations. Predictions from this hypothesis are that high genetic diversity of hosts and parasites are maintained and that evolution of new host and parasite genotypes is rapid. One way in which this rapid genetic change can take place is through a sexual lifestyle and so a further prediction is that the organisms are likely to be sexual. Another prediction is that the hosts and the parasites co-evolve and that this co-evolution evolves to associations that are highly specific between host and parasite genotypes.

In contrast to the red queen hypothesis, Law (1985) has considered that in mutualistic symbioses, selection pressures will not act against the most frequent genotype of host or symbiont if it receives a fitness benefit from the association. In contrast, they hypothesize that genotypes that cause the most benefit for their partner will be positively selected. An assumption in their hypothesis is that a given beneficial genotype, therefore, finds itself positively selected and, consequently, forms a stable strategy with its beneficial partner. The predictions of Law (1985) are that mutualistic organisms would not be under selection to form a sexual lifestyle, would not rapidly evolve new genotypes and would not evolve specificity. Law (1985) states that these predictions hold true for AMF in that they are asexual, they exhibit low rates of evolution and they are not specific. One scenario is that another more beneficial genotype could appear in a population that could replace the original genotype and that evolution and co-evolution could still proceed. Theoretical models based on "tit-for-tat" and "prisoners dilemma" predict that mutualistic or co-operative relationships between organisms can be stable (Nowak et al. 1995; Doebeli and Knowlton 1998), although they have not considered this scenario. In the case of the mycorrhizal symbiosis, however, plants can benefit more by forming associations with certain AMF. This could provide the selection pressure to form more specific symbioses. Plants that would form associations with particularly beneficial AMF would have a fitness advantage over their competitors. For Law's model (Law 1985) to apply to the interaction between plants and AMF, all AMF species would have to give an equal benefit to each plant species.

When Law (1985) produced his conceptual model, little or no genetic data were available on which to test whether the predictions were true for AMF. Therefore, predictions relied on morphological characterisation of the partners and the inferred rates of evolution. However, it is clear from detailed studies of other mutualistic interactions, namely the associations between figs and fig wasps that a considerable amount of radiation has occurred among the partners and that a high degree of specificity has also evolved or co-evolved (Herre et al. 1999). Also, several myco-heterotrophic plants associate only with a few fungal partners and, therefore, show a high degree of host specificity (Taylor et al., Chap. 15, this Vol.).

In the next sections, I examine why we do not already know whether specificity exists in the mycorrhizal symbiosis, the evidence supporting a lack of specificity in the arbuscular mycorrhizal symbiosis and the contrary evidence predicting the possible existence of specificity in the symbiosis.

16.6 Why Do We Not Already Know Whether Specificity Exists?

At first glance, it appears to be a relatively easy task to obtain descriptive field data showing whether there are specific interactions between certain plants and AMF. The identification of AMF in plant roots has, however, proved to be difficult. Morphological characterization of AMF inside roots has been attempted (Rosendahl et al. 1990; Merryweather and Fitter 1998). In both of these studies, morphological differences were observed among fungi in the roots but the fungi were not obtained into pot culture. As a result, it was not possible to compare the variation in morphology of each AMF within the roots of the different plant species, although this has been shown to vary (Lackie et al. 1987). Isozyme techniques also have been used to identify AMF in roots in pot experiments but they have had limited success (Hepper et al. 1988).

Since the demonstration by Simon et al. (1992a, 1993) that AMF can be distinguished on the basis of their rDNA sequences, identification of AMF in roots has concentrated on using DNA-based techniques. The techniques have now been developed to an extent where some data are being produced on AMF community structure and whether there is any specificity in the symbiosis. For a review of the techniques and the information that such techniques have so far yielded, see Clapp et al. (Chap. 8, this Vol.). Problems with using these techniques are that rDNA in AMF has been shown to be variable within individual AMF isolates and within individual spores (Sanders et al. 1995; Fig. 8.5 in Clapp et al., Chap. 8.5, this Vol.). Consequently, it is difficult to make primers that specifically amplify a given AMF species. Approaches with such primers have had limited success (van Tuinen et al. 1998). However, the variation in rDNA sequences means that such "assumed specific" primers may also amplify slightly different sequences of another AMF, or the primers may not amplify DNA from all individuals of the AMF of interest because some individuals may not contain that particular sequence, even though they are morphologically or functionally the same AMF. To be sure that PCR identification of an AMF is correct, the AMF also need to be isolated into pot culture so that sequences obtained from field roots can also be compared to sequences obtained from the cultured AMF. A way to avoid some of these problems is to sequence AMF rDNA from roots instead of relying on specific amplification. This approach has been taken by Helgasson et al. (1998; see also Clapp et al., Chap. 8, this Vol.), but it has the drawback of being time-consuming and expensive. A further disadvantage of this approach is that, although numerous different rDNA AMF sequences can be obtained from a single root, it is difficult to establish whether this variation comes from one AMF individual or a number of AMF individuals. One way of overcoming this problem may be to concentrate on other highly conserved functional genes that occur in low copy number and that may show less varia-

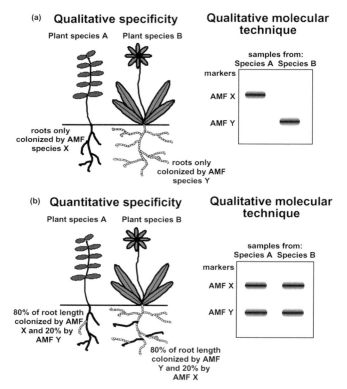

Fig. 16.2. Problems of detecting specificity in the AM symbiosis using qualitative molecular techniques. **a** Complete specificity exists in two hypothetical plant species *A* and *B* where they are colonized by hypothetical AMF species *X* and *Y* respectively (*left*). Qualitative specific molecular markers for *AMF X* and *Y* can be used to detect specificity in the symbiosis using PCR and by separating the PCR products on a gel (*right*). **b** Scenario where the symbiosis is specific because *AMF X* is able to colonize the roots of plant *A* better than *AMF Y* and vice versa for *AMF X* and *Y* colonizing plant *B* (*left*). The qualitative molecular identification technique does not reveal specificity in the symbiosis because positive amplifications show that *AMF X* and *Y* co-exist in plants *A* and *B* (*right*). A quantitative molecular technique would be needed in this case

tion within individual sequences. At present, sequences of other functional genes from AMF genomes have not been used for identification purposes. Currently, it is unknown whether different sequences can be obtained from the same functional gene. This clearly needs to be investigated in the future.

The other distinct drawback, so far, in the identification of AMF in roots of plants from natural ecosystems is that the methods that have been used successfully are qualitative but not quantitative. Quantitative PCR methods have been developed for identification of AMF in roots but these methods have not yet been effectively employed in the field (Simon et al. 1992b; Edwards et al. 1997). The development of these quantitative methods for use in the field is essential since specificity may not be reflected qualitatively (Fig. 16.2a) but quantitatively (Fig. 16.2b).

16.7 Evidence Supporting a Lack of Specificity

16.7.1 Arbuscular Mycorrhizal Fungi Have a Broad Host Range

Lack of specificity in the mycorrhizal symbiosis has been assumed because the roots of almost any plant species that is capable of forming the arbuscular mycorrhizal symbiosis (also called a mycotrophic species) can be colonized by almost any AMF isolate under carefully controlled conditions (Smith and Read 1997). Those conditions include a sterile substrate, host roots that have not been pre-colonized by native AMF from the soil in which it grows and the presence of no other AMF propagules in the soil. This simple demonstration that has been made many times, indicates that, under these artificial environmental conditions. the plant does not have the ability to prevent colonization by the fungus. Even in the case where the fungus provides no measurable growth benefit to the host, colonization is not eliminated, e.g. see data in Sanders (1993); van der Heijden et al. (1998a). It seems that plants do have the ability to regulate levels of colonization in cases where they have sufficient phosphate and this regulation has been hypothesized to occur by a reduced supply of carbohydrate to the fungus (Koide and Schreiner 1992). However, this is not conclusive evidence for whether specificity in the symbiosis occurs in natural communities. Such specificity in natural communities has been termed ecological specificity (Harely and Smith 1987) and there are several reasons why it could occur, even though experiments in artificial conditions indicate no specificity: Firstly, in such experiments, the fungus has no choice of host with which to display a preference. Secondly, certain fungal species may be quicker at colonizing certain plant species than others or may competitively exclude other AMF from the roots. Specificity could occur in the symbiosis in ecosystems if AMF have preferences for certain hosts, can colonize certain hosts better than other AMF species or can exclude other AMF following colonization (Wilson 1984).

16.7.2 A Systematic Perspective

Further evidence for the lack of specificity can be demonstrated by looking at the systematics of the partners involved in the arbuscular mycorrhizal symbiosis. As Law (1985) pointed out, AMF comprise a relatively small number of genera and species compared to the number of species of plant hosts. There are 6 known genera and approximately 150 described species of AMF (Morton and Benny 1990) and this constitutes a relatively low radiation, considering that AMF are ancient and occurred in roots of the first land plants from approximately 400 million years ago (Remy et al. 1995). Since there are many more host species than AMF species, it is clear that a single AMF species cannot be specific to one host plant but that it must form symbioses with many different plant species. This lack of specificity is also indicated by the fact that morphologically identical AMF species occur in different habitats comprising completely different plant species (Sanders 1993).

16.7.3 Repeatable Patterns of Arbuscular Mycorrhizal Fungal Community Structure

In plant and animal ecology there are repeatable and predictable patterns of species assemblages. In Europe, plant species assemblages have been studied in detail using various methods of vegetation classification (Ellenberg 1988). Give or take a few species, the presence of dominant and many of the subdominant plants in a given community are predictable. If a strict specificity were to exist among plant and AMF species, then repeatable patterns of AMF species should be seen that are predictable from knowing the species composition of plant communities. As outlined in the previous section, these data do not exist. Descriptions of AMF community structure have been given based on the relative abundance of different spores of AMF, although DNA-based investigations have shown that such measurements are unlikely to be accurate representations of AMF community structure. In order to accurately collect these data, the presence and absence of different AMF in the roots would have to be collected using methods other than counting the presence and absence of different AMF spores and reconstructing the AMF community structure on the basis of these data. Although investigations of the presence and absence of different spore types indicate that AMF community structure might change along gradients of vegetation or edaphic conditions (Johnson et al. 1991, 1992), there are no data indicating repeatable patterns showing that the same AMF species are found to be associated with the same plant species at different locations. The unpredictability in the distribution of AMF has been proposed as a possible mechanism preventing the evolution of specificity in the arbuscular mycorrhizal symbiosis. Hoeksema (1999) hypothesized that sufficient temporal or spatial variation or unpredictability in abundance of AMF could prevent the establishment of specificity between plants and AMF. Again, to know whether this is true with respect to AMF abundance will only be known when better descriptions are made on the distribution of AMF species.

16.7.4 Physiological Evidence

In an experiment where plant communities simulating calcareous grasslands were inoculated with native AMF, Grime et al. (1987) showed that labelled carbon was transferred from one of the dominant plants, *Festuca ovina*, to several of the sub-dominant plant species in the communities (see also Fig. 10.3 in van der Heijden, Chap. 10, this Vol.). The transfer of labelled carbon to sub-dominant plant species was negligible in the communities without AMF. The authors' interpretation was that the interspecific transfer of labelled carbon was due to the interconnection of plants by a common AMF hyphal network. This would mean that the same AMF species would connect the roots of different plant species and which, therefore, suggests that the fungi are not specific. The experiment of Grime et al. (1987) does not prove conclusively that the labelled carbon was transferred through the AMF hyphae. However, numerous other experiments demonstrate the movement of nutrients between plants of different species through an AMF hyphal network (see Simard et al., Chap. 2, this Vol.).

16.8 Evidence Supporting Specificity

16.8.1 Arbuscular Mycorrhizal Fungal Effects on Plant Fitness

Here, I take an evolutionary ecology approach to show the evidence existing that there could be selection towards specificity in the arbuscular mycorrhizal symbiosis. For many plant species, participation in the mycorrhizal symbiosis can clearly lead to changes in life-history traits that lead to fitness advantages, where fitness is defined as the probability of an individual passing its genes on to future generations. These advantages are manifested in terms of increased sexual reproductive output (Carey et al. 1992; Stanley et al. 1993), shorter time to reproductive maturity (Lu and Koide 1994; Shumway and Koide 1994) and increased clonal reproduction (Miller et al. 1987; Streitwolf-Engel et al. 1997). Interestingly, some of the fitness benefits have been shown to be independent of AMF effects on size, e.g. effects of AMF on clonal growth of *Prunella vulgaris* (Streitwolf-Engel et al. 1997) and increased phosphorus allocation to inflorescence of *Setaria lutescens* (Sanders and Koide 1994). This indicates that some fitness advantages of the AM symbiosis may not be explained by AMF effects on plant nutrition alone. Furthermore, some of those fitness benefits have been shown to actually occur in plant communities in the presence of other co-existing plant species (Sanders and Koide 1994; Streitwolf-Engel et al. 2001).

There are two prerequisites for specificity to evolve. A given plant species would have to obtain a fitness advantage by forming an association with a given AMF over that which could be obtained with other species of AMF that are present in the soil. Additionally, the most advantageous AMF species would not be the same for each plant species. This can be tested by inoculating different plant species with different AMF species that all originate from the same ecosystem, and testing for a plant species × AMF species interaction on a measure of plant fitness. Although many studies have shown that different AMF species have differential effects on plant growth, many of those studies are not ecologically relevant because they have not used different AMF and plant species originating from the same ecosystem (van der Heijden et al. 1998a). Recently, a few studies have been performed that show differential AMF isolate effects on plant growth using plants, AMF and soils that all originate from the same ecosystem (Streitwolf-Engel et al. 1997; van der Heijden et al. 1998a,b). In these studies, plant species benefited more or less, in terms of dry weight, according to which AMF occupied the roots. It is likely that this would lead to an increase in reproductive output, although in those studies this was not directly measured. Later investigations by Streitwolf-Engel et al. (2001) showed that some of the traits measured for *Prunella vulgaris* are fitness-related and are differentially affected by different AMF. Thus, selection pressures should exist for plants to form specific associations with AMF species.

16.8.2 Plant Species Effects on Fungal Fitness

The second prerequisite for the evolution of specificity in the arbuscular mycorrhizal symbiosis is that AMF receive a differential fitness benefit from different plant species. Hoeksema (1999) has proposed that the variation in benefit that AMF receive from their hosts is low and that this leads to low specificity on the fungal side of the symbiosis compared to the levels of specificity observed in ecto-mycorrhizal symbioses. This is particularly problematic to determine because it is difficult to define what the appropriate and measurable fitness-related traits are in AMF. Arbuscular mycorrhizal fungi have a clonal growth habit (Olsson et al., Chap. 4, this Vol.) and defining fitness in clonal organisms is often difficult, although possible where individual units of the organism can persist independently of the other (Harper 1977; Sackville-Hamilton et al. 1987; Charlesworth 1994). An AMF is obligately dependent on its host. Only after forming a symbiosis with a plant, can it then produce hyphae that grow out from the roots and have the ability to colonize new plants. One definition of fitness of an individual is the probability of passing on its genes to future generations. Therefore, one way that fitness may be measured could be as the success in terms of speed or ability to form new symbioses by colonizing new plant roots. Because extra-radical hyphae are the main vehicle for the colonization of roots, one useful measure of AMF fitness might be density of extra-radical hyphae. Unfortunately, many studies have concentrated on the effects of different hosts on the amount of colonization in the roots but have not measured external hyphal growth. Although measurements of AMF colonization inside the roots are likely to be a good measurement of direct host effects on AMF growth, they are not always correlated with external hyphal growth. In an experiment where *Prunella vulgaris* was inoculated with one of three different AMF, the relationship between percentage root length colonized by AMF and the external hyphal length was significantly negative for one of the AMF and not correlated at all for the other two AMF. However, there was a significant positive correlation between the number of vesicles produced in the roots and the density of external hyphae for each AMF isolate (R^2=0.85, 0.79 and 0.82, P=0.01; data previously unpublished but experiments described in Sanders et al. 1998). Thus, the level of AMF colonization inside the roots may not be a good predictor of the ability to colonize new plants. To my knowledge, no data exist on differences in external hyphal production among AMF isolates colonizing different host plant species, from the same place of origin.

Many studies have only measured spore production by AMF. In what are often short-term experiments, measurements of sporulation may not necessarily be a useful measure of AMF fitness. These spores, also termed azygospores are propagules of AMF that are usually produced on the termini of external hyphae in the soil. They are thought to be asexual products of the mother AMF and are capable of forming a new symbiosis with a plant. In temperate and tropical ecosystems, where living roots are available for colonization all year-round, the successful formation of new symbioses from spores may be relatively low compared with that formed by extra-radical hyphae. Indeed, many spores that can be extracted from soils are no longer viable (An et al. 1998). In arid and disturbed ecosystems, where a large number of plants are dead at certain times of the year and those which survive are sparsely distributed, the production of spores could be extremely important for the fitness of AMF for initiating new symbioses in periods when colonizable roots are available.

Despite the problems in defining measurements of AMF fitness, evidence exists for host species-induced changes in AMF life-history traits. Sanders and Fitter (1992) and Bever et al. (1996) have both shown that numbers of spores produced by different AMF were significantly and differentially affected by different host plant species. Thus, if sporulation was a good measure of fitness then selection pressures should exist for AMF to form specific associations with plant species. Data on AMF colonization of roots do not, however, support this. In the experiments by Streitwolf-Engel et al. (1997) and van der Heijden et al. (1998a), there were no AMF isolate × plant species interactions on AMF colonization by hyphae, even though main plant and AMF effects occurred. This means that a given AMF can benefit more by forming a symbiosis with a given plant species than with another species, however, other AMF will also benefit from forming a symbiosis with the same plant species. Thus, there would be no selection for specificity.

16.9 What Information Is Missing and Which Experiments Are Needed?

16.9.1 Arbuscular Mycorrhizal Fungal Benefit from Specific Hosts

Sanders (1993) hypothesized that if specificity occurs, then AMF from a given plant species would be able to colonize the roots of the specific host better than AMF from other plant species. Roots of four different plant species from a species-rich grassland, *Plantago lanceolata*, *Festuca rubra*, *Rumex acetosa* and *Trifolium pratense* were used to inoculate pot-grown *P. lanceolata*. Colonization by AMF differed according to inoculum source but the colonization by AMF originating from the roots of naturally occurring *P. lanceolata* were intermediate between the colonization levels observed with inoculum from *R. acetosa* and *T. pratense*. This experiment had a number of faults in the experimental design. One main fault was that the roots probably contained a mixture of different AMF. It could have been that some of the fungi in the roots of *P. lanceolata* were present in low levels in the field, but could form better symbioses with another plant species, e.g. with *T. pratense*. To test whether AMF really benefit most from the plant species that they colonise, such an experiment needs to be repeated after first isolating different AMF from the roots of the different plant species with which they normally form a symbiosis and not to carry out such an experiment with mixed inoculum.

16.9.2 Reciprocal Benefit Between Plant Species and Fungal Species

For specificity to evolve, plant and fungal combinations would be most favoured in selection if they provide reciprocal benefit. So far, few investigations have looked at this and evidence for reciprocal benefit is scarce. Several experiments indicate that AMF colonization is positively correlated with plant P concentration (Smith and Read 1997). However, in most, if not all, of those experiments, the plant and fungal material almost certainly came from different origins. In the experiments of Streitwolf-Engel et al. (1997, 2001) and Sanders et al. (1998), where plant and fungal material did have the same origin, there was no correlation between the percentage root length colonized by AMF or external AMF hyphal density and the growth of the plants. Although the contrary was found by van der Heijden et al. (1998a) for other plant-fungal combinations. Reciprocal benefit and its consequences in ecosystems are discussed by Bever et al. (Chap. 11, this Vol.). It is clear that more precise experiments are required to look for reciprocal benefits between specific plant and AMF combinations.

16.9.3 A Genetic Basis for Specificity in Plants

If specificity occurs, then there should be a genetic basis for its evolution. A recent study tested whether different genotypes of *P. vulgaris* exhibited different fitness as a result of inoculation with different AMF isolates (Streitwolf-Engel et al. 2001). All isolates originated from a species-rich calcareous grassland where the *P. vulgaris* genotypes were collected. There were no plant genotype × AMF isolate interactions for either the growth of the plants or the AMF. Thus, each AMF had an equal benefit, irrespective of which genotype it colonized, and plant genotypes did not have any advantage over other plant genotypes in forming specific symbioses with AMF isolates. These results thus suggest that there was no genetic basis for specificity in this particular plants species. The result of such experiments can only be attributed to genetic effects if all possible maternal effects are eliminated (Lynch and Walsh 1997). In the case of *P. vulgaris*, the maternal effects could be due to the different abiotic environments that each individual genotype locally experienced before collection, but would also include the maternal effects of the actual AMF that inhabited the roots of the different *P. vulgaris* genotypes in the field. Because AMF effects on the growth of *P. vulgaris* are strong and because AMF have been shown to confer maternal effects on offspring fitness in subsequent generations (Shumway and Koide 1994), the need to eliminate the maternal effects was essential. The maternal effects were reduced in this experiment by successively sub-culturing each *P. vulgaris* genotype in a homogeneous environment in meristem culture, without AMF. Reciprocal inoculation experiments in which different plant genotypes are inoculated with different AMF genotypes that were isolated from each of these plant genotypes in the field need to be done next.

16.9.4 A Genetic Basis for Specificity in Arbuscular Mycorrhizal Fungi

Specific differences in AMF genotypes have not yet been investigated. So far, different genotypes from plant populations have been tested for a differential response to AMF isolates. However, no serious attempts have been made to study differential effects of genetically different AMF from a population of one morphologically recognised AMF species (morphotype). Such experiments are clearly needed. The predictions that selection pressures exist favouring specificity because plants benefit from certain AMF are based on the assumption that differences in plant growth are caused by different AMF taxa. However, the variation in plant response to genetically different AMF from a population of one AMF morphotype has not been tested. If variation in plant growth is as great among plants inoculated with individual AMF from a population as among morphologically distinct AMF from the AMF community, then there may be no selection pressures favouring specific symbioses between plant species and recognisable AMF species/morphotypes. This has not been adequately addressed experimentally. Thus, specificity could evolve at the genotype level but would obscure any observable specificity at the species/morphotype level.

Evidence does, however, exist for considerable functional and genetic diversity within single AMF morphotypes. Feldmann (1998) has shown that sub-culturing of one AMF isolate into sub-populations can lead to great variation in the effect that the AMF sub-populations have on plant growth. Furthermore, Bever and Morton (1999) have shown that in a short time, e.g. a few generations, one AMF isolate or morphotype can be selected for a variety of different heritable characteristics. These results are consistent with evidence from molecular studies that unexpectedly high levels of genetic diversity exist in single AMF spores with respect to ribosomal DNA (Sanders et al. 1995; Lloyd-McGilp et al. 1996). It is, therefore, not unreasonable to hypothesise that considerable variation in plant growth may be caused by AMF but not at the morphologically recognizable AMF species level. Thus, selection pressures may exist for the formation of specificity between plants and certain AMF although, indeed, this may not be manifested at the level of what are currently accepted as different AMF species based on their morphology. This aspect of specificity in the AMF symbiosis has not previously been considered but it would mean that there is little or no relationship between morphologically based taxonomy of AMF and the ecology and specificity of AMF.

One clear bottleneck in understanding specificity in the AM symbiosis is a lack of knowledge concerning AMF population genetics. It has been hypothesized that AMF individuals may contain populations of genetically different nuclei (Sanders et al. 1996; Sanders 1999). At present, data that either support or refute this hypothesis are scarce. Hijri et al. (1999) have claimed that different rDNA sequences were present in one AMF isolate, *Scutellospora castanea*, and that they were segregated among nuclei. Hijri et al. (1999) diluted nuclear suspensions from single spores to an expected 'one-nucleus per sample' and performed PCR with specific primers that amplify the different rDNA sequences. Their results predicted that different rDNA sequences are carried on different nuclei. However, subsequent phylogenetic analyses of the r DNA sequence variants have cast doubt on this conclusion because some are identical to rDNA sequences of known ascomycete fungi and are, therefore, likely to be contaminants (Redecker et al. 1999). Trouvelot et al. (1999) also claim to have shown genetic differences among nuclei using fluorescent in situ hybridization although these data are also questionable. They have shown that the number of hybridization signals of a DNA probe to AMF nuclei differed among nuclei. They have interpreted this as a different copy number of rDNA among nuclei. The probe was raised by amplifying a specific piece of DNA. Therefore, differences in copy number do not represent sequence differences among nuclei, although varying copy number of a gene among nuclei would also mean that the genetic material among nuclei is not exactly the same. Caution must be given to the interpretation of such results because in situ techniques can easily produce such artifactual results, since the physical arrangement of the target sequences can affect whether probes are able to attach to them. Densely packed DNA material in a nucleus could prevent probes hybridizing to target sequences that are inside the nuclei rather than on the edge of the nucleus. Therefore, differences among nuclei in the density of the DNA could also give rise to such results.

Although the presence of different nuclei in fungal individuals has been well studied in basidiomycetes and ascomycetes, the situation of heterokaryosis in AMF would be quite different to those fungal groups. In ascomycetes and basidiomycetes, heterokaryosis occurs as the product of hyphal fusion between two haploid mycelia of opposite mating types. This results in nuclear fusion to form a diploid and subsequent segregation of genetic material into single haploid gametes contained in individual spores. Thus, the genetic diversity contained within an individual goes through a bottleneck at each generation. The situation in AMF would be quite different because there is no recognisable bottleneck where the genetic diversity in one individual is restricted.

If the hypothesis is true and AMF individuals contain populations of genetically different nuclei, then it could greatly alter how we approach the question of specificity in the symbiosis. Evolution of specific mycorrhizal associations may not easily be possible in an organism that contains multiple genomes. Let us consider the case where each genetically different AMF nucleus represents a genetic individual in a population. If certain AMF nuclear genotypes profit with certain plant species then selection for specific combinations of plants with certain nuclei could occur, but there may be no visible selection of plants for a given AMF morphotype. An outcome of such a situation would be that one AMF morphotype could connect several individual hosts of different species below ground but within that one morphologi-

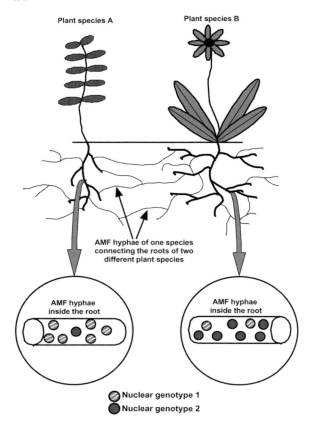

Fig. 16.3. A basis for genetic specificity in the mycorrhizal symbiosis with no observable specificity between plant species and AMF species. Two plant species, A and B, are colonized by the same AMF species and are connected by the hyphal network. The AMF contain a population of genetically different nuclei and selection for specificity acts at the nuclear level as a response to host environment, creating a patchy distribution of the nuclei in the hyphal network. In this example there are two different nuclear genotypes

cally recognizable AMF there could be distinct patches or clusters of nuclear genotypes occurring in the AMF in the roots of the different host species (Fig. 16.3). Whether this hypothesis is at all realistic could be experimentally tested when in situ hybridization techniques are specific and reliable enough to distinguish and count the frequency of genetically different nuclei in AMF growing in the roots of different plant species.

One further possibility regarding how the genetics of AMF may affect the ability to become specific concerns exchange of genetic information. Giovannetti et al. (1999) have shown that hyphae of the same isolate will readily anastomose with each other. This allows the possibility for genetic exchange. Frequent exchange of nuclei carrying different genetic information may reduce the likelihood of a genetic basis for specificity.

16.9.5 The Importance of the Hyphal Network for Understanding Specificity

The above hypothesis is, of course, only possible because of the existence of the hyphal network that connects plant individuals belowground. Such a hypothesis is difficult to explain at the genome level because it should be expected to result in considerable genetic conflict among nuclear genotypes. However, at the level of the whole organism, it could be a highly successful strategy. Retaining such an ability to profit with many different plant species could be highly beneficial. To an AMF, the variety of roots of different hosts represent a heterogeneous resource landscape. Different host roots may not only offer different amounts of resources, but may also simply offer different resources or a different quality of resources. Thus, the foraging strategy of AMF may make it more beneficial to not develop specificity as fitness may be maximized by foraging and exploiting different resources in different hosts (Fig. 16.4). The integrated nature of AMF would also mean that these

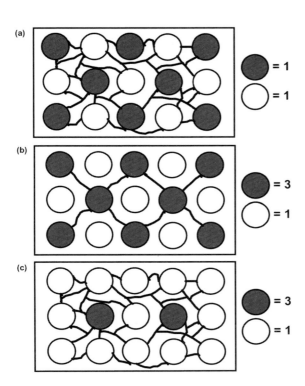

Fig. 16.4. Foraging by AMF could lead to the evolution of specificity or to an unspecific symbiosis depending on the comparative quality of the reward provided by different plant species and their relative abundance. a Two plant species in a community (shown by different *shaded circles*) give exactly the same reward (reward = 1 unit of benefit per host) to one AMF species. There is clearly no benefit for the AMF to forage for the best resource patches. b Hosts differ in quality (reward of 3 or 1) but have an equal abundance. It is beneficial for the AMF to evolve specificity and forage for the high resource patches. c Hosts differ in reward (reward of 3 or 1) but also in abundance. In this scenario, the AMF can obtain more resources from the community by colonizing any host, irrespective of species or reward

resources could be moved around in the network. A similar situation has been addressed in host-parasite ecology. Like AMF, the parasitic plant *Cuscuta* grows from plant to plant and one *Cuscuta* individual can parasitize many different plant species at the same time, while it is still physically and physiologically integrated. Experimental evidence exists that *Cuscuta* grows better when parasitizing two different host species than would be expected from the reward that it receives from growing on a monoculture of each of the two host species (Kelly and Horning 1999). Thus, *Cuscuta* can maximize its benefit by parasitizing different host species at the same time and this would also be expected to prevent the parasite from evolving specificity in this association. This may be a possible explanation for the existence of the AMF hyphal network and give a possible and testable explanation for why specificity may not evolve in the AM symbiosis.

16.10 Conclusions

The answer to whether there is any specificity in the AM symbiosis has not been satisfactorily answered. The existence of specificity in the symbiosis would have important consequences for our understanding of plant community ecology. There is clearly a long way to go before this question is answered. Despite its lack of appeal to many modern ecologists, detailed descriptive ecology of AMF communities is desperately needed to help answer this question. This will probably rely on reliable molecular identification methods, coupled with a clear understanding of AMF population genetics and the functioning of the hyphal network.

Acknowledgements. I wish to thank Alastair Fitter and two anonymous reviewers for helpful suggestions on how to improve this manuscript. This work was supported by a grant from the Swiss National Science Foundation No. 3100–050481.97.

References

An ZQ, Guo BZ, Hendrix JW (1998) Viability of soilborne spores of glomalean mycorrhizal fungi. Soil Biol Biochem 30:1133–1136

Bever JD, Morton J (1999) Heritable variation and mechanisms of inheritance of spore shape within a population of *Scutellospora pellucida*, an arbuscular mycorrhizal fungus. Am J Bot 86:1209–1216

Bever JD, Morton JB, Antonovics J, Schulz PA (1996) Host-dependent sporulation and species diversity of arbuscular mycorrhizal fungi in a mown grassland. J Ecol 84:71–82

Carey PD, Fitter AH, Watkinson AR (1992) A field study using the fungicide benomyl to investigate the effect of mycorrhizal fungi on plant fitness. Oecologia 90:550–555

Charlesworth B (1994). Evolution in age-structured populations, 2nd edn. Cambridge Univ Press, Cambridge
Clay K (1988) Clavicipitaceous fungal endophytes of grasses: coevolution and change from parasitism to mutualism. In: Pirozynski KL, Hawksworth DL (eds) Coevolution of fungi with plants and animals. Academic Press, London, pp 79–105
Doebeli M, Knowlton N (1998) The evolution of interspecific mutualisms. Proc Natl Acad Sci USA 95:8676–8680
Edwards SG, Fitter AH, Young JPW (1997) Quantification of an arbuscular mycorrhizal fungus, *Glomus mosseae*, within plant roots by competitive polymerase chain reaction. Mycol Res 101:1440–1444
Ellenberg H (1988) Vegetation ecology of central Europe, 4th edn. Cambridge Univ Press, Cambridge
Feldmann F (1998) The strain-inherent variability of arbuscular mycorrhizal effectiveness. II. Effectiveness of single spores. Symbiosis 25:131–143
Giovannetti M, Azzolini D, Citernesi AS (1999) Anastomosis formation and nuclear and protoplasmic exchange in arbuscular mycorrhizal fungi. Appl Environ Microbiol 65:5571–5575
Grime JP, Mackey JML, Hillier SH, Read DJ (1987) Floristic diversity in a model system using experimental microcosms. Nature 328:420–422
Harely JL, Smith SE (1987) Mycorrhizal symbiosis. Academic Press, London
Harper JL (1977) Population biology of plants. Academic Press, London
Helgasson T, Daniell TJ, Husband R, Fitter AH, Young JPW (1998) Ploughing up the wood-wide web? Nature 394:431–432
Hepper CM, Sen R, Azcon-Aguillar C, Grace C (1988) Variation in certain isozymes amongst different geographical isolates of the vesicular-arbuscular mycorrhizal fungi *Glomus clarum*, *Glomus monosporum* and *Glomus mosseae*. Soil Biol Biochem 20:51–59
Herre EA, Knowlton N, Mueller UG, Rehner SA (1999) The evolution of mutualisms: exploring the paths between conflict and cooperation. Trends Ecol Evol 14:49–53
Hijri M, Hosny M, van Tuinen D, Dulieu H (1999) Intraspecific ITS polymorphism in *Scutellospora castanea* (Glomales, Zygomycota) is structured within multinucleate spores. Fungal Genet Biol 26:141–151
Hoeksema JD (1999) Investigating the disparity in host specificity between AM and EM fungi: lessons from theory and better studied systems. Oikos 84:327–332
Johnson NC, Zak DR, Tilman D, Pfleger FL (1991) Dynamics of vesicular arbuscular mycorrhizae during old field succession. Oecologia 86:349–358
Johnson NC, Tilman D, Wedin D (1992) Plant and soil controls on mycorrhizal fungal communities. Ecology 73:2034–2042
Kelly CK, Horning K (1999) Acquisition order and resource value in *Cuscuta attenuata*. Proc Natl Acad Sci USA 96:13219–13222
Koide RT, Schreiner RP (1992) Regulation of the vesicular-arbuscular mycorrhizal symbiosis. Annu Rev Plant Physiol Plant Mol Biol 43:557–581
Lackie SM, Garriock ML, Peterson RL, Bowley SR (1987) Influence of host plant on the morphology of the vesicular-arbuscular mycorrhizal fungus *Glomus versiforme* (Daniels and Trappe) Berch. Symbiosis 3:147–158
Law R (1985) Evolution in a mutualistic environment. In: Boucher DH (ed) The biology of mutualism: ecology and evolution. Croom Helm, London, pp 147–170
Lloyd-McGilp SA, Chambers SM, Dodd JC, Fitter AH, Walker C, Young JPW (1996) Diversity of the ribosomal internal transcribed spacers within and among isolates of *Glomus mosseae* and related fungi. New Phytol 133:103–111
Lu XH, Koide RT (1994) The effects of mycorrhizal infection on components of plant growth and reproduction. New Phytol 128:211–218

Lynch M, Walsh B (1997) Genetics and analysis of quantitative traits. Sinauer Associates, Sunderland, Massachusetts

Margulis L, Fester R (eds) (1991) Symbiosis as a source of evolutionary innovation. MIT Press, Cambridge, Massachusetts

Maynard-Smith J, Szathmary E (1995) The major transitions in evolution. Freeman, San Francisco

Merryweather J, Fitter AH (1998) The arbuscular mycorrhizal fungi of *Hyacinthoides non-scripta*. I. Diversity of fungal taxa. New Phytol 138:117–129

Miller RM, Jarstfer AG, Pillai JK (1987) Biomass allocation in an *Agropyron smithii-Glomus* symbiosis. Am J Bot 74:114–122

Molina R, Massicotte H, Trappe JM (1992) Specificity phenomena in mycorrhizal symbioses: community-ecological consequences and practical implications. In: Allen MF (ed) Mycorrhizal functioning. Chapman Hall, New York, pp 357–423

Morton JB, Benny GL (1990) Revised classification of arbuscular mycorrhizal fungi (Zygomycetes): a new order, Glomales, two new suborders, Glomineae and Gigasporineae, and two new families, Acaulosporaceae and Gigasporaceae, with an emendation of Glomaceae. Mycotaxon 37:471–491

Nowak MA, May RM, Sigmund K (1995) The arithmetics of mutual help. Sci Am 272:50–55

Redecker D, Hijri M, Dulieu H, Sanders IR (1999) Phylogenetic analysis of a dataset of fungal 5.8S rDNA sequences shows that highly divergent copies of Internal Transcribed Spacers reported from *Scutellospora castanea* are of Ascomycete origin. Fungal Genet Biol 28:238–244

Remy W, Taylor TN, Hass H, Herp H (1995) Four-hundred-million-year-old vesicular-arbuscular mycorrhizae. Proc Natl Acad Sci USA 91:11841–11843

Rosendahl S, Rosendahl C, Søchting U (1990) Distribution of VA mycorrhizal endophytes amongst plants from a Danish grassland community. Agric Ecosyst Environ 29:329–336

Sackville-Hamilton NR, Schmid B, Harper JL (1987) Life-history concepts and the population biology of clonal organisms. Proc R Soc Lond B 232:35–57

Sanders IR (1993) Temporal infectivity and specificity of vesicular-arbuscular mycorrhizas in co-existing grassland species. Oecologia 93:349–355

Sanders IR (1999) No sex please, we're fungi. Nature 399:737–739

Sanders IR, Fitter AH (1992) Evidence for differential responses between host-fungus combinations of vesicular-arbuscular mycorrhizas from a grassland. Mycol Res 96:415–419

Sanders IR, Koide RT (1994) Nutrient acquisition and community structure in co-occurring mycotrophic and non-mycotrophic old-field annuals. Funct Ecol 8:77–84

Sanders IR, Alt M, Groppe K, Boller T, Wiemken A (1995) Identification of ribosomal DNA polymorphisms among and within spores of the Glomales: application to studies on the genetic diversity of arbuscular mycorrhizal fungal communities. New Phytol 130:419–427

Sanders IR, Clapp JP, Wiemken A (1996) The genetic diversity of arbuscular mycorrhizal fungi in natural ecosystems – a key to understanding the ecology and functioning of the mycorrhizal symbiosis. New Phytol 133:123–134

Sanders IR, Streitwolf-Engel R, van der Heijden MGA, Boller T, Wiemken A (1998) Increased allocation to external hyphae of arbuscular mycorrhizal fungi under CO_2 enrichment. Oecologia 117:496–503

Shumway DL, Koide RT (1994) Reproductive responses to mycorrhizal colonization of *Abutilon theophrasti* Medic. plants grown for two generations in the field. New Phytol 128:219–224

Shumway DL, Koide RT (1995) Size and reproductive inequality in mycorrhizal and non-mycorrhizal populations of *Abutilon theophrasti*. J Ecol 83:613–620

Simon L, Lalonde M, Bruns TD (1992a) Specific amplification of 18S fungal ribosomal genes from vesicular-arbuscular endomycorrhizal fungi colonizing roots. Appl Environ Microbiol 58:291-295

Simon L, Lévesque RC, Lalonde M (1992b) Rapid quantitation by PCR of endomycorrhizal fungi colonizing roots. PCR Methods Appl 2:76-80

Simon L, Bousquet J, Lévesque RC, Lalonde M (1993) Origin and diversification of endomycorrhizal fungi and coincidence with vascular land plants. Nature 363:67-69

Smith SE, Read DJ (1997) Mycorrhizal symbiosis. Academic Press, London

Stanley MR, Koide RT, Shumway DL (1993) Mycorrhizal symbiosis increases growth, reproduction and recruitment of *Abutilon theophrasti* Medic. in the field. Oecologia 94:30-35

Streitwolf-Engel R, Boller T, Wiemken A, Sanders IR (1997) Clonal growth traits of two *Prunella* species are determined by co-occurring arbuscular mycorrhizal fungi from a calcareous grassland. J Ecol 85:181-191

Streitwolf-Engel R, van der Heijden MGA, Wiemken A, Sanders IR (2001) The ecological significance of arbuscular mycorrhizal fungal effects on clonal reproduction in plants. Ecology 82(10):2846-2859

Trappe JM (1987) Phylogenetic and ecologic aspects of mycotrophy in the angiosperms from an evolutionary standpoint. In: Safir GR (ed) Ecophysiology of VA mycorrhizal plants. CRC Press, Boca Raton, pp 5-25

Trouvelot S, van Tuinen D, Hijri M, Gianinazzi-Pearson V (1999). Visualization of ribosomal DNA loci in spore interphasic nuclei of glomalean fungi by fluorescence in situ hybridization. Mycorrhiza 8:203-206

van der Heijden MGA, Boller T, Wiemken A, Sanders IR (1998a) Different arbuscular mycorrhizal fungal species are potential determinants of plant community structure. Ecology 79:2082-2091

van der Heijden MGA, Klironomos JN, Ursic M, Moutoglis P, Streitwolf-Engel R, Boller T, Wiemken A, Sanders IR (1998b) Mycorrhizal fungal diversity determines plant diversity, ecosystem variability and productivity. Nature 396:69-72

van Tuinen D Jacquot E, Zhao B, Gollotte A, Gianinazzi-Pearson V (1998) Characterisation of root colonisation profiles by a microcosm community of arbuscular mycorrhizal fungi using 25S rDNA-targeted nested PCR. Mol Ecol 7:879-887

Weiner J, Thomas SC (1986) Size variability and competition in plant monocultures. Oikos 47:211-222

Wilson JM (1984) Competition for infection between vesicular-arbuscular mycorrhizal fungi. New Phytol 97:427-435

Section F:
Conclusions

17 Mycorrhizal Ecology: Synthesis and Perspectives

MARCEL G.A. VAN DER HEIJDEN, IAN R. SANDERS

Contents

17.1	Introduction	441
17.2	Ecophysiology, Ecosystem Effects and Global Change	442
17.3	Biodiversity, Plant and Fungal Communities	446
17.4	Multitrophic Interactions	449
17.5	Host Specificity and Co-evolution	450
17.6	Conclusions	452
References		454

17.1 Introduction

There is an increasing awareness among ecologists and mycorrhizologists that mycorrhizal fungi are an integral part of ecosystems and that, therefore, their ecological function needs to be understood. Recently, many breakthroughs have pointed to this. It has, for instance, been shown that mycorrhizal fungi contribute to plant diversity, nutrient cycling, acquisition to nutrient sources previously thought not to be available to plants and finally to ecosystem functioning. Moreover, because of their abundance and potential ecological function at the community and ecosystem levels, mycorrhizal fungi could well contribute to community and ecosystem responses to global changes. Also, based on the effects of the symbiosis on plant fitness, the population ecology, dynamics and evolution of many plants, are unlikely to be fully understood without considering their fungal symbionts, although this area has sadly received little treatment in comparison to studies on physiological effects of the symbiosis. Another fascinating aspect of the mycorrhizal symbiosis is the occurrence of the so-called wood-wide web in the soil formed by mycorrhizal hyphae that connect individual plants together. Such a web can allow resources to flow from one individual to another. If this occurs at an ecologically significant level, and this is still controversial, then this would completely change our view on how communities function. We have taken a plant ecologist's approach to this subject but researchers must not forget the ecology and evolution of mycorrhizal fungi themselves, since the lack of information in this area is currently greatly inhibiting the understanding of how the fungi and the structure of the mycorrhizal fungal communities influence plant ecology. This Volume is devoted to mycorrhizal ecology. It gives a summary of the most recent advances and breakthroughs that have been made in this field. We hope that it allows ecologists and researchers of the mycorrhizal symbiosis to get a broad picture of how the symbiosis fits into the broader field of ecology.

Four levels of integration are considered in this Volume. These four levels, together with the introduction and this final chapter, correspond with the six sections of this volume. The mycorrhizal symbiosis is introduced in the first section (Chap. 1). The second section considers the ecophysiology of mycorrhizal plants and mycorrhizal fungi (Chaps. 2–6). Ecosystem effects of mycorrhizal fungi and interactions of the mycorrhizal symbiosis with global changes are included as well. The third section treats community level aspects (Chaps. 7–11). The impact of mycorrhizal fungi on the structure and diversity of plant communities is discussed. Factors that affect the structure and diversity of mycorrhizal fungal communities are discussed as well. This section also considers how feedback mechanisms might exist between plant and fungal populations and communities. Interactions among organisms are discussed in the fourth section (Chaps. 12–14). This part shows that mycorrhizal fungi interact with individuals and communities of many other organisms such as herbivores, saprotrophic fungi and soil invertebrates. Finally, the last section takes an evolutionary approach to whether host specificity occurs in the mycorrhizal symbiosis and what the consequences are for the plants and the fungi involved in the association (Chaps. 15 and 16)

In this chapter we summarize the major highlights presented in this Volume. We give a relatively broad summary and aim to present this chapter in such a form that it can be viewed as a short review on mycorrhizal ecology. We have also tried to include the latest breakthroughs that were not possible to include in individual chapters. We also refer to important recent work that is not covered in this Volume, but which is essential to consider in order to obtain a better overall picture of the contribution of the mycorrhizal symbiosis to ecology.

17.2 Ecophysiology, Ecosystem Effects and Global Change

As most authors of this Volume point out, the mycorrhizal symbiosis is abundant and widespread. It has been estimated that approximately 80% of all land plant species form associations with mycorrhizal fungi (Trappe 1987). The sheer abundance of mycorrhizal fungi must be enormous if one consid-

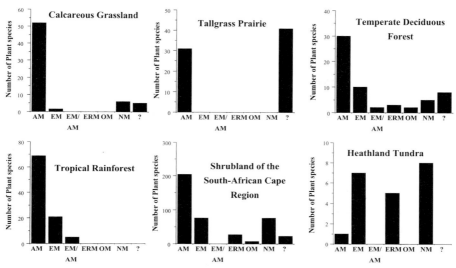

Fig. 17.1. Relative abundance of mycorrhizal plant species in different ecosystems. The number of plant species in a particular ecosystem that form arbuscular mycorrhizal (*AM*), ecto-mycorrhizal (*EM*), ericoid mycorrhizal (*ERM*) or orchid mycorrhizal (*OM*) associations are shown. Some plant species form no mycorrhizal associations (*NM*), some plants form a dual symbiosis with both ecto- mycorrhizal and arbuscular fungi (*EM/AM*) and for some species it is not known whether they form a symbiosis with mycorrhizal fungi (*?*). Plant species composition of a European calcareous grassland from Huovinen-Hufschmid and Körner (1998); of an American tall grass prairie from Hartnett and Wilson (1999) (kindly provided by G. Wilson); of European temperate deciduous forest from Stortelder et al. (1999); of a selection of 100 tree species in a tropical rain forest from Onguene and Kuyper (2001); of shrubland of the South-African Cape region from Allsopp and Stock (1993) and of arctic tundra vegetation from Michelsen et al. (1998). Mycorrhizal status of the plants from European ecosystems from Harley and Harley (1987) and of American tall grass prairies from Wilson and Hartnett (1998); for the other ecosystems both mycorrhizal status and plant species composition are given in the same reference

ers that most ecosystems are dominated by plants that form a symbiosis with these mutualistic soil fungi (Read 1991). For example, the majority of plants in a European calcareous grassland, an American tall grass prairie, temperate deciduous forests, a tropical rain forest and shrub land of the threatened South African Cape region associate with mycorrhizal fungi (Fig. 17.1). In contrast, plant communities of arctic tundra and alpine regions often contain a lower percentage of mycorrhizal plant species (Fig. 17.1; Smith and Read 1997). Furthermore, soils of most of such ecosystems contain large mycorrhizal hyphal networks. It is not unusual to find 20,000 km of hyphae per cubic metre soil. Mycorrhizal fungi are, therefore, both abundant and very widespread. Given their abundance and their effects on plant growth, it is not surprising that ecologists might expect them to play an important role in ecosystems. Read points out in Chapter 1 that there are several pitfalls in the path towards an understanding of mycorrhizal functioning in nature which need to be considered in order to investigate the ecological function of these fungi. Several attributes that enhance the relevance in studies of mycorrhizal involvement in plant population, community and ecosystem processes are summarized in Table 1.2.

Different types of mycorrhizal associations exist (Fig. 1.1, Chap. 1). The most common types are ecto-mycorrhizal associations between ecto-mycorrhizal fungi (basidomycetous fungi) and many tree species and arbuscular mycorrhizal associations between arbuscular mycorrhizal fungi (Zygomycetous fungi) and many grasses, forbs and tropical trees. The eco-physiology of these types is discussed respectively by Simard et al. (Chap. 2) and Jakobsen et al. (Chap. 3). Other types of mycorrhizal associations that have been less well investigated, and that receive less attention in this Volume, are associations between ericoid mycorrhizal fungi and plants of the Ericaceae family and between orchid mycorrhizal fungi and plants of the Orchidaceae. The biology of these different types is extensively discussed in Smith and Read (1997). Mycorrhizal fungi are not only abundant, they have important functions as well (Read, Chap. 1). It is well known that they have beneficial effects on plant growth (Simard et al., Chap. 2; Jakobsen et al., Chap. 3), especially when nutrient availability is low. This has mainly been attributed to improved plant nutrition (e.g. Table 2.1 in Simard et al., Chap. 2). The diameter of mycorrhizal hyphae is up to 60 times smaller than those of plant roots. Moreover, mycorrhizal fungi form extensive hyphal networks in the soil as shown in Figs. 4.1 and 4.2. This can explain why mycorrhizal fungi forage so effectively for nutrients in comparison to plant roots (they can enter small soil pores inaccessible to plant roots). Mycorrhizal fungi have, as a consequence, the ability to almost completely deplete nutrients from the soil (Table 3.1, in Jakobsen et al., Chap. 3). The roots of many ecto-mycorrhizal trees are completely encapsulated by a fungal mantle. Nutrients must pass this mantle before uptake by the plant occurs showing that it is not possible to understand plant nutrition without considering their fungal partner. The extent by which plants benefit from a symbiosis with mycorrhizal fungi varies depending on the identity of the plant and the fungus, the physiological state of the plant and environmental conditions. Some plants benefit greatly from mycorrhizal fungi while other species do not respond or are even negatively affected when colonized by mycorrhizal fungi (Figs. 9.5, 10.1). A number of traits, from both the plant and mycorrhizal fungi, determine the fungal impact on plant growth (Table 3.2 in Jakobsen et al., Chap. 3), pointing to the complexity of these plant-fungal interactions.

The multi-functional nature of these mutualistic soil fungi is starting to be revealed (e.g. Newsham et al. 1995). Mycorrhizal fungi not only improve plant nutrition, they can also protect plant roots against pathogens, contribute to soil structure and can reduce water stress. This area has been overlooked in comparison to nutritional effects of the symbiosis but is essential for understanding their ecological role.

Recent evidence suggests that mycorrhizal fungi have access to nutrients previously thought to be unavailable to plants. Both ecto-mycorrhizal fungi and ericoid mycorrhizal fungi exude a range of extracellular enzymes that are able to break down complex organic substances. These fungi have, consequently, access to organic nitrogen and organic phosphorus that can be transmitted to their hosts. This ability is especially important in nutrient-poor ecosystems such as heathlands, arctic or sub-arctic ecosystems and forests on acidic soils where organic nitrogen and phosphorus are the main forms of nitrogen and phosphorus (Aerts, Chap. 4). Moreover, it has been suggested that plants obtain animal-derived nitrogen through the predatory characteristics of their fungal associates (Klironomos and Hart 2001). Based on these findings, it has been suggested that models on nutrient cycling within forest ecosystems needs to be revised due to the ability of mycorrhizal fungi to use such organic nutrient sources (Figs. 14.1, 14.2 in Leake et al., Chap. 14). The suggestion that 'rock-eating mycorrhizal fungi' exist and that mycorrhizal fungi contribute to mineral weathering give further evidence for the observation that mycorrhizal fungi are able to extract nutrients from sources previously thought to be unavailable (Simard et al., Chap. 2; Landeweert et al. 2001). Recent evidence indicates that arbuscular mycorrhizal fungi exude enzymes as well and that they are able to use more complex organic nutrient sources as has originally been thought (Koide and Kabir 2000). Nevertheless, their ability is thought to be relatively small compared to ecto-mycorrhizal fungi and ericoid mycorrhizal fungi. This can also be derived from the ^{15}N values (an indicator for organically derived nitrogen) of their host plants. Plants associating with ericoid and ecto-mycorrhizal fungi have lower ^{15}N values compared to non-mycorrhizal plants or plants associating with arbuscular mycorrhizal fungi. Lower ^{15}N values indicate a better access to organic nitrogen (Table 4.1 in: Aerts, Chap. 4; Michelsen et al. 1998).

So far, most chapters have concentrated on the plant site of the mycorrhizal symbiosis. However, it is important not to forget the mycorrhizal fungal mycelium as well, for several reasons. Firstly, mycorrhizal hyphae are the primary structures that acquire nutrients. Secondly, the mycorrhizal mycelium acts as a sink for carbon (Simard et al., Chap. 2; Jakobsen et al., Chap. 3). Estimates suggest that 10–50 % of assimilated carbon is translocated to mycorrhizal roots. This is a considerable amount, in terms of the plant's carbon balance, but also in terms of the carbon cycle of whole ecosystems. Finally, due to their abundance, mycorrhizal hyphae interact with many organisms that live below ground. The characteristics of the mycorrhizal mycelium are discussed by Olsson et al. (Chap. 5). They show that mycorrhizal hyphae forage actively for nutrients and new host roots. The interactions of mycorrhizal hyphae with other organisms such as saprotrophic fungi and soil invertebrates are discussed in Chapters 13 and 14 and are summarized in Section 17.4. There are many gaps in our knowledge about the eco-physiology and functioning of the mycorrhizal mycelium.

Knowledge on the ecophysiology of mycorrhizal plants and mycorrhizal mycelium as presented in Chapters 2–5 is needed in order to predict how plants respond to current global changes. The influence of global changes such as elevated carbon dioxide, nitrogen deposition, ozone, UV radiation and climate change on mycorrhizal associations is discussed by Rillig et al. (Table 6.1; Fig. 6.1). Responses of mycorrhizal associations to global changes are positive as well as negative, depending on the identities of both the plant and the fungus. Further research is necessary to obtain a more complete picture as to how global changes affect the functioning of both the plants and the fungi.

17.3 Biodiversity, Plant and Fungal Communities

One of the major goals in ecology is to search for the mechanisms that determine biological diversity (Tilman 1988; Grime 2001). This is an important goal, considering the accelerating loss of species diversity that the earth is currently experiencing. Recently, several studies have pointed to the importance of mycorrhizal fungi as determinants of plant diversity and ecosystem functioning. These findings and the mechanisms by which mycorrhizal fungi enhance or reduce diversity are discussed by Hart and Klironomos (Chap. 9) and van der Heijden (Chap. 10). At this point, it must be made clear that the effects of mycorrhizal fungi on diversity are context dependent. Both nutrient availability and plant species composition determine the impact of mycorrhizal fungi on plant diversity (van der Heijden, Chap. 10, Fig. 10.2). In this chapter, it is hypothesized that the influence of AMF on plant diversity decreases with increasing productivity and nutrient availability. The presence of a mycorrhizal hyphal network in the soil might be of ultimate importance to promote plant diversity. The establishment of new seedlings (a main source of diversity) might be mediated by a hyphal web. It appears that seedlings of those plant species that are most dependent on AMF benefit most from a hyphal network (Fig. 10.3). It has been postulated that there is a flow of carbon and nutrients from one plant to another via mycelial networks and this may well provide a compelling mechanism for how seedlings might benefit from the symbiosis by tapping into the network to be fed. The ecological implications of this interplant resource transport are discussed by Simard et al. (Chap. 2).

It is not only the presence of mycorrhizal fungi that influences plant diversity. The species composition and diversity of mycorrhizal fungal communities also affect plant diversity and productivity (Hart and Klironomos, Chap. 9; van der Heijden, Chap. 10). A two-fold increase in plant diversity was observed when AMF species richness was increased from 1 to 16 species (Fig. 9.4). New results presented by Hart and Klironomos (Chap. 9) indicate that effects of AMF diversity on plant diversity and productivity are highest when AMF belonging to different genera are present (Fig. 9.2). Moreover, the influence of AMF diversity on plant diversity depends on plant species composition (e.g. see Fig. 10.7). Recent experiments by Jonsson et al. (2001) suggest that ecto-mycorrhizal diversity might also play a role in seedling establishment of boreal forest trees. They observed that the productivity of birch (*Betula pendula*) seedlings growing on nutrient-poor sterilized soil increased more than 200% when the number of ecto-mycorrhizal fungi increased from one to eight species. Birch is one of the keystone species of boreal forests, forests which are known to harbour a diverse and complex community of ecto-mycorrhizal fungi. The ecto-mycorrhizal diversity of these forests is in striking contrast to the inherent simplicity of the often species-poor stands of ecto-mycorrhizal trees (Erland and Taylor, Chap. 7). The results of Jonsson et al. (2001) thus provide a functional explanation about the importance of ecto-mycorrhizal diversity in boreal forests.

Hart and Klironomos and Van der Heijden take a largely plant-centric viewpoint of the symbiosis; analysing the effects of the fungi on the structure and diversity of plant communities. There is evidence showing that the reverse is true; that plants affect populations and communities of their fungal symbionts. Bever et al. (Chap. 11) take a population and community approach to look at how the fungi affect plant fitness but, at the same time, how plant species affect mycorrhizal fungal fitness. Bever et al. present the idea that we should not just study the one-way effects because both partners act on each other and consequently affect the fitness of the other partner. The outcome of this chapter is a prediction of the ecological outcome of both positive and negative feedbacks between plants and their mycorrhizal fungi.

Because of the differential effects of mycorrhizal fungal species on plant growth, it is essential to know which factors determine the diversity and composition of mycorrhizal communities. In addition, to better study the sort of feedbacks described by Bever, a better description of mycorrhizal fungal communities is necessary. The advances in the ability to describe mycorrhizal fungal communities are discussed for ecto-mycorrhizal fungi and arbuscular mycorrhizal fungi respectively by Erland and Taylor (Chap. 7) and Clapp et al. (Chap. 8). Both abiotic soil factors and biotic factors (such as plant species composition) affect the composition of mycorrhizal fungal communities (Chaps. 7 and 8). Perhaps the best studied is the effect of nitrogen fertilization on mycorrhizal fungi. Enhanced levels of soil available nitrogen, as caused by atmospheric nitrogen deposition, change the composition of ecto-mycorrhizal fungal communities. It also alters levels of root colonization of plants associating with either arbuscular mycorrhizal fungi, ecto-mycorrhizal fungi or ericoid mycorrhizal fungi (Table 5.1, Aerts, Chap. 5; Table 6.1, Rillig et al., Chap. 6; Table 7.1, Erland and Taylor, Chap. 7). It appears that nitrogen deposition and soil acidification reduce ecto-mycorrhizal diversity (Erland and Taylor, Chap. 7; Wallenda and Kotke 1998). This is often accompanied by a shift in community structure so that dominance by one fungus increases. Moreover, nitrogen deposition also leads to changes in nitrogen nutrition of ecto-mycorrhizal fungi. The proportion of species within the community that are specialised to utilise organic nitrogen sources (such as proteins) decline with increasing mineral nitrogen-availability (Fig. 7.2). This, in turn, can affect nutrient cycling within forests and disrupt the natural nitrogen cycle (Fig. 14.1). Moreover, it has been postulated that the negative impact of nitrogen deposition on ecto-mycorrhizal fungal diversity leads to reduced viability of many temperate and boreal forests. Much of the recent work describing ecto-mycorrhizal fungal communities that is covered by Erland and Taylor (Chap. 7) has been made with the use of DNA-based identification tools. Mycorrhizal fungi are small, unseen and their hidden nature makes it difficult to study them with standard techniques. Genetic studies provide an outcome to this problem and they have already greatly improved our knowledge of ecto-mycorrhizal fungal communities.

Genetic markers have also been used to identify arbuscular mycorrhizal fungi (both spores and in roots). This has started to improve our knowledge of arbuscular mycorrhizal fungal communities, but sadly still lags way behind in comparison to ecto-mycorrhizal fungal communities. Clapp et al. (Chap. 7) review the use of such genetic markers in arbuscular mycorrhizal ecology. They show that genetic studies are an essential tool for investigating the ecology and diversity of AMF communities. They also show that genetic studies can be used to describe AMF communities in the field (e.g. Fig. 8.4). The small amount of information that has so far been gained reveals intriguing aspects of these fungal communities. Clapp et al. make a strong claim that molecular techniques are needed because they have immediately shown the inadequacy of reconstructing mycorrhizal fungal communities based on spore counts. They also stress that no single molecular method detects all AMF present in soil and roots (Fig. 8.1). These molecular techniques will certainly contribute to a better insight into the mycorrhizal symbiosis in the future. However, Clapp et al. identify the poor understanding of the genetics and the maintenance of genetic variability as limiting the interpretation of current DNA-based approaches. In particular, within individual genetic variability, a phenomenon specific to the Glomales, is identified as reducing the precision of current molecular identification techniques and they call for more research towards a better understanding of the genetics of the Glomales.

17.4 Multitrophic Interactions

We have so far discussed direct contributions of the mycorrhizal symbiosis to plant ecology or the consequences of this on ecosystem functioning. However, it is clear from several chapters in this Volume that indirect effects of the mycorrhizal symbiosis can occur on other components of ecosystems, both above and below ground. Gehring and Whitham (Chap. 12) discuss the potential effects of the symbiosis on aboveground herbivory and on herbivore fitness (e.g. see Table 12.1).). Mycorrhizal fungi can affect herbivores in two ways, namely by altering food plant quality and by changing food plant quantity. Herbivores, in turn, affect mycorrhizal fungi as well, often resulting in decreasing levels of root colonization. The type of mycorrhizal fungi involved in multitrophic interactions are also highlighted in Chapter 12 (see Fig. 12.5). Aphid populations reared on ecto-mycorrhizal *Populus* plants were 6.9 times higher than those on plants colonized by arbuscular mycorrhizal fungi (Fig. 12.5). The influence of mycorrhizal fungi on aphid populations is especially interesting since aphids are keystone species. They have a great impact on ecosystem functioning through their effects on fixed carbon consumption (e.g. see Wimp and Whitham 2001). Moreover, different mycorrhizal fungal species differentially influence insect herbivore performance (Goverde et al. 2000; Gange 2001), stressing the importance of the composition of mycorrhizal fungal communities.

Another group of organisms that are affected by mycorrhizal fungi are soil organisms. The biomass and spatial distribution of many soil inhabitants, including earthworms, mites, collembola, nematodes, subterranean insects and saprotrophic soil fungi, are directly or indirectly affected by mycorrhizal fungi (Table 13.1 in Gange and Brown, Chap. 13; Leake et al., Chap. 14). Some of these soil inhabitants, such as mites and nematodes, feed on mycorrhizal hyphae and their abundance might be directly influenced by the availability of mycorrhizal mycelium. Other mycophagous animals such as collembola may also consume mycorrhizal hyphae, although recent feeding experiments suggest that they preferentially feed on non-mycorrhizal fungi (Gange 2001)

Mycorrhizal fungi also directly interact with saprotrophic soil fungi, and this can result in growth inhibition or even in competitive exclusion of one of the competing fungi. This has been nicely illustrated in Chapter 14 by Leake et al. (Fig. 14.3). In this study, competition between the ecto-mycorrhizal fungus *Paxillus involutus* and the saprotrophic wood-decay fungus *Phanerochaete velutina* was investigated by using autoradiography in microcosms (Fig. 14.3). Such interactions are of ecological interest, given the abundance of both saprotrophic fungi and ecto-mycorrhizal fungi in forest ecosystems and their contribution to carbon and nutrient cycling (e.g. Figs. 14.1, 14.2; also Simard et al., Chap. 2). Moreover, arbuscular mycorrhizal fungi interact with root pathogenic fungi within plant roots. By doing that, arbuscular mycorrhizal fungi can protect the plant against harmful effects of pathogens (Newsham et al. 1995). An important study by Setälä (2000) suggests that soil food web structure is affected by mycorrhizal fungi as well. In this study, soil food webs were experimentally manipulated with and without mycorrhizal fungi and the effects of these manipulations on plant growth were determined. Experimental studies such as those by Setälä (2000) need to be executed to understand the importance of the diversity and composition of the soil community for ecosystem functioning.

One interesting interaction that merits further study is that between mycorrhizal fungi and bacteria. Mycorrhizal fungi influence the composition of the bacterial community surrounding their hyphae or colonized plants roots (Timonen et al. 1998; Andrade et al. 1998). Bacteria, in turn, can promote mycorrhizal development (Garbaye 1994) and it is thought that bacteria which are associated with mycorrhizal fungi contribute to mineralization and solubilization of nutrients in the soil (Perotto and Bonfante 1997). Moreover, thousands of bacteria-like organisms (BLOs) belonging to the *Burkholderia* genus have been observed inside hyphae and spores of arbuscular mycorrhizal fungi (Bianciotto et al. 1996). It has also been suggested that ecto-mycorrhizal fungi contain endosymbiotic bacteria (Barbieri et al. 2000). Unfortunately, although it is clear that mycorrhizal fungi do have effects on soil micro-organisms and that soil organisms affect mycorrhizal fungi, much still needs to be investigated in this area. For example, very little is known about whether micro-organisms act differentially on different species of mycorrhizal fungi. This is an exciting field where intriguing discoveries are likely to be made. The studies mentioned above reveal the complexity and importance of belowground biological interactions. Examining these interactions is essential if one wishes to understand the patterns and processes we see in the world above ground.

17.5 Host Specificity and Co-evolution

The symbiosis between plants and mycorrhizal fungi is thought to be ancient. Fossil records and molecular clock dating suggest that the evolution of plants with arbuscular mycorrhizal fungi of the Glomales occurred approximately 400 million years ago (Simon et al. 1993; Remy et al. 1994). Fossils of Glomales-like fungi have also been found in the Ordovician, some 460 million years ago, when the land flora most likely comprised bryophytes (Redecker et al. 2000a). The protorhizoids of ancient plants might have been unable to obtain sufficient amounts of nutrients in the primitive soils where they occurred, without the help of fungi (Selosse and le Tacon 1998). Based on these findings and on the observation that mycorrhizal fungi associate with primitive plants such as liverworts, hornworts and lycopodia (Read et al. 2000; Schlüßler 2000;

Fig. 15.4 in Taylor et al., Chap. 15), it has been hypothesized that AMF enabled plants to colonize the land. Evidence for the occurrence of ancestral lineages of arbuscular mycorrhizal fungi has recently been found (Redecker et al. 2000b). Some of those 'ancient' mycorrhizal fungi are dimorphic, exhibiting spore stages of different genera within the Glomales (Morton et al. 1997). It has been hypothesized that these dimorphic fungi are precursors of 'modern' AMF. Despite their long evolutionary history, many mycorrhizal fungi appear to be non-host specific (Sanders, Chap. 16; Molina et al. 1992) and available evidence suggests that there is an ongoing parallel evolution of the partners in response to environmental change rather than co-evolution (Cairney 2000). For example, an interesting phylogenetic analysis by Hibbett et al. (2000) suggests that ecto-mycorrhizal fungi evolved repeatedly from saprotrophic precursors. Moreover, reversals of these symbiotic fungi into free-living saprotrophic fungi occurred, suggesting that the ecto-mycorrhizal associations are unstable, evolutionary dynamic associations (Hibbett et al. 2000).

As far as we know, arbuscular mycorrhizal fungi are completely dependent on their host for carbon while many ecto-mycorrhizal fungi can be cultured separately on agar plates without host roots. Despite their obligate nature, arbuscular mycorrhizal fungi are not thought to be host-specific. Most AMF species that have been isolated so far seem to be capable of associating with a wide range of plant species. The genetics, physiology and ecology of AMF might contribute to the absence of host specificity (Sanders, Chap. 16). AMF are multinucleate organisms. Hundreds, sometimes thousands, of nuclei have been observed within spores and hyphae of AMF (Bécard and Pfeffer 1993). It has been hypothesized that these nuclei are genetically diverse, given the genetic diversity occurring within single spores (see Clapp et al., Chap. 7; Sanders, Chap. 16). Sanders (Chap. 16) suggests that that the existence of populations of genetically diverse nuclei within AMF could prevent strong co-evolution or host specificity evolving (Fig. 16.3), but that this is dependent on either extensive nuclear migration through the hyphal network or frequent exchange of nuclei among hyphae. Hyphal fusion and anastomosis have been recently observed between hyphae (Giovanetti et al. 1999). Video-enhanced light microscopy even revealed that nuclear migration occurred, suggesting that genetic exchange may occur (Giovanetti et al. 1999). This lack of specificity is surprising since ecological experiments reveal that selection pressures exist that would favour the evolution of specificity. Sanders stresses that much basic descriptive ecology of AMF species in plant roots in natural ecosystems needs to be collected to help resolve the question of whether any ecological specificity exists in the arbuscular mycorrhizal symbiosis. Perhaps only those AMF with an extremely broad host range have been isolated so far. Clearly, there remains much to be done in this exciting field and many breakthroughs can be expected in coming years.

Most plant species are, like their fungal partners, not host-specific. Plant individuals can be colonized by different species of arbuscular mycorrhizal fungi (Clapp et al. 1995; van Tuinen et al. 1998) or ecto-mycorrhizal fungi (Jonsson et al. 2001). Dual colonization by both ecto-mycorrhizal fungi and arbuscular mycorrhizal fungi occurs also in some plant species (Jones et al. 1998), pointing to the absence of any host specificity. Moreover, a recent study by Streitwolf-Engel et al. (2001) provides some clues about whether or not host specificity is evolving in *Prunella vulgaris*, a plant species which is a good candidate to evolve host specificity since it is known to vary greatly in its response to different AMF species. This study showed that different genotypes of *Prunella* did not vary from each other in their growth response to different AMF species, pointing to the possibilities that either co-evolution is not occurring, it is proceeding very slowly or co-evolution has already occurred and is no longer proceeding. Some ecto-mycorrhizal trees show a certain degree of host specificity (Molina et al. 1992; Newton and Haigh 1998). The stability of host specificity is unclear, considering that reversals from mycorrhizal mutualists to saprotrophic fungi occur (Hibbett et al. 2000).

There is one major exception with regards to specificity in mycorrhizal symbioses; the myco-heterotrophic plants. Myco-heterotrophic plants are plants that depend on fungi for the supply of carbon and in which photosynthesis is completely or partly lost. Several myco-heterotrophic plants (orchids as well as monotropes) show tight specificity and are adapted to only one or a few mycorrhizal fungi (Table 15.1 in Taylor et al., Chap. 15). Spatial distribution of plants and fungi is, as a consequence of this specificity, often tightly linked. This is nicely illustrated in a study by Bidartondo et al. (2000). They observed that *Sarcodes sanguinea* individuals (Monotropoideae) occurred only at those patches where its host fungus, *Rhizopogon ellenae*, was present. Several myco-heterotrophic plants appear to show an extreme degree of host specialization with continuing co-evolution and host races between plants and fungi. This follows also from the observation that genotypes of the achlorophyllous orchid *Corrallorhiza maculata* appear to associate with their own clades of fungal species within the Russulaceae even when the different plant genotypes co-occur (Taylor et al., Chap. 15). Myco-heterotrophic plants are likely, due to their parasitic nature, to be forced to co-evolve continuously. The specialization of myco-heterotrophic plants, thus, contrasts with the absence of mycorrhizal specificity in many autotrophic plants.

17.6 Conclusions

The major advances and breakthroughs in the field of mycorrhizal ecology are presented in this Volume. We hope that this Volume gives ecologists and researchers of mycorrhizal symbiosis a picture of how mycorrhizae fit into the broader field of ecology.

The following conclusions emerge from this Volume:
1. The ecology of plants and plant communities cannot be understood without considering the ecology of their intimate root associates, the mycorrhizal fungi. Both arbuscular mycorrhizal fungi and ecto-mycorrhizal fungi provide plants with nutrients that are often limiting and/or inaccessible to the roots. This shows that mycorrhizal fungi need to be considered as biotic components of ecosystems that give plants access to resources.
2. The functioning of the mycorrhizal symbiosis has mostly been studied under simplified conditions in the lab or greenhouse. Ecological relevance is low in such study systems. Future research needs to investigate the ecological function of mycorrhizal symbiosis under ecologically realistic, multi-factorial conditions of the kind that prevail in nature.
3. Mycorrhizal fungi form dense and extensive mycelial networks that radiate into the soil and which forage for nutrients. Often, up to 80 % of the fungal biomass is allocated to these external hyphal networks. Despite this, not much is known about the functioning and ecology of mycorrhizal hyphal networks. Several of the authors have noted this and suggested that this large gap in our knowledge needs more attention.
4. Originally, most mycorrhizal researchers concentrated on the effects of mycorrhizal fungi on plants. Reviews in this Volume show the multi-trophic nature of the mycorrhizal symbiosis: mycorrhizal fungi act and interact with many organisms, ranging from small, unseen bacteria and decomposer fungi to big herbivores. Future studies will certainly increase our knowledge on the complexity and ecological importance of these interactions
4. Recently, several studies have shown that mycorrhizal fungi act as determinants of plant diversity and ecosystem functioning. These studies need to include a wider range of ecosystems at different latitudinal ranges and with varying levels of available nutrients. Feedbacks between plants and their mycorrhizal fungi may also be important in these processes and need to be further investigated. This is essential to obtain a more global view about the functional importance of mycorrhizal fungi at the ecosystem level.
5. Global changes such as nitrogen deposition and climate change affect the mycorrhizal symbiosis in various ways. The functionality and composition of mycorrhizal fungal communities are altered, for instance, due to increased levels of atmospheric nitrogen. It is hypothesized that future global changes may lead to changes in the composition of mycorrhizal fungal communities. These changes may, in turn, mediate global change impacts on plant communities.
6. Most studies suggest that specificity phenomena between plants and mycorrhizal fungi are relatively rare. There is one major exception. It is shown that a high degree of specificity exists between myco-heterotrophic plants and their fungal hosts. Unravelling whether, and to what extent, the symbiosis is specific or unspecific could help us to understand at what levels the symbiosis has an impact on ecosystem function.

References

Andrade G, Mihara KL, Linderman RG, Bethlenfalvay GJ (1998) Soil aggregation status and rhizo-bacteria in the mycorrhizosphere. Plant Soil 202:89–96

Allsopp N, Stock WD (1993) Mycorrhizal status of plants growing in the Cape Floristic Region, South-Africa. Bothalia 23:91–104

Barbieri E, Potenza L, Rossi I, Sisti D, Giomaro G, Rossetti S, Beimfohr C, Stocchi V (2000) Phylogenetic characterization and in situ detection of a Cytophaga-Flexibacter-Bacteroides phylogroup bacterium in *Tuber borchii* Vittad, ecto-mycorrhizal mycelium. Appl Environ Microbiol 66:5035

Bécard G, Pfeffer PE (1993) Status of nuclear division in arbuscular mycorrhizal fungi during in-vitro development. Protoplasma 174(1/2):62–68

Bianciotto V, Bandi C, Minerdi D, Sironi M, Tichy HV, Bonfante P, (1996) An obligately endosymbiotic mycorrhizal fungus itself harbors obligately intracellular bacteria. Appl Environ Microbiol 62:3005–3010

Bidartondo MI, Kretzer AM, Pine EM, Bruns TD (2000) High root concentration and uneven ecto-mycorrhizal diversity near *Sarcodes sanguinea* (Ericaceae): a cheater that stimulates its victims? Am J Bot 87:1783–1788

Cairney JWG (2000) Evolution of mycorrhiza systems. Naturwissenschaften 87:467–475

Clapp JP, Young JPW, Merryweather JW, Fitter AH (1995) Diversity of fungal symbionts in arbuscular mycorrhizas from a natural community. New Phytol 130:259–265

Gange (2000) Arbuscular mycorrhizal fungi, Collembola and plant growth. TREE 15:369–372

Gange AC (2001) Species specific responses of a root- and shoot feeding insect to arbuscular mycorrhizal colonization of its host plant. New Phytol 150:611–618

Garbaye J (1994) Helper bacteria – a new dimension to the mycorrhizal symbiosis. New Phytol 128:197–210

Giovannetti M, Azzolini D, Citernesi AS (1999) Anastomosis formation and nuclear and protoplasmic exchange in arbuscular mycorrhizal fungi. Appl Environ Microbiol 65:5571–5575

Goverde M, van der Heijden MGA, Wiemken A, Sanders IR, Erhard A (2000) Arbuscular mycorrhizal fungi influence life history traits of a lepidopteran herbivore. Oecologia 125:362–369

Grime JP (2001) Plant strategies, vegetation processes and ecosystem properties. Wiley, Chichester

Harley JL, Harley EL (1987) A check list of mycorrhiza in the British Flora. New Phytol 105 (Suppl):1–102

Hartnett DC, Wilson WT (1999) Mycorrhizae influence plant community structure and diversity in tall grass prairie. Ecology 80:1187–1195

Hibbett DS, Gilbert LB, Donoghue MJ (2000) Evolutionary instability of ecto-mycorrhizal symbioses in basidiomycetes. Nature 407:506–508

Huovinen-Hufschmid C, Körner C (1998) Microscale patterns of plant species distribution, biomass and leaf tissue quality in calcareous grassland. Bot Helvet 108:69–83

Jones MD, Durall DM, Tinker PB (1998) A comparison of arbuscular and ecto-mycorrhizal *Eucalyptus coccifera*: growth response, phosphorus uptake efficiency and external hyphal production. New Phytol 140:125–134

Jonsson LM, Nilsson MC, Wardle DA, Zackrisson O (2001) Context dependent effects of ecto-mycorrhizal species richness on tree seedling productivity. Oikos 93:353–364

Klironomos JN, Hart MM (2001) Animal nitrogen swap for plant carbon. Nature 410:651–652

Koide RT, Kabir Z (2000) Extraradical hyphae of the mycorrhizal fungus *Glomus intraradices* can hydrolyse organic phosphate. New Phytol 148:511–517

Landeweert R, Hoffland E, Finlay RD, Kuyper TW, van Breemen N (2001) Linking plants to rocks: ecto-mycorrhizal fungi mobilize nutrients from minerals. TREE 16(5):248–254

Michelsen A, Quarmby C, Sleep D, Jonasson S (1998) Vascular plant ^{15}N natural abundance in heath and forest tundra ecosystems is closely correlated with the presence and type of mycorrhizal fungi in roots. Oecologia 115:460–418

Molina R, Massicotte H, Trappe JM (1992) Specificity phenomena in mycorrhizal symbioses: community-ecological consequences and practical implications. In: Allen MF (ed) Mycorrhizal functioning. Chapman Hall, New York, pp 357–423

Morton JB, Bever JD, Pfleger FL (1997) Taxonomy of *Acaulospora gerdemannii* and *Glomus leptotichum*, synanamorphs of an arbuscular mycorrhizal fungus in Glomales. Mycol Res 101:625–631

Newsham KK, Fitter AH, Watkinson AR (1995) Multi-functionality and biodiversity in arbuscular mycorrhizas. Trends Ecol Evol 10:407–411

Newton AC, Haig JM (1998) Diversity of ecto-mycorrhizal fungi in Britain: a test of the species area relationship and the role of host specificity. New Phytol 138:619–627

Onguene NA, Kuyper TW (2001) Mycorrhizal associations in the rain forest of South Cameroon. For Ecol Manage 140:277–287

Perotto S, Bonfante P (1997) Bacterial associations with mycorrhizal fungi: close and distant friends in the rhizosphere. Trends Microbiol 5:496–501

Read D (1991) Mycorrhizas in ecosystems. Experientia 47:376–391

Read DJ, Duckett JG, Francis R, Ligrone R, Russell A (2000) Symbiotic fungal associations in 'lower' land plants. Philos Trans R Soc Lond Ser B Biol Sci 355:815–830

Redecker D, Kodner R, Graham LE (2000a) Glomalean fungi from the Ordovician. Science 289:1920–1921

Redecker D, Morton JB, Bruns TD (2000b) Ancestral lineages of arbuscular mycorrhizal fungi (Glomales). Mol Phylogenet Evol 14:276–284

Remy W, Taylor TN, Haas H, Kerp H (1994) Four hundred-million-year-old vesicular-arbuscular mycorrhizae. Proc Natl Acad Sci USA 91:11841–11843

Schlüßler A (2000) *Glomus claroideum* forms an arbuscular mycorrhiza-like symbiosis with the hornwort *Anthoceros punctatus*. Mycorrhiza 10:15–21

Selosse MA, Le Tacon F (1998) The land flora: a phototroph-fungus partnership? TREE 13:15–20

Setälä H (2000) Reciprocal interactions between Scots pine and soil food web structure in the presence and absence of ecto-mycorrhiza. Oecologia 125:109–118

Simon L, Bousquet J, Levesque RC, Lalonde M (1993) Origin and diversification of endomycorrhizal fungi and coincidence with vascular land plants. Nature 363:67–69

Smith SE, Read DJ (1997) Mycorrhizal symbioses, 2nd edn. Academic Press, London

Stortelder AHF, Schaminee JHJ, Hommel PWFM (1999) De vegetatie van Nederland. Deel 5 Plantengemeenschappen van ruigten, struwelen en bossen. Opulus Press, Leiden

Streitwolf-Engel R, van der Heijden MGA, Wiemken A, Sanders IR, (2001) The ecological significance of arbuscular mycorrhizal fungal effects on clonal plant growth. Ecology 82:2846–2859

Tilman D (1988) Plant strategies and the dynamics and structure of plant communities. Princeton Univ Press, Princeton

Timonen S, Jorgensen KS, Haahtela K, Sen R (1998) Bacterial community structure at defined locations of *Pinus sylvestris Suillus bovinus* and *Pinus sylvestris Paxillus involutus* mycorrhizospheres in dry pine forest humus and nursery peat. Can J Microbiol 44:499–513

Trappe JM (1987) Phylogenetic and ecological aspects of mycotrophy in the angiosperms from an evolutionary standpoint. In: Safir GR (ed) Ecophysiology of VA mycorrhizal plants. CRC Press, Boca Raton, pp 5–25

Van Tuinen D, Jacquot E, Zhao B, Gollotte A, Gianinazzi-Pearson V (1998) Characterisation of root colonisation profiles by a microcosm community of arbuscular mycorrhizal fungi using 25S rDNA-targeted nested PCR. Mol Ecol 7:879–887

Wallenda T, Kottke I (1998) Nitrogen deposition and ecto-mycorrhizas. New Phytol 139:169–187

Wilson GWT, Hartnett DC (1998) Interspecific variation in plant response to mycorrhizal colonisation in tallgrass prairie. Am J Bot 85:1732–1738

Wimp GM, Whitham TG (2001) Biodiversity consequences of predation and host plant hybridization on an aphid-ant mutualism. Ecology 82:440–452

Subject Index

A
abscisic acid   230
achlorophyllus plants   49, 50, 376, 390, 391, 402–405
acidification   184, 185
AFLP   212
age   43, 46, 47, 85
allocation of resources
– to mycorrhizal fungi   45–47, 78, 79
– to plants   35–39, 80–82
– within mycorrhizal fungi   105–109, 361
amino acids   42, 45, 46, 56, 120, 123, 129
ammonium   121, 124, 125, 127, 130
anamorph   379
anastomosis of hyphae   95, 99, 433, 451
ants   333
antagonistic interactions   176, 362–365
arbuscular mycorrhizal fungi (AMF)   4, 5, 76–81
– community structure   202, 203, 208–215, 255
– diversity and ecosystem functioning   211, 229–237, 255, 256
– ecophysiology   76–81
– effects on:
– – herbivores   305–310
– – plant diversity   233–235, 245–252, 333–339
– – plant nutrition   82–84
– genetics   215–217
– invertebrates   322–333
– molecular identification   203–207
– mycelium   102–104
– networks   52, 53, 229, 433
– phylogeny and taxonomy   213
– species composition   202, 203, 208–215, 255
– specificity   416–433

arbuscules   77, 78
arbutoid mycorrhizas   5, 48
Archeaosporaceae   207
arum-type mycorrhizas   77
Ascomycetes   5
auxins   38

B
bacteria   42, 43, 450
Basidiomycetes   5, 95, 347, 376, 379–388
benomyl   22, 327
bidirectional transfer   51–53, 60, 76
biochemical cycling   129, 346
biodiversity (see also diversity)   60, 119, 226–228, 446, 447
– ecosystem functioning   227–229
BLO's (bacteria like organisms)   450

C
^{14}C 51–53, 250, 361–364
C:N ratios   354, 355
C:P ratios   354
C_3 & C_4 species   11, 19
calcium   38, 39, 41
– oxalate   41
– phosphate   41
carbon (C)
– balance   167
– consumption/nutrition
– – by AMF   76–79
– – by EMF   45–47
– – interaction with herbivores   301, 302
– cycling   50, 53, 346
– dioxide elevated   137–143, 179, 180
– dynamics   399–405
– flow   106–108, 167

– flux   44–47
– transfer   48–50, 51–61, 250, 402–405
cellulase   354
chemo-trophic response   102
chitin   130
chloropicrin   21
chloropyrifos   335
clipping   56, 298, 301
^{13}C-NMR   76
coevolution   395–398, 419–421, 452, 453
coexistence   119, 125, 126, 252–254, 256, 257
coils   77
Coleoptera   333, 335
Collembola   328–330
colonization   142
– intensity of   78, 127, 142, 297–299, 301–304,
common mycorrhizal network (CMN)   47, 48, 58, 60, 98–104
community (see arbuscular and ecto-mycorrhizal fungi)
– mycorrhizal fungi   8–24, 172–187, 202–204, 211, 446, 447
– plants   8–24, 227–235, 245–249
community feedback (see feedback)
competition   42, 49, 52, 60, 117–133, 250, 251
– between fungi   353–358, 359–366, 175, 176
– between plants   125, 126, 252–255
concerted evolution   206
contaminant sequences   217, 218
contractile movement   43
copper   38
cost-benefit analysis   46, 47, 79, 85
cyst nematodes   339
cytoplasmic streaming   44

D

dark septate fungi   15
decomposition   129, 130, 346–348, 356–359
decomposer fungi   42, 346–367
defense chemicals   307
denitrification   356, 357
depletion zones   39, 40, 81
differential effects of different AMF species   81, 82, 231, 277, 428, 429
diffusion   43
Diptera   333, 335

dispersal of spores   323–225
diversity (see also species richness)   165, 226–228
– feedback mechanisms   271–274
– plants   231–235, 245–254, 256, 257, 333–340
– AMF   201–224, 231–235, 255, 256
– ecosystem functioning   227–229
– EMF   163–200
dying roots   55
dual mycorrhizal symbiosis
– herbivores   308–310
– effects on plants   15

E

earthworms   324, 325
ecophysiology   442–445
– of AMF   76–83
– of EMF   34–45
ecosystem functioning   4, 5, 34–45
ecto-mycorrhizal fungi (EMF)   4, 5, 34–45
– communities   168–171, 300, 446, 447
– ecophysiology   34–45
– effects on herbivores   305–310
– mycelium   100–102, 348–352,
– nutrient uptake   35–41, 347–356
– reversals to saprotrophs   451
– sporocarp counts   165–171
edaphic factors   172
endo-mycorrhizas   5
endophyte, fungal   248
enzymes   40–42, 77, 354, 355
ericoid mycorrhizal fungi   5, 119, 120, 125, 126, 128, 140, 443, 444
ergosterol   97
experiments
– constraints   8–23
– design of mycorrhizal experiments   8–23
exotic species   151
external hyphae (see also mycelium)
– of AMF   78, 96, 98–100, 102–104
– of EMF   95, 98–102, 348–352
extracellular enzymes   354, 355
exudates   46, 58
evolution   61, 105, 287, 399, 419–421
evolutionary maintenance of mutualism   287, 288, 419–421

Subject Index

F
feedback   257, 267–292
- dynamics   270–285
- implications   286–289
- mechanistic approach   274, 279–282
- phenomenological approach   275, 283–285
- testing   279–286
field studies   8, 9, 21–23, 333–340, 453
fire   177–179
fitness   106, 425, 426
foraging   100–109, 348, 362–366
- for inorganic nutrients   100–103
- for organic nutrients   101–104
- strategies   104, 105
forbs   337–339, 251
forest
- boreal   360
- coniferous   360
- management   179–187
- temperate, deciduous   347, 443
- tropical rainforest   443
fructose   77
fruiting bodies (see sporocarps)
functional groups
- plants   11, 18–20
- mycorrhizal fungi   13–15, 260
fungal regulation of transfer   58, 59, 61
fungal structures   5
- of AMF   77–79
- of EMF   100, 348–352
fungicides   21–23, 327, 334–340

G
gene conversion   206
generalist herbivore   305, 306, 314, 316
genetic organization of AMF   215–221, 430–432
genetic variation   215–220, 430
genotypes
- of fungi   13–15, 232, 396, 397, 429–431
- of plants   20
gibberelline   230
global change   135–160
- definition   136
- direct and indirect effects   151
- effects on symbionts   151, 152, 179–185
- factor interactions   149
- long term responses   150
- multiple factors   139

Glomales   209, 210, 450
glomalin   142
glutamine   45
glycine   82
glycosides   307
glycogen   77
glucose   77
glutamate   82
grassland
- calcareous   245–247, 443
- European   249
growth forms
- of plants   19
- of mycorrhizal fungi   94–100

H
H^+-ATPase   77
Hartig net   34, 393
heathlands   119, 122, 123, 128, 443
heavy metals   180, 181
herbivory   295–316
heterokaryosis   431
hexose uptake   77
humped-back curve   246, 247
humic residues   129, 356
humification   129, 357, 358
hydrophilic hyphae   98
hydrophobic hyphae   98
hyphae (see also mycelium)   39, 40, 43, 44, 46, 47, 94–100
- density   99
- extraradical/matrical   39–41, 94–99, 142
- length
-- of AMF   81
-- of EMF   39–41, 99
- links   47–53, 59–61, 402–405, 433
- networks   47, 48, 58, 60, 98–104
- turnover   46
hyphosphere   43

I
infection (see colonization)
inflow rates   39
inoculum types   15, 16
in situ hybridization   219, 432
intercellular hyphae of AMF   77
interface between plant and fungus
- of arbuscular mycorrhizas   77–79
- of ecto-mycorrhizas   34, 35
internal cycling   121

internal transcribed spacer (ITS) 205–207, 209, 216–220, 378, 384, 392, 393, 398
interplant transfer 50–59, 402–405
intrasporal sequence variation 215–217
insects 296–314, 323, 324; 328–332, 334–340
invertebrates 296–314, 322–333
iprodione 23, 335
iron 41
isoclines, plant growth 126, 252–254, 256
isoflavonoids 332
isotopes 40, 42, 47, 48, 50–53 (see also ^{15}N & ^{14}C)
isotope signatures 125
ITS sequences 216

K

keystone species 18, 19, 312, 313
kinetic parameters V_{max} and K_m 39, 40, 80, 86
K_m 39, 40

L

leaching 356, 357
Lepidoptera 333
life-history trades offs
– of mycorrhizal fungi 106
lignin 119, 129, 355
ligninase 355
lignocellulose 347, 348, 354, 356
liming 185, 186
lipids 77, 100
lipid signatures 78, 97, 100
litter 174
– chemistry 117, 119, 129, 130
– decomposition 117, 347, 355–359
luxury consumption 83

M

magnesium 38, 39, 41
mass flow 40, 43, 44, 56, 80
methyl-bromide 21
microbial communities 16, 17, 42, 43, 346, 450
microcosms 8–21
mineral nutrition 117
– by AMF 79–82
– by EMF 35–41

mineralization 119, 129, 130, 356–358
mites 327, 328
– mycophagous 327
molecular identification
– arbuscular mycorrhizal fungi 203–212
– ecto-mycorrhizal fungi 165–167
– quantitative versus qualitative 422
– orchidaceous mycorrhizal fungi 389, 390
monotropoid mycorrhizas 5, 382–385, 393, 394
morphological characterization
– arbuscular mycorrhizal fungi 212–215
– ectomycorrhizal fungi 165–167
– orchidaceous mycorrhizal fungi 389, 390, 392–394
morphotypes, EMF 175
multifunctionality 6, 444
– of AMF 230, 231
multiple infection 165, 204, 389
multitrophic interactions 296–314, 322–340, 449, 450
mutual interdependence 270–274
– evidence for 276–279
mycelium 98–105, 348–352
– carbon allocation 45–47, 76–79, 361–364
– cords 348, 349, 362, 363
– density 39, 40, 99
– disruption by invertebrates 323, 325
– energy storage 100
– function 350–352
– growth rates 98, 99
– – of AMF 78, 96, 98–100, 102–104
– – of EMF 95, 98–102
– nutrient transport 43, 44
– strands 98
– structure 98, 99, 349, 350
myco-heterotrophic plants 49, 50, 84, 375–413
mycorrhiza helper bacteria 450
mycorrhizal associations
– major types 5
– abundance 443
mycorrhizal dependency 8, 84–87, 246–251; 258, 259
mycorrhizal responsiveness (see mycorrhizal dependency)
mycorrhizal species sensitivity 231, 232, 258, 259
mycorrhizosphere 43

Subject Index

mycotrophy (see mycorrhizal dependency)

N

natural experiments 150
negative feedback 272–274, 280–282
nematicides 327
nematodes 325–327
– mycophagous 326, 327
– plant parasitic 325, 326
niche 172–174, 252, 286
nitrate 121
nitrification 121, 356, 357
nitrogen N 38, 40, 56
– ^{15}N natural abundance 124, 125
– $\partial^{15}N$ 124
– addition experiments 127
– cycle 125, 356, 357
– deposition 120, 127, 128, 143, 144, 181–184
– fertilization 181–184
– fixation 357
– fixing bacteria 43, 58
– fixing plants 54, 57, 58
– gradient 183
– inorganic 35–39, 81, 82, 100, 103, 120, 127, 355–357
– organic 38, 41, 42, 56, 101, 120, 127, 129, 356, 357
– stable isotope ^{15}N 124
– transfer 48, 53–55, 57–59
nutrient
– cycling 117–133, 356–359
– depletion 39, 80, 81
– retention 128
– sources 120, 121
– transfer 53–55, 77, 78
– translocation 39, 43, 44
– uptake 34–38, 42, 61, 79–84, 122, 123
nutrition (see also carbon or mineral nutrition) 38, 39, 46, 76–83, 119

O

optimal foraging (see also foraging)
– by mycorrhizal fungi 100–109
– theory 105–109
orchid mycorrhizas 5, 378–392, 443, 444
orchids 378–393
organic acids 41, 105

organic matter 101, 173, 357
organic soils 129
'organicization' 129, 130
ozone 137, 147, 148, 180

P

paris type mycorrhizas 77
Paraglomaceae 207
patchy environment 107
pathogens 38
PCR 205, 206
– generated microsatellite loci 212
– M13- primed 212
– microsatellite primed 212
– RFLP 49, 393, 394
peptides 121, 129
peroxidase 355
pH 176, 177
phenolics 129, 332
phosphatases 355, 356, 358
phosphorus (P) 38–41, 46, 79–84
– cycle 358, 359
– inorganic 80, 81, 102, 117, 118, 358
– organic 119, 120, 357–359
– orthophosphate 358, 359
– plant concentration 250
– plant content 234
– soil concentration 234
– transfer 53–55, 57–59
– transporters 82, 83
phylogeny 213
– trees 210, 216, 386, 387
physiology (see ecophysiology)
phytases 355
phytohormones 34
plant communities 47, 48
– structure 339
– diversity 231–235, 245–254, 256, 257
plant-soil feedbacks 129, 268–270
pollution 179–187
polyphenols 119, 357
positive feedback 271, 272
potassium 41, 80
precipitation 144, 146
primers
– family specific 208, 209
– glomalean specific 209–211
– species specific 208, 209
prisoners dilemma 420
productivity 234–236, 248
proteinases 355, 356

proteins 42, 121, 355
protein-tannin complexes 129, 130, 355

R
RAPD PCR 212
R-strategists 189
red queen hypothesis 419–421
reductionist approaches 4–8
redundancy hypothesis 227, 228, 235
regeneration (see seedling establishment)
resource
 – partitioning 172–174
 – competition 125, 126, 250, 251
respiration 46
RFLP 393
rhizobacteria 34, 38, 42, 43, 58
Rhizoctonia spp. 5, 14, 379, 381, 388
 – systematics 379–388
rhizomorph 43, 44, 94, 104, 348, 349
rhizosphere bacteria 42, 43
ribosomal genes 205–212, 215–221, 378, 386, 387
root
 – density 39, 40, 166
 – fibrousness 251
 – free compartment 81
 – root/shoot ratio 38
 – turnover 46, 191
runner hyphae 96, 104

S
sampling effect 19, 235, 236
sand dunes 99
saprotroph 356–359, 381, 400, 401
saprotrophic fungi 347–353, 356–367
SSCP 221, 402
S-strategists 189
sclerotia 98
seed germination 389, 391
seedling establishment 253, 254, 337–340, 353
sequence heterogeneity 215–220, 430–432
shrub land 443
small ribosomal subunit (SSU or 18S) 205–208, 210, 215
species composition
 – AMF 255, 256
 – EMF 168–171, 300, 446, 447
 – plant 8–24, 227–235, 245–249, 333–340

soil
 – aggregates 40
 – fauna 17, 323–333
 – invertebrates 322–333
 – minerals 41
 – moisture 175, 176
 – organic matter 101, 157, 173
 – pH 176, 177
 – pores 39, 41
 – spatial heterogeneity 173
 – sterilisation 12
 – structure 138, 142
 – weathering 41
source-sink 43, 55–59
spatial
 – distribution of AMF 233, 288
 – dynamics 288, 339
 – heterogeneity 173, 174
specificity 47–50, 60, 278, 279, 376–405, 415–437, 452, 453
 – definitions 378, 417
 – evidence against 423–425
 – evidence for 377–394, 425–428
 – evolution of 399, 415–437
 – functional 230, 425–429
 – genetic basis 377–397, 429–432
 – host 277–279, 425–428
 – importance of hyphal network 433, 434
 – influences on 395–399
 – in Orchidaceae 378–393
 – in Monotropoideae 393, 394
 – ontogenetic influences on 397–399
specialist herbivore 305, 306, 314, 316
species richness (see diversity)
spores
 – dispersal 323–325
 – diversity 201–224, 231–235, 255, 256
sporocarps 45–47, 165, 166, 168–171
stability 227
succession 237, 238, 334–338
sucrose 44, 45, 77
sulphur 38

T
tallgrass prairie 245, 248, 443
tannins, 129
teleomorph 379
temperature 144–146, 177
terpenoids 332
tit-for-tat 420

transfer
- carbon   48–53, 55–61, 402–405
- net   50–53, 60
- nitrogen   48, 53–55, 57–59
- nutrient   48, 50, 53–55, 57–61
- phosphorus   53–55, 57–59
- regulation   55–59
transporters   82
trehalose   77
T-RFLP   221
trophic growth   95
trophic niches   399–405
TTGE   221
tundra   443

U
uptake
- kinetics   40, 129
- specific rates   38, 39
UV-B   137, 148, 149
unequal crossing over   206

V
vertebrates   297, 298
V_{max} 39, 40

W
water availability   146, 147
weathering   41
wildfire   177, 178
wood ash   101, 186, 187
wood decay fungi   354, 364

X
xylem loading   38

Z
zinc   38
Zygomycetes   5, 96, 202

Taxonomic Index

A
Abies
– *amabilis* 42
– *lasiocarpa* 41
– *magnifica* 50
Acaulospora
– *denticulata* 216, 232, 233
– *gerdemannii* 213, 221
– *koskei* 210
– *laevis* 210
– *morrowiae* 284
– *rugosa* 210
– *scrobiculata* 204, 210, 216
– *spinosa* 210
– *trappei* 214, 221, 284
Acer
– *pseudoplatanus* 214
– *saccharum* 148, 329
Acianthus
– *caudatus* 380
– *exsertus* 380
– *reniformis* 380
Afzelia africana 37
Agropyron desertorum 297, 303
Agropyron spicatum 303
Albatrelus ovinus 181
Allium
– *porrum* 233, 328, 329
– *vinale* 280–283
Allotropa virgata 382, 394, 395
Alnus
– *glutinosa* 48
– *incanca* 49
– *rubra* 48
– *viridis* 48
Anthoxanthum odoratum 246, 283
Amanita spp. 98, 175
– *franchetii* 166
– *muscaria* 36
Amerorchis rotundifolia 380

Ammophila breviligulata 326
Amphinema byssoides 182, 186
Andropogon
– *gerardii* 334
– *scoparius* 334
Aphelenchus
– *avena* 327
– *composticola* 327
Aphelenchoides spp. 327
Apophysomyces elegans 210
Arabis hirsuta 246, 254
Arbutus spp. 48
Archeaosporaceae 207
Arctia caja 297, 306, 307, 314
Arctostaphylos spp. 48
Armillaria spp. 383, 390
Ascomycetes 5
Atriplex gardneri 147

B
Basidiomycetes 5, 95, 347, 376, 379–388
Betula 50
– *pendula* 36, 179, 360–365, 402, 404
– *papyrifera* 51, 52, 56, 179, 180
Boletus spp. 98
Bouteloua gracilis 297
Briza media 246
Bromus erectus 246, 248, 250, 254
Brachypodium pinnatum 246, 250, 254
Burkholderia 450

C
Caladenia spp. 380
Calluna vulgaris 119, 125, 126
Calypso bulbosa 380
Campanula rotundifolia 246
Carex 19

Carex flacca 246, 254
Cenococcum
– *geophilum* 173, 175, 176, 182, 185, 190, 191, 385
– *graniforme* 98
Centauria nigra 246
Centaurium erythrea 246, 254
Cephalanthera austinae 382, 392, 395, 402
Ceratobasidium spp. 381, 396
Chamerion angustifolium 232
Chaitophorus populicola 306, 308, 309, 312, 313
Choristoneura occidentalis 297
Cinara pinea 306
Cirsium arvense 306
Clitocybe squamulosa 385
Coleoptera 333, 335
Collembola 328–330
Corallorhiza
– *maculata* 382, 392, 395–399, 402, 452
– *mertensiana* 382, 392
– *odontorhiza* 398
– *striata* 382
– *trifida* 50, 382, 392, 398, 401, 402
Cortinarius spp. 175
Cryptothallus mirabilis 376, 404, 405
Cypripedium
– *calceolus* 398
– *candidum* 380
– *parviflorum* 380
– *passerinum* 398, 399
cyst nematodes 339

D
dark septate fungi 15
Dactylorhiza
– *majalis* 393
– *purpurella* 380
Dactylis glomerata 246
Deschampsia flexuosa 119
Didymoplexis minor 390
Diptera 333, 335
Diuris spp. 380
Douglas fir (see *Pseudotsuga menziessi*)

E
earthworms 324, 325
Elaphomyces muricatus 385
Endogone pisiformis 210

Entrophospora spp. 210
– *colombiana* 213, 232, 308
– *infrequens* 213
Epilachnis verivestis 306, 308
Epipactis helleborine 390
Erica tetralix 119
Erthryomyces crocicreas 383, 389, 390, 401
Eucalyptus
– *coccifera* 99
– *dives* 277
– *viminalis* 277

F
Fagus spp. 14
– *sylvatica* 37, 42, 176
Festuca
– *ovina* 53, 246, 250, 251, 254, 425
– *rubra* 246, 428
Folsomia candida 328
Fomes spp. 390
– *mastoporus* 383, 401
Fragaria X ananassa 306, 330
Fusarium spp. 216

G
Galeola altissima 382, 389, 390, 401
– *hydra* 390
– *septentrionalis* 382, 390, 401
Galium verum 246
Ganoderma australe 383, 389
Gastrodia
– *cunninghamii* 382, 401
– *elata* 390
– *minor* 382
– *sesamoides* 382, 401
Geopora cooperi 299
Geosiphon pyriforme 210, 213
Gigaspora spp.
– *gigantea* 144, 210, 232
– *margarita* 144, 148, 206, 210, 212, 215, 216
– *rosea* 219, 220
Gilpinia pallida 306
Glomales 209, 210, 450
Glomus spp. 58, 99, 210, 308
– *aggregatum* 144, 232
– *brasilianum* 221
– *caledonium* 81, 82, 216, 277
– – runner hyphae 96
– *claroideum* 210, 215–217, 232

Taxonomic Index

– *clarum* 308, 327
– *constrictum* 217
– *coronatum* 216, 217
– *diaphanum* 279
– *etunicatum* 206, 210, 231–233, 308
– *fasciculatum* 210, 216, 331
– *fistulosum* (= *G. claroideum*) 215, 216
– *geosporum* 144, 210, 216, 217
– *intraradices* 81, 82, 103, 206, 209, 216, 219, 231–33, 277, 310, 330
– – external hyphae 96
– *invermaium* 81
– *leptotichum* (= *Archaeospora leptoticha*) 144, 210, 216, 221
– *macrocarpum* 328
– *mosseae* 81, 141, 144, 208, 210, 216, 217, 219, 232, 279, 310, 330, 331
– *occultum* 144, 213, 221
– *versiforme* 80, 210, 216
– *vesiculiferum* 210
Glycine max 297, 306
Goodyera
– *oblongifolia* 380
– *repens* 380
Gutierrezia sarothrae 142, 149

H
Hebeloma spp. 98
– *crustiniforme* 36, 37, 47, 147
– *cylindrosporum* 14, 36
– *sacchariolens* 180
Helianthemum 14
Heliothis zea 306
Hemitomes congestum 382
Heterobasidion annosum 147
Heteroderma trifolii 339
Hexalectris spicata 392, 401
Hieracium pilosella 246, 250, 254
Hyacinthoides non-scripta 208, 211, 213, 214
Hydnellum 98
Hydnum rufescens 182
Hysterangium 98
Hypholoma fasciculare 359, 360

J
Juncus spp. 19

L
Laccaria spp. 180
– *bicolor* 36, 37, 139, 176
– *laccata* 37, 47, 98, 139
– *proxima* 37
Lactarius spp. 98
– *rufus* 182, 183
Larix kaempferi 174
Leccinum scabrum 181
Leontodon hispidus 246
Lepidoptera 333
Lepista nuda 102
Lolium perenne 339
Lotus corniculatus 246, 254, 306, 310
Loweporus tephroporus 383, 389
Lumbricus terrestris 325
Luzula spp. 19
Lycoperdon perlatum 383
Lygus rugulipennis 306

M
Marasmius coniatus 390
Matsucoccus acalyptus 299, 300, 306
Medicago truncatula 82
Meloidogyne incognita 326
Microporus affinus 383, 389
Microtis parviflo 380
Molinia caerulea 119, 125, 126
Monotropa
– *hypopitys* 384, 394, 395
– *uniflora* 382, 384, 394
Monotropastrum spp. 384
Mycena thuja 383
Mucor indicus 210
Myzus ascalonicus 306

N
Neottia nidus-avis 382, 392, 398, 401

O
Onychiurus armatus 17
Otiorhynchus sulcatus 306, 310, 330

P
Panicum
– *sphaerocarpon* 283
– *virgatum* 279
Paraglomaceae 207

Paxillus spp. 98
- *involutus* 17, 36, 37, 40, 42, 48, 49, 98, 100, 139, 177, 179, 180, 182, 191, 359–365
Phanerochaete velutina 358, 361–365
Phaeosphaeria nodorum 216
Phellinus spp. 389
Picea
- *abies* 182, 185
- *mariana* 36, 37, 38
- *rubens* 175
- *sitchensis* 36
Piceirhiza
- *gelatinosa* 190
- nigra 186
Piloderma spp. 95
- *croceum* 177, 187
Pinirhiza rosea 186
Pinus spp. 14
- *banksiana* 178
- *contorta* 36, 48, 173, 402
- *edulis* 299, 300, 302, 306, 314
- *muricata* 121, 166
- *palustris* 179
- *pinaster* 36
- *ponderosa* 48, 147, 149
- *rigida* 36, 37
- *strobus* 179
- *sylvestris* 36–40, 49, 147–149, 179, 185–187, 298, 306, 355, 359
- *taeda* 180
pinyon pine (see *Pinus edulis*)
Pisolithus
- *arhizus* 36, 174, 177
- *tinctorius* 36, 37. 149, 176, 308
Plantago lanceolata 53, 141, 142, 231, 232, 246, 280–283, 297, 306–308, 314, 335, 428
Platanthera
- *hyperborea* 380, 398, 399
- *leucophaea* 380
- *obtusata* 380
Pleuricospora fimbriolata 384
Poa pratensis 246
Pogonia ophioglossoides 380
Polyommatus icarus 306, 310
Populus spp. 449
- *angustifolia* X *Populus fremontii* 306, 308
- *balsamifera* 298
- *tremuloides* 149
Prunella
- *grandiflora* 246, 254

- *vulgaris* 246, 250, 254, 452, 426, 429
Pseudotsuga menziesii 48, 51, 52, 56, 149, 297, 302
Pterospora andromedea 384, 393–395
Pterostylis
- *aff. Rufa* 380
- *barbata* 380
- *nana* 380, 396

Q

Quercus
- *petraea* 214
- *robur* 37, 139, 149
- *rubra* 181

R

Rhizanthella
- *gardneri* 382, 400
- *slateri* 400
Rhizoctonia spp. 5, 14, 379, 381, 388
- *monilioides* 381
- *repens* 381, 388, 391, 400
- *solani* 381, 391
- *subtilis* 381
- systematics 379–388
Rhizopogon 48
- *ellenae* 50, 385, 394, 396, 452
- *subcaerulescens* 385, 393
- *vinicolor* 383
Rhizopus oryzae 210
Rhizomucor variabilis 210
Robinia pseudoacacia 139
Rumex acetosa 246, 428
Russula
- *amoenolens* 166
- *ochroleuca* 175, 185

S

Sanguisorba minor 246, 254
Saksenaea vasiformis 210
Salix spp. 37, 50, 298
- *repens* 405
- *viminalis* 37
Sarcodes sanguinea 50, 384, 393, 394, 396, 452
Scabiosa columbaria 246
Scleroderma 37
Scutellospora spp.
- *calospora* 82, 144, 216, 232, 233, 278, 284, 430

Taxonomic Index 469

– *castanea*   215, 216, 219, 220
– *dipapillosa*   210
– *dipurpurescens*   210
– *pellucida*   210, 280–282, 284, 288
Schizaphis graminum   306
Schizolachnus pineti   306
Sebacina vermifera   381, 383, 401
Setaria lutescens   426
Silene nutans   246
Solidago missouriensis   279
Sorgastrum nutans   334
Sorghum spp.   306
Spiranthes
– *romanzoffiana*   399
– *sinensis*   380, 391, 395, 398
Spodoptera frugiperda   306
Suillus spp.   98, 100
– *bovinus*   37, 42, 100, 139, 363
– *luteus*   36
– *pungens*   98, 166
– *variegatus*   37, 359, 360

T
Taraxacum officinale   306, 330
Thanatephorus
– *cucumeris*   381, 400
– *gardneri*   383
– *pennatus*   381, 392
Thelephora spp.   98
– *terrestris*   36, 37, 42, 102, 180
Thuja plicata   51
Tomentella sublilicina   173

Tricholoma
– *magnivelare*   394
– *terreum*   299
Trifolium spp.
– *pratense*   246, 254, 329, 428
– *repens*   96, 142, 326, 339
– *subterraneum*   148
Truncocolumella citrina   385
Tsuga
– *canadensis*   179
– *heterophylla*   48
Tuber puberulum   186
Tulasnella
– *calospora*   381, 388
– *cruciata*   381
Tylospora fibrillosa   173, 175, 183

U
Urophora carduii   306

X
Xerocomus badius   182

Y
Yoania spp.   382

Z
Zygomycetes   5, 96, 202

Ecological Studies
Volumes published since 1997

Volume 125
Ecology and Conservation of Great Plains Vertebrates (1997)
F.L. Knopf and F.B. Samson (Eds.)

Volume 126
The Central Amazon Floodplain: Ecology of a Pulsing System (1997)
W.J. Junk (Ed.)

Volume 127
Forest Decline and Ozone: A Comparison of Controlled Chamber and Field Experiments (1997)
H. Sandermann, A.R. Wellburn, and R.L. Heath (Eds.)

Volume 128
The Productivity and Sustainability of Southern Forest Ecosystems in a Changing Environment (1998)
R.A. Mickler and S. Fox (Eds.)

Volume 129
Pelagic Nutrient Cycles: Herbivores as Sources and Sinks (1997)
T. Andersen

Volume 130
Vertical Food Web Interactions: Evolutionary Patterns and Driving Forces (1997)
K. Dettner, G. Bauer, and W. Völkl (Eds.)

Volume 131
The Structuring Role of Submerged Macrophytes in Lakes (1998)
E. Jeppesen et al. (Eds.)

Volume 132
Vegetation of the Tropical Pacific Islands (1998)
D. Mueller-Dombois and F.R. Fosberg

Volume 133
Aquatic Humic Substances: Ecology and Biogeochemistry (1998)
D.O. Hessen and L.J. Tranvik (Eds.)

Volume 134
Oxidant Air Pollution Impacts in the Montane Forests of Southern California (1999)
P.R. Miller and J.R. McBride (Eds.)

Volume 135
Predation in Vertebrate Communities: The Białowieża Primeval Forest as a Case Study (1998)
B. Jędrzejewska and W. Jędrzejewski

Volume 136
Landscape Disturbance and Biodiversity in Mediterranean-Type Ecosystems (1998)
P.W. Rundel, G. Montenegro, and F.M. Jaksic (Eds.)

Volume 137
Ecology of Mediterranean Evergreen Oak Forests (1999)
F. Rodà et al. (Eds.)

Volume 138
Fire, Climate Change and Carbon Cycling in the North American Boreal Forest (2000)
E.S. Kasischke and B. Stocks (Eds.)

Volume 139
Responses of Northern U.S. Forests to Environmental Change (2000)
R. Mickler, R.A. Birdsey, and J. Hom (Eds.)

Volume 140
Rainforest Ecosystems of East Kalimantan: El Niño, Drought, Fire and Human Impacts (2000)
E. Guhardja et al. (Eds.)

Volume 141
Activity Patterns in Small Mammals: An Ecological Approach (2000)
S. Halle and N.C. Stenseth (Eds.)

Volume 142
Carbon and Nitrogen Cycling in European Forest Ecosystems (2000)
E.-D. Schulze (Ed.)

Volume 143
Global Climate Change and Human Impacts on Forest Ecosystems: Postglacial Development, Present Situation and Future Trends in Central Europe (2001)
J. Puhe and B. Ulrich

Volume 144
Coastal Marine Ecosystems of Latin America (2001)
U. Seeliger and B. Kjerfve (Eds.)

Volume 145
Ecology and Evolution of the Freshwater Mussels Unionoida (2001)
G. Bauer and K. Wächtler (Eds.)

Volume 146
Inselbergs: Biotic Diversity of Isolated Rock Outcrops in Tropical and Temperate Regions (2000)
S. Porembski and W. Barthlott (Eds.)

Volume 147
Ecosystem Approaches to Landscape Management in Central Europe (2001)
J.D. Tenhunen, R. Lenz, and R. Hantschel (Eds.)

Volume 148
A Systems Analysis of the Baltic Sea (2001)
F.V. Wulff, L.A. Rahm, and P. Larsson (Eds.)

Volume 149
Banded Vegetation Patterning in Arid and Semiarid Environments (2001)
D. Tongway and J. Seghieri (Eds.)

Volume 150
Biological Soil Crusts: Structure, Function, and Management (2001)
J. Belnap and O.L. Lange (Eds.)

Volume 151
Ecological Comparisons of Sedimentary Shores (2001)
K. Reise (Ed.)

Volume 152
Global Biodiversity in a Changing Environment: Scenarios for the 21st Century (2001)
F.S. Chapin, O. Sala, and E. Huber-Sannwald (Eds.)

Volume 153
UV Radiation and Arctic Ecosystems (2002)
D.O. Hessen (Ed.)

Volume 154
Geoecology of Antarctic Ice-Free Coastal Landscapes (2002)
L. Beyer and M. Bölter (Eds.)

Volume 155
Conserving Biological Diversity in East African Forests: A Study of the Eastern Arc Mountains (2002)
W.D. Newmark

Volume 156
Urban Air Pollution and Forests: Resources at Risk in the Mexico City Air Basin (2002)
M.E. Fenn, L. I. de Bauer, and T. Hernández-Tejeda (Eds.)

Volume 157
Mycorrhizal Ecology (2002)
M.G.A. van der Heijden and I.R. Sanders (Eds.)

Volume 158
Diversity and Interaction in a Temperate Forest Community: Ogawa Forest Reserve of Japan (2002)
T. Nakashizuka and Y. Matsumoto (Eds.)

Volume 159
Big-Leaf Mahogany: Genetic Resources, Ecology and Management (2003)
A. E. Lugo, J. C. Figueroa Colón, and M. Alayón (Eds.)

Volume 160
Fire and Climatic Change in Temperate Ecosystems of the Western Americas (2003)
T. T. Veblen et al. (Eds.)

Volume 161
Competition and Coexistence (2002)
U. Sommer and B. Worm (Eds.)

Volume 162
How Landscapes Change: Human Disturbance and Ecosystem Fragmentation in the Americas (2003)
G.A. Bradshaw and P.A. Marquet (Eds.)

Volume 163
Fluxes of Carbon, Water and Energy of European Forests (2003)
R. Valentini (Ed.)

Volume 164
Herbivory of Leaf-Cutting Ants: A Case Study on *Atta colombica* in the Tropical Rainforest of Panama (2003)
R. Wirth, H. Herz, R.J. Ryel, W. Beyschlag, B. Hölldobler

Printing: Saladruck, Berlin
Binding: Stürtz, Würzburg